Lecture Notes in Computer Science 16350

Founding Editors

Gerhard Goos
Juris Hartmanis

Editorial Board Members

Elisa Bertino, *Purdue University, West Lafayette, IN, USA*
Wen Gao, *Peking University, Beijing, China*
Bernhard Steffen, *TU Dortmund University, Dortmund, Germany*
Moti Yung, *Columbia University, New York, NY, USA*

The series Lecture Notes in Computer Science (LNCS), including its subseries Lecture Notes in Artificial Intelligence (LNAI) and Lecture Notes in Bioinformatics (LNBI), has established itself as a medium for the publication of new developments in computer science and information technology research, teaching, and education.

LNCS enjoys close cooperation with the computer science R & D community, the series counts many renowned academics among its volume editors and paper authors, and collaborates with prestigious societies. Its mission is to serve this international community by providing an invaluable service, mainly focused on the publication of conference and workshop proceedings and postproceedings. LNCS commenced publication in 1973.

Silvia Bonomi · Partha Sarathi Mandal ·
Peter Robinson · Gokarna Sharma ·
Sebastien Tixeuil
Editors

Stabilization, Safety, and Security of Distributed Systems

27th International Symposium, SSS 2025
Kathmandu, Nepal, October 9–11, 2025
Proceedings

Editors
Silvia Bonomi
University of Rome La Sapienza
Rome, Italy

Peter Robinson
Augusta University
Augusta, GA, USA

Sebastien Tixeuil (iD)
Sorbonne University
Paris, France

Partha Sarathi Mandal
Indian Institute of Technology Guwahati
Guwahati, Assam, India

Gokarna Sharma
Kent State University
Kent, OH, USA

ISSN 0302-9743   ISSN 1611-3349 (electronic)
Lecture Notes in Computer Science
ISBN 978-3-032-11126-5   ISBN 978-3-032-11127-2 (eBook)
https://doi.org/10.1007/978-3-032-11127-2

Preface

The papers in this volume were presented at the 27th International Symposium on Stabilization, Safety, and Security of Distributed Systems (SSS 2025), held during October 9–11, 2025, in Kathmandu, Nepal.

SSS is an international forum for researchers and practitioners in the design and development of distributed systems with a focus on systems that are able to provide guarantees on their correctness, performance, and/or security in the face of an adverse operational environment. Research in distributed systems continues to evolve rapidly, driven by the proliferation of dynamic and heterogeneous environments such as autonomous robotic swarms, edge-enabled sensor networks, decentralized AI systems, blockchain-based infrastructures, and multi-cloud federations. The rise of 5G/6G connectivity, IoT ecosystems, and quantum-aware architectures has further expanded the scope and complexity of distributed computing. Modern applications–including AI-driven edge analytics, zero-trust cloud services, decentralized finance (DeFi), and autonomous cyber-physical systems–demand that distributed computations be not only scalable and efficient but also resilient, privacy-preserving, and secure against increasingly sophisticated threats. The integration of AI-enhanced security mechanisms, secure multiparty computation, and post-quantum cryptography is becoming essential to ensure trustworthiness and self-healing capabilities across distributed platforms.

SSS started as the Workshop on Self-Stabilizing Systems (WSS), the first two of which were held in Austin in 1989 and in Las Vegas in 1995. Starting in 1995, the workshop was held biennially; it was held in Santa Barbara (1997), Austin (1999), and Lisbon (2001). As interest grew and the community expanded, in 2003 the title of the forum was changed to the Symposium on Self-Stabilizing Systems (SSS). SSS was organized in San Francisco in 2003 and in Barcelona in 2005. As SSS broadened its scope and attracted researchers from other communities, significant changes were made in 2006. It became an annual event, and the name of the conference was changed to the International Symposium on Stabilization, Safety, and Security of Distributed Systems (SSS). From then, SSS conferences were held in Dallas (2006), Paris (2007), Detroit (2008), Lyon (2009), New York (2010), Grenoble (2011), Toronto (2012), Osaka (2013), Paderborn (2014), Edmonton (2015), Lyon (2016), Boston (2017), Tokyo (2018), Pisa (2019), virtually (2020, 2021), Clermont-Ferrand (2022), Jersey City (2023), and Nagoya (2024).

This year the program was organized into three tracks reflecting major trends related to distributed systems: (A) Stabilization and Locality in Distributed Computing (chaired by Peter Robinson), (B) Time, Safety, and Security in Distributed Computing (chaired by Silvia Bonomi), and (C) Moving and Computing (chaired by Partha Sarathi Mandal). We received 59 submissions from 19 countries.

Each submission was reviewed in a double-blind process by at least three Program Committee members with the help of external reviewers. Out of the submitted papers, 21 were selected for presentation as regular papers. The symposium also included two keynotes given by John Augustine and Sonia Ben Mokhtar, seven brief announcements, and four invited papers. Selected extended papers from the symposium will be published in a special issue of the journal *Theoretical Computer Science*. The committee also made awards to the following papers:

- **Best paper award**: *On the Computational Power of Mobile Robots under Sequential Schedulers*, by Caterina Feletti, Paola Flocchini, and Nicola Santoro.
- **Best student paper award**: *A Poly-Log Approximation for Transaction Scheduling in Fog-Cloud Computing and Beyond*, by Ramesh Adhikari, Costas Busch, and Pavan Poudel.

On behalf of the Program Committee, we extend our heartfelt thanks to all authors who submitted their work to SSS. We deeply appreciate the significant time and intellectual effort contributed by the Program Committee members in shaping the symposium. Our sincere gratitude also goes to the external reviewers for their thoughtful and insightful evaluations, and to EasyChair for streamlining both the review process and the preparation of the proceedings.

We are especially thankful to the SSS Steering Committee for their invaluable guidance throughout. Finally, we warmly acknowledge the Organizing Committee, whose dedication and coordination made this edition of the conference truly memorable.

September 2025

Silvia Bonomi
Partha Sarathi Mandal
Peter Robinson
Gokarna Sharma
Sebastien Tixeuil

Organization

General Co-chairs

Gokarna Sharma   Kent State University, USA
Sébastien Tixeuil   Sorbonne Université, France

Track A. Stabilization and Locality in Distributed Computing

Peter Robinson (Chair)   Augusta University, USA
Bogdan Chlebus   Augusta University, USA
Varsha Dani   Rochester Institute of Technology, USA
Stéphane Devismes   Université de Picardie Jules Verne, France
Fabien Dufoulon   Lancaster University, UK
Leszek Gasieniec   University of Liverpool, UK
Taisuke Izumi   Osaka University, Japan
Dominik Kaaser   TU Hamburg, Germany
Valerie King   University of Victoria, Canada
Christoph Lenzen   CISPA Helmholtz Center for Information Security, Germany
Yannic Maus   TU Graz, Austria
Avery Miller   University of Manitoba, Canada
William K. Moses Jr.   Durham University, UK
Thomas Nowak   ENS Paris-Saclay, France
Shreyas Pai   IIT Madras, India
Gopal Pandurangan   University of Houston, USA
Sriram Pemmaraju   University of Iowa, USA
Joel Rybicki   Humboldt University of Berlin, Germany
Christian Scheideler   Paderborn University, Germany

Track B. Time, Safety, and Security in Distributed Computing

Silvia Bonomi (Chair)   Sapienza University of Rome, Italy
Leonardo Aniello   University of Southampton, UK
Quentin Bramas   University of Strasbourg, France
Jérémie Decouchant   Delft University of Technology, Netherlands
Antonella Del Pozzo   CEA, France

Xavier Defago	Institute of Science Tokyo, Japan
Giovanni Farina	Niccolò Cusano University, Italy
Letterio Galletta	IMT, Italy
Maurice Herlihy	Brown University, USA
Moti Medina	Bar-Ilan University, Israel
Mikhail Nesterenko	Kent State University, USA
Sathya Peri	Indian Institute of Technology Hyderabad, India
Valerio Schiavoni	University of Neuchâtel, Switzerland
Elad Michael Schiller	Chalmers University of Technology, Sweden

Track C. Moving and Computing

Partha Sarathi Mandal (Chair)	Indian Institute of Technology Guwahati, India
Evangelos Bampas	Université Paris-Saclay, France
Doina Bein	California State University, Fullerton, USA
Anaïs Durand	Université Clermont Auvergne, France
Paola Flocchini	University of Ottawa, Canada
Konstantinos Georgiou	Toronto Metropolitan University, Canada
Barun Gorain	Indian Institute of Technology Bhilai, India
Sayaka Kamei	Hiroshima University, Japan
Ajay Kshemkalyani	University of Illinois Chicago, USA
Euripides A. Markou	University of Ioannina, Greece
Kaushik Mondal	Indian Institute of Technology Ropar, India
Lata Narayanan	Concordia University, Canada
Alfredo Navarra	University of Perugia, Italy
Debasish Pattanayak	University of Ottawa, Canada
Giuseppe Prencipe	University of Pisa, Italy
Anisur Rahaman Molla	Indian Statistical Institute, Kolkata, India
Masahiro Shibata	Kyushu Institute of Technology, Japan
Yuichi Sudo	Hosei University, Japan

Additional Reviewers

Adhikari, Ramesh	Leonard, Andrew
Capozzi, Gianluca	Manaswini, Piduguralla
Das, Bibhuti	Oglio, Joseph
Demore, Joseph	Poudel, Pavan
Francis, Maria	Pramanick, Subhajit
Kumar, Manish	Wada, Koichi

Local Organization Committee

Shree Bhadra Wagle (Chair)	Kantipur Engineering College, Nepal
Keshar Prasain	Kantipur Engineering College, Nepal
Rabindra Khati	Kantipur Engineering College, Nepal
Prahlad Chapagain	Kantipur Engineering College, Nepal
Sahadeep Thapa	Kantipur Engineering College, Nepal
Saroj Sharma	Kantipur Engineering College, Nepal
Suman Shrestha	Kantipur Engineering College, Nepal

Treasurer

Kamal Khadga   Kantipur Engineering College, Nepal

Proceedings Chair

Pavan Poudel   University of Houston-Clear Lake, USA

Registration Chair

Ramesh Adhikari   Augusta University, USA

Steering Committee

Anish Arora	Ohio State University, USA
Stéphane Devismes	Université de Picardie Jules Verne, France
Shlomi Dolev	Ben-Gurion University of the Negev, Israel
Sayaka Kamei	Hiroshima University, Japan
Sandeep Kulkarni	Michigan State University, USA
Toshimitsu Masuzawa	Osaka University, Japan
Elad Michael Schiller	Chalmers University of Technology, Sweden
Sébastien Tixeuil (Chair)	Sorbonne Université, France

Advisory Committee

Sukumar Ghosh	University of Iowa, USA
Franck Petit	Sorbonne Université, France
Ted Herman	University of Iowa, USA

In Memory of

Ajoy Kumar Datta
Edsger W. Dijkstra
Mohamed Gouda

Contents

Keynote: Resilient Distributed Computing on External Data

John Augustine[✉]

Indian Institute of Technology Madras, Chennai, India
augustine@iitm.ac.in

Abstract. Distributed computing via message passing [6] has blossomed over the years, thanks in large part to the anchoring provided by the LOCAL model, the CONGEST model, and other related models like the congested clique, and the massively parallel computation model. Alongside, we have also seen the development of fault tolerant and Byzantine resilient models of computing [5]. In all of these models, the data is assumed to be within the network with each node holding a portion of the data. Consequently, the study of fault tolerance is somewhat stifled [4] because the data contained within the faulty nodes are lost at best (under crash failures) or maliciously misrepresented at worst (under Byzantine failures).

We explore several real-world contexts where the data is, in fact, external to the network and accessible to all nodes through queries and API calls – often at a cost that must be optimized. Inspired by these contexts, we will present a recent Byzantine resilient distributed computing model on external data [1]. We will then focus on the download problem that requires all good nodes in the network to optimally learn all the external data (represented as an array of n bits that can be individually queried) through careful collaboration despite β fraction (out of the k nodes) being Byzantine. The download problem is fundamental because all other conceivable problems can be solved locally after the download.

It is easy to see that n/k is a lower bound for the average number of queries per node even when the fraction of Byzantine nodes β is 0. Somewhat surprisingly, we will see that the download problem can be solved in the synchronous setting with at most a $\mathrm{polylog}(n)$ factor overhead for any fixed $\beta < 1$ [2,3]. We will end with some discussion on what might be interesting directions that can be explored further.

Acknowledgements. This is based on joint works with Jeffin Biju, Soumyottam Chatterjee, Valerie King, Manish Kumar, Shachar Meir, David Peleg, Srikkanth Ramachandran, and Aishwarya Thiruvengadam.

References

1. Augustine, J., et al.: Byzantine resilient distributed computing on external data. In: 38th International Symposium on Distributed Computing (DISC) (2024)

2. Augustine, J., et al.: Brief announcement: distributed download from an external data source in asynchronous faulty settings. In: 39th International Symposium on Distributed Computing (DISC) (2025)
3. Augustine, J., et al.: Distributed download from an external data source in byzantine majority settings. In: 39th International Symposium on Distributed Computing (DISC) (2025)
4. Augustine, J., Molla, A.R., Pandurangan, G., Vasudev, Y.: Byzantine connectivity testing in the congested clique. In: 36th International Symposium on Distributed Computing, DISC (2022)
5. Lamport, L., Shostak, R., Pease, M.: The byzantine generals problem. ACM Trans. Program. Lang. Syst. 4(3), 382–401 (1982)
6. Peleg, D.: Distributed Computing: A Locality-Sensitive Approach. Society for Industrial and Applied Mathematics (SIAM), Philadelphia, PA (2000)

Keynote: On the Safety and Security of Decentralised Machine Learning

Sonia Ben Mokhtar[✉]

CNRS, LIRIS, Lyon, France
`sonia.ben-mokhtar@cnrs.fr`

Abstract. There is a strong momentum towards data-driven services at all layers of society and industry. This started from large scale web-based applications such as Web search engines (e.g., Google, Bing), social networks (e.g., Facebook, TikTok, Twitter, Instagram) and recommender systems (e.g., Amazon, Netflix) and is becoming increasingly pervasive thanks to the adoption of handheld devices and the advent of the Internet of Things. Recent initiatives such as Web 3.0 are coming with the promise of decentralising such services for empowering users with the ability to gain back control over their personal data, and prevent a few economic actors from over concentrating decision power. However, decentralising online services calls for decentralising the data and the machine learning algorithms on which they heavily rely. While Federated Learning allows training machine learning models over decentralised data, it still relies on the centralised computation of model aggregations. In this presentation, I will present recent research works targeting the decentralisation of machine learning beyond the well know Federated Learning concept. A particular focus will be given on recent advances and open challenges for enforcing safety and security in decentralised machine learning.

S. Bonomi et al. (Eds.): SSS 2025, LNCS 16350, p. 3, 2026.
https://doi.org/10.1007/978-3-032-11127-2_2

Near-Optimal Stability for Distributed Transaction Processing in Blockchain Sharding

Ramesh Adhikari[(✉)][iD], Costas Busch[iD], and Dariusz R. Kowalski[iD]

Augusta University, Augusta, GA 30912, USA
`{radhikari,kbusch,dkowalski}@augusta.edu`

Abstract. In blockchain sharding, n processing nodes are divided into s shards, and each shard processes transactions in parallel. A key challenge in such a system is to ensure system stability for any "tractable" pattern of generated transactions; this is modeled by an adversary generating transactions with a certain rate of at most ρ and burstiness b. This model captures worst-case scenarios and even some attacks on transactions' processing, e.g., DoS. A stable system ensures bounded transaction queue sizes and bounded transaction latency. It is known that the absolute upper bound on the maximum injection rate for which any scheduler could guarantee bounded queues and latency of transactions is $\max\left\{\frac{2}{k+1}, \frac{2}{\lfloor\sqrt{2s}\rfloor}\right\}$, where k is the maximum number of shards that each transaction accesses. Here, we first provide a single leader scheduler that guarantees stability under injection rate $\rho \leq \max\left\{\frac{1}{16k}, \frac{1}{16\lceil\sqrt{s}\rceil}\right\}$. Moreover, we also give a distributed scheduler with multiple leaders that guarantees stability under injection rate $\rho \leq \frac{1}{16c_1 \log D \log s} \max\left\{\frac{1}{k}, \frac{1}{\lceil\sqrt{s}\rceil}\right\}$, where c_1 is some positive constant and D is the diameter of shard graph G_s. This bound is within a poly-log factor from the optimal injection rate, and significantly improves the best previous known result for the distributed setting [2].

Keywords: Blockchain Sharding · Transaction Scheduling · Partial Synchrony · Adversarial Stability · Locality-based Clustering

1 Introduction

Blockchain provides several unique features to transaction processing and storing, such as transparency, immutability, and fault tolerance [25], and has been used in many applications [8,21]. However, despite these advantages, it suffers from scalability issues, low throughput, and high latency. One of the methods to improve scalability is the sharding technique [18,20,28], which divides the entire blockchain network into smaller groups of nodes, called *shards*, and allows them to process transactions in parallel.

S. Bonomi et al. (Eds.): SSS 2025, LNCS 16350, pp. 4–20, 2026.
https://doi.org/10.1007/978-3-032-11127-2_3

The goal of this work is to improve the stability of transaction processing within sharded blockchain systems. Stability of a system is defined by maintaining bounded buffers (queues) and bounded latency of transactions' processing in an arbitrarily long run. We focus on worst-case scenarios and thus consider an adversarial model of transactions' generation: new transactions could be generated (also called injected or arriving) at any time as long as certain injection rate ρ and burstiness b are preserved [2,10]. This model also captures Denial of Service (DoS) attacks [22,23], in blockchain sharding systems. The goal is to guarantee stability (i.e., bounded queues and latency) for as large injection rate ρ as possible and for any fixed burstiness b.

Similarly to [1,2,4], we consider the blockchain sharding system with n nodes that are divided into s shards. Each shard has its own local blockchain and holds a subset of the accounts and states related to those accounts. We consider the partial-synchronous model [16], which is more practical than the synchronous communication model considered in [2,4]. We assume that a transaction T_i might be generated in any shard $S_i \in s$, which we call the *home shard* of transaction T_i. Moreover, considering the sharding model [1,2,4], each transaction T_i could access accounts from at most k shards, which we call the *destination shards* for T_i. Transaction T_i can be split into subtransactions $T_{i,j}$, where each subtransaction is sent to the respective *destination shard* that holds the corresponding account for the commitment. Similarly to [2,4], the maximum distance between the home shard and destination shards of a transaction is at most $d \leq D$ in the shard graph G_s, where D is the diameter of G_s.

We assume that shards have queues that store pending transactions. A shard picks transactions from its queue, splits them into subtransactions, and sends them to destination shards for processing in parallel. A conflict occurs when shards pick transactions that access the same account in destination shards. Such conflicts prevent the transactions from being committed concurrently, and forces the conflicting transactions to be committed in a serialized order. Our proposed scheduler coordinates with the shards to determine a schedule with a sequence of conflict-free sets of transactions that commit concurrently, which ultimately imposes an efficient serialization of conflicting transactions.

Contributions. We improve the best-known transaction injection rates [2], which guarantee stability, and we achieve this by employing a specific event-driven approach in designing the schedulers. Another advantage of our approach is that it does not require full synchrony. Moreover, the previous work [2] assumes a clique graph with unit distances for a single leader scheduler, whereas we consider a general graph with a diameter D. For such graphs, the scheduler in [2] would stabilize for D times smaller injection rate. Similarly, in a multi-leader setting, our scheduler allows stability for d times higher injection rate of transactions than in [2]. The summary of our contributions and their comparison to [2] is shown in Table 1, which we also describe briefly as follows:

- First, we provide a single leader scheduler in a partially-synchronous setting, which guarantees stability under injection rate $\rho \leq \max\{\frac{1}{16k}, \frac{1}{16\lceil\sqrt{s}\rceil}\}$. The queue size of this scheduler's pending transactions is bounded by $2(2b + \rho \cdot$

Table 1. Comparison of our schedulers with the best-known results in [2], where $d \leq D$ denotes the maximum distance of any transaction from its home shard to the shards it will access, $\mathfrak{D}$ denotes an upper bound on the local processing time and the communication delay between any two shards in our proposed protocol p(is at least d), and c', c_1 are some constants implied by analysis.

	Previous Work [2]	Proposed Results
	Synchronous	Partially-synchronous
	Single Leader Scheduler	
Txn injection rate ρ	$\rho \leq \max\left\{\frac{1}{18k}, \frac{1}{\lceil 18\sqrt{s}\rceil}\right\}$	$\rho \leq \max\left\{\frac{1}{16k}, \frac{1}{16\lceil\sqrt{s}\rceil}\right\}$ (Th 1)
Pending queues size	$\leq 4bs$	$\leq 2(2b + \rho \cdot 48\mathfrak{D})s$ (Th 1)
Latency	$\leq 36b \cdot \min\{k, \lceil\sqrt{s}\rceil\}$	$\leq 32b \cdot \min\{k, \lceil\sqrt{s}\rceil\} + 96\mathfrak{D}$ (Th 1)
Shards network	Clique graph with unit distance	General graph with diameter D
	Multiple Leaders Scheduler	
Txn injection rate ρ	$\rho \leq \frac{1}{c'd\log^2 s} \cdot \max\left\{\frac{1}{k}, \frac{1}{\sqrt{s}}\right\}$	$\rho \leq \frac{1}{16c_1\log D\log s}\max\left\{\frac{1}{k}, \frac{1}{\lceil\sqrt{s}\rceil}\right\}$ (Th 3)
Pending queues size	$\leq 4bs$	$\leq 2(2b + \rho \cdot 48c_1\mathfrak{D}\log D\log s)s$ (Th 3)
Latency	$\leq 2 \cdot c'bd\log^2 s \cdot \min\{k, \lceil\sqrt{s}\rceil\}$	$32c_1 b\log D\log s \cdot \min\{k, \sqrt{s}\}$ $+96c_1\mathfrak{D}\log D\log s$ (Th 3)
Shards network	General graph with diameter D	

$48\mathfrak{D})s$ and the latency is bounded by $32b \cdot \min\{k, \lceil\sqrt{s}\rceil\} + 96\mathfrak{D}$, where $\mathfrak{D}$ denotes an upper bound on the combined delay of local processing (including consensus in a shard) and communication between any shards.

- Next, we provide a multi-leader distributed scheduler, where multiple non-overlapping shard leaders process the transactions in parallel. This scheduler guarantees stability for injection rate $\rho \leq \frac{1}{16c_1\log D\log s}\max\left\{\frac{1}{k}, \frac{1}{\lceil\sqrt{s}\rceil}\right\}$, for which queue sizes are bounded by $2(2b + \rho \cdot 48c_1\mathfrak{D}\log D\log s)s$ and transaction latency is bounded by $32c_1 b\log D\log s \cdot \min\{k, \sqrt{s}\} + 96c_1\mathfrak{D}\log D\log s$, where c_1 is a positive constant coming from the analysis.

Note that in [2], for multiple leaders, the injection rate ρ is suboptimal because it inversely depends (linearly) on the parameter d. In our result, Theorem 3, the injection rate ρ does not depend linearly on d, but only on $\log D$; hence, our bound on injection rate ρ is near-optimal (within poly-log factors).

Paper Organization. Sect. 2 provides the related work, and Sect. 3 describes the preliminaries for this study and the sharding model. Section 4 presents a single leader scheduler. Section 5 generalizes the techniques to a multi-leader scheduler in the distributed setting. We provide the conclusion in Sect. 6. Missing proofs and other details are deferred to the full version, see [3].

2   Related Work

Several blockchain sharding protocols [18,20,26,28] have been proposed to solve the scalability issue of the blockchain. Although these protocols provide better throughput and process the transactions with minimum latency, none of these protocols provide stability analysis under adversarial transaction generation. Similarly, there is recent work on transaction scheduling in blockchain sharding [4], but this paper does not provide stability analysis; they only focus on the performance-oriented scheduling problem.

The only other known previous work that provides a stability analysis for blockchain sharding appears in [2]. However, the communication model considered in their protocol is synchronous, which may not accurately represent real distributed blockchain scenarios. Moreover, their transaction injection rate is not close to an optimal one for the distributed setting (i.e. it deviates from the optimal injection rate by a factor of $1/d$). In this work, we consider a partially-synchronous model, and we also provide near-optimal transaction injection rate bounds (that do not depend on d) for which the system remains stable. Hence, we significantly improve on the previous work. Moreover, in [2], the single leader model was analyzed under a synchronous setting with a clique graph of unit distance. However, such assumptions may not be practical in real-world blockchain networks. Therefore, we extend the model here not only by allowing partially synchronous communication but also by considering a general graph structure with a diameter D and maximum communication delay $\mathfrak{D}$, where shards can be connected to each other in a more flexible topology.

The adversarial model we consider here was first introduced in the context of adversarial queuing theory by Borodin *et al.* [10] for stability analysis in routing algorithms, where packets are continuously generated over time. Adversarial queuing theory [10] provides a framework for establishing worst-case injection rate bounds, and it was recently used to analyze blockchain sharding environments with unpredictable transaction injections [2]. Adversarial queuing theory has also been applied to various dynamic tasks in communication networks and channels [9,14,15]. Moreover, in [11], the authors introduced a stable scheduling algorithm for software transactional memory systems under adversarial transaction generation. Their model assumes a synchronous communication model, it allows objects to move across nodes, and transactions are executed once the required objects become available. However, in the blockchain sharding model, objects are static in shards, and transaction commitment requires confirmation from all the respective involved shards. Therefore, the results in [11] do not directly apply to our transaction scheduling model.

3 Technical Preliminaries

Similar to the previous works [1,2], we consider a blockchain system with n participating nodes, which are divided into s shards $S_1, S_2, \cdots S_s$, where each shard S_i is a subset of nodes, i.e., $S_i \subseteq \{1, \ldots, n\}$. The shards are pairwise disjoint, forming a partition of the n nodes, that is, for any $i \neq j$, $S_i \cap S_j = \emptyset$, and $n = \sum_i |S_i|$. Similarly to [2], we assume that shards communicate via message passing, where all non-faulty nodes within a shard must reach consensus on each message before sending it to another shard. To achieve this, each shard executes a consensus algorithm before transmitting messages (e.g., using PBFT [13] in the shard). We denote $n_i = |S_i|$ is the total number of nodes in shard S_i (the size of S_i), and by f_i is the number of faulty (Byzantine) nodes in S_i. For Byzantine fault tolerance and security guarantees, we assume that each shard S_i satisfies the condition $n_i > 3f_i$. Moreover, similarly to [1,2,4,18], we assume that the system consists of a set of shared accounts $\mathcal{O}$ (we also refer to them as objects). The set $\mathcal{O}$ is partitioned into disjoint subsets $\mathcal{O}_1, \ldots, \mathcal{O}_s$ where $\mathcal{O}_i$ represents the collection of objects handled by shard S_i. Each shard S_i maintains its own local blockchain (ledger) based on the subtransactions it processes for $\mathcal{O}_i$.

Transactions and Subtransactions. For a transaction T_i the respective *home shard* S_i is where it gets injected into. Similar to [1,2,4,18], a transaction T_i is represented as a collection of subtransactions $T_{i,a_1}, \ldots, T_{i,a_j}$ where subtransaction T_{i,a_l} accesses only objects in $\mathcal{O}_{a_l}$. Thus, subtransaction T_{i,a_l} has a respective *destination shard* S_{a_l} where it will ultimately be appended in the local ledger. An intra-shard transaction is a special case where the transaction accesses only accounts in the home shard. If a transaction accesses accounts outside the home-shard, then we refer to it as a *cross-shard* transaction. In [28], it was observed that cross-shard transactions are most frequent (99.98% of all transactions for 16 shards), thus we focus on the general case of cross-shard transactions.

Two transactions T_i and T_j *conflict* if they access a common object $o_k \in \mathcal{O}$ concurrently and at least one of them modifies (updates) the value of o_k. For safe transaction processing, the corresponding subtransactions of T_i and T_j must be serialized in the same order across all involved (destination) shards.

Communication Model, Emit, and Events. We consider a partially synchronous communication model [16], which assumes that there exists a Global Stabilization Time (GST) after which all messages are guaranteed to be delivered within a known bounded delay ($\mathfrak{D}$), and this delay is known to all shards. Similar to previous works in [2,5,18], we adopt cluster sending protocols (see the details in the extended version [3]) from *the fault-tolerant cluster-sending problem* [19] and Byshard [18], where shards run consensus before sending a message and assume that sending a message between two shards incurs a delay of at most $\mathfrak{D}$ (after GST). Transactions that are generated before GST are accounted in the burstiness parameter b. The parameter $\mathfrak{D}$ is determined by two delays: (i) the local consensus worst time (δ_{cons}) within a shard, which we normalize to be at most one time unit (i.e., $\delta_{cons} = 1$); and (ii) the worst network delay (δ_{comm}), network delay is the time it takes for a message (e.g., transactions or state request) to travel from one shard to another in the shard graph G_s, which can be arbitrary depending on the network delays. Hence, $\mathfrak{D} = \delta_{cons} + \delta_{comm}$.

We define two key terms describing how messages and actions propagate within our scheduling algorithm: (i) *Emit*: Represents the act of sending a message or request asynchronously. E.g., if shard S_i receives a new transaction T_k and there is some shard, say S_j, destined to process the transaction T_k, then shard S_i emits a send request T_k to shard S_j. (ii) *Event*: Represents the reception of an emitted message. This triggers an appropriate response. Events occur when messages reach their destination after a bounded delay (at most $\mathfrak{D}$) since their emit. In our example, upon receipt of the message, the destination shard S_j performs specified action to put the transaction into the processing schedule.

Adversarial Model of Transactions' Generation. The adversarial model [2, 10] we consider here models the process in which users generate and inject transactions into the sharded blockchain system continuously and arbitrarily with transaction injection rate at most ρ and burstiness b, where $0 < \rho \le 1$ and $b > 0$. We assume that each injected transaction accesses at most k shards, and adds a unit congestion to each accessing destination shard. The injected transactions are stored in the pending queue of the shard, and we examine and analyze the combined pending queue size throughout the system. The adversary model is constrained in such a way that over any contiguous time interval of length $t > 0$, the congestion at any shard (number of transactions that access accounts in the shard) does not exceed $\rho t + b$ transactions. Here, ρ represents the maximum transaction injection rate per unit time at each destination shard, and b represents the adversary's ability to generate a burst (maximum number) of transactions within any unit of time, including transactions generated before the Global Stabilization Time (GST) due to partial synchrony.

pIn the literature, there are basically two adversarial models: one is called a window type adversary (see e.g., [10]), and the one considered in our work is called a leaky bucket adversary, see e.g., [7]. As discussed in [15], the former is a logically restricted form of the latter; however, in most considered applications of this adversarial framework, they are equivalent (see e.g., [24]). An additional reason why we chose the leaky bucket model over the window adversarial model is the ability to easily account for transactions injected before the Global Stabilization Time (GST) as part of the burst (which needs to be resolved in the considered period after the GST). One of the advantages of having b in our performance formulas is, among others, that we can see how the transactions injected before the GST may influence system performance even after the GST.

Similar to the literature in this topic [2], our current analysis focuses on the congestion of transaction processing capacity within shards. Note that transactions are processed only in the home and destination shards (or in the leader shard), but not in transit. The delays based on the communication congestion, caused by transiting transactions between the home and destination shards, are accounted for by the parameter $\mathfrak{D}$ (see above, Communication Model). The parameter $\mathfrak{D}$ upper bounds both processing and network delays. Notation-wise, the delay caused by communication congestion is included in δ_{comm}, which is the part of the parameter $\mathfrak{D}$. There was a vast number of papers, started by [7, 10], analyzing the stability and delays caused by communication congestion. For instance, [29] applied adversarial models to capture phenomena related to the routing of packets with varying priorities and failures in networks. Authors in [6] addressed the impact of link failures on the stability of communication algorithms by way of modeling them in adversarial terms.

Depending on the communication model and setting (e.g., failures), our parameter δ_{comm} could be replaced by the bounds obtained in those papers. In this view, using a generic parameter δ_{comm} makes our paper more general, because it is not dependent on specific communication model – as long as there exists a bound on δ_{comm} in that model.

4   Event-Driven Scheduler with Single Leader

In this section, we design and analyze an event-driven single leader transaction scheduler that operates in a partial-synchronous communication model.

High-level Idea: In our approach, after a transaction T_i is generated at the home shard, then, the home shard forwards the transaction to the leader shard S_{ldr}, which is responsible for scheduling the transactions. Upon receiving transaction information from multiple home shards, the leader shard determines the schedule length, i.e., the time required to process and commit these transactions. S_{ldr} constructs a transaction conflict graph G_T and uses a greedy vertex coloring algorithm to determine schedule length and conflict-free commit order. If the computed schedule length is at least as long as the previously computed schedule length, the leader initiates the scheduling process. In order to properly determine the schedule length, the leader shard first requests the latest account states from the destination shards that hold the relevant accounts. Once the leader shard receives the account state information, the leader pre-commits the transactions according to the color they get in G_T and creates a batch order for each destination shard; this is possible because the leader has gathered all required account state information from the destination shards. This pre-committed batch is then sent to the respective destination shards. After receiving the pre-committed batch of subtransactions, each destination shard runs locally a consensus algorithm to reach an agreement on the received order and to append the respective transactions into their local blockchain.

Algorithm Details: Our Event-Driven Single-Leader Scheduler (EDSLSCHEDULER) Algorithm 1 is structured into three key phases: (i) transaction initiation at home shards, (ii) scheduling at the leader shard, and (iii) finalization at destination shards. Let us describe each phase in detail.

(i) At the home shards: When a new transaction T_i is generated, it is immediately sent to the designated leader shard S_{ldr} by emitting an transaction sending event *transaction_send(T_i)*. We assume that all shards are aware of the leader's ID.

(ii) At the leader shard: At the leader shard S_{ldr} the transactions are maintained in two queues. The first one is a pending transaction queue (PQ_{ldr}) that stores received but unscheduled transactions. The second one is a scheduled transaction queue (SQ_{ldr}) that tracks transactions that are scheduled but awaiting for final commitment. Upon receiving a new transaction T_i due to the emitting an event *transaction_send(T_i)* from a home shard, the leader appends transaction T_i to PQ_{ldr}. As the pending queue may contain transactions from multiple shards, the leader calculates the schedule length λ, which represents an upper bound on the total processing time required for all transactions in PQ_{ldr}. To determine λ and conflict-free commit order, leader shard S_{ldr} constructs a transaction conflict graph G_T from transactions PQ_{ldr} using

greedy vertex coloring algorithm. If λ is greater than or equal to the last recorded schedule length $LastEventSchLength$, the leader triggers a scheduling event. The leader then increments the version number $\mathcal{V}$ of this scheduling event (initially set to 0), moves transactions from PQ_{ldr} to SQ_{ldr}, forms a transaction batch $BatchTxn(\mathcal{V})$, and set the version for $G_{\mathcal{T}}$ to $G_{\mathcal{T}\mathcal{V}}$. It updates the last event schedule length $LastEventSchLength$ to λ and determines the destination shards for each transaction. Next, the leader requests the latest account states from the destination shards by emitting an event $account\_state\_request(BatchTxn(\mathcal{V}))$.

The versioning mechanism is introduced to handle the partial-synchronous and event-driven nature of our scheduling algorithm. Each scheduling event is associated with a version number $\mathcal{V}$ which is tagged in transactions, batch requests, and final committed transactions. This versioning is crucial for tracking the sequence of events and ensuring correct order processing. Since scheduling events can be triggered at any point in time, the leader shard may request account states from destination shards at different moments. Therefore, versioning helps the scheduler determine the correct state for each transaction batch. For instance, when a scheduling threshold (with respect to λ) is reached, the leader requests account states from destination shards. Without versioning, it would be ambiguous which batch the response corresponds to, potentially leading to inconsistencies. Additionally, versioning prevents conflicts by ensuring that account state modifications are correctly sequenced across multiple scheduling events.

Upon receiving the account state responses from the destination shards due to emit $account\_state\_response(BatchTxn(\mathcal{V})$ from destination shards, the leader shard pre-commit the transactions according to the execution order they get in colored conflict transaction graph $G_{\mathcal{T}\mathcal{V}}$. Transactions within the same color group C_k are checked for pre-commit feasibility simultaneously based on account balance constraints (because same-color transactions are non-conflicting transactions, as they access different accounts and can be committed at the same time). This pre-commit in a leader is possible because the leader now has the latest account state information. Pre-committed transactions are divided into subtransactions according to the shards they access. Then, the leader creates pre-committed subtransaction batches $PrecommitSubTxnBatch(\mathcal{V}, S_j)$ for each destination shard S_j. Once all transactions in $G_{\mathcal{T}\mathcal{V}}$ are processed, the leader sends the pre-committed batch orders to the respective destination shards by emitting the event $send\_precommit\_batch(PrecommitSubTxnBatch(\mathcal{V}, S_j))$.

(iii) At the Destination Shards: Upon receiving an account state request from the leader shard, the event $account\_state\_request(BatchTxn(\mathcal{V}))$ gets triggered. If the requested account is available, this indicates that there is no other transaction in progress (the account state is read but has not yet finally committed). Then the destination shard set the account to *occupied* to prevent conflicting access and responds with the current account state by emitting the event $account\_state\_response(BatchTxn(\mathcal{V}))$. Otherwise, if an ongoing transaction is

---

Algorithm 1: EDSLSCHEDULER

---

1 S_{ldr}: Leader shard; PQ_{ldr}: Pending transactions queue in leader shard;
2 SQ_{ldr}: Scheduled transactions queue in leader shard;
3 $\mathcal{V}$: Version number of scheduling event, initially $\mathcal{V} = 0$, and it increments;
4 $LastEventSchLength$: Last scheduled event length, initially, 0;
5 $BatchTxn(\mathcal{V})$: Batch transaction of version $\mathcal{V}$;
6 $PrecommitSubTxnBatch(\mathcal{V}, S_j)$: Pre-committed subtransactions batch for shard S_j;

7 /* **Event on Home Shards:** */
8 **Upon event: *new transaction T_i generated on any home shard S_i***
9   Shard S_i **Emits** *transaction_send(T_i)* for T_i to be received by leader shard S_{ldr};

10 /* **Events on Leader Shard S_{ldr}:** */
11 **Upon event: *transaction_send(T_i)***
12 S_{ldr} appends T_i to pending queue PQ_{ldr};
13 S_{ldr} constructs transaction conflict graph $G_{\mathcal{T}}$ and colors it using greedy vertex coloring to determine schedule length λ for transactions in PQ_{ldr};
14   **if** $\lambda \geq LastEventSchLength$ **then**
15   // Scheduling Event Triggered; S_{ldr} does the following:
16   Update the version number: $\mathcal{V} \leftarrow \mathcal{V} + 1$;
17   Move transactions from PQ_{ldr} to SQ_{ldr} with version $\mathcal{V}$;
18   Create batch transaction $BatchTxn(\mathcal{V})$;
19   Assign the version $\mathcal{V}$ to the transaction conflict graph $G_{\mathcal{T}}$ as $G_{\mathcal{T}\mathcal{V}}$;
20 $LastEventSchLength \leftarrow \lambda$;
21   Determine destination shards for each $T_i \in BatchTxn(\mathcal{V})$;
22   **Emit** *account_state_request($BatchTxn(\mathcal{V})$)* to be received by respective destination shards;

23 **Upon event: *account_state_response($BatchTxn(\mathcal{V})$)***
24   **if** *All account states received for $BatchTxn(\mathcal{V})$* **then**
25   According to the commit order, which was set previously in the transaction conflict graph $G_{\mathcal{T}\mathcal{V}}$ of $BatchTxn(\mathcal{V})$, S_{ldr} does the following;
26   **foreach** *color group C_k in $G_{\mathcal{T}\mathcal{V}}$* **do**
27   Pre-commit or abort transactions $T_i \in C_k$ by checking transactions and account state conditions according to the commit order determined by coloring;
28   If T_i is pre-committed, split T_i into subtransactions and create pre-committed subtransactions batch $PrecommitSubTxnBatch(\mathcal{V}, S_j)$ for each destination shard S_j;
29   **Emit** *send_precommit_batch($PrecommitSubTxnBatch(\mathcal{V}, S_j)$)* to the corresponding destination shards S_j, for each shard S_j;

30 **Upon event: *final_commit_response($T_{i,j}, \mathcal{V}$)***
31   If for all subtransactions $\overline{T}_{i,j}$ of T_i, *final_commit_response($T_{i,j}, \mathcal{V}$)* was emitted, remove T_i from SQ_{ldr};
32   If $T_i \in LastEventSchLength$ then $LastEventSchLength \leftarrow LastEventSchLength - |T_i|$;

33 /* **Events on Destination Shards:** */
34 **Upon event: *account_state_request($BatchTxn(\mathcal{V})$)***
35   **if** *Requested account is available in shard S_j* **then**
36   **Emit** *account_state_response($BatchTxn(\mathcal{V})$)* to S_{ldr} and set respective account to *occupied*;
37   **else**
38   Wait until the account is available and Emit account state response after wait;

39 **Upon event: *send_precommit_batch($PrecommitSubTxnBatch(\mathcal{V}, S_j)$)***
40 S_j reaches consensus on $PrecommitSubTxnBatch(\mathcal{V}, S_j)$ and appends it to the local blockchain; and sets respective account to *available*;
41 S_j **Emit** *final_commit_response($T_{i,j}, \mathcal{V}$)* to S_{ldr} shard;

---

modifying the account, the destination shard delays its response until the current state is finalized.

Similarly, when a destination shard receives a pre-committed batch from the leader, the event *send_precommit_batch(PrecommitSubTxnBatch(V, S_j))* is triggered. The destination shard then executes a consensus on the received batch and appends the finalized transactions to its local blockchain. After finalizing the transactions (subtransactions), the shard resets the account to available, allowing new transactions to access the account. The destination shard then emits a final commit response *final_commit_response($T_{i,j}$, V)* to the leader, confirming successful commitment. Once the leader receives the final commit responses for all subtransactions of a given transaction T_i, it removes T_i from SQ_{ldr}. If T_i belonged to the most recent scheduled batch, the leader reduces the last recorded schedule length by the number of committed transactions.

4.1 Analysis for Algorithm 1

Recall that each transaction accesses at most $k \geq 1$ out of $s \geq 1$ shards, and the burstiness is $b \geq 1$. Suppose $\mathfrak{D}$ is the upper bound on the local processing time and the communication delay between any two shards as described in Sect. 3. For the sake of analysis, we divide time into a sequence of intervals $I_1, I_2, \ldots$, each of duration τ as defined below. To simplify the analysis, we also introduce an artificial interval I_0 preceding the start of time, during which no transactions are present. Let the maximum transaction injection rate be ρ', as defined below, and let ζ denote the maximum number of transactions that can be generated in an interval:

$$\tau := 16b \cdot \min\{k, \sqrt{s}\} + 48\mathfrak{D} \ , \quad \rho' := \max\left\{\frac{1}{16k}, \frac{1}{16\lceil \sqrt{s}\rceil}\right\} \ , \quad \zeta := (2b + \rho \cdot 48\mathfrak{D})s \ .$$

Lemma 1. *In Algorithm 1, for transaction generation rate $\rho \leq \rho'$ and $b \geq 1$, in any interval I_z of length τ, where $z \geq 1$, there are at most ζ new transactions generated within I_z.*

Proof. According to the definition of the adversary, during any interval I_z of length τ, the maximum congestion added to any shard is:

$$\rho \cdot \tau + b \leq b + \rho \cdot 48\mathfrak{D} + b = 2b + \rho \cdot 48\mathfrak{D} \ . \tag{1}$$

Moreover, since each transaction accesses at least one shard and there are at most s shards, the total number of new transactions generated in I_z is at most $(2b + \rho \cdot 48\mathfrak{D})s = \zeta$.

Lemma 2. *In Algorithm 1, for transaction generation rate $\rho \leq \rho'$ and $b \geq 1$ in any interval I_z of length τ (where $z \geq 1$), if there are at most ζ pending transactions, then it takes at most $\frac{\tau}{4}$ time units to commit all ζ transactions.*

Lemma 3. *In Algorithm 1, for transaction generation rate $\rho \leq \rho'$ and $b \geq 1$, in any interval I_z of length τ, where $z \geq 1$, it holds that:*
(i) transactions that are scheduled at the last event at I_z will be committed by at most the middle of I_{z+1}, and (ii) all the transactions generated in the interval I_z will be committed or aborted by the end of I_{z+1}.

Proof. (Sketch). We prove this lemma by induction on the intervals of length τ. The core idea is that each scheduling event processes a bounded number of transactions (i.e. ζ), which can be committed within a fraction (i.e. $\frac{\tau}{4}$) of the interval length. Let us consider the last scheduling event in interval I_z. All transactions generated up to this event are bounded by ζ (from Lemma 1) and require at most $\frac{\tau}{4}$ time units to commit (by Lemma 2). Additionally, if an earlier event occurred in the same interval, it also contributes at most another $\frac{\tau}{4}$ time units. Therefore, the total time unit required to commit all transactions scheduled by the last event in I_z is at most $\frac{\tau}{2}$. This proves property (i). Moreover, any additional transactions generated in I_z after the last event require at most $\frac{\tau}{4}$ time unit (from Lemma 2), and can be processed together with new transactions from the next interval (i.e. I_{z+1}) which is also bounded by $\frac{\tau}{4}$ (also from Lemma 1, and Lemma 2). Therefore, the total schedule time required is $\frac{\tau}{2} + \frac{\tau}{4} + \frac{\tau}{4} = \tau$, as the length of I_{z+1} is τ which is sufficient to process (commit or abort) all the transactions from interval I_z, this proof the property (ii). $\square$

Theorem 1 (Stability of EDSLSCHEDULER). *In Algorithm 1, for transaction generation rate* $\rho \leq \max\left\{\frac{1}{16k}, \frac{1}{16\lceil\sqrt{s}\rceil}\right\}$ *and burstiness* $b \geq 1$, *the number of combined pending transactions throughout the system at any given time is at most* $2(2b + \rho \cdot 48\mathfrak{D})s$, *and the transaction latency is at most* $32b \cdot \min\{k, \sqrt{s}\} + 96\mathfrak{D}$.

Proof To estimate the number of pending transactions during any time unit, consider a time unit within an interval I_z. From the proof of Lemma 1, the maximum number of stale transactions during any time of I_z is $\zeta \leq (2b + \rho \cdot 48\mathfrak{D})s$. Within I_z, there can be at most $(2b + \rho \cdot 48\mathfrak{D})s$ newly generated transactions. Therefore, the upper bound on pending transactions during any time unit is $2(2b + \rho \cdot 48\mathfrak{D})s$. For estimating transaction latency, we rely on the fact that a transaction generated in an interval will be processed by the end of the next interval, see Lemma 3 property (ii). Consequently, the transaction latency is bounded by twice the duration of the maximum interval length τ:

$$2\tau = 2(16b \cdot \min\{k, \lceil\sqrt{s}\rceil\} + 48\mathfrak{D}) = 32b \cdot \min\{k, \lceil\sqrt{s}\rceil\} + 96\mathfrak{D} \ . \qquad \square$$

Theorem 2 *Our proposed scheduling Algorithm 1 provides safety and liveness.*

5   Event-Driven Scheduler with Multiple Leaders

A central authority, as considered in Sect. 4, might not exist in some blockchain sharding since, in principle, each transaction could be generated independently in a distributed manner in any shard, and gathering knowledge about them could be time- and resource-consuming. Here, we discuss a fully distributed scheduling approach using a clustering technique that allows the transaction schedule to be computed in a decentralized manner without requiring a single central authority. Each cluster has its own leader, and within each cluster, the distributed approach invokes the Event-Driven Single-Leader Scheduling Algorithm 1 from Sect. 4.

5.1   Shard Clustering

Similarly to [2], we consider a multi-layer clustering [17] of shard graph G_s, which is calculated before the algorithm starts and known to all the shards. This clustering scheme has been previously used to optimize transaction scheduling and execution [2,4,12,27]. A key advantage of this approach is locality-awareness, where transactions accessing nearby shards are prioritized over those that involve more distant shards [2,4]. The cluster is formed by dividing the shard graph G_s into $H_1 = \lceil \log D \rceil + 1$ layers of clusters (logarithms are base 2, and D is the diameter of G_s). Each layer consists of multiple clusters, and each cluster contains a set of shards. Layer i, where $0 \leq i < H_1$, is a sparse cover of G_s, which has the following properties: (i) Each cluster at layer i has a diameter of at most $O(2^i \log s)$; (ii) Each shard belongs to at most $O(\log s)$ different clusters within layer i; (iii) For $i \geq 1$, for every shard S_k, at least one cluster at layer i contains the entire 2^{i-1}-neighborhood of S_k.

The layer 0 is a special case that has each shard as a cluster on its own. Each layer $i \geq 1$ is further subdivided into $H_2 = O(\log s)$ partitions, following the sparse cover construction in [17]. These partitions, referred to as *sub-layers*, are labeled from 0 to $H_2 - 1$. A shard may be present in all H_2 sub-layers, ensuring that at least one of these clusters contains its complete 2^{i-1}-neighborhood.

Within each cluster at layer $i \geq 1$, a leader shard is designated such that its 2^{i-1}-neighborhood is entirely contained within the cluster. To systematically identify clusters across layers and sub-layers, we introduce the concept of height, represented as a tuple (i, j), where i refers to the layer and j refers to the sublayer. Following [2,12,27], cluster heights are lexicographically ordered to maintain a structured hierarchy. For a cluster C let $height(C) = (i, j)$ (where $0 \leq i < H_1$ represents the layer, and $0 \leq j < H_2$ represents the sublayer), denote its height.

The *home cluster* of a transaction T_i is determined based on its home shard S_i and the destination shards it needs to access. Let z be the maximum distance between S_i and the destination shards that transaction T_i accesses. The home cluster of S_i is defined as the lowest-layer and sub-layer (with lowest lexicographic order) cluster that contains the z-neighborhood of S_i. Each home cluster has a designated leader shard responsible for handling all transactions originating from shards within that cluster. When a transaction T_i is generated at its home shard, the transaction is sent to the leader shard of the corresponding home cluster, where the scheduling process is determined.

5.2   Multi-leader Scheduling Algorithm

The pseudocode of the Multi-leader Scheduler is deferred to the full version of this paper [3]. Consider a cluster C with $height(C) = (i, j)$. Within cluster C we use Algorithm 1 to schedule and commit transactions. The leader S_{ldr} of C first appends any transactions it receives into its pending queue PQ_{ldr}. If the schedule length λ of PQ_{ldr} transactions is at least the previously computed schedule length, denoted as $LastEventSchLength$, then the leader schedules and processes those transactions. To schedule transactions, S_{ldr} first acquires *scheduleControl*. Once it has *scheduleControl* it moves transactions from PQ_{ldr} to SQ_{ldr} and processes those transactions similar to Algorithm 1.

The acquisition of *scheduleControl* is coordinated with "parent" and "child" clusters. For a cluster C with $height(C) = (i, j)$ let (i', j') and (i'', j'') denote its immediate higher and lower heights, namely, $(i', j') > (i, j) > (i'', j'')$ (assume that (i, j) is not the top-most or bottom-most height, as those can be treated as simpler special cases). A *parent* cluster of C is any cluster C' at the immediate above height (i', j') that has common shards with C. A *child* cluster of C is any cluster C'' at the immediate below height (i'', j'') that has common shards with C. Note that C may have multiple parents or children.

The communication between C and its parent/child clusters occurs through event-driven requests and responses between their leaders. If C is at the bottom-most height, $(i, j) = (0, 0)$, then C has by default the scheduling control, namely, $scheduleControl(C) = true$. However, if C is not at the bottom-most height then it initially does not have the scheduling control and it must acquire it from its children. The *schedule_control_request* event is used to request scheduling control from C to all children clusters. Once all the children have returned a positive *schedule_control_response*, then $scheduleControl(C) = true$. When this happens, the leader S_{ldr} of C proceeds with executing locally Algorithm 1. After that, C relinquishes its control back to its children by sending *schedule_control_release* which pass it down to their own children until it reaches the bottom-most height.

Each time when the leader shard S_{ldr} of C has pending transactions PQ_{ldr} the schedule length λ reaches the minimum $LastEventSchLength$, then it requests the scheduling control from its children. However, C may request to obtain control on behalf of a parent. If any of its parent C' requests the scheduling control from C, then if C currently has the scheduling control, it responds positively to C'. Otherwise, C forwards the control acquisition request down to its children. When all the children respond positively, then it passes the control to the parent C'. While trying to obtain control for its parent, C may use the control to process its own transactions if new transactions have arrived in leader S_{ldr}, before passing it to the parent C'.

If cluster C requests control acquisition for processing its own new transactions, then it only needs to involve the children which intersect with the destination shards of the transactions in the queue PQ_{ldr} of its leader S_{ldr}. This assures that there will be no interference between leaders that have no common intersecting destination shards in the transactions of their pending queues. When a control acquisition request is forwarded by C to children, we include in the request the involved destination shards, in order to pass the request only to the children clusters that contain such destination shards.

5.3   Analysis for Multi-leader Scheduler

The multi-leader scheduler extends the single-leader scheduling algorithm (Algorithm 1) while introducing an additional overhead cost due to its hierarchical clustering structure. This overhead is $O(\log D \log s)$, where $O(\log D)$ accounts for the layered hierarchy and $O(\log s)$ comes from the sublayers. Here, D represents the diameter of shard graph G_s, and s denotes the total number of shards. Moreover, each cluster C at level (i, j) has its own diameter $\mathfrak{D}_i \leq \mathfrak{D}$, where $\mathfrak{D}$ is the maximum diameter among all clusters, which represents the upper bound on local processing time and communication delay between any two shards. This upper bound $\mathfrak{D}$ is defined in Sect. 3.

For the sake of analysis of the multi-leader scheduling algorithm, we divide time into a sequence of intervals $I_1, I_2, \ldots$, each of duration τ' as defined below. To simplify the analysis, we also have an artificial interval I_0 before time begins with no transactions. Let the maximum transaction injection rate be ρ'', as defined below, where c_1 is some combined positive constant that comes from layers and sublayers, and let ζ' denote the maximum number of transactions that can be generated in an interval:

$$\tau' := 16c_1 b \log D \log s \cdot \min\{k, \sqrt{s}\} + 48c_1 \mathfrak{D} \log D \log s \,,$$

$$\rho'' := \frac{1}{16c_1 \log D \log s} \max\left\{\frac{1}{k}, \frac{1}{\lceil \sqrt{s} \rceil}\right\} , \zeta' := (2b + \rho 48c_1 \mathfrak{D} \log D \log s)s \,.$$

Lemma 4 *In Multi-Leader Scheduler, for transaction generation rate $\rho \leq \rho''$ and $b \geq 1$, in any interval I_z of length τ', where $z \geq 1$, there are at most ζ' new transactions generated within I_z.*

Proof According to the definition of the adversary, during any interval I_z of length τ', the maximum congestion added to any shard is:

$$\rho \cdot \tau' + b \leq b + \rho \cdot 48c_1 \mathfrak{D} \log D \log s + b = 2b + \rho \cdot 48c_1 \mathfrak{D} \log D \log s \,.$$

Since each transaction accesses at least one shard and there are at most s shards, the total number of new transactions generated in I_z is at most $(2b + \rho \cdot 48c_1 \mathfrak{D} \log D \log s)s = \zeta'$.

Lemma 5 *In Multi-Leader Scheduler, for transaction generation rate $\rho \leq \rho''$ and $b \geq 1$ in any interval I_z of length τ' (where $z \geq 1$), if there is only one cluster C and there are at most ζ' pending transactions, then it takes at most $\frac{\tau'}{4c_1 \log D \log s}$ time units to commit all ζ' transactions.*

Lemma 6 *In Multi-Leader Scheduler, for transaction generation rate $\rho \leq \rho''$ and $b \geq 1$ in any interval I_z of length τ' (where $z \geq 1$), if there are at most ζ' pending transactions considering all layers and sublayers, then it takes at most $\frac{\tau'}{4}$ time units to commit all ζ' transactions.*

We now restate Lemma 3 for the Multi-Leader Scheduler. The proof follows the same structure as for Algorithm 1, with one change: Every time when the scheduling event occurs, instead of using Lemma 2, we use Lemma 6. With this change, the rest of the proof remains the same, and we get the following:

Lemma 7 *In Multi-Leader Scheduler, for transaction generation rate $\rho \leq \rho''$ and $b \geq 1$, in any interval I_z of length τ', where $z \geq 1$, it holds that: (i) transactions that are scheduled at the last event at I_z will be committed by at most the middle of I_{z+1}, and (ii) all the transactions generated in the interval I_z will be committed or aborted by the end of I_{z+1}.*

Theorem 3 (Multi-leader Stability). *In the multi-leader scheduler, for injection rate $\rho \leq \frac{1}{16c_1 \log D \log s} \cdot \max\left\{\frac{1}{k}, \frac{1}{\sqrt{s}}\right\}$, with burstiness $b \geq 1$, the number of combined pending transactions throughout the system at any given time unit is at most $2(2b + \rho \cdot 48c_1 \mathfrak{D} \log D \log s)s$, and the transaction latency is upper bounded by at most $32c_1 b \log D \log s \cdot \min\{k, \sqrt{s}\} + 96c_1 \mathfrak{D} \log D \log s$.*

Proof To estimate the number of pending transactions in any time unit, consider a time unit within an interval I_z. From the proof of Lemma 4, the maximum number of stale transactions during any time of I_z is $(2b + \rho \cdot 48c_1 \mathfrak{D} \log D \log s)s$. Within I_z, there can be at most $(2b + \rho \cdot 48c_1 \mathfrak{D} \log D \log s)s$ newly generated transactions. Therefore, the upper bound on pending transactions during any time unit is $2(2b + \rho \cdot 48c_1 \mathfrak{D} \log D \log s)s$. For estimating transaction latency, we rely on the fact that a transaction generated in an interval will be processed by the end of the next interval, see Lemma 7. Consequently, the transaction latency is bounded by twice the duration of the maximum interval length τ':

$$2\tau' = 32c_1 b \log D \log s \cdot \min\{k, \sqrt{s}\} + 96c_1 \mathfrak{D} \log D \log s.$$

Theorem 4 *There is no deadlock in the Multi-Leader Scheduling Algorithm.*

Corollary 1 Our proposed Multi-Leader Scheduling Algorithm provides safety and liveness.

6   Conclusions

In conclusion, we designed a stable transaction scheduling algorithm for the blockchain sharding system. Our algorithm significantly improves the stable transaction injection rate compared to the best-known result in [2]. Moreover, our proposed scheduler works in a partially-synchronous model, which is more practical than the synchronous model considered in [2]. To our knowledge, the bound we establish here provides the most efficient stable transaction rate for blockchain sharding systems. Blockchain designers can use our results to build systems that are resilient to DoS attacks.

Our proposed schedulers reduce the communication between the leader and destination shards in [2] by pre-fetching account states and performing local pre-commitment at the leader shard. This change is one of the reasons we obtain a near-optimal injection rate. In future work, we plan to simulate our proposed scheduling algorithm to validate our theoretical results. Further research directions could be dynamic shard clustering for a multi-leader scheduler, as our current model assumes that clusters are precomputed.

Acknowledgements. This paper is supported by NSF grant CNS-2131538.

References

1. Adhikari, R., Busch, C.: Lockless blockchain sharding with multiversion control, pp. 112–131. Springer-Verlag, Berlin, Heidelberg (2023)
2. Adhikari, R., Busch, C., Kowalski, D.R.: Stable blockchain sharding under adversarial transaction generation. In: Proceedings of the 36th ACM Symposium on Parallelism in Algorithms and Architectures, pp. 451–461 (2024)
3. Adhikari, R., Busch, C., Kowalski, D.R.: Near-optimal stability for distributed transaction processing in blockchain sharding. arXiv preprint arXiv:2509.02421 (2025), to be appear in the 27th International Symposium on Stabilization, Safety, and Security of Distributed Systems
4. Adhikari, R., Busch, C., Popovic, M.: Fast transaction scheduling in blockchain sharding. arXiv preprint arXiv:2405.15015 (2024)
5. Adhikari, R., Busch, C., Popovic, M.: On the efficiency of dynamic transaction scheduling in blockchain sharding. arXiv preprint arXiv:2508.07472 (2025), proceedings of the 39th International Symposium on Distributed Computing
6. Alvarez, C., Blesa, M., Serna, M.: The impact of failure management on the stability of communication networks. In: Proceedings. Tenth International Conference on Parallel and Distributed Systems, 2004. ICPADS 2004. pp. 153–160 (2004). https://doi.org/10.1109/ICPADS.2004.1316091
7. Andrews, M., Awerbuch, B., Fernandez, A., Leighton, T., Liu, Z., Kleinberg, J.: Universal-stability results and performance bounds for greedy contention-resolution protocols. J. ACM **48**(1), 39–69 (2001)
8. Azzi, R., Chamoun, R.K., Sokhn, M.: The power of a blockchain-based supply chain. Comput. Industr. Eng. **135**, 582–592 (2019)
9. Bender, M.A., Farach-Colton, M., He, S., Kuszmaul, B.C., Leiserson, C.E.: Adversarial contention resolution for simple channels. In: Proc. of the 17th ACM SPAA'05, pp. 325–332 (2005)
10. Borodin, A., Kleinberg, J., Raghavan, P., Sudan, M., Williamson, D.P.: Adversarial queuing theory. J. ACM (JACM) **48**(1), 13–38 (2001)
11. Busch, C., Chlebus, B.S., Kowalski, D.R., Poudel, P.: Stable scheduling in transactional memory. In: Algorithms and Complexity: 13th International Conference, CIAC 2023, Larnaca, Cyprus, June 13–16, 2023, Proceedings, pp. 172–186. Springer (2023)
12. Busch, C., Herlihy, M., Popovic, M., Sharma, G.: Dynamic scheduling in distributed transactional memory. Distrib. Comput. **35**(1), 19–36 (2022)
13. Castro, M., Liskov, B., et al.: Practical byzantine fault tolerance. In: OsDI. **99**, 173–186 (1999)
14. Chlebus, B.S., Kowalski, D.R., Rokicki, M.A.: Maximum throughput of multiple access channels in adversarial environments. Distributed Comp. **22**, 93–116 (2009)
15. Chlebus, B.S., Kowalski, D.R., Rokicki, M.A.: Adversarial queuing on the multiple access channel. ACM Trans. Algorithms (TALG) **8**(1), 1–31 (2012)
16. Dwork, C., Lynch, N., Stockmeyer, L.: Consensus in the presence of partial synchrony. J. ACM (JACM) **35**(2), 288–323 (1988)
17. Gupta, A., Hajiaghayi, M.T., Räcke, H.: Oblivious network design. In: Proc. of the 17th ACM-SIAM Symp. on Discrete Algorithms (SODA'06), pp. 970–979 (2006)
18. Hellings, J., Sadoghi, M.: Byshard: sharding in a byzantine environment. Proc. VLDB Endowment **14**(11), 2230–2243 (2021)
19. Hellings, J., Sadoghi, M.: The fault-tolerant cluster-sending problem. In: Varzinczak, I. (ed.) Foundations of Information and Knowledge Systems, pp. 168–186. Springer International Publishing, Cham (2022)

20. Luu, L., Narayanan, V., Zheng, C., Baweja, K., Gilbert, S., Saxena, P.: A secure sharding protocol for open blockchains. In: Proceedings of the 2016 ACM SIGSAC Conference on Computer and Communications Security, pp. 17–30 (2016)
21. McGhin, T., Choo, K.K.R., Liu, C.Z., He, D.: Blockchain in healthcare applications: research challenges and opportunities. J. Netw. Comput. Appl. **135**, 62–75 (2019)
22. Nguyen, T., Thai, M.T.: Denial-of-service vulnerability of hash-based transaction sharding: attack and countermeasure. IEEE Trans. Comp. **72**(3), 641–652 (2022)
23. Raikwar, M., Gligoroski, D.: Dos attacks on blockchain ecosystem. In: European Conference on Parallel Processing, pp. 230–242. Springer (2021)
24. Rosén, A.: A note on models for non-probabilistic analysis of packet switching networks. Inf. Process. Lett. **84**(5), 237–240 (2002)
25. Sankar, L.S., Sindhu, M., Sethumadhavan, M.: Survey of consensus protocols on blockchain applications. In: Proc. of the 4th ICACCS'17, pp. 1–5 (2017)
26. Secure, A.: The zilliqa project: a secure, scalable blockchain platform (2018)
27. Sharma, G., Busch, C.: Distributed transactional memory for general networks. Distrib. Comput. **27**(5), 329–362 (2014). https://doi.org/10.1007/s00446-014-0214-7
28. Zamani, M., Movahedi, M., Raykova, M.: Rapidchain: scaling blockchain via full sharding. In: Proceedings of the 2018 ACM SIGSAC, pp. 931–948 (2018)
29. Àlvarez, C., Blesa, M., Díaz, J., Serna, M., Fernández, A.: Adversarial models for priority-based networks. Networks **45**(1), 23–35 (2005)

A Poly-log Approximation for Transaction Scheduling in Fog-Cloud Computing and Beyond

Ramesh Adhikari[1], Costas Busch[1], and Pavan Poudel[2]

[1] Augusta University, Augusta, GA 30912, USA
{radhikari,kbusch}@augusta.edu
[2] University of Houston-Clear Lake, Houston, TX 77058, USA
poudel@uhcl.edu

Abstract. Transaction scheduling is crucial to efficiently allocate shared resources in a conflict-free manner in distributed systems. We investigate the efficient scheduling of transactions in a network of fog-cloud computing model, where transactions and their associated shared objects can move within the network. The schedule may require objects to move to transaction nodes, or the transactions to move to the object nodes. Moreover, the schedule may determine intermediate nodes where both objects and transactions meet. Our goal is to minimize the total combined cost of the schedule. We focus on networks of constant doubling dimension, which appear frequently in practice. We consider a batch problem where an arbitrary set of nodes has transactions that need to be scheduled. First, we consider a single shared object required by all the transactions and present a scheduling algorithm that gives an $O(\log n \cdot \log D)$ approximation of the optimal schedule, where n is the number of nodes and D is the diameter of the network. Later, we consider transactions accessing multiple shared objects (at most k objects per transaction) and provide a scheduling algorithm that gives an $O(k \cdot \log n \cdot \log D)$ approximation. We also provide a fully distributed version of the scheduling algorithms where the nodes do not need global knowledge of transactions.

Keywords: Distributed systems · shared object · fog-cloud computing · transaction scheduling · communication cost · doubling dimension graph

1 Introduction

There are distributed systems that process a large number of concurrent transactions in industry sectors like FinTech, e-commerce, social media, telecommunications, fog-cloud computing, etc. [9,42,44]. A coordination problem arises when multiple transactions attempt to simultaneously read from and write to the same shared objects, such as common accounts. To prevent inconsistencies while accessing shared objects, each transaction should be executed in an atomic

S. Bonomi et al. (Eds.): SSS 2025, LNCS 16350, pp. 21–39, 2026.
https://doi.org/10.1007/978-3-032-11127-2_4

way. Traditionally, locks are used to coordinate the actions of transactions and prevent inconsistencies while accessing shared objects [43]; however, if locks are not handled properly, that leads to a deadlock and priority inversion. To address these issues, we need to efficiently schedule the execution of transactions ensuring that each shared object is accessed by only one transaction at a time.

Consider a distributed system consisting of n processing nodes interconnected in a network represented as graph G. A set of transactions $\mathcal{T}$ and a set of shared objects $\mathcal{O}$ that are required to be accessed by the transactions are initially located at different nodes of the graph G. We consider the case where multiple transactions require to access the shared objects concurrently. In order for a transaction to execute, it needs to have an exclusive access to all the required objects. This can be achieved by either moving the objects to the transaction node (data-flow model [18,37]), or moving the transaction to the object nodes (control-flow model [33,36]). Recently, Busch $et\ al.$ [4] considered a more flexible transaction execution model called $dual\text{-}flow$ model where objects and transactions meet at arbitrary nodes of G during the execution.

Busch $et\ al.$ [4] studied transaction scheduling problem in $Trees$ with the objective of minimizing total communication cost. However, trees are inherently fragile (i.e., they may easily get disconnected on edge removals), lack redundancy, and are inefficient for modeling general networks (e.g., distances may not be preserved and congestion may increase). In contrast, graphs with a constant doubling dimension [15] provide a compelling alternative, particularly in the context of scalable, fault-tolerant, and real-time systems, by introducing redundancy, robustness, and flexibility needed for real-world high-performance distributed systems. Typical examples of graphs with constant doubling dimension are 2-dimensional grids [1,12] and randomly distributed unit disk graphs [28]. Such graphs are used in various distributed and routing systems, such as solving communication and graph problems [24,26], resource management [10,12,40], vehicular routing [19], routing and location services in networks [1,11,25].

In this paper, we studied transaction scheduling in a distributed system modeled as a graph G with a constant doubling dimension, aiming to minimize the total communication cost for executing transactions. Similar to [4], we also adopted the dual-flow model for transaction execution. We assume that moving an object along a unit-length edge in G costs α, while moving a transaction along the unit-length edge costs $\beta < \alpha$. (The case $\alpha \leq \beta$ corresponds to the well-studied data-flow model [6–8,37,38].) We consider a synchronous communication model, where time is divided into discrete steps [18]. At each time step, a node may perform one of the following actions: receive objects or transactions from neighboring nodes; execute transactions that have gathered all required objects; or forward objects or transactions to neighboring nodes.

Contributions. The primary goal of this paper is to provide an efficient schedule for the execution of transactions in a network G accessing shared objects. We consider G as a graph of constant doubling dimension. Table 1 compares our results with the most related work [4]. We provide the following contributions:

Table 1. Comparison of our proposed scheduler with the most related work [4], where n is the total number of nodes, D is the diameter of graph G, and k is the maximum number of objects accessed by each transaction.

Source	Scheduling Model	Metric	Communication cost approximation	
			Single object	**Multiple objects** (at most k)
Busch *et al.* [4]	Centralized	Tree	$O(1)$	O(k)
This paper	Centralized and distributed	Constant doubling dimension graph	$O(\log n \cdot \log D)$	$O(k \cdot \log n \cdot \log D)$

- Assuming each node has global knowledge of all transactions and each transaction accesses a single shared object, we propose a global-aware distributed scheduling algorithm with an $O(\log n \cdot \log D)$ approximation in communication cost, where n is total number of nodes and D is the network diameter.
- For transactions accessing up to k shared objects, we present a global-aware distributed scheduling algorithm with an $O(k \cdot \log n \cdot \log D)$ approximation in communication cost.
- When nodes do not have global knowledge of transactions, we introduce a fully distributed scheduling algorithm for the single-object case and explain how it can be extended to handle multiple shared objects.

Techniques. When there is a single shared object o to be accessed by all the transactions in $\mathcal{T}$, our scheduling algorithm consists of two main steps: (1) Find a set of nodes S_f in G where transactions and the object o can meet together with minimal cost, and (2) Make the object o visit each node in S_f and execute the transactions in order. Each node in S_f contains at least one transaction. When the object visits a node in S_f, respective transaction(s) in the node will execute sequentially. The overall schedule provides a global order of all transactions in $\mathcal{T}$. We show that the schedule of our algorithm has $O(A \log D)$ approximation, where A be the approximation of the optimal TSP tour for the nodes in S_f.

To calculate the set S_f, we use a hierarchical sparse partition H of G, as in [20]. Then the object follows an approximate TSP tour for visiting the nodes of S_f. We can use any of the two ways to calculate the tour.

- *Universal TSP*: As outlined in [20], the hierarchy H can be used to provide a global order of all the nodes in G which is called a *universal TSP tour*. For any subset of G, the respective order of nodes in H gives an $A = O(\log n)$ approximation of the optimal TSP tour. When considering S_f, using the respective TSP tour, we obtain the $O(\log n \cdot \log D)$ approximation.
- *MST*: Alternatively, we can first calculate a minimum weight spanning tree (MST) with the nodes of S_f, which can be used to approximate the tour with $A = 2$. Thus, this approach gives an overall approximation of $O(\log D)$ for our proposed algorithm.

We would like to note that although the MST approach gives a better approximation, the distributed version of the approach could require more messages. In particular, assuming that the hierarchical partition H is given, the number of messages to compute the tour with H is $O(n \log D)$. While using the MST approach for the tour, it involves $O(n^2)$ messages.

When transactions are required to access multiple shared objects, we extend the algorithm for a single shared object. First, a set of nodes S in G is calculated with respect to each individual object and the transactions requiring that object where the transactions and the object can meet together with minimal cost. Next, for each transaction T, a common node s' is found where all the required objects for T will gather with minimal additional cost. s' is added to the final set of nodes S_f and T will also move to s' for the execution. Following a TSP tour, objects will move to their respective nodes in S_f, and the transactions in each node execute sequentially when all the required objects gather at that node.

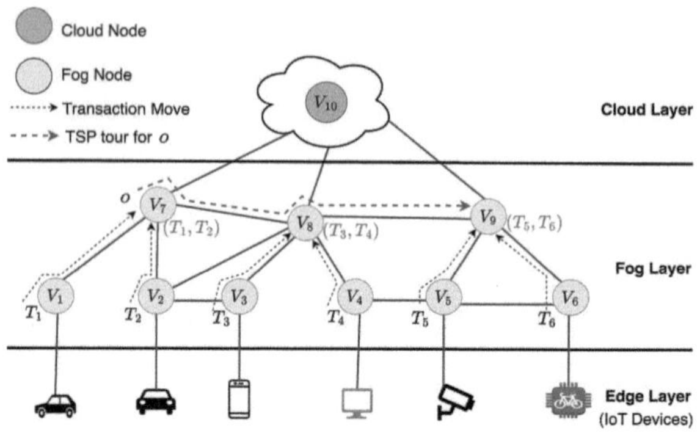

Fig. 1. Illustration of fog-cloud computing model.

Applications. Our scheduling technique is applicable in many distributed applications such as transactional memory (TM) [17,39], IoT and fog-cloud computing [13,29,30,42], financial systems [14], and Big Data [21]. In the following, we discuss the applicability in fog-cloud computing in detail.

Fog-cloud computing: IoT applications often rely on centralized cloud computing nodes. However, there exists a high communication delay between cloud and IoT devices [22,30]. Distributed fog computing addresses this challenge by providing resources at the edge of the network, closer to end devices that reduce network delay [27,30]. In fog-cloud computing model, transactions are generated from user IoT devices and sent to the fog layer. If the fog node has sufficient resources and the object it requires to execute, then the transaction is scheduled and executed in the fog layer; otherwise, it is routed to a cloud server for execution [22,27,30]. Figure 1 shows a simple example of the fog-cloud computing.

As a practical example, consider a vehicle-tracking scenario where IoT devices, like traffic cameras, transmit vehicle data as transactions to the fog layer. Each transaction creates, reads, or updates a vehicle-related object. Due to vehicle mobility, the object may be distant from the transaction node, leading to communication overhead. To reduce this cost, the object, the transaction, or both can be moved. Our proposed algorithm efficiently enables such movements for better transaction scheduling and execution.

Paper Organization: The rest of the paper is organized as follows: We discuss related work in Sect. 2 and model and preliminaries in Sect. 3. We present global-aware distributed scheduling algorithms in Sect. 4, followed by fully-distributed algorithms in Sect. 5. We conclude our paper in Sect. 6. Some of the pseudocodes, proofs, and other details are omitted due to space constraints.

2 Related Work

Several studies have explored the scheduling of transactions within distributed systems considering fog-cloud computing [30,31]. Nikoui et al. [31] employed a genetic algorithm (GA) to minimize energy consumption in task scheduling for green cloud computing systems. However, their approach assumes a central cloud broker utilizing the GA to allocate tasks across a set of virtual machines. Later, Nikoui et al. [30] introduced a cost-aware genetic-based task scheduling algorithm for fog-cloud environments, also relying on a centralized fog node, termed a fog broker, to handle transaction scheduling. Peixoto et al. [32] proposed a multilevel fog-cloud architecture for transaction scheduling. All of these algorithms depend on a single fog broker, and none of these works considers efficient communication analysis, such as determining when to move the object, the transaction, or both to intermediate nodes, to optimize communication costs and improve the overall efficiency of transaction scheduling.

Extensive research [5,7,8,23,33–35] has been done on scheduling transactions in distributed transactional memory systems. In transactional memory, each transaction requires access to specific objects for execution. Coordination between objects and transactions is often necessary for accessing objects and executing transactions. The majority of TM research is based on the data-flow model [18,41], in which transactions remain static while objects move between nodes to reach the locations where the transactions are held. In contrast, several studies adopt the control-flow model [33,36], where transactions move across nodes to access static objects. Lately, the dual-flow model [4,16] has been introduced in which both objects and transactions are moved to some intermediate node to minimize total communication cost. Additionally, there are transaction scheduling techniques in the context of blockchain sharding [2,3]. However, in blockchain sharding, objects are static and only transactions send messages, without a focus on reducing communication costs.

The most closely related work is [4], where the authors provide two variants of transaction scheduling algorithms for minimizing communication cost in transactional memory. However, their results are restricted to trees and do not apply to graphs with constant doubling dimension that we consider here. Moreover, the algorithms proposed in [4] rely on a centralized model where nodes have access to global information about transactions and objects. In contrast, this paper also presents a fully distributed version of the algorithms that operate without requiring global knowledge of transactions. The novelty of our algorithms lies in the use of hierarchical sparse partition H to compute the schedule.

3   Technical Preliminaries and Model

We model the network as a connected weighted graph $G = (V, E, w)$. The n vertices in the set V represent the processing nodes that may hold object(s) and transaction(s). Communication links between nodes are represented by edges in the set $E \subseteq V \times V$, and each edge is associated with a weight assigned by the function $w : E \to \mathbb{R}^+$ to denote the distance between the two nodes. The minimum distance between a pair of nodes is 1. A *path* p in G is a sequence of nodes with respective edges between adjacent nodes. There is a path between each pair of nodes and the distance between two nodes varies from 1 to D, where D is the diameter of G.

For a node v, the y neighborhood $N_y(v)$, where $y \geq 0$, is the set of nodes that are at a distance at most y from v (we also include v in $N_y(v)$). For a set of nodes S, the y-neighborhood of S is $N_y(S) = \bigcup_{v \in S} N_y(v)$, which is all the nodes that are at a distance at most y from some node in S. The length of a path p in G, denoted as $|p|$, is the sum of the weights of its edges; if the path is just a single node, then its length is trivially 0.

Similar to the previous work in [4], we adopt dual-flow model, allowing both shared objects and transactions the flexibility to move between the network nodes. The cost of moving an object of size α across a unit-weight edge is represented by α. Similarly, the cost of moving a transaction across a unit-weight edge is denoted by β where $\alpha > \beta$. The scheduling algorithm is responsible for determining the execution schedule $\mathcal{E}$ of the transactions, considering the movements of both objects and transactions in G.

We consider graphs with constant doubling dimension ($2^\delta = \Theta(1)$) as described in [40]. In the following definition, a *ball of radius* r refers to the $N_r(v)$-neighborhood of some node v.

Definition 1 (Doubling-dimension of Graph *[40]. The doubling dimension of a graph G is the smallest value of δ such that every ball of radius r in G can be covered by the union of at most 2^δ balls of radius $r/2$. If δ remains constant, we say G has a constant doubling dimension.*

3.1   Partition Hierarchy

Given a weighted graph $G = (V, E, w)$ with diameter $D \leq n$, we can build a partition hierarchy H with $O(\log D)$ levels (or layers), similar to the construction in [20]. A (r, σ, I)-partition divides V into a group of sets $\mathcal{X} = \{X_1, X_2, \dots\}$ such that each set $X_i \in \mathcal{X}$ has diameter at most $r \cdot \sigma$ and for each node $v \in V$, the neighbourhood $N_r(v)$ intersects with at most I sets in the partition; namely, $|\{X_i : X_i \in \mathcal{X} \wedge X_i \cap N_r(v) \neq \emptyset\}| \leq I$.

H consists of $h + 2$ levels, where $h = \lceil \log_\rho D \rceil$, where $\rho = 4\sigma$. Each level $l \geq 0$ of H, is a (r_l, σ, I)-partition $\mathcal{P}_l$ computed with $r_l = \min(D, \rho^l)$. In each set X_i in $\mathcal{P}_l$, an arbitrary node is designated as leader node $leader(X_i)$. Let $\mathcal{P}_{-1}$ be the trivial partition where each node of G is a cluster by itself. The maximum level is h, and at that level, the whole graph is a single cluster. For a leader ℓ at level $l < h$, the parent cluster (and respective parent leader) is the cluster at layer $\mathcal{P}_{l+1}$ that contains ℓ.

For graphs with a constant doubling dimension, the parameters ρ, σ, and I are all constant values [20]. For any subset S', let G' be the complete graph consisting of only the nodes in S', such that the edge weight between two nodes in G' is the same as the distance between them in G. Then, the order of the nodes for S' in the TSP tour of G gives a κ-factor approximation for the TSP tour of S' in G' (by connecting the last to the first node in S'), where $\kappa = O(\log n)$. In [20], the authors describe a method to transform the hierarchy H into to a κ-universal TSP tour for G. Figure 2 illustrates a partition hierarchy on a constant doubling dimension of graph G with $n = 7$ nodes where the left figure shows a partitioning scheme and the right figure shows leader nodes of each cluster and their parents (connected through a virtual link) in the graph G.

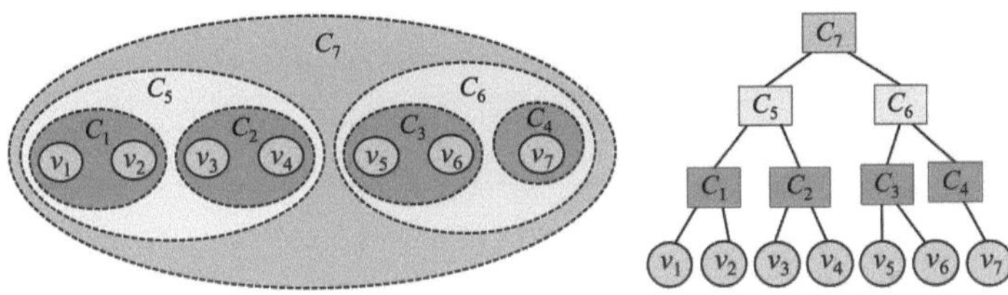

Fig. 2. Hierarchical partitioning of graph G with constant doubling dimension.

4   Global-Aware Distributed Scheduling

In this section, we provide two basic transaction scheduling algorithms, GLOBALAWARE_SINGLEOBJ and GLOBALAWARE_MULTIPLEOBJS, for graphs with a constant doubling dimension, assuming that the nodes have global knowledge of transactions and objects they access.

4.1   Single Object

We consider a single shared object o of size $\alpha > 1$ and a set of transactions $\mathcal{T} = \{T_1, T_2, \dots\}$ initially positioned at the nodes of G. All the transactions in $\mathcal{T}$ require object o for execution. We provide a basic scheduling algorithm denoted as GLOBALAWARE_SINGLEOBJ for the execution of transactions accessing object o and its pseudocode is given as Algorithm 1.

Algorithm 1: GLOBALAWARE_SINGLEOBJ

Input : Graph G with (r, σ, I)-partition H (with $h + 2$ levels), and a set of transactions $\mathcal{T}$

1 $v' \leftarrow$ initial home node for object o;
2 $\alpha, \beta \leftarrow$ cost of moving object o and a transaction over a unit weight edge of G, respectively;
3 Initialize $S \leftarrow \emptyset$ and $\gamma \leftarrow \lceil \frac{\alpha}{\beta} \rceil$;
4 **for each** *level l from 0 to $h + 1$ in H with partition $\mathcal{P}_l$* **do**
5   **for each** *cluster $X \in \mathcal{P}_l$* **do**
6         $T(X) \leftarrow$ set of transactions contained by the nodes in X that have not been assigned a dedicated super-leader yet;
7   **if** $|T(X)| \geq 2\gamma$ **then**
8   Add $leader(X)$ to the set of super-leaders S;
9   Assign $leader(X)$ as the dedicated super-leader for all the transactions in $T(X)$;

10 **for each** $T_i \in \mathcal{T}$, if T_i doesn't have a dedicated super-leader, **then** move T_i to v';
11 $S_f \subseteq S \leftarrow$ final set of dedicated super-leaders;
12 **for each** *level $l \geq 0$ in H with partition $\mathcal{P}_l$* **do**
13   Let $\widehat{L}_l$ be the set of dedicated super-leaders at level l;
14     $\widehat{T}_l \leftarrow$ set of transactions contained by the dedicated super-leaders at level l;
15   **if** $|\widehat{T}_l| < 8I\alpha$ **then**
16   Move all the transactions in $\widehat{T}_l$ to v';
17   Update $S_f \leftarrow S_f \setminus \widehat{L}_l$;

18 Move each transaction $T_i \in \mathcal{T}$ to its corresponding dedicated super-leader;
19 Calculate TSP tour on the set of dedicated super-leaders $S_f \cup \{v'\}$;
20 Object o traverses the ordered nodes of the TSP tour with transactions at the respective node being executed;

A hierarchical partition H of a given graph G can be computed as described in Sect. 3.1. Now, our task is to find a set of special nodes in H (which we call *super-leaders*) to move object o for providing access to the transactions. The concept of super-leaders introduced here is derived from the notion of supernodes utilized in trees, as outlined in [4]. A leader node v of a cluster (set) X of a partition at some level $l \geq 0$ becomes a super-leader if the cluster contains sufficiently large number of transactions such that the cost of moving object o from v to any of the transaction nodes contained in X is less than the total cost of moving all the transactions contained in X to v. The idea is that the movement of an object incurs high communication cost compared to the movement of a transaction, thus if some cluster contains trivially less number of transactions, then moving an object to the leader of that cluster results in greater communication cost than transferring the transactions from that cluster to the next-level cluster. Overall, we attempt to reduce the length of object movement path at each level by computing the super-leaders. For each transaction $T_i \in \mathcal{T}$, we find a super-leader at which T_i is executed and we call it a *dedicated super-leader* for T_i. If a transaction does not have a dedicated super-leader, then it is moved and executed at the node where object o is initially located.

Let v' be the initial home node for object o in H. At each level $l \geq 0$ in H, if a cluster (set) X of the partition $\mathcal{P}_l$ at level l contains at least 2γ (where $\gamma = \lceil \frac{\alpha}{\beta} \rceil$) transactions that are not assigned a dedicated super-leader yet, then $leader(X)$ becomes a super-leader and each transaction $T_i \in X$ is assigned $leader(X)$ as a dedicated super-leader. Let S be the set of such super-leaders. Each super-leader $s \in S$ also maintains the set of transactions (nodes) $\widehat{T}(s)$ for which s is a dedicated super-leader. If some transactions do not have a dedicated super-leader yet, then those transactions are moved directly to v' for the execution.

Next, we prune the dedicated super-leaders. Let $S_f \subseteq S$ be the final set of dedicated super-leaders. Initially, each dedicated super-leader $s \in S$, where $|\widehat{T}(s)| \geq 2\gamma$, is added to S_f. At each level $l \geq 0$ in H, if the sum of total number of transactions contained by the dedicated super-leaders at level l is less than $8I\alpha$, then those transactions will be relocated to the object node for the execution and all the dedicated super-leaders at level l are removed from S_f. This threshold is needed in order to minimize the total communication cost as shown later in the proof of Lemma 3 that if the number of transactions is less than $8I\alpha$, then moving transactions to the node where the object is located is cheaper than moving the object to the transactions node.

In the final stage, each transaction is sent to its assigned dedicated super-leader in the set S_f. The goal is to move object o to each super-leader in S_f and execute the corresponding transactions there. This is achieved by computing a TSP tour starting from v' that visits all nodes in S_f. To compute the tour, we can use the Universal TSP approach based on the hierarchy H as in [20].

Figure 3 illustrates an execution of Algorithm 1. The figure in the left shows the set of dedicated super-leaders highlighted in orange color. Assuming that the object o is initially positioned at node v_1, the dashed line on the right figure traces the TSP tour for o moving from v_1 to the dedicated super-leaders executing the transactions in each node.

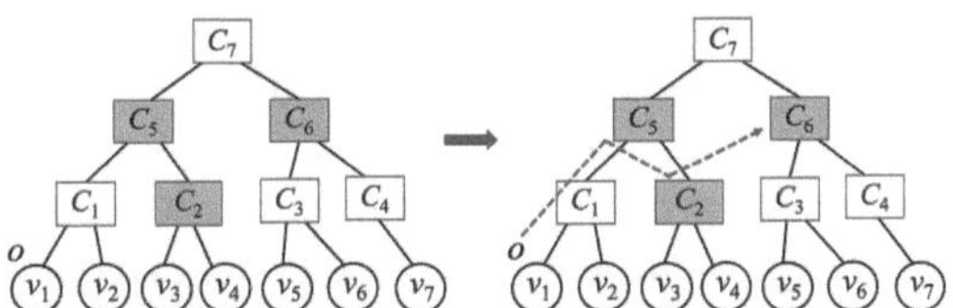

Fig. 3. Illustration of dedicated super-leaders and the TSP tour for moving object along the dedicated super-leaders in H by Algorithm 1.

Analysis. Without loss of generality, assume that $\alpha > 1$, and $\beta = 1$. Hence, $\gamma = \alpha/\beta = \alpha$, and the criterion of a leader becoming a super-leader is $2\gamma = 2\alpha$.

Each involved node in V holds at most one transaction. Let $\widehat{T}_i \subseteq \mathcal{T}$ denotes the set of transactions whose dedicated super-leader is at level i. Let $\widehat{L}_i$ denotes the set of dedicated leaders at level i (i.e., each $v \in \widehat{L}_i$ has a transaction that picked it as dedicated). Let $Tour^*(\widehat{L}_i)$ denotes the optimal length TSP tour for the nodes $\widehat{L}_i$, which also includes the original position of the object o. Let C^* be the optimal cost of executing all the transactions.

For a set of Z transactions, let $C_{\mathrm{direct}}(Z)$ be the cost of a schedule that sends all Z transactions to the object's node (object does not move in this schedule).

Lemma 1. *For $Z > 0$ transactions, $C_{direct}(Z) \leq 4(|Z|/\alpha + \log_2 D)C^*$.*

Proof. Let u be the node with the object o. Let $Z_i \subseteq Z$, $i \geq 0$, be the set of transactions whose distance from u is in the range $[2^i, 2^{i+1})$. Let $C(Z_i)$ be the cost of moving the transactions Z_i to u. We have $C(Z_i) < |Z_i|2^{i+1}$.

In the best case scenario, all Z_i transactions are gathered at a single node v at distance 2^i from u. Ideally, the optimal schedule would move o in an intermediate node x between u and v (or x may coincide with u or v). Let d_1 be the distance between u and x and d_2 be the distance between x and v, where $d_1 + d_2 = 2^i$, and $d_1, d_2 \geq 0$. The cost of moving object o to x is $d_1\alpha$. The cost of moving the Z_i transactions to x is $d_2|Z_i|$. Thus, $C^* = d_1\alpha + d_2|Z_i|$. We examine two cases:

- $|Z_i| < \alpha$: then $C^* > d_1|Z_i| + d_2|Z_i| = 2^i|Z_i|$. Hence, $C(Z_i) < |Z_i|2^{i+1} < 2C^* < 2(|Z_i|/\alpha + 1)C^*$.
- $|Z_i| \geq \alpha$: then $C^* \geq d_1\alpha + d_2\alpha = 2^i\alpha$. Hence, $C(Z_i) < |Z_i|2^{i+1} = |Z_i|2^{i+1}\alpha/\alpha \leq 2|Z_i|C^*/\alpha < 2(|Z_i|/\alpha + 1)C^*$.

Hence, $C(Z_i) \leq 2(|Z_i|/\alpha+1)C^*$. Since $0 \leq i \leq \lceil \log_2 D \rceil$, and $\sum_{i=0}^{\lceil \log_2 D \rceil} |Z_i| = |Z|$, we have:

$$C_{\text{direct}}(Z) \leq \sum_{i=0}^{\lceil \log_2 D \rceil} C(Z_i) \leq \sum_{i=0}^{\lceil \log_2 D \rceil} 2(|Z_i|/\alpha + 1)C^* = 2\,C^* \sum_{i=0}^{\lceil \log_2 D \rceil} (|Z_i|/\alpha + 1)$$

$$= 2\,C^*(|Z|/\alpha + 1 + \lceil \log_2 D \rceil) \leq 4(|Z|/\alpha + \log_2 D)C^*. \qquad \square$$

Let q be the path of the object in an optimal schedule for all the transactions in $\mathcal{T}$ which results in the optimal schedule with cost C^*. We can recursively decompose q to a sequence of edge-disjoint subpaths $q_1, q_2, \ldots, q_k$, $k \geq 1$, with parameter ξ as follows. Let q_1 be the shortest prefix subpath of q such that its length $|q_1|$ is at least $|q_1| \geq \xi$. If such q_1 does not exist, then $q_1 = q$. Let q' be the remaining subpath of q where the first node of q' is the same with the last node of q_1. If q' consists of at least two nodes, repeat the same process recursively on q'. We have the following properties for optimal path q:

- The first and last node of q is the first of q_1, and the last of q_k, respectively.
- The first node of q_j is the same with the last node of q_{j-1}, where $j > 1$.
- For $1 \leq j < k$, each q_j has a length at least $|q_j| \geq \xi$ such that the removal of the last edge splits q_j into prefix q_j' and suffix q_j'' with $|q_j'| < \xi$.
- For the last subpath q_k, if it consists of three or more nodes, then for its prefix q_j', it holds $|q_j'| < \xi$; otherwise, $|q_k| > 0$.
- $k \leq |q|/\xi + 1$: Since each of the first $k - 1$ segments has length at least ξ, we have $|q| \geq (k - 1)\xi$. Therefore, $k - 1 \leq |q|/\xi$, which implies $k \leq |q|/\xi + 1$.

Lemma 2. *For a path q_j, $1 \leq j \leq k$, with v_1 and v_2 being the first and last nodes of q_j, any $c \geq 0$, $N_c(q_j) \subseteq N_{2c}(\{v_1, v_2\})$.*

Proof. This holds immediately if q_j consists only of v_1 and v_2, since $N_c(q_j) = N_c(\{v_1 \cup v_2\}) \subseteq N_{2c}(\{v_1, v_2\})$. Suppose now that q_j consists of three or more nodes. Let v' be the node of q_j which is adjacent to v_2. We have that $dist_G(v_1, v') \leq c$. Thus, all the nodes of q_j between v_1 and v' are at distance less than ξ from v_1; hence, the ξ-neighbors of all these nodes are at a distance less than 2ξ from v_1. Thus, $N_c(q_j) \subseteq N_{2c}(v_1) \cup N_c(v_2) \subseteq N_{2c}(\{v_1, v_2\})$. $\qquad \square$

Lemma 3. *For $|\widehat{T_i}| \geq 8I\alpha$, $C^* \geq |\widehat{T_i}|\rho^{i-1}/(16I)$.*

The proof of Lemma 3 is omitted due to space constraints.   □

Lemma 4. *The number of ρ^i-neighborhoods that are required to cover a $8\sigma\rho^i$-neighborhood of G is at most $\zeta := 2^{\delta \log(8\sigma)}$.*

Proof. Let G be a graph with doubling dimension δ. Consider a ball of radius $8\sigma\rho^i$ centered at a vertex v in G, denoted as $B(v, 8\sigma\rho^i)$. We want to cover this ball with ρ^i-neighborhoods. From the definition, any ball of radius $2r$ can be covered by at most 2^δ balls of radius r. Here, $r = 4\sigma\rho^i$. Therefore, the ball $B(v, 8\sigma\rho^i)$ can be covered by at most 2^δ balls of radius $4\sigma\rho^i$. Similarly, to cover $4\sigma\rho^i$, we need 2^δ balls of radius $2\sigma\rho^i$ and so on. We can estimate the total number of balls as $\prod_{i=1}^{k} 2^\delta = (2^\delta)^k$, where $\frac{8\sigma\rho^i}{2^k} = \rho^i$. Thus, $k = \log(8\sigma)$. Hence, $2^{\delta k} = 2^{\delta \log(8\sigma)}$.   □

Lemma 5. $\widehat{L_i} \subseteq N_{4\sigma\rho^i}(q)$.

Proof. Suppose that there is $v \in \widehat{L_i}$, such that $v \notin N_{4\sigma\rho^i}(q)$. Let X be the cluster of v at partition $\mathcal{P}_i$. Let d be the distance of v to the closest node in q; note that $d > 4\sigma\rho^i$. The distance between any node in X to its closest node in q is at least $d - \sigma\rho^i$. Let $T(X)$ be the transactions with home node in X. Since v is a super-leader, $|T(X)| \geq 2\alpha$. The cost c_1 of moving these transactions to q is

$$c_1 \geq (d - \sigma\rho^i)2\alpha = (2d - 2\sigma\rho^i)\alpha > (d + 4\sigma\rho^i - 2\sigma\rho^i)\alpha = (d + 2\sigma\rho^i)\alpha .$$

On the other hand, the cost c_2 of gathering the $T(X)$ transactions to v and then having the object move from q to v is $c_2 \leq 2\alpha\sigma\rho^i + d\alpha < c_1$. Therefore, the path q is not optimal; a contradiction. Thus, $\widehat{L_i} \subseteq N_{4\sigma\rho^i}(q)$.   □

Using Lemmas 2, 4, and 5, we obtain the following two results:

Lemma 6. *For $|\widehat{T_i}| \geq 4I\alpha$ and $|q| \geq \rho^{i-1}/2$, $Tour^*(\widehat{L_i}) \leq 74\zeta I\sigma\rho|q|$.*   □

Lemma 7. *For $|\widehat{T_i}| \geq 4I\alpha$ and $|q| < \rho^{i-1}/2$, $Tour^*(\widehat{L_i}) \leq 36C^*\rho\zeta\sigma/\alpha$.*   □

Let R be the set of indices in range $0, \ldots, h$ such that for each $i \in R$, $|\widehat{T_i}| \geq 8I\alpha$ and $\widehat{T} = \bigcup_{i \in R} \widehat{T_i}$. We have a set of dedicated super-leaders $S_f = \bigcup_{i \in R} \widehat{L_i}$. Let $Tour^*(S_f)$ denote the cost of the optimal TSP tour covering all the nodes in S_f, including the object's initial position, and $Tour(S_f)$ be the actual cost of the object tour incurred by Algorithm 1. (Without loss of generality, assume that $Tour^*(S_f) > 0$, since otherwise we would only have the direct move cost for the transactions.) We establish the following three Lemmas (proofs are omitted for brevity).

Lemma 8. $\alpha \cdot Tour^*(S_f) \leq 74(h+1)\zeta I\sigma\rho C^*$.   □

Lemma 9. *For transactions $\widehat{T}$, cost of Algorithm 1 is $C(\widehat{T}) \leq 74\frac{Tour(S_f)}{Tour^*(S_f)}(h+1)\zeta I\sigma\rho C^*$.*   □

Lemma 10. *For transactions $\mathcal{T} \setminus \widehat{T}$, cost of Algorithm 1 is $C_{direct}(\mathcal{T} \setminus \widehat{T}) \leq 36(h+2)IC^* + 4C^* \log_2 D$.*   □

Theorem 1. *For executing all transactions in $\mathcal{T}$, the total communication cost of Algorithm 1 is $C = O\left(C^* \cdot \frac{Tour(S_f)}{Tour^*(S_f)} \cdot \log D\right)$.*

Proof. For the cost C of Algorithm 1 we have, $C = C(\widehat{T}) + C_{\text{direct}}(\mathcal{T} \setminus \widehat{T})$. Thus, from Lemmas 9 and 10, we get:

$$C \leq 74 \frac{Tour(S_f)}{Tour^*(S_f)}(h+1)\zeta I \sigma \rho C^* + 36(h+2)IC^* + 4C^* \log_2 D.$$

The result follows since $h = O(\log D)$, and σ, I, ρ, ζ are constants (see Lemma 4 for ζ being a constant).   □

When using a TSP tour based on the minimum weight spanning tree, we get $Tour(S_f)/Tour^*(S_f) \leq 2$. Thus, from Theorem 1, we get the following result.

Corollary 1. *Using a MST based TSP to implement the tour of the object, the approximation of Algorithm 1 is $O(\log D)$.*   □

Using a universal TSP tour based on H, we get $Tour(S_f)/Tour^*(S_f) = O(\log n)$. Thus, from Theorem 1, we get the following result.

Corollary 2. *Using a universal TSP to implement the tour of the object, the approximation of Algorithm 1 is $O(\log n \cdot \log D)$.*   □

4.2   Multiple Objects

We consider a set of shared objects $\mathcal{O} = \{o_1, o_2, \ldots\}$, all initially positioned at some node v' of G. Each transaction $T_i \in \mathcal{T}$ accesses at most k objects from $\mathcal{O}$. The set of objects accessed by transaction T_i is denoted as $objs(T_i) \subseteq \mathcal{O}$. We assume that nodes have global knowledge of transactions and provide a scheduling algorithm GLOBALAWARE_MULTIPLEOBJS for executing the transactions in $\mathcal{T}$. The pseudocode is provided in Algorithm 2.

In Algorithm 2, the objective is to ensure synchronized access to the objects at minimal cost for executing the transactions. We accomplish this by first using Algorithm 1 to calculate the dedicated super-leaders for each object individually. We then combine some of the dedicated super-leaders of the objects in a set of nodes S_f. Finally, we use a single TSP tour over $S_f \cup \{v'\}$, where the objects move synchronously from node to node along the tour, executing the corresponding transactions at each visited node. Note that an object visits only the nodes in the TSP tour where it is needed for a transaction, skipping all others.

The set of dedicated super-leaders S_f is determined as follows. Each transaction $T_i \in \mathcal{T}$ accesses up to k objects, potentially with different dedicated super-leaders for each such object in $objs(T_i)$. For each T_i, we select the nearest super-leader among those assigned to its objects, and designate it as the T_i's

---

Algorithm 2: GLOBALAWARE_ MULTIPLEOBJS

Input : Graph $G = (V, E, w)$ with (r, σ, I)-partition H, and a set of transactions $\mathcal{T}$

1   $v' \leftarrow$ initial home node for all objects in $\mathcal{O}$; $S_f \leftarrow \emptyset$;

2   // Find super-leaders w.r.t. each object
3   **for each** *object $o_j \in \mathcal{O}$* **do**
4   Use Algorithm 1 to find the dedicated super-leaders for object o_j;

5   // Find dedicated super-leader for T_i
6   **for each** *transaction $T_i \in \mathcal{T}$* **do**
7   **if** *T_i has at least one dedicated super-leader assigned above for any of the objects in $objs(T_i)$* **then**
8   Find the closest super-leader s in H among all the dedicated super-leaders assigned to T_i and mark it as the dedicated super-leader for T_i;
9   Move T_i to s and add s to S_f;

10   **else** move T_i to the initial home node of objects, v';

11   // Build and traverse TSP tour
12   Calculate TSP tour for $S_f \cup \{v'\}$;
13   Each object $o_j \in \mathcal{O}$ follows the order of the TSP tour for visiting the nodes where o_j is required to execute respective transactions;

---

dedicated super-leader s. Transaction T_i moves to s, and s is added to S_f. During the TSP traversal of S_f, all objects in $objs(T_i)$ will be brought to the leader node s, and T_i is executed. If T_i has no dedicated super-leader (i.e., none of its objects do), then it moves to v' and is executed when v' is visited in the tour.

Analysis. We continue with the analysis of Algorithm 2. Consider an object $o_j \in \mathcal{O}$. If a super-leader s' of o_j, as returned by Algorithm 1, is not used in the set S_f calculated by Algorithm 2, then another super-leader is being used in S_f which is closer to the transactions that were going to move to s'. Thus, since $\alpha > \beta$, the cost of such a change of super-leader does not incur an additional cost with respect to object o_j (where cost of o_j is as calculated in Theorem 1).

Now consider the case where object o_j is accessed by a transaction T_i, originally located at some node u, and T_i does not have a dedicated super-leader for o_j under Algorithm 1, but does have a dedicated super-leader s under Algorithm 2. If the distance from u to s is no greater than the distance from u to v', then the cost of accessing o_j remains unaffected (as calculated in Theorem 1).

If the distance from u to s is more than the distance from u to v', then there is additional cost due to the transaction T_i traveling a longer distance in Algorithm 2. However, since T_i has s as a dedicated super-leader, there must exist some object $o_z \in \mathcal{O}$ which is accessed by T_i for which s is a dedicated super-leader for T_i. The additional cost with respect to object o_j is no more than the cost C' of moving T_i to s, as counted in the analysis of object o_z (in Theorem 1). Since T_i can access at most $k - 1$ other objects similar to o_j, the additional cost for moving these objects to s is at most $(k - 1)C'$.

Hence, using S_f increases the total cost by at most a factor of k compared to the sum of the costs of individual objects. As a result, the approximation of Theorem 1 scales by a factor of k, giving the following theorem and corollaries:

Theorem 2. *For executing all transactions in $\mathcal{T}$, the total communication cost of Algorithm 2 is $C = O\left(k \cdot C^* \cdot \frac{Tour(S_f)}{Tour^*(S_f)} \cdot \log D\right)$.* $\square$

Corollary 3. *Using a MST based TSP to implement the tour of the object, the approximation of Algorithm 2 is $O(k \cdot \log D)$.* $\square$

Corollary 4. *Using a universal TSP to implement the tour of the object, the approximation of Algorithm 2 is $O(k \cdot \log n \cdot \log D)$.* $\square$

5   Fully Distributed Scheduling

In this Section, we provide fully distributed transaction scheduling algorithms assuming that the nodes in G do not have the global knowledge of transactions. First, we discuss the algorithm in detail for the case where transactions access a single shared object. Later, we provide a high-level idea for extending the algorithm for multiple shared objects.

5.1   Single Object

As in Sect. 4.1, we consider a single shared object o of size $\alpha > 1$ and a set of transactions $\mathcal{T} = \{T_1, T_2, \dots\}$ initially positioned at arbitrary nodes of G. Each transaction $T_i \in \mathcal{T}$ requires access to object o for the execution. We provide a fully distributed scheduling algorithm FULLYDISTRIBUTED_SINGLEOBJ for the execution of transactions in $\mathcal{T}$ accessing the object o. The algorithm operates in three phases: (1) Cluster leaders share transaction information with their parent leaders and identify super-leaders; (2) Each transaction selects its dedicated super-leader and moves to that node for execution; (3) A TSP tour is computed to move the object to each dedicated super-leader for transaction execution. We now describe each phase in detail.

Phase 1: In this phase, transaction information is propagated through the levels of the hierarchy H to identify the super-leaders. The process begins with each node simultaneously sharing its transaction data with the leader of the corresponding cluster at level $l = 0$ in H. At each level $l \geq 0$, the leader of each cluster (set) X ($leader(X)$) in the partition $\mathcal{P}_l$ counts the transactions within its cluster. If the count reaches at least 2γ, the cluster leader becomes a super-leader and notifies all child leaders recursively down to the level 0. It also forwards the transaction count to the parent cluster at level $l + 1$, continuing this process up to the root of the hierarchy H. At the end of Phase 1, a set S of all possible super-leaders is computed, and each cluster leader has the knowledge of S.

Phase 2: In this phase, each node independently determines the dedicated super-leader for its associated transaction $T_i \in \mathcal{T}$. Let X be the lowest-level cluster in H that contains T_i, and S be the set of super-leaders whose information was received by $leader(X)$ from its parent clusters. Then, T_i selects a dedicated super-leader $s \in S$ such that s is the closest super-leader to T_i among all in S. If S is empty (i.e., $leader(X)$ didn't receive any super-leader information from its parents), then T_i is moved to the node where the object o is initially located.

The set of dedicated super-leaders is further pruned as follows. At each level $l \geq 0$ in H, the algorithm checks the total number of transactions across all dedicated super-leaders. If this total is less than $8I\alpha$, then the associated transactions are moved to the initial home node of object o and the corresponding dedicated super-leaders are unmarked. To compute the total at level l, a reference leader node s_z is chosen at the level l, and all dedicated super-leaders at that level send their transaction counts to s_z. Consequently, s_z sums all the counts of the number of transactions it received and sends the calculated sum to all the dedicated super-leaders of that level.

Phase 3: In this phase, the algorithm determines the schedule to move the object o to each dedicated super-leader and executes the transactions. This can be done in two ways: by using a universal TSP tour or an MST-based tour. Pseudocode and descriptions of the approaches are omitted due to space constraints.

Analysis. In the fully distributed algorithm FULLYDISTRIBUTED_SINGLEOBJ, each node initially knows only its own transactions. The key challenge is to compute the set of super-leaders, which is handled in Phase 1 of the algorithm. After that, the algorithm proceeds similarly to Algorithm 1. Thus, any additional communication cost comes only from Phase 1 of the distributed algorithm. Nevertheless, we show that FULLYDISTRIBUTED_SINGLEOBJ also achieves the same approximation as of GLOBALAWARE_SINGLEOBJ (i.e., Algorithm 1).

Let C be the total communication cost of Algorithm 1 that involves the movement of transactions from current node to the respective dedicated super-leaders (say C_1) and the movement of object between the dedicated super-leaders in H (say C_2). That means, $C = C_1 + C_2$.

Let C' be the total communication cost of FULLYDISTRIBUTED_SINGLEOBJ, which involves three phases. Let C'_{p_1}, C'_{p_2}, and C'_{p_3} be the communication cost of Phase 1, Phase 2, and Phase 3, respectively. Then, $C' = C'_{p_1} + C'_{p_2} + C'_{p_3}$. Since the execution of Phase 2 and Phase 3 of the algorithm is equivalent to Algorithm 1, we have, $C'_{p_2} + C'_{p_3} = C$. That means, $C' = C'_{p_1} + C$.

Phase 1 runs only once throughout the execution. Three types of messages are exchanged between different nodes of H during the execution of Phase 1.

i. Each node sends its transaction information to its cluster leader at level $l = 0$ in H. This cost is at most C_1.
ii. Each cluster leader at level $l \geq 0$ sends the count of total transactions in its cluster to the leader of parent cluster at level $l + 1$. This cost is at most C_2.
iii. When a cluster leader becomes a super-leader, it recursively notifies all child leaders down to the level 0. This cost is also at most C_2.

Hence, $C'_{p_1} \leq C_1 + 2C_2 < 2C$, which implies, $C' < 2C + C < 3C$. Therefore, the following theorem is immediate:

Theorem 3. *Algorithm 2 achieves $O(\log n \cdot \log D)$–approximation in communication cost when using universal TSP tour, and $O(\log D)$–approximation when using universal TSP tour of the object.*

5.2 Multiple Objects

In this section, we briefly describe how the fully distributed algorithm for single object can be extended to handle the transactions accessing multiple shared objects. Consider a set of shared objects $\mathcal{O} = \{o_1, o_2, \ldots\}$, and each transaction $T_i \in \mathcal{T}$ accesses at most k objects from $\mathcal{O}$. As in FULLYDISTRIBUTED_SINGLEOBJ, transaction information is first shared through the levels of the hierarchy H. Then, we calculate the dedicated super-leaders with respect to each object by running FULLYDISTRIBUTED_SINGLEOBJ. After that, the set of dedicated super-leaders for each transaction $T_i \in \mathcal{T}$ are determined by combining the object-specific super-leaders as in Algorithm 2. Finally, we calculate the TSP tour on the dedicated super-leaders. Each object $o_j \in \mathcal{O}$ visits the respective dedicated super-leader following the TSP tour and transactions are executed at each dedicated super-leader when the required objects gather.

6 Conclusions

In this paper, we studied transaction scheduling problem in distributed systems modeled as constant doubling dimension graphs. We used a dual-flow transaction execution model in which both objects and transactions can move within the network to minimize communication cost. We proposed centralized and fully distributed versions of the algorithm for transaction scheduling that achieve polylogarithmic approximations in communication cost. In the future, it would be interesting to extend the algorithms and analysis for general and arbitrary graphs. Experimental evaluation of the designed algorithms against different application benchmarks in a practical setting would also be interesting.

Acknowledgments. This paper is supported by NSF grant CNS-2131538.

References

1. Abraham, I., Gavoille, C., Goldberg, A.V., Malkhi, D.: Routing in networks with low doubling dimension. In: 26th IEEE International Conference on Distributed Computing Systems (ICDCS'06), pp. 75–75. IEEE (2006)
2. Adhikari, R., Busch, C.: Lockless blockchain sharding with multiversion control. In: International Colloquium on Structural Information and Communication Complexity, pp. 112–131. Springer (2023)
3. Adhikari, R., Busch, C., Kowalski, D.R.: Stable blockchain sharding under adversarial transaction generation. In: Proceedings of the 36th ACM Symposium on Parallelism in Algorithms and Architectures, pp. 451–461 (2024)

4. Busch, C., Chlebus, B.S., Herlihy, M., Popovic, M., Poudel, P., Sharma, G.: Flexible scheduling of transactional memory on trees. Theoret. Comput. Sci. **978**, 114184 (2023)
5. Busch, C., Chlebus, B.S., Kowalski, D.R., Poudel, P.: Stable scheduling in transactional memory. In: International Conference on Algorithms and Complexity, pp. 172–186. Springer (2023)
6. Busch, C., Herlihy, M., Popovic, M., Sharma, G.: Impossibility results for distributed transactional memory. In: Proceedings of the 2015 ACM Symposium on Principles of Distributed Computing, pp. 207–215 (2015)
7. Busch, C., Herlihy, M., Popovic, M., Sharma, G.: Fast scheduling in distributed transactional memory. In: Proceedings of the 29th ACM Symposium on Parallelism in Algorithms and Architectures, pp. 173–182 (2017)
8. Busch, C., Herlihy, M., Popovic, M., Sharma, G.: Dynamic scheduling in distributed transactional memory. Distrib. Comput. **35**(1), 19–36 (2022)
9. Cao, Y., Fan, W., Ou, W., Xie, R., Zhao, W.: Transaction scheduling: from conflicts to runtime conflicts. Proc. ACM Manag. Data **1**(1), 1–26 (2023)
10. Ceccarello, M., Pietracaprina, A., Pucci, G., Upfal, E.: Mapreduce and streaming algorithms for diversity maximization in metric spaces of bounded doubling dimension. Proc. VLDB Endowment **10**(5), 469–480 (2017)
11. Chan, T.H.H., Gupta, A., Maggs, B.M., Zhou, S.: On hierarchical routing in doubling metrics. ACM Trans. Algorithms (TALG) **12**(4), 1–22 (2016)
12. Gao, J., Guibas, L., Milosavljevic, N., Zhou, D.: Distributed resource management and matching in sensor networks. In: 2009 International Conference on Information Processing in Sensor Networks, pp. 97–108 (2009)
13. Goudarzi, M., Palaniswami, M., Buyya, R.: Scheduling IoT applications in edge and fog computing environments: a taxonomy and future directions. ACM Comput. Surv. **55**(7), 1–41 (2022)
14. Gramoli, V., Lu, Z., Tang, Q., Zarbafian, P.: Aoab: optimal and fair ordering of financial transactions. In: 2024 54th Annual IEEE/IFIP International Conference on Dependable Systems and Networks (DSN), pp. 377–388. IEEE (2024)
15. Gupta, A., Krauthgamer, R., Lee, J.R.: Bounded geometries, fractals, and low-distortion embeddings. In: 44th Symposium on Foundations of Computer Science, FOCS 2003, Cambridge, October 11-14, 2003, Proceedings, pp. 534–543. IEEE Computer Society (2003)
16. Hendler, D., Naiman, A., Peluso, S., Quaglia, F., Romano, P., Suissa, A.: Exploiting locality in lease-based replicated transactional memory via task migration. In: DISC, pp. 121–133 (2013)
17. Herlihy, M., Moss, J.E.B.: Transactional memory: architectural support for lock-free data structures. In: ISCA, pp. 289–300 (1993)
18. Herlihy, M., Sun, Y.: Distributed transactional memory for metric-space networks. Distrib. Comput. **20**, 195–208 (2007)
19. Jayaprakash, A., Salavatipour, M.R.: Approximation schemes for capacitated vehicle routing on graphs of bounded treewidth, bounded doubling, or highway dimension. ACM Trans. Algorithms **19**(2), 1–36 (2023)
20. Jia, L., Lin, G., Noubir, G., Rajaraman, R., Sundaram, R.: Universal approximations for tsp, steiner tree, and set cover. In: Proceedings of the Thirty-seventh Annual ACM Symposium on Theory of Computing, pp. 386–395 (2005)
21. Kang, Y., Pan, L., Liu, S.: Job scheduling for big data analytical applications in clouds: a taxonomy study. Futur. Gener. Comput. Syst. **135**, 129–145 (2022)
22. Khiat, A., Haddadi, M., Bahnes, N.: Genetic-based algorithm for task scheduling in fog-cloud environment. J. Netw. Syst. Manage. **32**(1), 3 (2024)

23. Kim, J., Ravindran, B.: On transactional scheduling in distributed transactional memory systems. In: Symposium on Self-Stabilizing Systems, pp. 347–361. Springer (2010)
24. Kitamura, N., Kitagawa, H., Otachi, Y., Izumi, T.: Low-congestion shortcut and graph parameters. Distrib. Comput. **34**(5), 349–365 (2021). https://doi.org/10.1007/s00446-021-00401-x
25. Konjevod, G., Richa, A.W., Xia, D.: Dynamic routing and location services in metrics of low doubling dimension. In: Proceedings of the Twenty-seventh ACM Symposium on Principles of Distributed Computing, pp. 417–417 (2008)
26. Kuhn, F., Moscibroda, T., Wattenhofer, R.: On the locality of bounded growth. In: Proceedings of the Twenty-fourth Annual ACM Symposium on Principles of Distributed Computing, pp. 60–68 (2005)
27. Mokni, M., Yassa, S., Hajlaoui, J.E., Omri, M.N., Chelouah, R.: Multi-objective fuzzy approach to scheduling and offloading workflow tasks in fog-cloud computing. Simul. Model. Pract. Theory **123**, 102687 (2023)
28. Mulzer, W., Willert, M.: Compact routing in unit disk graphs. In: 31st International Symposium on Algorithms and Computation (ISAAC 2020), pp. 16–1. Schloss Dagstuhl–Leibniz-Zentrum für Informatik (2020)
29. Narman, H.S., Hossain, M.S., Atiquzzaman, M., Shen, H.: Scheduling internet of things applications in cloud computing. Ann. Telecommun. **72**(1), 79–93 (2017)
30. Nikoui, T.S., Balador, A., masoud Rahmani, A., Bakhshi, Z.: Cost-aware task scheduling in fog-cloud environment. In: 2020 CSI/CPSSI International Symposium on Real-Time and Embedded Systems and Technologies (RTEST), pp. 1–8 (2020)
31. Nikoui, T.S., Jabbehdari, S., Bagheri, A.: Providing a cloud broker-based approach to improve the energy consumption and achieve a green cloud computing. Int. J. Comput. Appl. **138**(1), 42–49 (2016)
32. Peixoto, M.L.M., Genez, T.A., Bittencourt, L.F.: Hierarchical scheduling mechanisms in multi-level fog computing. IEEE Trans. Serv. Comput. **15**(5), 2824–2837 (2021)
33. Poudel, P., Rai, S., Guragain, S.: Ordered scheduling in control-flow distributed transactional memory. Theor. Comput. Sci. **993**, 114463 (2024). https://doi.org/10.1016/J.TCS.2024.114463
34. Poudel, P., Rai, S., Sharma, G.: Processing distributed transactions in a predefined order. In: ICDCN '21: International Conference on Distributed Computing and Networking, Virtual Event, January 5-8, 2021, pp. 215–224. ACM (2021). https://doi.org/10.1145/3427796.3427819
35. Poudel, P., Sharma, G.: Graphtm: an efficient framework for supporting transactional memory in a distributed environment. In: ICDCN, pp. 11:1–11:10 (2020)
36. Saad, M.M., Ravindran, B.: Snake: control flow distributed software transactional memory. In: Défago, X., Petit, F., Villain, V. (eds.) Stabilization, Safety, and Security of Distributed Systems, pp. 238–252. Springer, Berlin Heidelberg, Berlin, Heidelberg (2011)
37. Sharma, G., Busch, C.: Distributed transactional memory for general networks. Distrib. Comput. **27**(5), 329–362 (2014). https://doi.org/10.1007/s00446-014-0214-7
38. Sharma, G., Busch, C.: A load balanced directory for distributed shared memory objects. J. Parallel Distr. Comput. **78**, 6–24 (2015)
39. Shavit, N., Touitou, D.: Software transactional memory. In: Proceedings of the Fourteenth Annual ACM Symposium on Principles of Distributed Computing, pp. 204–213 (1995)

40. Srinivasagopalan, S., Busch, C., Iyengar, S.: An oblivious spanning tree for single-sink buy-at-bulk in low doubling-dimension graphs. IEEE Trans. Comput. **61**(5), 700–712 (2011)
41. Tilevich, E., Smaragdakis, Y.: J-Orchestra: automatic java application partitioning. In: Magnusson, B. (ed.) ECOOP 2002. LNCS, vol. 2374, pp. 178–204. Springer, Heidelberg (2002). https://doi.org/10.1007/3-540-47993-7_8
42. Tran-Dang, H., Kim, D.S.: Disco: distributed computation offloading framework for fog computing networks. J. Commun. Netw. **25**(1), 121–131 (2023)
43. Usui, T., Behrends, R., Evans, J., Smaragdakis, Y.: Adaptive locks: combining transactions and locks for efficient concurrency. J. Parallel Distr. Comput. **70**(10), 1009–1023 (2010)
44. Zhang, Q., et al.: Efficient distributed transaction processing in heterogeneous networks. Proc. VLDB Endowment **16**(6), 1372–1385 (2023)

Efficient Distributed Algorithms for Shape Reduction via Reconfigurable Circuits

Nada Almalki[1]([✉)] ⓘ, Siddharth Gupta[2] ⓘ, Othon Michail[1] ⓘ,
and Andreas Padalkin[3] ⓘ

[1] Department of Computer Science, University of Liverpool, Liverpool, UK
`{n.almalki,othon.michail}@liverpool.ac.uk`
[2] Department of Computer Science & Information Systems, BITS Pilani, K K Birla
Goa Campus, Goa, India
`siddharthg@goa.bits-pilani.ac.in`
[3] Paderborn University, Paderborn, Germany
`andreas.padalkin@upb.de`

Abstract. In this paper, we study the problem of efficiently reducing geometric shapes into other such shapes in a distributed setting through size-changing operations. We develop distributed algorithms using the *reconfigurable circuit* model to enable fast node-to-node communication. We study the *connectivity* graph model. Let n denote the number of agents and k the number of turning points in the initial shape. We show that any tree-shaped configuration can be reduced to a single agent using only *shrinking* operations in $O(k \log n)$ rounds w.h.p., and to its *incompressible* form in $O(\log n)$ rounds w.h.p. given prior knowledge of the incompressible nodes, or in $O(k \log n)$ rounds otherwise. When both *shrinking* and *growth* operations are available, we give an algorithm that transforms any tree to a topologically equivalent one in $O(k \log n + \log^2 n)$ rounds w.h.p. On the negative side, we show that one cannot hope for $o(\log^2 n)$-round transformations for all shapes of $\Theta(\log n)$ turning points.

Keywords: growth process · shrinking process · collision avoidance · programmable matter

1 Introduction

Reconfiguration and autonomous formation of networks and geometric structures is an active area of research, with applications ranging from swarm and reconfigurable robotics to dynamic networks and self-assembly. A central problem in this area of reconfigurable systems is for a set of agents to transform from an initial shape into a target shape efficiently, and without violating structural and other constraints.

S. Gupta was supported, in part, by the Anusandhan National Research Foundation (ANRF) grant ANRF/ECRG/2024/006849/PMS.
A. Padalkin was supported by the DFG Project SCHE 1592/10-1. The full version of the paper with all omitted details is available on arXiv [2].

Much of the existing work has focused on *size-preserving* transformations, which guide the system through a sequence of configuration changes without altering its overall size. Other work, particularly in algorithmic self-assembly and distributed computing, has studied passive *size-changing* dynamics, whose accumulation leads to structural changes over time. Recent research has begun to investigate transformations that can actively form or deform a configuration by locally controlling the addition or removal of individual agents. These processes are motivated by biological and chemical systems, such as embryonic growth in multicellular organisms, and by dynamical systems, like cellular and self-reproducing automata. They are also related to recent work on algorithmic reconfiguration of graphs and networks [6,10,12,13].

To the best of our knowledge, the paper by Woods *et al.* [16] is the only study of a distributed model (called the *nubot* model) for exponentially fast size-changing transformations. In terms of its dynamics, their model uses a combination of size-changing and size-preserving rules. An example of the latter is local rotations causing relative movement between agents. The application of rules that would cause collisions or other structural violations is effectively excluded by the model. As an active self-assembly model, it also makes a set of specific molecular modeling assumptions, such as its evolution over time being a continuous-time Markov process (as in stochastic CRNs) and Brownian motion-style agitation.

Motivated by [16] and [12], Almalki and Michail [3] proposed a centralized geometric model that is more abstract than [16], aiming to understand what transformations can or cannot be performed (with the main focus being on in time polylogarithmic in the size of the target shape) if *only* size-changing rules are available. They studied restricted types of growth dynamics, conditioned to affect specific parts of the shape (e.g., whole columns) and defined in a way that ensures no collisions occur. Almalki *et al.* [1] generalized this to parallel growth rules that can be applied to any subset of the agents. These are not guaranteed to be collision-free *a priori*, so collision avoidance becomes the responsibility of the algorithm designer. They show that some classes of shapes can be grown efficiently (in polylogarithmic time), while others require superpolylogarithmic time.

In this work, we take this investigation further, to a fully distributed setting, remaining focused on understanding the structural and (distributed) algorithmic properties of size-changing dynamics. Communication is a central challenge in taking this step. Indeed, global communication based on classical distributed models such as message passing is exponentially slower than the speed of parallel reconfiguration available, thus cannot be useful. Other issues include the lack of common orientation or chirality, and constraints stemming from the decentralization and limited computational and communication resources available to the agents. To address these challenges, we make use, for the first time in this class of problems, of the *reconfigurable circuit* [9] communication model.

Parallel transformation processes can be exponentially fast and are highly dynamic, which makes it difficult to avoid structural violations like collisions

[11]. It is therefore important to have fast and reliable communication within the system—especially over long distances. In order to overcome these issues, Feldmann *et al.* [9] proposed the deployment of *reconfigurable circuits* which allow to connect subsets of agents. Each agent can send simple signals (beeps) through any circuit it is connected to, which are instantaneously received by all agents connected to the same circuit. Reconfigurable circuits have been shown to enable polylogarithmic-time solutions (in the number of agents n) to various problems, such as shape recognition/containment [5,9], spanning tree [15], and shortest path [14]. Reconfigurable circuits have so far only been used in the context of polylog-time distributed computations (primarily in the *amoebot* model of programmable matter and for graph problems) and size-preserving reconfiguration, and have not yet been applied to exponentially fast parallel reconfiguration.

In this work, we leverage reconfigurable circuits as a communication model for rapid, distributed size-changing transformations. In particular, we study distributed algorithms that must efficiently reduce a given shape into a target shape (sometimes a single node) through shrinking or by combining shrinking and growth, while avoiding collisions. We focus on *shape reduction* because it aligns naturally with *distributed* algorithms, especially in uniform models where agents act as *finite-state* automata. Unlike growth, which starts from a single agent and propagates outward— typically forming regular or simple structures— reduction does not require prior knowledge of the final shape. Achieving complex structures via *growth* would require each agent to store a complete map of the final shape or receive information proportional to its size, both are infeasible for limited-memory agents. This work thus offers a new application domain for reconfigurable circuits.

1.1   Contribution

We aim to understand the conditions under which a set of agents (also called *nodes* throughout) can efficiently reduce their geometric arrangement to another such arrangement (referred to as *shape* throughout), using either only *shrinking* operations or a combination of *shrinking* and *growth* operations. A set of *shrinking* operations reduces a shape by absorbing nodes into neighboring nodes, while a set of *growth* operations expands a shape by adding nodes adjacent to existing nodes.[1] Our transformations result from distributed algorithms executed by limited agents, each of which operates with constant memory and is computationally equivalent to a finite-state machine. For communication, we employ the *reconfigurable circuit* model, as its speed can support the fast dynamics of the transformations under consideration. Reconfigurable circuits enable efficient global coordination by allowing the nodes to exchange limited information instantly (in constant time) across a dynamically changing structure.

---

[1] As will become clear, shrinking in this work involves translation, making it fundamentally different from the shrinking rule of [16] which is merely node deletion.

The challenge lies in designing efficient distributed algorithms that perform these operations while maintaining structural integrity and avoiding collisions— undesirable structural violations where nodes overlap, intersect, or whose movement breaks the structure. Understanding the distributed complexity and developing distributed algorithms for size-changing transformation tasks is important for engineering self-deploying systems and to begin understanding some of the algorithmic properties involved in formation processes of nature. Our work is also relevant to research on extremely weak models of distributed computing, such as population protocols [4] and variants, beeping models [7], and programmable matter [8]. In contrast to the considerable amount of work on size-preserving dynamics (e.g., in reconfigurable robotics), algorithmically controlled size-changing dynamics are less understood and the development of appropriate models is still in its infancy. We aim to take a step in this direction by proposing a novel such model and initiating its study.

We consider connected geometric shapes formed by nodes in a two-dimensional square grid in the *connectivity* graph model, in which edges form only when new nodes are created. Our goal is to determine complexity bounds and design distributed algorithms for different reduction objectives, aiming to optimize time efficiency without violating structural constraints. Our main results are the *BFS shrinking* algorithm, which reduces any tree to a single node in $O(k \log n)$ rounds w.h.p.[2], where the parameter k represents the number of turning points and n the number of nodes in the initial shape, and the *incompressible tree* algorithm, which reduces any tree to its incompressible form in $O(\log n)$ rounds w.h.p. when the incompressible nodes are known in advance, or in $O(k \log n)$ rounds w.h.p. otherwise. Moreover, by combining these approaches, we can shrink any tree to a single node through its incompressible tree in $O(\log n + k \log k)$ rounds. These algorithms use only shrinking operations. The *target tree* algorithm, by combining shrinking and growth operations, ensures the reduction of any tree to any *topologically equivalent* tree in $O(k \log n + \log^2 n)$ rounds w.h.p. We also give centralized lower bounds (implying equivalent distributed lower bounds): (*i*) There exists an infinite family of pairs of *geometrically equivalent* paths of $k = \Theta(\sqrt[3]{n})$ turning points each such that, if only shrinking is available, $\Omega(k \log k)$ rounds are required to reduce one path to the other. (*ii*) There exists an infinite family of pairs of *geometrically equivalent* paths of $k = \Theta(\log n)$ turning points each such that, if only shrinking is available, $\Omega(\log^2 n)$ rounds are required to reduce one path to the other. These results are summarized in Table 1. In the full version of the paper [2], we also study an alternative graph model, the *adjacency model*, where edges are added between nodes as soon as they become adjacent.

The remainder of the paper is organized as follows. In Sect. 2, we formally introduce our *distributed* computational model and define the growth and shrinking operations. In Sect. 2.1, we define the *reduction* problem and various variants that we consider in this work. In Sect. 3, we present key algorithmic primitives of

---

Table 1. A summary of our results. All upper bounds require a preprocessing phase that requires $O(\log n)$ rounds w.h.p. that we omit in the bounds.

Reduction to	Bound
Single node	$O(k \log n)$ w.h.p.
Single node (known incompressible nodes)	$O(\log n + k \log k)$ w.h.p.
Geometrically equivalent tree $(k = \Theta(\sqrt[3]{n}))$	$\Omega(k \log k)$
Geometrically equivalent tree $(k = \Theta(\log n))$	$\Omega(\log^2 n)$
Incompressible tree	$O(k \log n)$
Incompressible tree (known incompressible nodes)	$O(\log n)$
Topologically equivalent tree	$O(k \log n + \log^2 n)$

previous work that serve as the basis for our algorithms. In Sect. 4, we develop distributed algorithms for different shape reduction problems under the *connectivity* model. In Sect. 5, we discuss potential directions for future research, including optimizing our bounds and extending our methods to other models and problems.

2   Models and Problems

We study a system composed of computational agents (we refer to the agents as *nodes* for the rest of the paper) that initially form a connected shape $S = (V, E)$ of size $|V| = n$ on a two-dimensional square grid. Each node $u \in V$ occupies a distinct point (u_x, u_y) of the grid. We adopt the connectivity model of [1]. In this model, $E \subseteq \{uv \mid u, v \in V$ and u, v are adjacent$\}$ is a set of edges between pairs of adjacent nodes, where two nodes $u = (u_x, u_y)$ and $v = (v_x, v_y)$ are *adjacent* if $u_x \in \{v_x - 1, v_x + 1\}$ and $u_y = v_y$ or $u_y \in \{v_y - 1, v_y + 1\}$ and $u_x = v_x$, that is, their orthogonal distance on the grid is one. We illustrate nodes as maximal circles inscribed within grid cells to represent their positions; however, our results hold for any geometry of individual nodes that does not trivially make nearby nodes intersect. In other words, while our model allows for an arbitrary geometry of individual nodes, we require that adjacent nodes are arranged so that their boundaries do not intersect. A shape is *connected* if the graph that defines it is a connected graph. Throughout the paper, we restrict attention to connected shapes.

Nodes operate as anonymous (and randomized) finite-state automata, executing actions in discrete synchronous rounds. Each node has a compass orientation and a chirality. Initially, the nodes may not agree on a common compass orientation or chirality.

The nodes are able to communicate through *reconfigurable circuits* as follows [9]. Each edge in E consists of a constant number c of *links*. The constant c is equal for all edges. In this paper, we will assume $c = 2$ which is sufficient and necessary for all results. Each incident node owns one endpoint of each link which we call *pins*. Each node labels its pins of each incident edge from 1 to c. If the two nodes incident to the same edge share a chirality, then the pin with label i is connected to a pin with label $c - i + 1$. Otherwise, the pin labels are matching, i.e., each pin with label i is connected to a pin with label i. Each node partitions its pins to disjoint *partition sets*. This partitioning is called the *circuit configuration* of the node. Let P denote the set of all partition sets of all nodes. Consider the graph $C = (P, L)$, where two partition sets in P are connected if they share pins of the same link. We call the connected components of C *circuits*. In each round, each node can reconfigure its circuit configuration, i.e., it can change its partitioning, and send a simple signal (a *beep*) on an arbitrary number of its partition sets. This beep is received by all nodes connected to the same circuit at the beginning of the next round. However, they do not obtain any information about the origin of the beeps or the number of beeps in case multiple nodes decide to beep on the same circuit.

After describing the static structure of the system and the communication process, we now define the operations that nodes can perform. In each round, nodes process incoming signals, transition to new states accordingly, and may execute one of the two operations—growth or shrinking—or take no action and remain in their current state. A *growth* operation applied on a node u generates a new node in one of the points adjacent to u and possibly translates some part of the shape. A *shrinking* operation, on the other hand, is applied on a node u to absorb an adjacent node v, and possibly translate v's neighbors towards u. In this context, translation is the outcome of the applied operation, where the addition of new nodes or removal of existing nodes within the shape pushes or pulls its parts, causing a shift in position. One or more growth or shrinking operations applied in parallel to nodes of a shape S either cause a *collision* or yield a new shape S'. There are two types of collisions: *node collisions* and *cycle collisions*. When describing the outcome of these operations, we assume there is an *anchor* node which is stationary, and other nodes move relative to it. This is without loss of generality (w.l.o.g.) under the assumption that the constructed shapes are equivalent up to translations.

Temporarily disregarding collisions, we begin by defining the effect of a single operation on a tree shape—first for growth, then for shrinking. Collisions will be handled later for the general case, where one or more operations are applied in parallel. Let $T = (V, E)$ be a tree and $u_0 \in V$ its anchor. We set u_0 to be the root of T. A single growth operation is applied on a node $u \in V$ toward a point (x, y) adjacent to u. There are two cases that are free from collisions: (i) there is no edge between u and (x, y), (ii) (x, y) is occupied by a node v and $uv \in E$. We first define the effect in each case. In case (i), the growth operation generates a node u' at point (x, y) and connects it to u. In case (ii), assume w.l.o.g. that u is closer to u_0 in T than v. Let $T(v)$ denote the subtree of T rooted at node v. Then, the operation generates a node u' between u and v, connected to both, which translates $T(v)$ by one unit away from u along the axis parallel to uv. After this, u' occupies (x, y) and uv has been replaced by $\{uu', u'v\}$. We refer to a *neighbor handover* as the process by which a node u

transfers a neighbor w, located perpendicular to the direction of an operation (e.g., growth), to another node u'. This happens by a unit translation of $T(w)$ or $T(u)$ along the axis parallel to uu', depending on which of u, w, respectively, is closer to u_0 in T.

We now define the effect of a single shrinking operation on a tree shape. A single *shrinking* operation is applied on a node u to absorb an adjacent node v, located either horizontally or vertically next to u. Assume that u is at (u_x, u_y) and v at (v_x, v_y). Let $\langle d_x, d_y \rangle = \langle x_u - x_v, y_u - y_v \rangle \in \{\langle \pm 1, 0 \rangle, \langle 0, \pm 1 \rangle\}$ denote the unit vector from v toward u. There are two cases: (*i*) if v has no neighbors (other than u), the operation removes v and deletes the edge uv; (*ii*) if v has neighbors in the direction orthogonal to or opposite of $\langle d_x, d_y \rangle$, each of these neighbors is translated by one unit in the direction $\langle d_x, d_y \rangle$ pulling them towards u. For example, if v is the east or west of u in a horizontal shrinking (i.e., $d_y = 0$) where a node at (v_x, v_{y+1}) is moved to $(v_x + d_x, v_{y+1})$, one at (v_{x+1}, v_y) is moved to $(v_{x+1} + d_x, v_y)$, and one at (v_x, v_{y-1}) is moved to $(v_x + d_x, v_{y-1})$. These translated nodes are then connected to u, replacing their previous connection to v. The shrinking operation is applied only when all target positions for the translated nodes are unoccupied, ensuring the process is collision-free.

Let Q be a set of operations to be applied *in parallel* to a connected shape S, each operation on a distinct pair of nodes or a node and an unoccupied point. We assume that all operations in such a set are applied *concurrently*, have the same *constant execution speed*, and their *duration* is equal to one *round*. A *node collision* occurs if the trajectories of any two nodes meet. A *cycle collision* occurs when two parts of a cycle grow unequally. A set of operations is said to be *collision free* if it does not cause any node or cycle collisions.

A growth (shrinking) process σ starts from an initial shape S_I and in each round $r \geq 1$, applies a set of parallel growth or shrinking operations—possibly a single operation—on the current shape S_{r-1} to give the next shape S_r, until a final shape S_F is reached at a round r_f. In the case of a growth process, we say that, σ *grows* S_F *from* S_I *in* r_f *rounds*; in the case of a shrinking process, we say that σ *shrinks* S_I *to* S_F *in* r_f *rounds*. We also assume that parallel operations have the same cardinal direction and that a node gets at most one operation per round. These assumptions simplify the description of algorithms and can be easily dropped.

A node $u \in V$ of a shape S is a leaf if and only if its degree is one, i.e., $\delta(u) = 1$. Let $L \subseteq V$ denote the set of all leaves. A node $u \in V$ of a shape S is called a *turning point* if either $u \in L$ or if there are two neighbors v_1 and v_2 of u such that $v_1 u$ is perpendicular to $u v_2$. For example, in the special case of a path shape $P = (u_1, u_2, \ldots, u_n)$, a node u_i is a turning point if either $i \in \{1, n\}$ or $u_{i-1} u_i$ is perpendicular to $u_i u_{i+1}$. For uniformity of our arguments, we add the endpoints of a path to the set of its turning points. We denote the set of all turning points by $TP \subseteq V$, and by definition, $L \subseteq TP$. A node $u_j \in P$ is a segment node if it lies between any two consecutive turning points and satisfies $\delta(u) = 2$. The sequence of segment nodes between two turning points forms a segment s of length ℓ, where ℓ is the number of nodes in the segment, including both turning points.

We define the *relative position* of $v = (v_x, v_y)$ with respect to $u = (u_x, u_y)$ by $r(u,v) = (r_H(u,v), r_V(u,v)) \in \{\rightarrow, 0, \leftarrow\} \times \{\uparrow, 0, \downarrow\}$ such that, $r_H(u,v) = \{\rightarrow$ if $u_x < v_x$, 0 if $u_x = v_x$, $\leftarrow$ if $u_x > v_x\}$, and $r_V(u,v) = \{\uparrow$ if $u_y < v_y$, 0 if $u_y = v_y$, $\downarrow$ if $u_y > v_y\}$. We call any two trees T and T' *geometrically equivalent* if and only if there exists a one-to-one correspondence between their turning points such that only the segment lengths and the relative positions of the turning points may differ. We call two *geometrically equivalent* trees T and T' additionally *topologically equivalent* if and only if the relative positions of all corresponding turning points are the same.

A *column* or *row of a shape S* refers to a row or column of the grid that contains at least one node of S. We call a column or row of a shape S *compressible* if all nodes in that row or column are segment nodes, i.e., the row or column contains no turning points. Otherwise, it is *incompressible*. We call the nodes of incompressible columns or rows *incompressible nodes*. We define the incompressible form of a shape S, denoted by $i(S)$, as the shape obtained by removing all compressible rows and columns from S. The resulting shape is incompressible and therefore topologically equivalent to S [1].

2.1   Problems

We study the problem of reducing an initial shape $S_I = (V_I, E_I)$ to a target shape $S_F = (V_F, E_F)$, where $|V_F| \leq |V_I|$. We assume that the constructed shapes are equivalent up to translations. In what follows, we restrict attention to S_I and S_F being trees, denoted T_I and T_F, respectively. Throughout, n denotes the number of nodes and k the number of turning points in T_I. For any pair of geometrically or topologically equivalent trees T_I, T_F, let s_i denotes the same segment with lengths $\ell(s_i^I)$ and $\ell(s_i^F)$ in T_I and T_F, respectively. For such trees, the input convention assumes that in T_I each segment s_i^I stores a binary representation of $\ell(s_i^I)$ and $\ell(s_i^F)$. We study the following problems:

SINGLE NODE REDUCTION. Let T_I be an initial tree. The system must reduce itself from T_I to $T_F = (\{u_0\}, \emptyset)$ while avoiding collisions.

GEOMETRICAL REDUCTION. As in SINGLE NODE REDUCTION, with T_F now being a tree geometrically equivalent to T_I, satisfying $\ell(s_i^F) \leq \ell(s_i^I)$ for all segments s_i.

TOPOLOGICAL REDUCTION. As in GEOMETRICAL REDUCTION, with T_F additionally being topologically equivalent to T_I.

INCOMPRESSIBLE REDUCTION. A restricted version of TOPOLOGICAL REDUCTION, where $T_F = i(T_I)$.

We also consider a variant of SINGLE NODE REDUCTION and INCOMPRESSIBLE REDUCTION where each node additionally knows in advance if it is an incompressible node.

3   Preliminaries

In this section, we highlight fundamental primitives presented in previous work, which form the basis for our proposed algorithms. Feldmann *et al.* [9] have proposed algorithms that allow us to elect a unique leader node, to align the compass orientation of all nodes, and to obtain an agreement about the chirality of all nodes. Each of these algorithms requires $O(\log n)$ rounds w.h.p. Our algorithms utilize (some of) these algorithms in a preprocessing step which we will explicitly state for each algorithm.

In case we perform some subroutine with varying runtimes in different parts of the shape in parallel, we use the synchronization mechanism of [15] to keep them synchronized. This only adds a constant factor to the runtime. In the following, we summarize the key primitives used throughout the paper. Additional technical details, including binary representation and the *primary and secondary circuit (PASC)* algorithm, are presented in the full version of the paper [2].

Lemma 1 (Feldmann et al. *[9],* **Padalkin** et al. *[15]).* *Let s be a segment of m nodes. PASC algorithm computes the length of s and stores it within s in $O(\log m)$ rounds. Given an $m' \leq m$, the PASC algorithm also identifies the first m' nodes of s in $O(\log m)$ rounds.*

Let $u = (u_x, u_y)$ and $v = (v_x, v_y)$ be two nodes. Let $|u_x - v_x|$ $(|u_y - v_y|)$ denote the *horizontal (vertical) distance* of u to v. The PASC algorithm can be adapted to compute the horizontal (vertical) distance of each node u_j to a reference node u_r. We refer to [15] for more details.

Lemma 2 (Padalkin et al. *[15]).* *Let S be a shape of m nodes and u_r an arbitrary reference node of S. The spatial PASC algorithm computes the sign of the horizontal (vertical) distance of u_r to each node u_j and stores it in u_j in $O(\log m)$ rounds.*

4   Connectivity Model

We start our examination in the connectivity model. In each of the following subsections, we study one of the four reduction problems: SINGLE NODE REDUCTION, GEOMETRICAL REDUCTION, INCOMPRESSIBLE REDUCTION, and TOPOLOGICAL REDUCTION.

4.1   Single Node Reduction

In this section, we present the *BFS shrinking* algorithm, which reduces any arbitrary tree T to a single node. We assume common chirality or establish it in a preprocessing phase (see Sect. 3). Note that we do not apply the synchronization mechanism from [15], as operations progress uniformly across the shape. The following presents this in detail.

BFS Shrinking. This algorithm consists of three subroutines, *segment detection, segment coloring and shrinking*, and *final segment shrinking*. First, we identify segments starting from the leaves and progressing toward their turning points. Once a segment is identified, it proceeds to *segment coloring and shrinking* subroutine, where it is colored and shrunk by parity-based operations. This recursive process continues until T reduces to a single node or segment. In the latter case, the *final segment shrinking* subroutine handles simultaneous leaf formation at both endpoints. The algorithm completes in $O(k \log n)$ rounds.

Segment Detection. This subroutine allows nodes to detect leaves in their segment and determine their direction by establishing two parallel circuits along each segment s_i. For simplicity, we assume all nodes share the same chirality (clockwise or counterclockwise). Otherwise, we establish this assumption in a preprocessing phase . A complete description of the parallel circuits construction and the processing of signals is given in the full version of the paper [2].

Now, there are three possible scenarios for each segment s_i. First, if there is no signal (beep) in the segment s_i, neither endpoints of s_i is a leaf. In this case, the segment remains inactive, waiting for one of its endpoints to become a leaf. Second, if there is a beep on only one circuit, exactly one endpoint is a leaf. The segment then proceeds to the *segment coloring and shrinking* subroutine. Third, if there is a beep on both circuits (i.e., two leaves send signals simultaneously), both endpoints of the segment are leaves. In this case, we proceed with the *final segment shrinking* subroutine where the leaves perform a leader election.

Lemma 3. *Each segment agrees on a common orientation once one of its endpoints is a leaf. This process requires $O(1)$ rounds for all segments.*

After applying the *segment detection* subroutine, we proceed with *segment coloring and shrinking*. To compute a coloring, we use the first round of the PASC algorithm by Padalkin et al. [15], which compute a parity of each node in a segment based on its distance from the a reference node. Nodes are then colored blue (even parity) or green (odd parity), and shrinking proceeds accordingly.

Lemma 4. *For any segment s_i, each iteration of the* segment coloring *subroutine completes in $O(1)$ rounds.*

Lemma 5. *The* segment coloring and shrinking *subroutine completes in $O(\log \ell)$ rounds, where ℓ is the length of the segment.*

Final Segment Shrinking. When T is shrunk to a single segment, it is not possible to determine an orientation during the segment detection subroutine, which implies that we cannot decide which endpoint of the segment will perform the PASC for the coloring and shrinking process. We resolve this via a applying leader election between the leaves.

Lemma 6. *The final segment agrees on a common orientation through a leader election process, which requires $O(\log n)$ rounds w.h.p.*

By Lemmas 3, 4, 5, 6, we conclude the following:

Theorem 1. *After an $O(\log n)$-round preprocessing (w.h.p.), the* BFS shrinking *shrinks any tree T to a single node in $O(k \log n)$ rounds w.h.p.*

4.2   Geometrical Reduction

In this section, we consider the reduction of a tree T_I into a geometrically equivalent tree T_F. We will show some lower bounds for the case that we are only allowed to perform shrinking operations. Note that in general, it is not always possible to perform such a reduction with only shrinking operations.

Theorem 2. *Let $\mathcal{A}$ be an algorithm that reduces a path T_I to a geometrically equivalent path T_F using only shrinking operations. Then, there exists an infinite family of pairs of paths of $k = \Theta(\sqrt[3]{n})$ turning points each, for which $\mathcal{A}$ requires $\Omega(k \log k)$ rounds.*

The construction used in the proof of the above theorem appears in the full version on arXiv [2]. The following corollary shows that we cannot hope for an $o(\log^2 n)$-rounds algorithm even for small $k = \Theta(\log n)$.

Corollary 1. *Let $\mathcal{A}$ be an algorithm that reduces a path T_I to a geometrically equivalent path T_F using only shrinking operations. Then, there exists an infinite family of pairs of paths of $k = \Theta(\log n)$ turning points each, for which $\mathcal{A}$ requires $\Omega(\log^2 n)$ rounds.*

4.3   Incompressible Reduction

In this section, we present the *incompressible tree* algorithm which reduces an initial arbitrary tree T_I to its incompressible form $i(T_I)$. We assume that we have a leader node given and that the nodes share a common compass orientation and chirality. Otherwise, we establish these assumptions in a preprocessing phase (see Sect. 3). This allows us to define a direction for each segment (w.l.o.g. $\rightarrow$ and $\downarrow$) and with that an order.

Incompressible Tree Reduction. In the following, we present the *incompressible tree* algorithm. For now, we assume that each node knows whether it is an incompressible node and with that whether it belongs to a compressible column and row. Later, we show how to remove this assumption. Almalki *et al.* [1] have proven that in order to obtain the incompressible shape, we have to compress all compressible columns and rows.

The nodes of a maximal sequence of consecutive compressible columns (rows) form sets of parallel horizontal (vertical) subsegments of the same length [1] which we call *compressible subsegments*. Hence, in order to shrink the compressible columns (rows), we need to shrink the compressible subsegments. For that, we apply the *segment coloring and shrinking* subroutine to compress all compressible columns and rows (see Lemma 5). Since we cannot shrink a segment to length 0, we add the preceding incompressible node to each compressible subsegment and shrink the resulting subsegments to 1. We proceed as follows. First, we apply the subroutine on all horizontal compressible subsegments in parallel. Then, we repeat the procedure with all vertical compressible subsegments.

Theorem 3. *After an $O(\log n)$-round preprocessing (w.h.p.), and given the incompressible nodes, the* incompressible tree *algorithm reduces the initial tree T_I to its incompressible form $i(T_I)$ in $O(\log n)$ rounds.*

Note that an incompressible tree contains at most $O(k^2)$ nodes since each column and row must contain at least one of the k turning points. Hence, by running the incompressible tree algorithm prior to the BFS shrinking algorithm (see Theorem 1), we can improve the runtime.

Corollary 2. *Given the incompressible nodes, the* incompressible tree *algorithm combined with the* BFS shrinking *algorithm reduces the initial tree to a single node in $O(\log n + k \log k)$ rounds w.h.p.*

Incompressible Node Computation. If we do not have the incompressible nodes given, we need to compute them before applying the *incompressible tree* algorithm. First, note that given an arbitrary node $u \in S$, we can apply the spatial PASC algorithm to compute for each node $v \in S$ its relative position to u, i.e., $r(u, v)$.

Corollary 3. *Let $u \in S$. The* spatial PASC algorithm *computes $r(u, v)$ for each $v \in S$ in $O(\log n)$ rounds.*

We now explain how to compute all incompressible nodes and with that all compressible subsegments. Recall that compressible columns (rows) do not contain any turning points. This is the case if the horizontal (vertical) distance of the column (row) to all turning points is not 0. Hence, we simply iterate through all turning points $u \in TP$ and apply the spatial PASC algorithm such that each node $v \in S$ can check whether its column (row) contains any turning points, i.e., whether for all $u \in TP$, $r_H(u, v) \neq 0$ ($r_V(u, v) \neq 0$).

In order to iterate through all turning points, we utilize the *election primitive*[3] for trees of [14] which requires a single round.

Lemma 7. *We compute all incompressible nodes and with that all compressible segments in $O(k \log n)$ rounds.*

Corollary 4. *Given no additional assumptions, after an $O(\log n)$-round preprocessing (w.h.p.), the* incompressible tree *algorithm reduces the initial tree to its incompressible shape in $O(k \log n)$ rounds.*

Note that in case no additional assumptions are given, i.e., the incompressible nodes are not given, the *incompressible tree* algorithm does not help to reduce the runtime of the BFS shrinking algorithm since their runtimes match.

4.4 Topological Reduction

In this section, we employ both shrinking and growth operations to achieve the reduction objectives. We begin by presenting the *target tree* algorithm, which reduces an arbitrary tree T_I to another topologically equivalent tree T_F.

---

[3] Note that the election primitive is not a leader election algorithm. It utilizes a leader node to elect a node from a given set. In case we apply the primitive in a segment, we can use the first node for that.

Recall that $\ell(s_i^I)$ $(\ell(s_i^F))$ denote the length of segment s_i in the initial (target) tree. We assume that for each i, $\ell(s_i^F) \leq \ell(s_i^I)$. Furthermore, we assume that initially, each segment s_i stores a binary representation of $\ell(s_i^F)$ (see Sect. 3) which is possible due to the first assumption. We also assume that we have a leader node given and that the nodes share a common compass orientation and chirality. Otherwise, we establish these assumptions in a preprocessing phase (see Sect. 3). Note that $\ell(s_i^F) \leq \ell(s_i^I)$ for all i is not sufficient for T_F to be a topologically equivalent tree of T_I. We are able to test whether T_I and T_F are topologically equivalent.

Topological Equivalence Test. For each turning point u, we apply the spatial PASC algorithm on both T_I and T_F, so that all other turning points can compare their relative positions to u in T_I and T_F, respectively. However, we are not able to apply the subroutine on T_F directly. Instead, we will simulate T_F in T_I by identifying a unique node in T_I for each node in T_F. All other nodes will simply forward the signals between their neighbors. This is effectively the same as if those nodes were not present in the execution. Note that each turning point in T_F already has a corresponding node in T_I. Hence, we only have to identify $\ell(s_i^F) - 2$ further nodes in each segment s_i. We use the PASC algorithm to identify the first $\ell(s_i^F) - 2$ segment nodes in each s_i.

Lemma 8. *Let $\mathcal{A}$ be any algorithm that does not apply any growth or shrinking operations. If for each i, $\ell(s_i^F) \leq \ell(s_i^I)$, T_I can simulate $\mathcal{A}$ on T_F after a preprocessing time of $O(\log n)$ rounds.*

The *topological equivalence subroutine* proceeds as follows. We iterate through all turning points by utilizing the election primitive of [14] as before. For each turning point u, we apply the spatial PASC algorithm on T_I and T_F. This allows each of the other turning points, to compare its relative position to u in T_I and T_F. Finally, after comparing all relative positions, we perform a notification round where we check whether there is a turning point that has identified a violation, i.e., a difference of its relative position to another turning point. For that, we first establish a circuit connecting all nodes. Then, each turning point that has identified a violation beeps on that circuit. T_F is topologically equivalent to T_I if and only if there was no beep.

Lemma 9. *The* topological equivalence *subroutine checks whether T_F is topologically equivalent to T_I in $O(k \log n)$ rounds.*

Target Tree Reduction. We start by outlining the challenges of reducing the initial tree to a target tree. The first idea one may have is to apply the *segment coloring and shrinking* subroutine (see Lemma 5) on all segments in parallel until each segment has reached its target length. However, even if T_I and T_F are topologically equivalent, this approach does not guarantee that intermediate trees stay topologically equivalent which may cause collisions.

To always avoid collisions, we will apply the *segment coloring and shrinking* subroutine on the compressible subsegments instead of the tree segments. By uniformly changing the length of the compressible subsegments belonging to the same sequence of consecutive compressible columns (rows), we can make sure that all intermediate trees are topologically equivalent and with the guarantee that no collisions can occur. Let $\ell(c_j^I)$ $(\ell(c_j^F))$ denote the length of the compressible subsegment c_j in the initial (target) tree. However, the assumption that for each i, $\ell(s_i^F) \leq \ell(s_i^I)$ does not imply that for each j, $\ell(c_j^F) \leq \ell(c_j^I)$, i.e., some compressible subsegments grow. It is even possible that $\ell(c_j^F)/\ell(c_j^I) = \Omega(n)$ which makes it impossible to store $\ell(c_j^F)$ within c_j.

In the following, we will explain how to deal with all these challenges. We first outline our *target tree* algorithm. It consists of three phases: *compressible subsegment computation* phase, *growth* phase, and *shrinking* phase. In the *compressible subsegment computation* phase, we compute the lengths of all compressible subsegments of T_I and T_F. Then, we grow or shrink all compressible subsegments in the other two phases. In the *growth* phase, we grow all compressible subsegments c_j with $\ell(c_j^F) > \ell(c_j^I)$ and in the *shrinking* phase, we shrink all compressible subsegments c_j with $\ell(c_j^F) < \ell(c_j^I)$.

Compressible Subsegment Computation phase. We compute the lengths of all compressible subsegments of T_I and T_F, respectively. For that, we first compute all compressible columns and rows of T_I and T_F (see Lemma 7). Note that we need to simulate the subroutine for T_F (see Lemma 8). Then, we apply the PASC algorithm on each compressible subsegment c_j^I and c_j^F in parallel to compute its length $(+1)$ which we store within the segment itself. Consider the compressible subsegments c_j of a segment s_i. For each c_j, s_i now stores $\ell(c_j^I)+1$ and $\ell(c_j^F)+1$. However, these lengths are not necessarily aligned since the incompressible nodes may differ between T_I and T_F. Hence, we apply the transfer primitive to shift $\ell(c_j^I)$ and $\ell(c_j^F)$ of each c_j to a subsegment c_j' of s_i.

Lemma 10. *In the* compressible subsegment computation *phase, the* target tree *algorithm computes* $\ell(c_j^I) + 1$ *and* $\ell(c_j^F) + 1$ *for each* c_j *and stores it in a segment* c_j' *in* $O(k \log n)$ *rounds.*

Growth and Shrinking phases. In the growth (shrinking) phase, we grow (shrink) all compressible subsegments c_j with $\ell(c_j^F) > \ell(c_j^I)$ $(\ell(c_j^F) < \ell(c_j^I))$. Observe that it does not matter which nodes exactly perform a growth (shrinking) operation in a segment as long as the number of operations is the same. The idea is therefore to first compute a subsegment c_j'' for each c_j and then to perform the growth (shrinking) operations of c_j in c_j''. Note that we cannot use c_j' for c_j since it may not have enough nodes, i.e., $\ell(c_j') < \ell(c_j^I) + 1$.

We split the phase into two subphases: *segmentation* subphase and *transformation* subphase. In the *segmentation* subphase, we compute a subsegment c_j'' for each c_j and in the *transformation* subphase, we apply the growth (shrinking) operations. In both subphases, we will only consider the c_j's relevant for the phase, i.e., the c_j's with $\ell(c_j^F) > \ell(c_j^I)$ $(\ell(c_j^F) < \ell(c_j^I))$ for the growth (shrinking) phase. Note that each c_j' is able to compare $\ell(c_j^F)$ and $\ell(c_j^I)$ to determine in which phase it is participating. We first perform the *segmentation* subphase, followed by the *transformation* subphase. For details, see [2].

Lemma 11. *In the* segmentation *subphase, the* target tree *algorithm computes a disjoint subsegment* c_j'' *with length* $\max\{\ell(c_j^I), \ell(c_j^F)\} + 1$ *and stores* $\ell(c_j^F)$ *and* $\ell(c_j^I)$ *in it in* $O(k \log n)$ *rounds.*

Lemma 12. *In the* transformation *subphase, the* target tree *algorithm grows (shrinks) all compressible subsegments* c_j *with* $\ell(c_j^F) > \ell(c_j^I)$ *(*$\ell(c_j^F) < \ell(c_j^I)$*) to a length of* $\ell(c_j^F)$ *in* $O(\log^2 n)$ *rounds.*

Finally, we apply the operations. In the *growth* phase, each marked node performs a growth operation. In the *shrinking* phase, we apply one iteration of the *segment coloring and shrinking* subroutine on the marked nodes. Note that the growth operations may split up the binary representation of the stored values. In this case, we deal with them in the same way as with the excess nodes during the simulation of T_F (see Lemma 8): The newly grown nodes do not participate in the computation of m_j (or any other future computation, e.g., in the next segmentation subphase) and simply forward the signals which has the same effect as removing them again. However, they do participate in the final PASC algorithm to mark nodes to grow. We have a similar problem in the *shrinking* phase. If we apply a shrinking operation on two nodes that store c bits, the absorbing node has to store $2c$ bits. Since we may need to apply up to $O(\log n)$ shrinking operations, the absorbing node can end up with cn bits which it is not capable of storing. In order to resolve this issue, we redistribute the stored values within the subsegment by using the transfer primitive. We proceed to the next phase (if we are in the *growth* phase) or terminate the algorithm (if we are in the *shrinking* phase) once $m_j = 0$ for all c_j in all s_i.

Theorem 4. *After an* $O(\log n)$*-round preprocessing (w.h.p.), the* target tree *algorithm reduces* T_I *to* T_F *in* $O(k \log n + \log^2 n)$ *rounds.*

5   Open Problems

Several open problems remain to be addressed. One problem is to study the symmetric shrinking mechanisms that avoid reliance on leaders or predefined orientations. Another problem is encoding the target shape and transforming between shapes, particularly under memory constraints, for example, growing from a single node to a target shape or transforming a larger shape to a smaller one, or a smaller to a larger. Mixed models that allow both shrinking and growing also raise questions about how to preserve memory. Also, exploring alternative communication methods, such as local visibility, offers the potential for improving coordination in these systems. An exciting broader research direction opened by this work is the integration of rapid communication schemes, such as reconfigurable circuits, into fast, parallel transformation processes, potentially combining size-preserving and size-changing operations.

References

1. Almalki, N., Gupta, S., Michail, O.: On the exponential growth of geometric shapes. In: ALGOWIN. LNCS, vol. 15026, pp. 16–30. Springer (2024)
2. Almalki, N., Gupta, S., Michail, O., Padalkin, A.: Efficient distributed algorithms for shape reduction via reconfigurable circuits. CoRR abs/2503.18663 (2025)
3. Almalki, N., Michail, O.: On geometric shape construction via growth operations. Theor. Comput. Sci. **984**, 114324 (2024)
4. Angluin, D., Aspnes, J., Diamadi, Z., Fischer, M.J., Peralta, R.: Computation in networks of passively mobile finite-state sensors. Distributed Comput. **18**(4), 235–253 (2006)
5. Artmann, M., Padalkin, A., Scheideler, C.: On the shape containment problem within the amoebot model with reconfigurable circuits. CoRR abs/2501.16892 (2025)
6. Avin, C., Schmid, S.: Toward demand-aware networking: a theory for self-adjusting networks. Comput. Commun. Rev. **48**(5), 31–40 (2018)
7. Cornejo, A., Kuhn, F.: Deploying wireless networks with beeps. In: DISC. LNCS, vol. 6343, pp. 148–162. Springer (2010)
8. Derakhshandeh, Z., Dolev, S., Gmyr, R., Richa, A.W., Scheideler, C., Strothmann, T.: Brief announcement: amoebot - a new model for programmable matter. In: 26th ACM Symposium on Parallelism in Algorithms and Architectures, SPAA '14, pp. 220–222. ACM (2014)
9. Feldmann, M., Padalkin, A., Scheideler, C., Dolev, S.: Coordinating amoebots via reconfigurable circuits. J. Comput. Biol. **29**(4), 317–343 (2022)
10. Götte, T., Hinnenthal, K., Scheideler, C., Werthmann, J.: Time-optimal construction of overlay networks. Distributed Comput. **36**(3), 313–347 (2023)
11. Gupta, S., van Kreveld, M.J., Michail, O., Padalkin, A.: Collision detection for modular robots - it is easy to cause collisions and hard to avoid them. In: ALGO-WIN. LNCS, vol. 15026, pp. 76–90. Springer (2024)
12. Mertzios, G.B., Michail, O., Skretas, G., Spirakis, P.G., Theofilatos, M.: The complexity of growing a graph. J. Comput. Syst. Sci. **147**, 103587 (2025)
13. Michail, O., Skretas, G., Spirakis, P.G.: Distributed computation and reconfiguration in actively dynamic networks. Distributed Comput. **35**(2), 185–206 (2022)
14. Padalkin, A., Scheideler, C.: Polylogarithmic time algorithms for shortest path forests in programmable matter. In: PODC, pp. 65–75. ACM (2024)
15. Padalkin, A., Scheideler, C., Warner, D.: The structural power of reconfigurable circuits in the amoebot model. Nat. Comput. **23**(4), 603–625 (2024)
16. Woods, D., Chen, H., Goodfriend, S., Dabby, N., Winfree, E., Yin, P.: Active self-assembly of algorithmic shapes and patterns in polylogarithmic time. In: ITCS, pp. 353–354. ACM (2013)

Block Transactional Memory: A Complexity Study

Parwat Singh Anjana[1(✉)] and Srivatsan Ravi[1,2]

[1] Supra Research, Kelowna, USA
{p.anjana,s.ravi}@supra.com
[2] University of Southern California, Los Angeles, CA, USA

Abstract. In the smart contract ecosystem each transaction contains code to be executed, and all nodes that execute the block of transactions must arrive at the same final state. Consequently, it becomes necessary to maximize the parallel execution of transactions within a node in settings where execution time is a bottleneck (e.g., under heavier workloads, larger blocks, or in future scaling designs) while still allowing all nodes to reach a consistent final state resulting from the execution of a set of transactions. Traditional software transactional memory has been identified as a possible abstraction for the parallel execution of transactions within a block. However, in the smart contract setting, the execution order must follow the preset order of the transactions in the block. This paper presents a complexity study of this model for in-memory transactions (or block transactional memory) and investigates the fundamental bounds that might exist. To address this, we formalize the model of block transactional memory and identify the safety property that is required. We then present algorithmic designs for single-version and multi-version block transactional implementations that provide preset serializability. Finally, we show some fundamental limitations of the block transactional memory model, which establishes the limits of parallelism.

Keywords: Blockchain · Smart Contract · Parallel Execution

1 Introduction

A virtual machine (VM) in blockchain systems is a deterministic transaction execution environment that runs smart contracts, for example, the EVM on Ethereum [17] or the MoveVM on Aptos [6]. For higher execution throughput, smart contract ecosystems rely on the parallel execution of a *block* of user-specified *transactions* that each perform a sequence of *reads* and *writes* on VM states [7,18–20,28,36]. A *block proposer* node takes as input a *block* consisting of n transactions and a *preset* total order $T_1 \rightarrow T_2 \ldots, \rightarrow T_n$. The node attempts to execute transactions (n) in parallel so that the state resulting from the parallel execution must be the same as the state resulting from the *sequential* execution of the transactions $T_1 \cdot T_2 \cdots T_n$.

S. Bonomi et al. (Eds.): SSS 2025, LNCS 16350, pp. 56–75, 2026.
https://doi.org/10.1007/978-3-032-11127-2_6

As with traditional shared memory, we may consider *coarse-grained* or *fine-grained locks* or *atomic transactions* for parallel execution of user-specified transactions. While safety for traditional shared memory allows the parallel execution to be *equivalent* (here, equivalence means the two executions contain the same transactions, performing the same transactional operations and receiving the same return values from those operations [21,22]) to *some* sequential execution (known as *serializability*), safety for smart contract transactions requires that the parallel execution be equivalent to the exact sequential execution that respects the preset order (transaction order) of the user-specified transactions in the block.

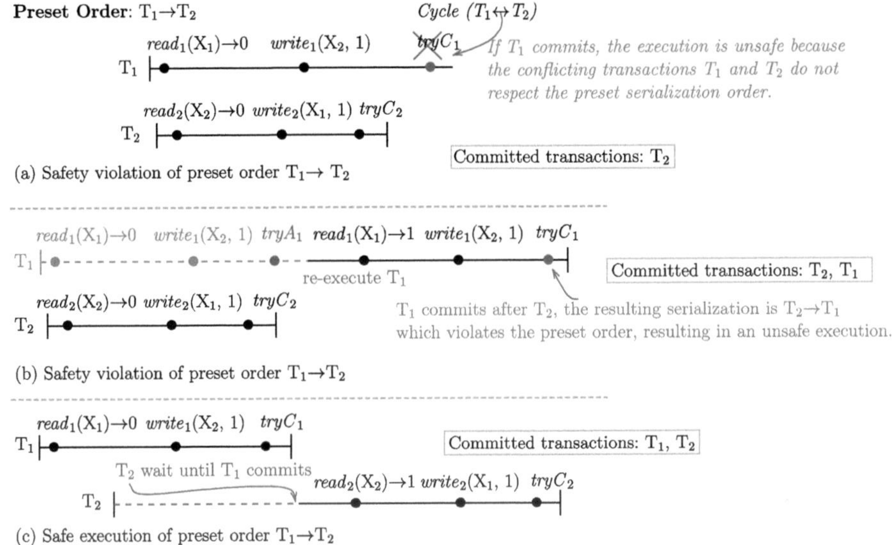

Fig. 1. Safe execution in preset order: sub-figure (a) shows an unsafe execution where conflicting transactions T_1 and T_2 commit in arbitrary order; (b) shows an unsafe execution where transactions commit in a serialization order other than the preset order; and (c) shows a safe execution where T_2 waits for T_1 to commit, preserving the preset order. An execution is considered correct if it behaves like some legal sequential execution that preserves the preset serialization, i.e., same transactions, same reads/writes, and same return values. Unlike traditional TM safety definitions such as opacity [21] and virtual-world consistency [13,25], block transactional memory enforces preset order explicitly, making preset serializability strictly stronger than standard serializability and its variants [20,22].

This paper studies the complexity of *transactional memory (TM)* algorithms for safe parallel execution of user-specified transactions [20]. Intuitively, if the execution must respect the preset order, it seemingly restricts the concurrency significantly. For instance, if we consider that transaction T_1 must precede T_2 in the preset order (denoted $T_1 \rightarrow T_2$), then without *a priori* knowledge of the read-write conflicts on the set of VM states accessed, committing T_2 prior to the commit of T_1 may result in a safety violation. To illustrate this, consider the following execution: T_1 performs a read of a VM state X_1 (reading value 0), followed by T_2 reading a VM state X_2 (reading value 0), and then writing a new value 1 to VM state X_1, as shown in Fig. 1. Observe in Fig. 1a that if T_2 commits at this point in the execution and T_1 later attempts to write a new value 1 to X_2 after the commit of T_2, then the resulting execution does not respect the preset order in any extension. Clearly, if transaction T_1 commits, the resulting execution is not equivalent to any sequential execution. Alternatively, if T_1 aborts and re-starts the execution, any subsequent read of X_2 will return the value 1 written by T_2, thus violating the preset order again, as illustrated in Fig. 1b. Consequently, Fig. 1c shows that the only way to avoid this violation is for T_2 to wait until T_1 commits. This observation motivates our study of the complexity of block transactional memories.

Contributions. In this paper, we formalize the model of *block transactional memory (BTM)* [20] (à la traditional *transactional memory (TM)* [33]). While traditional TMs adopt *serializability and its variants for safety*, block TMs need the serialization to respect a pre-specified preset order. Since smart contract execution in blockchain occurs within the deterministic environment of a VM, safety requirements like opacity [21] and virtual-world consistency [13,25], which ensure consistent reads even for aborted transactions, are generally not a concern in smart contract VMs [16,20]. Hence, we precisely define the property of preset serializability and ask: what are the fundamental limitations of parallelism in BTM? We present the design of single-version and multi-version block transactional implementations that satisfy preset serializability. We then present a proposition on the *longest conflict chain* which imposes a fundamental limit on the maximum achievable speedup. Finally, we present the costs associated with the number of versions of each VM state that must be maintained. We show that, under natural restrictions of liveness and the assumption that reading transactions are *invisible*, it is necessary in BTM implementations to maintain a large number of versions for each VM state.

Roadmap. Section 2 briefly discusses related parallelization techniques. Section 3 provides the technical preliminaries. Section 4 presents our algorithms and Sect. 5 presents results on the cost of parallelism with fundamental lower bounds. We conclude with the takeaways in Sect. 6.

2   Related Work at a Glance

In this section, we briefly discuss smart contract parallel execution models, existing algorithms, and how traditional TM models relate to those introduced in this paper.

Parallel Execution Models. Four fundamentally different execution models have emerged for the parallel execution of block transactions. First, the *Ethereum model* [17]—the proposer speculatively executes the block transactions and attaches dependency information to the block for the validators. Second, the *Aptos model* [6]—all nodes first agree on a transaction order and then run a local executor in parallel. Third, the *Solana model* [34]—transactions carry explicit read–write sets that drive parallel scheduling. Finally, the *Sui model* [35]— object ownership (exclusive vs. shared) identifies independent transactions. These models fall into two broad classes: *read-write-oblivious* executors rely only on runtime speculation, whereas *read-write-aware* executors exploit client-supplied access hints.

Read-Write-Oblivious (Optimistic) Execution. Pioneering work by Dickerson et al. [16] used Pessimistic ScalaSTM plus a proposer-appended happens-before graph for efficient execution of Ethereum blocks at validator nodes. Optimistic, multi-version variants followed [3–5,38], while Saraph & Herlihy [31] introduced a bin-based two-phase schedule that rolls back conflicting transactions and replays them sequentially. Aptos proposed Block-STM [20], a block transactional memory engine that enforces a fixed preset serialization and leverages multi-version concurrency control; every validator re-executes the proposer's transaction order in the block (called the preset order) locally to reach consensus on the VM state.

Read-Write-Aware Execution. When read–write sets are known, static or dynamic schedulers can eliminate conflicts *a priori*. ParBlockchain pre-computes access sets by static analysis or optimistic execution, and appends a dependency graph to the block [2]. Solana's SeaLevel executor iteratively acquires locks on disjoint access sets, running conflict-free subsets in parallel [36,37]. Hyperledger Sawtooth introduced a tree scheduler; a faster DAG variant was later proposed in [30]. Hay and Friedman [23] proposed a batch-schedule-execute approach that leverages conflict graphs and graph coloring for deterministic scheduling. Sui leverages its object model to bypass consensus for exclusively owned objects and executes shared-object transactions sequentially through consensus [18,19].

Block-STM as Modern BTM. Block-STM inherits decades of transactional memory research yet imposes a stricter requirement: executions must be equivalent to a globally agreed preset order, substantially constraining concurrency. Each transaction maintains per-incarnation read and write sets, with incarnations resulting from multiple executions of a transaction due to abort; validation double-reads the read set to detect inconsistencies, and initiates cascade aborts to later transactions in the order [20]. Our lower bounds (Theorem 1) formalize this "ordering curse" and match Block-STM's behavior.

Links to Classical Transactional Memory. Conventional TM systems (e.g., TL2 [15], NOrec [14], DSTM [32]) guarantee equivalence to *some* sequential schedule. Ethereum-style proposers emulate this by exporting a DAG of their speculative schedule, but such algorithms and bounds do not directly apply to the Block-STM setting. Nevertheless, Perelman et al. showed that *mv-permissive* TMs cannot be online space-optimal [29]; Kuznetsov et al. proved that wait-free read-only transactions require all historical versions [27]. We adapt these results to derive our versioning bounds for block transactional memory, although more generally, it might be possible to derive the bounds in [27,29] for the block transactional memory model via a *reduction*.

Take-Away. Existing approaches trade off read-write-oblivious flexibility (no client hints) against read-write-aware efficiency (explicit access sets). However, there has been no attempt to formalize block transactional memory within the asynchronous shared memory model, as has been done with traditional TM systems [22]. This paper represents the first such formalization, which is leveraged to address the limits of parallelism in block transactional memory.

3   Transactional Memory: Model and Definitions

In this section, we formally define the model and properties of block transactional memory. We present this as an extension of the formalism associated with traditional transactional memory [22] models. Throughout this paper, we consider the setting in which a *block* of n smart contract transactions is executed concurrently on an asynchronous shared memory machine.

3.1   Definitions and Technical Preliminaries

Transactional Memory (TM). A *transaction* is a sequence of *transactional operations*, which include *reads* and *writes* performed on a set of *objects* (alternatively, an object can represent a *VM state*), followed by a termination operation: either *tryC* (represents a commit request) or A (represents a abort request). A TM *implementation* provides a set of concurrent *processes* with deterministic algorithms that implement reads and writes on VM states using a set of *shared objects*. More precisely, for each transaction T_k, a TM implementation must support the following operations: $read_k(X)$, where X is a VM state, that returns a value in a domain V or a special value $A_k \notin V$ (*abort*), $write_k(X, v)$, for a value $v \in V$, that returns *ok* or A_k, and $tryC_k$ that returns $C_k \notin V$ (*commit*) or A_k. The transaction T_k completes when any of its operations return A_k or C_k.

Configurations and Executions. A *configuration* of a TM implementation specifies the state of each location and each process. In the *initial* configuration, each location has its initial value and each process is in its initial state. An *event* (or *step*) of a transaction invoked by some process is an invocation of an operation, a response of a operation, or an atomic *primitive* operation applied to a location along with its response. An *execution fragment* is a (finite or infinite) sequence of events $E = e_1, e_2, \dots$. An *execution* of a TM implementation M is an execution fragment that starts from the initial configuration, such that events are issued and responses returned as specified by M.

For any finite execution E and execution fragment E', $E \cdot E'$ denotes the concatenation of E and E', and we say that $E \cdot E'$ is an *extension* of E. For every transaction identifier k, $E|k$ denotes the subsequence of E restricted to events of transaction T_k. If $E|k$ is non-empty, we say that T_k *participates* in E, and let $txns(E)$ denote the set of transactions that participate in E. Two executions E

and E' are *indistinguishable* to a set $\mathcal{T}$ of transactions, if, for each transaction $T_k \in \mathcal{T}$, $E|k = E'|k$. A transaction $T_k \in txns(E)$ is *complete* if $E|k$ ends with A_k or C_k, i.e., T_k ends with a response event; otherwise, T_k is *incomplete*. The execution E is *complete* if all transactions in $txns(E)$ are complete in E.

History. A TM *history* is the subsequence of an execution consisting of the invocation and response events of operations. Two histories H and H' are *equivalent* if $txns(H) = txns(H')$ and for every transaction $T_k \in txns(H)$, $H|k = H'|k$.

We assume that shared memory locations (VM states) are accessed with *read-modify-write* primitives. A read-modify-write primitive event on a location is said to be *trivial* if, in any configuration, its application does not change the state of the location; otherwise, it is considered *nontrivial*.

We say that an execution fragment E is *step contention-free* for *operation* op_k if the events of $E|op_k$ are contiguous in E. We say that an execution fragment E is *step contention-free* for T_k if the events of $E|k$ are contiguous in E.

Sequential History. A history H is said to be *sequential* (denoted H_{seq}) if the events of each transaction $T_k \in txns(H)$ appear contiguously in H. A sequential history is obtained by executing transactions in a fixed order, in our model, this is the preset order, such that: $\forall T_i, T_j \in txns(H_{\text{seq}})$, $T_i \to T_j \Rightarrow$ All $E|i$ precede those of $E|j$ in H_{seq}.

Transaction Access Sets. For a transaction T_k, we denote all the VM states accessed by its *reads* and *writes* as $Rset(T_k)$ and $Wset(T_k)$, respectively. We denote all the *operations* of a transaction T_k as $Dset(T_k)$. The *read set* (resp., the *write set*) of a transaction T_k in an execution E, denoted $Rset_E(T_k)$ (and resp. $Wset_E(T_k)$), is the set of VM states that T_k attempts to read (and resp. write) by issuing a read (and resp. write) invocation in E (for brevity, we sometimes omit the subscript E from the notation). The *data set* of T_k is $Dset(T_k) = Rset(T_k) \cup Wset(T_k)$. T_k is called *read-only* if $Wset(T_k) = \emptyset$; *write-only* if $Rset(T_k) = \emptyset$ and *updating* if $Wset(T_k) \neq \emptyset$.

Conflict. Two transactions T_i and T_j are said to be in *conflict* if they access a common VM state and at least one of them performs an update operation.

$$(Wset(T_i) \cap Wset(T_j)) \cup (Wset(T_i) \cap Rset(T_j)) \cup (Rset(T_i) \cap Wset(T_j)) \neq \emptyset \quad (1)$$

Legal Reads. Let H be a sequential history. For every operation $read_k(X)$ in H, we define the *latest written value* of X as follows: if T_k contains a $write_k(X, v)$ that precedes $read_k(X)$, then the latest written value of X is the value of the latest such write to X. Otherwise, the latest written value of X is the value of the argument of the latest $write_m(X, v)$ that precedes $read_k(X)$ and belongs to a committed transaction in H. This write is well-defined since H can be assumed to start with an initial transaction (T_{init}) writing to all VM states. So, a successful read $read_k(X)$ (i.e., $v \neq A_k$) in S is said to be *legal* if there exists a transaction T_j that wrote v to account X and *committed* before $read_k(X)$.

Legal Histories. We say that a sequential history S is *legal* if every read of a VM state returns the *latest written value* of this VM state in S. It means that sequential history S is legal if all its reads are legal.

3.2   Block Transactional Memory (BTM)

Read-Write-Oblivious Block Transactional Memory. Unless otherwise specified, this paper considers BTM implementations in which, for every execution E, $Dset_E(T_k)$ is unknown (to T_k and every other transaction in E) at the start of transaction T_k in E. More specifically, let E be an execution such that $T_k \notin txns(E)$ and let $E \cdot e$ be an extension in which T_k performs an invocation associated with some VM state X_i. Then, we consider that $Dset_E(T_k)$ only contains X_i in $E \cdot e$ and other VM states that might be accessed by T_k in any extension of $E \cdot e$ are unknown both to T_k or any other transaction in $E \cdot e$. More generally, if we consider E to be an execution in which T_k has only invoked m operations involving at most m VM states, then we consider that $Dset_E(T_k)$ only contains these m VM states in this execution.

Given a set of n transactions with a *preset* order $T_1 \rightarrow T_2 \rightarrow \cdots \rightarrow T_n$, we need a deterministic parallel execution protocol that efficiently executes block transactions using preset serialization and always leads to the same state even when executed sequentially. The objective is to execute the block transactions in parallel with the given preset order. However, without knowing the transactions' read and write sets at the start of the transaction invocation, we also need parallel execution to always produce a deterministic final state equivalent to the state produced by sequential execution in the preset order.

We refer to a read-write-oblivious BTM implementation with a *preset* order $T_1 \rightarrow T_2 \rightarrow \cdots \rightarrow T_n$ as BTM[1,..., n].

Definition 1 (*Preset* **Serializability**). *Let H be a history of a BTM[1,..., n] implementation M. We say that H is* preset serializable *if H is equivalent to a legal sequential history S that is $H_1 \cdots H_i \cdot H_{i+1} \cdots H_n$ where H_i is the complete history of transaction T_i. We say that a BTM[1,..., n] implementation M is* preset serializable *if every history of M is preset serializable.*

Read-Write-Aware Block Transactional Memory. As taxonomized in Sect. 2, block transactional memory implementations can also provide with the read and write set of VM states right at the start of the transaction itself. This is indeed the model adopted by smart contract ecosystems like Solana [34]. For pedagogical purposes, we articulate this model, although this paper is primarily concerned about the read-write-oblivious block transactional memory model.

We say that a BTM implementation is *read-write-aware* if for every execution E, $Rset_E(T_k)$ (and resp. $Wset_E(T_k)$) are known at the start of the transaction T_k in E. Thus, unlike a read-write-oblivious BTM, a transaction T_k may be fully aware of its entire $Dset_E(T_k)$ before invoking the first operation and may write this to shared memory for other transactions to read.

We refer to a read-write-aware BTM implementation with a *preset* order $T_1 \rightarrow T_2 \ldots \rightarrow T_n$ as $\mathrm{BTM}[(Rset(T_1),\ Wset(T_1));\ \ldots;\ (Rset(T_n),\ Wset(T_n))]$.

4 Block-TM Algorithms

This section presents algorithms for the design of single-version and multi-version read-write-oblivious block transactional memory implementations. The lines specific to the single-version variant are highlighted in red and lines specific to the multi-version are highlighted in blue in Algorithm 1.

4.1 A Single-Version Block Transactional Memory Algorithm

We now present the design of a preset serializable $\mathrm{BTM}[1,\ldots, n]$ implementation that maintains exactly one version of every VM state.

Implementation State. For every VM state X_j, we maintain a memory location v_j that stores the value of X_j. Additionally, for a BTM implementation $\mathrm{M}[1,\ldots, n]$, for each transaction T_k, we maintain a memory location Entry[k] which is a binary variable indicating if T_k has completed.

Read Implementation. Consider any transaction T_k executed by process p_k. The implementation of $read_k(X_j)$ first checks if the VM state X_j is already contained in $Wset(T_k)$ (line 2). If X_j is in $Wset(T_k)$, it adopts that value (line 8). Otherwise, reads the value from v_j and adds it to its read set (line 4).

Write Implementation. The $write_k(X, v)$ implementation simply stores the value v locally (line 14) or updates it (line 17) if previously written to the same VM state, deferring the actual update in shared memory to $tryC_k$.

Commit Implementation. During $tryC_k$, each transaction T_k checks (line 34) whether Entry[$k - 1$] $= true$; then it performs various validation. If T_k is an updating transaction, $tryC_k$ performs a validation of the read set of T_k (line 36) and aborts it (line 37), if validation fails. Finally, on successful validation, T_k updates its write set to shared memory (line 39) and sets Entry[k]$= true$ (line 41). The very first transaction, T_1, can be simply write to shared memory without validation, since there will be no transaction preceding it in preset, and T_2 cannot commit until the previous $Entry[1]$ is set to $true$. Note that the abort due to read set invalidation at line 37, can only occur because of a transaction T_i preceding T_k in the preset order and not due to one succeeding T_k in the preset order.

Algorithm 1: Strict-serialisable BTM $[1..n]$ implementation L for transaction T_k—single-version and multi-version

```
     Data:  − v_j, for each VM state X_j (single entry)
            − a preset-ordered version list ⟨v_j⟩ for each VM state X_j
 1  fun read_k(X_j):
 2      if X_j ∉ Wset(T_k) then
            // X_j is not in Wset_k:
            // Read from shared memory
 3          ov_j := read(v_j)
 4          Rset(T_k) ← Rset(T_k) ∪ {X_j, ov_j}
            // Read largest version of X_j
                written by some T_i → T_k
 5          [ov_j, i] := read_lvp(T_k, X_j)
 6          Rset(T_k) ← Rset(T_k) ∪ {X_j, [ov_j, i]}
 7      else
            // X_j is already in Wset(T_k)
 8          ov_j := Wset(T_k).locate(X_j)
 9          [ov_j, k] := Wset(T_k).locate(X_j)
10      return ov_j

11  fun write_k(X_j, v):
12      nv_j := v ;
13      if X_j ∉ Wset(T_k) then
14          Wset(T_k) ← Wset(T_k) ∪ {X_j, nv_j}
15          Wset(T_k) ← Wset(T_k) ∪ {X_j, [nv_j, k]}
16      else
            // Update current value to v
17          Wset(T_k).update(X_j, nv_j)
18          Wset(T_k).update(X_j, [nv_j, k])
19      return ok

20  fun read_lvp(T_k, X_j):
        // Traverse X_j's version list to
            find largest version by some T_i < T_k
        // The version list of X_j reflects the
            preset order
21      [ov, i] ← [0, 0];
22      forall [ov_j, i] ∈ X_j do
23          if k > i then
                [ov, i'] ← [ov_j, i]
24      return [ov, i]

25  fun tryC_k():
26      if Wset(T_k) = ∅ then
            // Validation for read-only
                transactions
27          forall [ov_j, i] ∈ Rset(T_k) do
28              if [ov_j, i] ≠ read_lvp(T_k, X_j)
                then
29                  if i > k then
30                      Rset(T_k).[ov_j, i] :=
                            read_lvp(T_k, X_j)
31                  else
                        return A_k
32          return C_k
33      else
            // Ensure commit order
34          if Entry[k−1] = false then
35              wait until Entry[k − 1] = true
            // Validation of Rset(T_k)
36          if ∃X_j ∈ Rset(T_k):
                ov_j ≠ read(v_j)  [ov_j, k] ≠
                read_lvp(T_k, X_j) then
37              return A_k
            // Update shared memory from
                Wset(T_k)
38          forall X_j ∈ Wset(T_k) do
                // Write back to shared
                    memory
39              write(X_j, v_j)
                // Append new version
                    [v_j, k] in the
                    preset-ordered version
                    list of X_j
40              write(X_j, [v_j, k])
41          write(Entry[k], true);
42          return C_k
```

4.2 Single-version to multi-version

How Can Multi-versioning Help?. Multi-versioning helps in general with two transactions writing to the same VM state, i.e., write-write conflict. The two writing transactions can execute concurrently by updating their respective "versions" of the VM state in a multi-versioned data structure. For a BTM implementation $M[1, \ldots, n]$, multi-versioning can help in the following way.

Consider the scenario shown in Fig. 2, where T_2 is an updating transaction that writes the value 1 to VM states X_1 and X_2. The execution begins with T_1 reading X_1, which returns the initial value 0. After this read, T_2 performs its writes of value 1 to X_1 and X_2, and then commits. Observe that if we extend this execution with T_1 performing a read of X_2, maintaining multiple versions allows this read of X_2 to return the old value 0, thus preserving preset serializability;

this is illustrated in Fig. 2b. On the other hand, a single-version implementation (Fig. 2a) will necessarily need to force transaction T_2 to *wait* until T_1 finishes and additionally, perform a shared memory-write to indicate the presence of the read-only transaction.

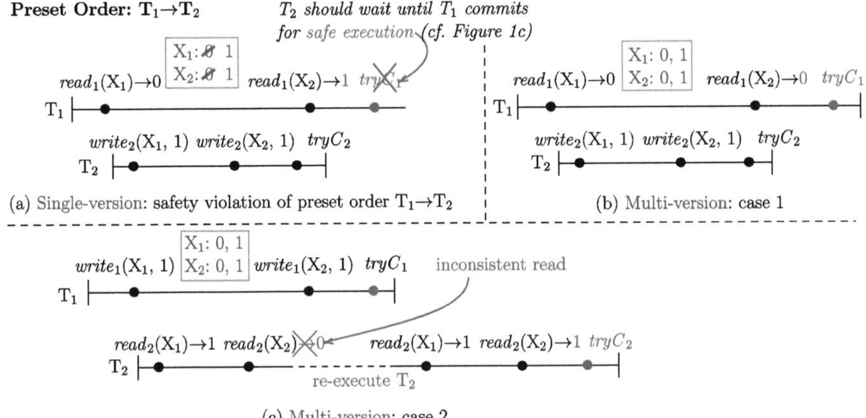

Fig. 2. Single-version to multi-version: with two transactions T_1 and T_2 and a preset order $T_1 \rightarrow T_2$, sub-figure (a) shows that when VM states X_1 and X_2 have a single version in the shared memory, transaction T_2 must wait for T_1 to commit first for safe execution; when executing transactions safely in the preset order, sub-figures (b) and (c) illustrate cases where multi-versioning is useful, and where it is not, respectively.

Now consider the modification of the above execution in Fig. 2 c, where we have T_2 performing a read of X_1 (returning 0) followed by the updating transaction T_1 running to completion and then T_2 resumes with a read of X_2. Note that in this scenario, it is possible for T_2 to validate the read set during the read of X_2, and then retry T_2 and return the last written values of X_1 and X_2 by updating transaction T_1.

Multi-version Block Transactional Memory Algorithm. We now present the design of a preset serializable multi-version block transactional memory $BTM[1,\ldots, n]$, that maintains multiple versions for every VM state.

Implementation State. For the multi-version block transactional memory, for each VM state X_j, we maintain a memory location v_j that stores a set of tuples $([v_1, k], [v_2, k'], \ldots)$. Each tuple $[v, k]$ represents the value v of X_j and k is the transaction T_k that wrote that value. Further, the version list of X_j follows the preset serialization order for storing versions.

Read Implementation. The implementation of $read_k(X_j)$ first checks at line 2 if VM state X_j is already contained in $Wset(T_k)$ and if so, adopts that value, at line 9. Otherwise, T_k reads the value of X_j from the largest version by a transaction T_i such that $T_i \rightarrow T_k$ in preset order and adds it to its read set

$(Rset(T_k))$, at lines 5–6. The method $read\_lvp()$, defined at lines 20-24, finds and returns the value of the largest version created by a preceding transaction, along with the transaction identifier. The version list will never be empty since an initial version for VM state X_j is created during initialization by T_{init}.

Write Implementation. The $write_k(X, v)$ implementation stores the version $[v, k]$ locally at line 15, or updates it in the $Wset(T_k)$ at line 18 if it has previously written to the same VM state, while deferring the actual update in the shared memory to $tryC_k$.

Commit Implementation. If T_k is a read-only transaction (line 26), it performs its read set validation at line 28 ; if any read becomes invalid due to a lower-order transaction, then it returns A_k at line 31; otherwise, it reads the updated value and returns C_k at line 32. If T_k is an updating transaction, it performs a validation of the read set at line 36 by re-reading the corresponding versions of each VM state it reads and aborts the transaction (at line 37) if validation fails. A read set validation fails when at least one preceding transaction T_i (such that $T_i \rightarrow T_k$) updates a new version of X_j after the version read by T_k. When read set validation fails for a transaction, there are generally two possible choices: (i) abort and re-execute the transaction from scratch, or (ii) resume execution from the point where the inconsistency was detected. We adopt the former strategy, as it is simpler to implement and easier to reason about correctness. Finally, on successful validation, T_k updates its write set to the shared memory at line 40 and sets the Entry$[k]=$ *true* at line 41.

Algorithmic Optimizations. Observe that in Algorithm 1, multi-versioning is only exploited by the read-only transaction. While this is done only for presentation succinctness, extending this to any updating transaction T_i is fairly straightforward since all versions of the VM state (including those possibly written by a transaction T_j; $T_i \rightarrow T_j$) are maintained by the implementation.

We remark that the linear (in the size of the transaction's read set) validation cost can be mitigated in some executions by employing a *global timestamp*, as used by traditional TM implementations like TL2 [15]. The read validation is performed only if the global timestamp has changed since the start of the transaction, thus indicating the presence of a concurrent updating transaction that has committed. However, this might affect performance on Non-Uniform Memory Architectures due to a lack of disjoint-access parallelism [9,26]. This is because the timestamp will need to be updated even by transactions that do not have a read-write conflict over their respective datasets.

5   The Cost of Parallelism in Block Transactional Memory

This section analyzes the limits of parallelism in blockchain (Sect. 5.1), establishes the fundamental versioning cost and derives the lower bound implications for read-write-aware block transactional memory (Sect. 5.2).

5.1 The Limits of Parallelism

The *longest conflict chain*, referred to as the critical path or span, is a well-established limiting factor in parallel execution. Blumofe and Leiserson in [11] propose a foundational framework of the conflict chain, which formalizes the notions of work and span in multithreaded computations. Building on this, they also proposed work-stealing schedulers [12] to demonstrate how the length of the critical path fundamentally limits parallel execution speedup. Herlihy and Moss [24] explore the impact of conflicts in the context of transactional memory systems, showing how cascading aborts and serialization bottlenecks can arise due to conflicting accesses. Berger et al. [10] emphasize the cost of serialization introduced by conflict chains in speculative parallelism, reinforcing the broader applicability of critical path analysis across concurrency models.

In the context of parallel execution on blockchains, the longest conflict chain offers a fundamental upper bound on the maximum achievable speedup in BTMs. This chain represents a sequence of transactions with interdependent state accesses that must be executed sequentially to ensure preset serializability. Block-STM [20], which utilizes multi-versioning, also cannot bypass this performance bound, since the transactions in the longest conflict chain must still commit in order. In contrast, Ethereum EVM, which lacks explicit read-write specifications, may serialize more transactions than necessary, while Solana's read-write-aware execution can better exploit available parallelism. However, in all models, the longest conflict chain forms a critical path, placing a hard limit on throughput regardless of how aggressively parallel execution is pursued.

We now formalize the longest conflict chain as a graph-theoretic problem, allowing a precise analysis of its implications for parallel execution in blockchains. Let $\mathcal{T} = \{T_1, T_2, \ldots, T_n\}$ denote the set of transactions in a block with a preset order $T_1 \to T_2 \to \cdots \to T_n$.

Definition 2 (Conflict Graph). *A conflict graph is defined as a directed graph $G = (V, E)$, in which every vertex $v_i \in V$ represents a transaction $T_i \in \mathcal{T}$, and an edge $(v_i, v_j) \in E$ represents a conflict between T_i and T_j (as defined in Eq. 1).*

Definition 3 (Longest Conflict Chain). *A sequence of transactions $(T_{i_1}, T_{i_2}, \ldots, T_{i_k}) \subseteq \mathcal{T}$ is a conflict chain if, for all $1 \leq j < k$, there is an edge $(T_{i_j}, T_{i_{j+1}}) \in E$. Since each transaction in the chain depends on the previous one, the chain must be executed in order to preserve the preset serializability defined in Definition 1.*

Now, let L denote the length of the longest conflict chain in G. Formally, $L = \max_{chains\ C \subseteq G} |C|$, where C ranges over all conflict chains in G.

Proposition 1 (Upper Bound on Parallel Speedup). *Let $n = |\mathcal{T}|$ be the total number of transactions in the block, and let L be the length of the longest conflict chain. Then the maximum parallel speedup achievable, $S_{\max}$, is bounded by $S_{\max} \leq \frac{n}{L}$*

Proof Assume that each transaction in $\mathcal{T}$ requires one unit of time under sequential execution. Then, the total sequential execution time is $\mathcal{T}_{\mathrm{seq}} = n$. Let $\mathcal{T}_{\mathrm{par}}$ denote the parallel execution time. For any parallel execution, the transactions forming the longest conflict chain must be executed sequentially in preset order due to dependency constraints, so the parallel execution time is bounded by the length of this chain, i.e., $\mathcal{T}_{\mathrm{par}} \geq L$. Hence, the maximum speedup is $S_{\max} = \frac{\mathcal{T}_{\mathrm{seq}}}{\mathcal{T}_{\mathrm{par}}} \leq \frac{n}{L}$.

Thus, the *longest conflict chain* defines the minimum execution time under any parallel execution strategy, and bounds the maximum speedup by $\frac{n}{L}$, where n is the number of transactions and L is the length of the longest conflict chain.

As shown in Fig. 3, let T_i denote a transaction with a $Rset(T_i)$ (VM states read by T_i) and a $Wset(T_i)$ (VM states written by T_i). We now illustrate this limitation using the following set of transfer transactions $\{T_1, T_2, \cdots, T_6\}$ accessing VM states $\{X_1 \ldots X_{10}\}$ with the preset order $\{T_1 \to T_2 \cdots \to T_6\}$. For simplicity, we consider transfer transactions with small read and write sets, although in practice these sets can be much larger, depending on the smart contracts and blockchain network. In the *read-write-aware* model, each transaction declares read-write sets upfront before execution starts. The *read-write-oblivious* model computes these sets at run-time, observing accesses as they occur to detect and resolve conflicts during parallel execution.

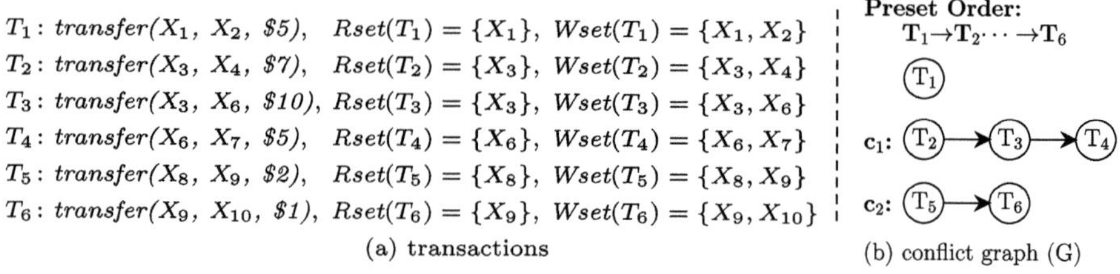

$T_1: transfer(X_1, X_2, \$5),\quad Rset(T_1) = \{X_1\},\ Wset(T_1) = \{X_1, X_2\}$
$T_2: transfer(X_3, X_4, \$7),\quad Rset(T_2) = \{X_3\},\ Wset(T_2) = \{X_3, X_4\}$
$T_3: transfer(X_3, X_6, \$10),\ Rset(T_3) = \{X_3\},\ Wset(T_3) = \{X_3, X_6\}$
$T_4: transfer(X_6, X_7, \$5),\quad Rset(T_4) = \{X_6\},\ Wset(T_4) = \{X_6, X_7\}$
$T_5: transfer(X_8, X_9, \$2),\quad Rset(T_5) = \{X_8\},\ Wset(T_5) = \{X_8, X_9\}$
$T_6: transfer(X_9, X_{10}, \$1),\ Rset(T_6) = \{X_9\},\ Wset(T_6) = \{X_9, X_{10}\}$

(a) transactions

(b) conflict graph (G)

Fig. 3. Transactions $\{T_1, T_2, \ldots, T_6\}$ with read-write sets and conflict graph.

From Definition 2, we compute the longest chain of conflicting transactions in the block by constructing a conflict graph $G = (V, E)$, as shown in Fig. 3 b, where each vertex $v_i \in V$ represents a transaction T_i, and there is an edge $(v_i, v_j) \in E$ if T_i and T_j conflict (cf. Eq. 1). The conflict chains in the conflict graph are computed using Definition 3. There are two conflict chains, c_1: $T_2 \to T_3 \to T_4$ and c_2: $T_5 \to T_6$. Then the longest chain is c_1, because $|c_1| > |c_2|$, from Definition 3. Hence, the upper bound on the maximum parallel speedup for this block is $S_{\max} \leq \frac{n}{L} = \frac{6}{3} \approx 2\times$.

5.2   On Space Complexity

In this section, we establish some fundamental versioning costs that are inherent to block transactional memory.

First, we specify the following progress condition. We say that a Block-STM implementation BTM$[1,\ldots,n]$ provides *sequential TM progress* if, for every finite execution E of M, every transaction T_k that runs sequentially commits.

Read Invisibility [9,27]. Informally, in a BTM using invisible reads, a transaction cannot reveal any information about its read set to other transactions, and cannot reveal any information about the status of its execution (such as committed or aborted) to other transactions. Formally, a BTM implementation $M[1,\ldots,n]$ uses *invisible reads* if, for every execution E of M: for every read-only transaction $T_k \in txns(E)$, no event of $E|k$ is nontrivial in E; and for every other transaction T_i, $1 \leq i \leq n$, T_k does not participate in E or any extension of E.

First, we observe that, under this definition of invisible reads, read-only transactions are not allowed to maintain a state (as in Algorithm 1) that informs other transactions in the preset order whether they have started executing or have completed execution. Given this restriction, it is easy to observe the challenge of implementing a BTM implementation $M[1,\ldots,n]$ that uses invisible reads and ensures sequential TM progress.

Consider an execution that starts with the transaction T_1 reading VM state X_1 and returning the initial value 0. Now we extend this execution with a transaction T_2, which also reads X_2, returning the initial value 0, and then writes the value 1 to X_1. Given invisible reads and the sequential progress assumption, T_2 must commit. If we further extend the execution so that T_1 writes 1 to X_2, the earlier read by T_2 becomes inconsistent. The only way to avoid this inconsistency is for the effects of T_1 to be *visible* to T_2, allowing T_2 to wait for T_1 to commit before reading a valid version of X_2.

We now derive how preset serialization necessitates retaining multiple versions of every VM state that is being updated by updating transactions. In other words, given a block of n transactions, there exist executions in which $\Omega(n)$ distinct versions of every VM state might need to be maintained in shared memory, or read-only transactions might need to write to shared memory to inform updating transactions of their presence.

Theorem 1. *Consider any implementation of BTM$[1,\ldots,n]$ with sequential TM progress and invisible reads. Then, M has an execution in which it uses $\Omega(n)$ versions of each VM state accessed by some read-only transaction.*

Proof. Without loss of generality, we consider an execution involving 4 transactions $T_1 \to T_2 \to T_3 \to T_4$ in which each transaction accesses only two VM states X_1 and X_2. Consider an execution E that consists of a read-only transaction T_1 that performs a read of VM state X_1 returning the initial value 0 of X_1. We assume that the read-only transaction T_1 is invisible. Immediately following the read, consider the following extension:

- The complete execution of an updating transaction T_2 that writes the value 1 to VM state X_1 and writes value 1 to VM state X_2. Since T_1 is invisible (by assumption), transaction T_2 must commit since this execution is indistinguishable to T_2 from an execution in which T_1 does not participate (by assumption of sequential progress).

- Now we extend this execution with a transaction T_3 that performs a read of X_1; this must return the value 1.
- Following this, we introduce a new updating transaction T_4 that writes value 2 to VM states X_1 and X_2.
- Now we extend the read-only transaction T_3 that performs a read of X_2. Since T_3 precedes T_4 in the preset order, we require that the value returned by the read of X_2 be the latest written value of X_2 in the sequential history $T_1 \cdot T_2 \cdot T_3$. Consequently, the only value that the read of X_2 by T_3 can return in any extension of this execution is the value 1 (observe that this is true even if T_3 aborts and re-runs in an execution without any contention with concurrent transactions).
- Immediately following the complete execution of T_4, transaction T_1 resumes execution and reads X_2.

Since T_1 is the first transaction in the preset order, the only value that this read can return is the initial value 0. Observe that in this execution (cf. Fig. 4), we require both versions of X_2 (0 and 1) to be maintained. We can iteratively extend this execution E such that $\Omega(c)$ versions of X_1 and X_2 must be maintained, where c is the cardinality of the updating transactions each performing a write of a new version to X_1 and X_2.

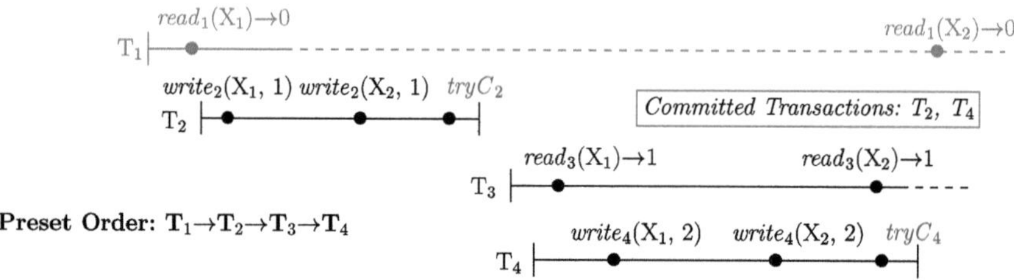

Fig. 4. Invisible read and safe execution: an example showing 4 transaction T$_1$−T$_4$ and a preset order T$_1 \rightarrow$T$_2 \rightarrow$T$_3 \rightarrow$T$_4$, where T$_1$ and T$_3$ are read-only transactions, while T$_2$ and T$_4$ are updating transactions. T$_1$ is an invisible read transaction, while T$_2$ is a committed transaction. The execution illustrates the values read and written by transactions during safe execution in the preset order.

Implication for Read-Write-Aware BTM. We now argue that the lower bound on versioning applies also to a restricted class of read-write-aware BTM implementations. We consider the set of *strictly data-partitioned* implementations [22] which intuitively require two transactions which access disjoint sets of VM states to not share a common memory location. Intuitively, this restriction when applied to a $BTM[(Rset(T_1),\ Wset(T_1));\ \ldots;\ (Rset(T_n),\ Wset(T_n))]$ makes this only as useful as the read-write-oblivious model.

Formally, a BTM implementation M is *strictly data-partitioned* [22], if for every VM state X, there exists a set of memory locations $Loc_M(X)$ such that (i) for any two VM states X_1, X_2; $Loc_M(X_1) \cap Loc_M(X_2) = \emptyset$, (ii) for every execution E of M and every transaction $T \in txns(E)$, every memory location accessed by T in E is contained in $Loc_M(X)$ for some $X \in Dset(T)$, (iii) for all executions E and E' of M, if $E|\neg X = E'|\neg X$ for some VM state X, then the configurations after E and E' only differ in the states of the memory locations in $Loc_M(X)$. Here, $E|\neg X$ denote the subsequence of the execution E derived by removing all events associated with X.

To understand how the construction in Theroem 1 can be applied to strictly data-partitioned block transactional memory implementations, consider the execution involving two transactions, T_1 and T_2: T_1 reads VM state X_1, then T_2 performs writes of new values to VM states X_1 and X_2. Observe that, by the assumption of invisible reads, such an execution exists. By the assumption of strict data-partitioning, the read of X_1 by T_1 cannot access memory locations associated with X_2 and thus, cannot *pre-read* the values associated with X_2. Consequently, when we extend this execution with the read of X_2 by T_1, the initial value of X_2 (prior to the update by T_2) must be returned. Thus, intuitively, both versions of X_2 must be maintained. Extending this argument to the construction of the multi-phased execution in Theroem 1, the proof follows.

Remarks. We remark that, while the definition of strict data-partitioning might seem overly restrictive (e.g., global timestamps or process-specific shared memory locations might not be used), we conjecture that it is possible to extend the equivalence to a more practical model of block transactional memory implementations (analogous to those developed for traditional in-memory transactions [1,8]—that could incorporate the delineation of the "value" of the VM state from the "metadata" associated with shared memory synchronization).

6   Discussion

Reduced Hardware Transactions. Intel shared memory architectures support hardware transactions that allow the *atomic* execution of a set of transactional reads and writes. Specifically, current CPUs include instructions to mark a block of memory accesses as transactional, allowing them to be executed *atomically* in hardware. Hardware transactions typically provide automatic conflict detection at cacheline granularity, thus ensuring that a transaction will be aborted if it experiences memory contention on a cacheline. Typically, if the hardware transaction is executed sequentially, i.e., in the absence of step-contention, the transaction will commit. In the read-write-oblivious block transactional memory model, the "ordering curse" that arises from the preset order prevents a transaction T_j from committing before transaction T_i when $T_i \rightarrow T_j$ (as explained in Fig. 1). Thus, the execution of a set of transactional writes may effectively need to be carried out sequentially, thus perfectly supporting the use of hardware transactions. The writes of the values in a transaction's write set are carried out by a hardware transaction and should succeed with high probability. It remains an open question whether hardware transactions can be exploited in a more comprehensive way for block transactional memory, akin to attempts with traditional TMs [15,22,33].

Relaxing Preset Serializability. The definition of preset serializability (Definition 1) requires reads of VM states to return the latest written value to that state respecting the preset order. However, certain workloads in smart contract ecosystems might not require such restrictions, allowing for the possibility of relaxed state semantics. This flexibility allows the development of different algorithmic techniques and yield block transactional memory implementations with better performance.

Equivalence. For any parallel execution approach in both the *read-write-aware* and *read-write-oblivious* settings, identifying the conflicts between transactions and resolving them efficiently is the key to increased transaction throughput [7,18–20,28,36]. As we have articulated in Sect. 2, smart contract parallel execution models seek to provide hints about transactions to the parallel execution engine to enable potentially faster execution, thus overcoming the restrictions imposed by the preset order. Both Solana [34] and Sui [35] equip transactions with *a priori* state access information about the read-write sets along with potential transactional conflicts (in the case of Sui). Whether these models yield concrete complexity separation results compared to the read-write-oblivious model remains an outstanding open question.

References

1. Alistarh, D., Kopinsky, J., Kuznetsov, P., Ravi, S., Shavit, N.: Inherent limitations of hybrid transactional memory. In: Moses, Y. (ed.) DISC 2015. LNCS, vol. 9363, pp. 185–199. Springer, Heidelberg (2015). https://doi.org/10.1007/978-3-662-48653-5_13
2. Amiri, M.J., Agrawal, D., El Abbadi, A.: ParBlockchain: leveraging transaction parallelism in permissioned blockchain systems. In: 2019 IEEE 39th International Conference on Distributed Computing Systems (ICDCS), pp. 1337–1347. IEEE, Los Alamitos (2019). https://doi.org/10.1109/ICDCS.2019.00134
3. Anjana, P.S., Attiya, H., Kumari, S., Peri, S., Somani, A.: Efficient concurrent execution of smart contracts in blockchains using object-based transactional memory. In: Georgiou, C., Majumdar, R. (eds.) NETYS 2020. LNCS, vol. 12129, pp. 77–93. Springer, Cham (2021). https://doi.org/10.1007/978-3-030-67087-0_6
4. Anjana, P.S., Kumari, S., Peri, S., Rathor, S., Somani, A.: An efficient framework for optimistic concurrent execution of smart contracts. In: 27th Euromicro International Conference on Parallel, Distributed and Network-Based Processing (PDP), pp. 83–92. IEEE (2019). https://doi.org/10.1109/EMPDP.2019.8671637
5. Anjana, P.S., Kumari, S., Peri, S., Rathor, S., Somani, A.: OptSmart: a space efficient optimistic concurrent execution of smart contracts. Distrib. Parallel Databases **42**(2), 245–297 (2024). https://doi.org/10.1007/s10619-022-07412-y

6. The Aptos blockchain: safe, scalable, and upgradeable web3 infrastructure. https://aptosfoundation.org/whitepaper/aptos-whitepaper_en.pdf. Accessed 11 Aug 2022
7. Arcology: a blockchain ecosystem with unlimited scalability (2021). https://doc.arcology.network/arcology-concurrency-control. Accessed 12 Jan 2025
8. Attiya, H., Hillel, E.: The cost of privatization in software transactional memory. IEEE Trans. Comput. **62**(12), 2531–2543 (2013). https://doi.org/10.1109/TC.2012.159
9. Attiya, H., Hillel, E., Milani, A.: Inherent limitations on disjoint-access parallel implementations of transactional memory. Theory Comput. Syst. **49**(4), 698–719 (2011). https://doi.org/10.1007/s00224-010-9304-5
10. Berger, E.D., Yang, T., Liu, T., Novark, G.: Grace: safe multithreaded programming for C/C++. In: Proceedings of the 24th ACM SIGPLAN Conference on Object-Oriented Programming Systems Languages and Applications (OOPSLA), pp. 81–96. ACM (2009). https://doi.org/10.1145/1640089.1640097
11. Blumofe, R.D., Leiserson, C.E.: Scheduling multithreaded computations by work stealing. In: Proceedings of the 35th Annual Symposium on Foundations of Computer Science (FOCS), pp. 356–368. IEEE (1994). https://doi.org/10.1109/SFCS.1994.365680
12. Blumofe, R.D., Leiserson, C.E.: Scheduling multithreaded computations by work stealing. J. ACM **46**(5), 720–748 (1999). https://doi.org/10.1145/324133.324234
13. Crain, T., Imbs, D., Raynal, M.: Read invisibility, virtual world consistency and permissiveness are compatible. Research Report, ASAP - INRIA - IRISA - CNRS : UMR6074 - INRIA - Institut National des Sciences Appliquées de Rennes - Université de Rennes I (2010). http://hal.inria.fr/inria-00533620/en/
14. Dalessandro, L., Spear, M.F., Scott, M.L.: NOrec: streamlining STM by abolishing ownership records. SIGPLAN Not. **45**(5), 67–78 (2010). https://doi.org/10.1145/1837853.1693464
15. Dice, D., Shalev, O., Shavit, N.: Transactional locking II. In: Dolev, S. (ed.) DISC 2006. LNCS, vol. 4167, pp. 194–208. Springer, Heidelberg (2006). https://doi.org/10.1007/11864219_14
16. Dickerson, T., Gazzillo, P., Herlihy, M., Koskinen, E.: Adding concurrency to smart contracts. In: Proceedings of the ACM Symposium on Principles of Distributed Computing, PODC 2017, , pp. 303–312. ACM, New York (2017). https://doi.org/10.1145/3087801.3087835
17. Ethereum (ETH): open-source blockchain-based distributed computing platform. https://www.ethereum.org/. Accessed 5 Jan 2025
18. Fouda, M.: The case for parallel processing chains (2022). https://medium.com/alliancedao/the-case-for-parallel-processing-chains-90bac38a6ba4. Accessed 10 Jan 2025
19. Foundation, S.: All about parallelization (2024). https://blog.sui.io/parallelization-explained/. Accessed 8 Jan 2025
20. Gelashvili, R., et al.: Block-STM: scaling blockchain execution by turning ordering curse to a performance blessing. In: Proceedings of the 28th ACM SIGPLAN Annual Symposium on Principles and Practice of Parallel Programming, PPoPP 2023, pp. 232–244. Association for Computing Machinery, New York (2023). https://doi.org/10.1145/3572848.3577524
21. Guerraoui, R., Kapalka, M.: On the correctness of transactional memory. In: Proceedings of the 13th ACM SIGPLAN Symposium on Principles and Practice of Parallel Programming, PPoPP 2008, pp. 175–184. ACM, New York (2008). https://doi.org/10.1145/1345206.1345233

22. Guerraoui, R., Kapalka, M.: Principles of Transactional Memory, Synthesis Lectures on Distributed Computing Theory. Morgan and Claypool (2010). https://doi.org/10.1007/978-3-031-02002-5
23. Hay, Y., Friedman, R.: Batch-schedule-execute: on optimizing concurrent deterministic scheduling for blockchains. In: Proceedings of the 43rd IEEE Symposium on Reliable Distributed Systems (SRDS) (2024). https://doi.org/10.1109/SRDS64841.2024.00025
24. Herlihy, M., Moss, J.E.B.: Transactional memory: architectural support for lock-free data structures. In: Proceedings of the 20th Annual International Symposium on Computer Architecture (ISCA), pp. 289–300. ACM (1993). https://doi.org/10.1145/165123.165164
25. Imbs, D., de Mendivil, J.R., Raynal, M.: Brief announcement: virtual world consistency: a new condition for STM systems. In: Proceedings of the 28th ACM Symposium on Principles of Distributed Computing, PODC 2009, pp. 280–281. Association for Computing Machinery, New York (2009). https://doi.org/10.1145/1582716.1582764
26. Israeli, A., Rappoport, L.: Disjoint-access-parallel implementations of strong shared memory primitives. In: PODC, pp. 151–160. Association for Computing Machinery, New York (1994). https://doi.org/10.1145/197917.198079
27. Kuznetsov, P., Ravi, S.: On partial wait-freedom in transactional memory. In: Proceedings of the 2015 International Conference on Distributed Computing and Networking, ICDCN 2015, Goa, India, 4–7 January 2015, p. 10. Association for Computing Machinery, New York (2015). https://doi.org/10.1145/2684464.2684473
28. Labs, M.: Parallel execution & monad. https://medium.com/monad-labs/parallel-execution-monad-f4c203cddf31. Accessed 10 Jan 2025
29. Perelman, D., Fan, R., Keidar, I.: On maintaining multiple versions in STM. In: PODC, pp. 16–25. Association for Computing Machinery, New York (2010). https://doi.org/10.1145/1835698.1835704
30. Piduguralla, M., Chakraborty, S., Anjana, P.S., Peri, S.: Dag-based efficient parallel scheduler for blockchains: hyperledger sawtooth as a case study. In: Cano, J., Dikaiakos, M.D., Papadopoulos, G.A., Pericàs, M., Sakellariou, R. (eds) Euro-Par 2023. Lecture Notes in Computer Science, vol. 14100, pp. 184–198. Springer, Heidelberg (2023). https://doi.org/10.1007/978-3-031-39698-4_13
31. Saraph, V., Herlihy, M.: An empirical study of speculative concurrency in ethereum smart contracts. In: International Conference on Blockchain Economics, Security and Protocols (Tokenomics 2019), pp. 4:1–4:15. OpenAccess Series in Informatics (OASIcs), Schloss Dagstuhl–Leibniz-Zentrum fuer Informatik, Dagstuhl (2019). https://doi.org/10.4230/OASIcs.Tokenomics.2019.4
32. Scherer, III, W.N., Scott, M.L.: Advanced contention management for dynamic software transactional memory. In: Proceedings of the Twenty-fourth Annual ACM Symposium on Principles of Distributed Computing, PODC 2005, pp. 240–248. ACM, New York (2005). https://doi.org/10.1145/1073814.1073861
33. Shavit, N., Touitou, D.: Software transactional memory. In: Proceedings of the Fourteenth Annual ACM Symposium on Principles of Distributed Computing, PODC 95, pp. 204–213. Association for Computing Machinery, New York (1995). https://doi.org/10.1145/224964.224987
34. Solana documentation. https://docs.solana.com/. Accessed 13 Jan 2025
35. Sui documentation: discover the power of sui through examples, guides, and concepts. https://docs.sui.io. Accessed 10 Jan 2024
36. Umbraresearch: lifecycle of a solana transaction. https://www.umbraresearch.xyz/writings/lifecycle-of-a-solana-transaction. Accessed 15 Jan 2025

37. Yakovenko, A.: Sealevel - parallel processing thousands of smart contracts (2019). https://medium.com/solana-labs/sealevel-parallel-processing-thousands-of-smart-contracts-d814b378192. Accessed 23 Jan 2025
38. Zhang, A., Zhang, K.: Enabling concurrency on smart contracts using multiversion ordering. In: Cai, Y., Ishikawa, Y., Xu, J. (eds.) APWeb-WAIM 2018. LNCS, vol. 10988, pp. 425–439. Springer, Cham (2018). https://doi.org/10.1007/978-3-319-96893-3_32

Brief Announcement: The Steiner Shortest Path Tree Problem

Omer Asher[1]($\boxtimes$), Yefim Dinitz[1]($\boxtimes$), Shlomi Dolev[1]($\boxtimes$), Li-on Raviv[2]($\boxtimes$), and Baruch Schieber[3]($\boxtimes$)

[1] Department of CS, Ben-Gurion University of the Negev, Beer-Sheva, Israel
`omeras@post.bgu.ac.il`, {`dinitz,dolev`}`@cs.bgu.ac.il`
[2] Gilat, Petah Tikva, Israel
`LionR@gilat.com`
[3] Department of CS, New Jersey Institute of Technology, Newark, NJ, USA
`sbar@njit.edu`

Abstract. We introduce and study a novel problem of computing a shortest path tree with the minimum number of non-terminals. It can be viewed as an (unweighted) *Steiner Shortest Path Tree* (SSPT) that spans a given set of terminal vertices by shortest paths from a given source while minimizing the number of nonterminal vertices included in the tree. This problem is motivated by applications where shortest-path connections from a source are essential, and where reducing the number of intermediate vertices helps limit cost, complexity, or overhead.

We show that the SSPT problem is NP-hard. To approximate it, we introduce and study the *shortest path subgraph* of a graph. Using it, we show an approximation-preserving reduction of SSPT to the uniform vertex-weighted variant of the Directed Steiner Tree (DST) problem, termed UVDST. Consequently, the algorithm of [Grandoni et al., 2023] approximating DST, implies a quasi-polynomial polylog-approximation algorithm for SSPT. We present a *polynomial* polylog-approximation algorithm for UVDST, and thus for SSPT for a restricted class of graphs.

1 Introduction

Given an undirected or directed graph with nonnegative edge weights and a source vertex s, a shortest path tree from s can be found by classic algorithms. We introduce the problem of finding a shortest path tree that spans a given set of vertices, called terminals, while minimizing the number of nonterminal vertices in the tree. This problem is closely related to the classical unweighted Steiner Tree problem, but differs in that it designates a root vertex and enforces that all paths from the root to each terminal are shortest. Due to this relation, we term the problem the Steiner Shortest Path Tree problem (SSPT). This problem

Partially supported by the Rita Altura Trust Chair in Computer Science, Frankel Center for Computer Science, BGU-NJIT Inst. for Future Technologies, the Israeli Smart Transportation Research Center (ISTRC), and Israeli Science Foundation (Grant No. 465/22).

is motivated by scenarios where shortest-path connections from a source are essential and where reducing the number of intermediate vertices helps limit cost, complexity, or overhead.

SSPT is a fundamental graph problem that has some obvious applications. One such application is multicasting from a source (root) to a given set of terminal nodes, while simultaneously minimizing latency by restricting paths to be shortest and reducing the total number of bits transmitted by limiting the use of nonterminal nodes. In the context of secure communication, it is desirable to minimize the number of nonterminal nodes exposed to the multicast message while still achieving the fastest possible delivery. An additional optimization application is in *hierarchical supply chain management* [9]. There, designing the most cost-effective hierarchical supply chain from a given set of potential hubs gives rise to the SSPT problem.

We prove that SSPT is NP-Hard. To obtain an approximation, we first introduce and study the *subgraph of shortest paths* of a weighted graph, which is the union of all the shortest paths from the source. Using this concept, we show an approximation-preserving reduction of SSPT to the uniform vertex-weighted variant of the Directed Steiner Tree (DST) problem [2], termed UVDST. Thus, any approximation for UVDST implies a similar approximation for SSPT.

The DST algorithm of [2] achieves an $O(k^\varepsilon)$-approximation in polynomial time for any fixed $\varepsilon > 0$, where k is the number of terminals. The DST algorithm of [8] achieves an $O(\log^2 k / \log\log k)$-approximation in quasi-polynomial time. Thus, the same results hold for SSPT.

We suggest a *polynomial* polylog-approximation algorithm solving UVDST on polylogarithmically shallow graphs. For logarithmically shallow graphs, our algorithm achieves an $O(\log^2 k)$-approximation ratio. Note that random graphs, expander graphs, and Small World graphs (i.e., the Internet network) are polylogarithmically shallow. We extend these results to SSPT, requiring that there exists a polylogarithmically (or logarithmically) shallow shortest path tree.

Our problem shares conceptual connections with various Steiner tree variants. While shallow-light tree problems [6] focus on bounding stretch and minimizing total edge weight, SSPT minimizes the number of nonterminal vertices under strict path constraints, which sets it apart from the shallow-light family. The *hop-constrained Steiner tree* problem [7] introduces a constraint on the number of edges allowed in the path from the root to each terminal. SSPT is more restrictive in this regard, as it permits only the shortest paths.

The full version of this paper can be found in the CORR archive [1].

2   SSPT Problem Definition and Hardness

Let us define the Steiner Shortest Path Tree (SSPT) problem. Its instance (G, s, X) is an undirected or directed nonnegatively weighted graph $G = (V, E, w)$, $w : E \to \mathbb{R}^+$, where a source vertex s and a subset of vertices X, $s \notin X$, called terminals, are distinguished. The goal is to find a shortest path tree w.r.t. w from s to the terminals that has a minimum number of nonterminal vertices.

Next, we prove that SSPT is NP-Hard by a reduction from the classic Set Cover problem. Any instance $I = (U, \mathcal{S})$ of the Set Cover problem consists of a universal set U and a set of its subsets $\mathcal{S}$. The goal is to find a minimum cardinality subset of $\mathcal{S}$ that covers U. The corresponding (either undirected or directed) SSPT instance is constructed as follows. Consider the usual representation of I by the bipartite graph with the sides $\mathcal{S}$ and U and the edges (S, x) for all $S \in \mathcal{S}, x \in S$. We add to it the vertex s with the edges (s, S) for all $S \in \mathcal{S}$. All edges are of weight 1. Observe that any tree rooted at s with edges from s to $\mathcal{S}$ and from $\mathcal{S}$ to U only is a shortest path tree in G, and vice versa. Thus, the reduction provides a natural one-to-one correspondence between the feasible solutions of the instances of Set Cover and SSPT. Since it maintains the objective function value, it keeps the optimality of solutions, as required.

3   Shortest Path Subgraph

Consider a weighted, undirected or directed graph $G = (V, E, w)$ with a vertex s distinguished as its source. A reasonable approach to pruning the input of some of shortest path problems is to introduce the *shortest path subgraph* rooted at s. The layered network data structure of the *unweighted* shortest paths from s to a target vertex t was introduced by Dinitz in [4] and used as the basis of Dinitz's network flow algorithm [4] and subsequent network flow algorithms.

The generalization to the case of *weighted* graphs was suggested by Dinitz et al. in [5] for the shortest paths from s to a fixed vertex t. Here, we further generalize the approach of [5] to the shortest paths from s to all other vertices of G.

We consider G to be a directed graph (digraph). This also covers the case of undirected graphs, by the usual reduction that replaces every undirected edge by the pair of oppositely directed edges between the same vertices. For any $x, y \in V$, we denote by $d(x, y)$ the *distance*, that is, the length of the shortest path from x to y in G. We define the *shortest path subgraph* rooted at s, $\tilde{G}(s)$, as the union of all shortest paths from s to all other vertices in G. (We stress that the shortest path subgraph $\tilde{G}(s)$ is a *digraph* even if G is undirected.)

Theorem 1.   *1. The subgraph $\tilde{G}(s)$ consists of all vertices reachable from s in G and all edges $(u, v) \in E$ satisfying $d(s, u) + w(u, v) = d(s, v)$.*
  2. Any path from vertex x to vertex y in $\tilde{G}(s)$ has weight $d(s, y) - d(s, x)$ and is a shortest path from x to y in G.

Let us describe some properties of the shortest path subgraph $\tilde{G}(s)$. It can be constructed by running a shortest path algorithm and a simple scan of G. If the edge weights in G are *nonnegative*, the only cycles in $\tilde{G}(s)$ consist of zero-weight edges only; thus, in their absence, it is *acyclic*. Given a set of *terminals* $X \subseteq V$, we can consider a similar subgraph $\tilde{G}(s, X) \subseteq \tilde{G}(s)$ which is the union of all the shortest paths from s to the vertices in X.

By Theorem 1(1), the condition on a path P from s to be shortest in G is equivalent to the condition that P is in $\tilde{G}(s)$. This relation can simplify the shortest path-related problems. In particular, the SSPT problem instance (G, s, X) is equivalent to finding, in the *unweighted* digraph $\tilde{G}(s)$ (or $\tilde{G}(s, X)$), a tree rooted at s and spanning X with the minimum number of nonterminal vertices.

4   Approximations of the UVDST and SSPT Problems

We define the minimum non-terminals directed Steiner tree (UVDST) problem as follows: given a digraph, a source vertex s, and a set of terminals X, $s \notin X$, we seek a tree rooted at s and spanning X with the minimum number of non-terminal vertices. Note that UVDST is the uniform vertex-weighted version of the directed Steiner tree problem studied in [2,8] obtained by assigning weights 1 to all edges entering non-terminals and 0 to those entering terminals.

By the analysis in Sect. 3, the SSPT problem on (G, s, X) is equivalent to the UVDST problem on $(\tilde{G}(s), s, X)$ (or on $(\tilde{G}(s, X), s, X)$), keeping the objective function. Therefore, any α-approximation of the latter problem is an α-approximation of the former problem. We thus arrive at the following reduction yielding a family of approximation algorithms for the SSPT problem.

Theorem 2. *Let $\mathcal{A}$ be any approximation algorithm solving the UVDST problem with approximation ratio α. Then, given an instance $I = (G, s, X)$ of the SSPT problem, running $\mathcal{A}$ on the instance $(\tilde{G}(s), s, X)$ (or $(\tilde{G}(s, X), s, X)$) of UVDST returns a tree which is an α-approximation to the optimum for I.*

In particular, the quasi-polynomial $O(\log^2(|X|)/\log\log|X|)$-approximation algorithm in [8] yields a quasi-polynomial $O(\log^2(|X|)/\log\log|X|)$-approximation algorithm for the SSPT problem.

In what follows, we suggest a *polynomial* polylog-approximation algorithm for the UVDST problem on polylogarithmically shallow graphs (to be defined precisely below). This implies, by Theorem 2, a similar result for the SSPT problem. By saying that T is a tree, we mean a tree rooted at s whose edges are directed along the paths from s. For any tree T, let $nt(T)$ denote the number of non-terminals in T.

Let $G[X]$ be the subgraph of G induced by X.

Lemma 1. *For any tree T^X in $G[X]$ rooted at t and any tree T in G spanning t, there exists a tree T' in G spanning all terminals in T^X with $nt(T') = nt(T)$.*

Let $\mathcal{S}$ be the set of strongly-connected components of $G[X]$ that are not reachable in $G[X]$ from any other strongly-connected component.

Lemma 2. *For any tree T spanning at least one terminal in each component in $\mathcal{S}$, there exists a tree T' spanning X with $nt(T') = nt(T)$.*

Proof. (Sketch) Let F be a DFS forest in $G[X]$ from all terminals spanned by T. Tree T' is T together with all edges in F leading to the terminals not in T. $\square$

Observe that in any tree T feasible for I, that is, spanning X, the set of edges from non-terminals to terminals "covers" $\mathcal{S}$, in a sense. We denote by $Pre\mathcal{S}$ the set of non-terminals $\{v \in V \setminus X : \exists(v, t) \in E, t \in S, S \in \mathcal{S}\}$ (that is, the vertices in $Pre\mathcal{S}$ are the non-terminals in G with outgoing edges to $\mathcal{S}$). For any $v \in Pre\mathcal{S}$, let $N(v) = \{S \in \mathcal{S} : \exists(v, t) \in E, t \in S\}$. We denote by I^{SC} the instance of Set Cover defined by the subsets $\{N(v) \subseteq \mathcal{S} : v \in Pre\mathcal{S}\}$.

Lemma 3. *For any instance I of UVDST, $OPT(I^{SC})$ lower bounds $OPT(I)$.*

Consider the following *algorithm* $\mathcal{A} = \mathcal{A}(I)$ for the UVDST problem, described by stages **(1)** to **(6)**: **(1)** Construct I^{SC} based on I. **(2)** Find an $O(\log|\mathcal{S}|)$-approximation of I^{SC} by the algorithm in [3], denoted P. Note that P is defined by $V_{SC} \subseteq Pre\mathcal{S}$ such that $P = \{N(v) : v \in V_{SC}\}$. **(3)** Construct an edge set E_{SC} by choosing for each $S \in \mathcal{S}$, a single edge $(v, t) \in E$ with $t \in S$ and $v \in V_{SC}$. **(4)** Build a BFS tree T_{BFS} from s to V_{SC} in G. **(5)** Build a tree T by adding to T_{BFS} all edges in E_{SC} leading to the terminals not in T_{BFS}. **(6)** Return the tree T' obtained from T by expanding it as in the proof of Lemma 2.

A digraph G with a source vertex s is *R-shallow* if for every $v \in V$, there exists a path from s to v that has no more than R edges. We say that a digraph G with a source vertex s is *polylogarithmically, resp., logarithmically shallow* if it is R-shallow where R is polylogarithmic, resp., logarithmic in $|X|$.

Theorem 3. *The algorithm $\mathcal{A}$ is polynomial and returns an $O(R \cdot \log(|X|))$-approximation to the UVDST problem on R-shallow digraphs. Thus, algorithm $\mathcal{A}$ computes a polylog-approximation of UVDST on polylogarithmically shallow digraphs, and an $O(\log^2|X|)$-approximation on logarithmically shallow digraphs.*

A weighted graph G (either undirected or directed) with a source vertex s is *R-shortest path shallow* if for every $v \in V$ there exists a *shortest* path from s to v with no more than R edges. We say that G with a source vertex s is *polylogarithmically, resp., logarithmically shortest path shallow* if it is R-shortest path shallow where R is polylogarithmic, resp., logarithmic in $|V|$.

Corollary 1. *There exists a polynomial algorithm for the (undirected or directed) SSPT problem on instances (G, s, X) that provides a polylog-approximation when G is polylogarithmically shortest path shallow, and an $O(\log^2|X|)$-approximation when G is a logarithmically shortest path shallow.*

References

1. Asher, O., Dinitz, Y., Dolev, S., Raviv, L.O., Schieber, B.: The steiner shortest path tree problem. CoRR (2025). arxiv preprint arxiv:2509.06789
2. Charikar, M., et al.: Approximation algorithms for directed steiner problems. J. Algor. **33**(1), 73–91 (1999)
3. Chvatal, V.: A greedy heuristic for the set-covering problem. Math. Oper. Res. **4**(3), 233–235 (1979)
4. Dinic, E.A.: An algorithm for the solution of the max-flow problem with the polynomial estimation. Soviet Math. Dokl. **11**(5), 1277–1280 (1970)
5. Dinitz, Y., Dolev, S., Kumar, M.: Polynomial time prioritized multi-criteria k-shortest paths and k-disjoint all-criteria-shortest paths. CoRR arxiv:2101.11514 (2021)
6. Elkin, M., Solomon, S.: Steiner shallow-light trees are exponentially lighter than spanning ones. SIAM J. Comput. **44**(4), 996–1025 (2015)
7. Gouveia, L., Simonetti, L., Uchoa, E.: Modeling hop-constrained and diameter-constrained minimum spanning tree problems as steiner tree problems over layered graphs. Math. Program. **128**(1), 123–148 (2011)

8. Grandoni, F., Laekhanukit, B., Li, S.: $O(\log^2 k/\log\log k)$-approximation algorithm for directed steiner tree: A tight quasi-polynomial time algorithm. SIAM J. Comput. **52**(2), STOC19-298–STOC19-322 (2023)
9. Miller, T.: Hierarchical Operations and Supply Chain Planning. Springer, Heidelberg (2002)

Brief Announcement: Searching for an Eventually-Emerging Black Hole in Rings

François Bonnet[1], Quentin Bramas[2], and Anissa Lamani[2(✉)]

[1] Institute of Science Tokyo, Tokyo, Japan
[2] University of Strasbourg, CNRS, ICUBE, Strasbourg, France
`alamani@unistra.fr`

Abstract. We study a novel variant of the Black Hole Search (BHS) problem where the black hole, a node that silently destroys visiting agents, can appear at any time during execution, rather than being present initially as assumed in some of the previous works. Our focus is on ring networks using synchronous agents. We provide two solutions: a 4-agent algorithm for rings without the knowledge of n but when only a single link leading to the black hole is found and a 3-agent algorithm assuming known ring size. We also prove that the problem is not solvable with 2-agents (even with whiteboard and a home base that remains safe), making our 3-agent algorithm optimal.

1   Introduction

The Black Hole Search (BHS) problem is a well-studied theoretical problem in distributed computing and algorithm design. It focuses on finding a dangerous node (black hole) in a network where any agent entering the node gets destroyed, and no information is returned.

This paper investigates a variant of the Black Hole Search problem by synchronous agents within a ring topology. Whereas what is commonly assumed in the literature, in the newly proposed variant of the problem, the black hole is not necessarily present initially and hence, **it can appear anytime during the execution**.

The first paper introducing the Black Hole Search problem was [5] followed by several papers extending the problem to other topologies and settings (e.g., [1,10]).

In these papers, different conditions have been considered, such as prior knowledge of the network size, the synchrony level of the agents (synchronous or asynchronous), communication mechanisms (e.g., whiteboard/pebbles), and whether the agents are initially scattered or start from a single node called a base station. In these previous works, the main goal is to first identify the conditions that allow a team of agents to successfully solve the BHS problem, and then propose distributed solutions that are optimal in terms of the number of mobile agents required and the move complexity.

S. Bonomi et al. (Eds.): SSS 2025, LNCS 16350, pp. 82–87, 2026.
https://doi.org/10.1007/978-3-032-11127-2_8

Variants of the BHS problem include having multiple black holes [4], gray/Byzantine holes [9], rendezvous in dangerous networks (e.g., [6]), exploration with fault tolerance (e.g., [7]) and decontamination of mobile threats [3].

Recent investigations have extended the study of the problem to dynamic graphs(e.g., [2,8]), primarily focusing on ring networks with 1-interval connectivity. The goal is to determine how the computational complexity of finding a black hole changes when the graph is dynamic.

We address the aforementioned problem on ring networks using synchronous agents, but, contrary to the previous work, we assume that the black hole may appear after the beginning of the execution. We present two algorithms, one using four agents that are not aware of the size of the ring, capable only of finding one link leading to the black hole, and one using three agents knowing the size of the ring, able to identify both links leading to the black hole.

2   Model

We consider a team $\mathcal{A}$ of k agents located on an anonymous ring of n connected nodes denoted $u_0, u_1, \ldots, u_{n-1}$. The ring is consistently oriented *i.e.*, all the agents agree on the clockwise direction. All the agents start at the same node called *the home base (HB)*. Each agent has a unique identifier in the interval $[1, k]$ and unbounded memory. In this work, we consider the *FaceToFace* model in which agents can communicate directly when they are collocated on the same node. All agents execute the same algorithm that takes as input the current state of the agent and the states of the agents that are collocated. Agents do not have pebbles (or tokens) and nodes do not have whiteboards. They are synchronous and the time is discretized in rounds.

A *black hole* is a node that destroys every agent located on it. When an agent is destroyed, it disappears from the system. We say that a black hole appears in node u at time t if starting from time t, node u becomes a black hole. The agents that are not on u are not impacted; the agents in node u at time $t - 1$ that do not leave u, or arriving on u at time $t' \geq t$ are killed. We denote by $\mathcal{K}_t$ the set of agents that are not killed at time t.

In this paper, we consider a single eventually-emerging black hole *i.e.*, there exists a time $t \geq 0$ and a node u such that a black hole appears on u at time t. Of course, we consider that the black hole cannot appear at the home base at time $t = 0$. However, it can appear at the home base at any time $t \geq 1$.

The Eventually-Emerging Black Hole Search (EBHS) Problem: We say an algorithm solves the EBHS problem if, in any execution with an eventually-emerging black hole, an agent remains alive and all surviving agents eventually correctly identify the location of the black hole and terminate. In the literature, finding the location of the black hole either means identifying (or marking) all the links leading to the black hole, or simply finding a safe path leading to the black hole. In our setting, we say an algorithm solves the *weak* EBHS problem if it identifies a single link leading to the black hole, and solves the *strong* EBHS problem if it identifies all the direct links leading to the black hole. Note that when the size of the ring is known, every algorithm that solves the weak version of the problem also solves its strong version.

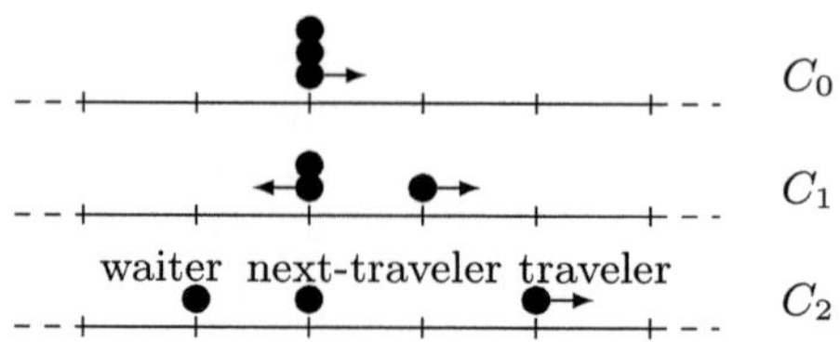

Fig. 1. The first three configuration of Algorithm $\mathcal{A}_2$.

3   Solving the EBHS Problem

We first show that the problem is not solvable with only two agents as it is the case when the black hole is present from the initial configuration. This result holds even with a whiteboard, safe home base, and the knowledge of n.

Theorem 1. *There is no algorithm solving the EBHS problem with only two synchronized agents, even with a safe home base, whiteboards, the knowledge of n, and implicit termination.*

We now present two deterministic algorithms that solve the EBHS problem under different hypotheses. The first algorithm $\mathcal{A}_1$, solves the weak EBHS problem with four agents that are not aware of the size of the ring. The second algorithm, $\mathcal{A}_2$, solves the strong EBHS problem with three synchronous agents which know the size of the ring.

Algorithm $\mathcal{A}_1$: We present a deterministic algorithm $\mathcal{A}_1$ that solves the EBHS problem using four agents using only face-to-face communications. The size of the ring n is unknown and the HB is assumed unsafe. This first algorithm consists in running an algorithm that works when the black hole is initially present by simply executing the cautious walk. The idea is to create two groups of agents, each of them executing such an algorithm in the same direction while keeping a distance between them. Since the two groups never meet, the appearance of the black hole can only impact one of the two groups. Hence, the correctness of the algorithm follows from the correctness of the cautious walk strategy.

Consequently, the following theorem holds:

Theorem 2. *Algorithm $\mathcal{A}_1$ solves the weak EBHS problem with 4 synchronous agents.*

Algorithm $\mathcal{A}_2$: We now present Algorithm $\mathcal{A}_2$ that solves the EBHS problem using three synchronous agents and assuming that the size of the ring n is known and the home base is unsafe. The algorithm works as follows: Initially, one agent has a role denoted as *traveler* and moves clockwise and the two others stay idle. Then, the agents repeatedly perform the same phase. In a phase, the traveler moves $n-2$ times clockwise. Another agent takes the role of the *next traveler* and stays idle for $n-1$ rounds. The last agent takes the role of the *waiter*, it moves once counterclockwise, and then stays idle for $n-2$ rounds. The first three configurations of each phase C_0, C_1, C_2 are shown in Fig. 1.

After $n-2$ rounds in the phase, the traveler agent reaches the waiter agent. Then each agent stays idle for one round and then they repeat the phase after exchanging their role: the traveler becomes the waiter, the waiter becomes the next-traveler, and the next-traveler becomes the traveler, as shown in Fig. 2. Observe that at each phase the configuration is similar but rotated by one node counterclockwise.

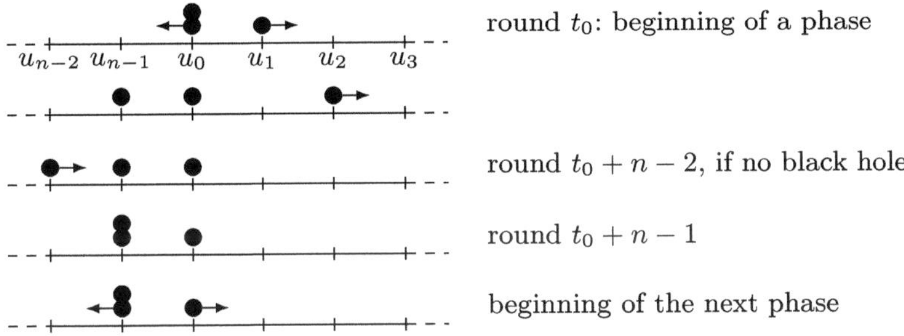

Fig. 2. Execution of Algorithm $\mathcal{A}_2$ when no black hole is detected.

This phase is repeated until a black hole appears. If a black hole kills the traveler agent while it performs the tour of the ring, then, after $n-2$ rounds, the waiter agent detects that and moves clockwise to warn the third agent and starts a cautious walk to find the black hole, as shown in Fig. 3.

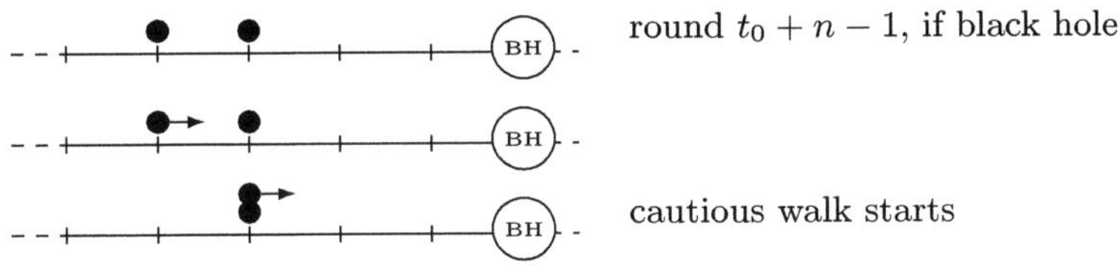

Fig. 3. Execution of Algorithm $\mathcal{A}_2$ when the traveler agent is killed.

If a black hole appears and kills two collocated agents, then this happens at the beginning of the phase and the traveler is still alive. When the traveler

round $t_0 + n - 1$, if black hole

Fig. 4. Execution of Algorithm $\mathcal{A}_2$ when two agents are killed.

agent finishes its tour, it detects that the waiter agent is not there waiting for his arrival. Hence, it knows that the adjacent node contains a black hole, as shown in Fig. 4. Finally, observe that since the size of the ring n is known, every agent that identifies one direct link leading to the black hole can also identify the second direct link leading to it. Hence, the following theorem holds:

Theorem 3. *Algorithm $\mathcal{A}_2$ solves the strong EBHS with three synchronous agents that know the size n of the ring.*

4 Conclusion

In this work, we introduced a new version of the Black Hole Search problem in which the black hole is not present initially but can emerge at any time $t \geq 1$. We presented two deterministic algorithms assuming synchronous robots starting from an unsafe home base. Future work could focus on either further exploring other models or investigating strategies for handling multiple emerging black holes or dynamic networks.

References

1. Balamohan, B., Dobrev, S., Flocchini, P., Santoro, N.: Exploring an unknown dangerous graph with a constant number of tokens. Theor. Comput. Sci. **610**, 169–181 (2016). https://doi.org/10.1016/j.tcs.2014.07.013
2. Bhattacharya, A., Italiano, G.F., Mandal, P.S.: Searching for a black hole in a dynamic cactus. J. Graph Algor. Appl. **29**(2), 127–166 (2025). https://doi.org/10.7155/jgaa.v29i2.3042
3. Cai, J., Flocchini, P., Santoro, N.: Decontaminating a network from a black virus. Int. J. Network. Comput. **4**(1), 151–173 (2014). https://doi.org/10.15803/ijnc.4.1_151
4. Cooper, C., Klasing, R., Radzik, T.: Searching for black-hole faults in a network using multiple agents. In: Shvartsman, M.M.A.A. (ed.) OPODIS 2006. LNCS, vol. 4305, pp. 320–332. Springer, Heidelberg (2006). https://doi.org/10.1007/11945529_23
5. Dobrev, S., Flocchini, P., Prencipe, G., Santoro, N.: Mobile search for a black hole in an anonymous ring. Algorithmica **48**(1), 67–90 (2007). https://doi.org/10.1007/s00453-006-1232-z
6. Dobrev, S., Flocchini, P., Prencipe, G., Santoro, N.: Asynchronous gathering in a dangerous ring. Algorithms **16**(5), 222 (2023). https://doi.org/10.3390/a16050222
7. D'Emidio, M., Frigioni, D., Navarra, A.: Exploring and making safe dangerous networks using mobile entities. In: Cichoń, J., Gebala, M., Klonowski, M. (eds.) ADHOC-NOW 2013. LNCS, vol. 7960, pp. 136–147. Springer, Heidelberg (2013). https://doi.org/10.1007/978-3-642-39247-4_12

8. Flocchini, P., Kellett, M., Mason, P.C., Santoro, N.: Searching for black holes in subways. Theory Comput. Syst. **50**(1), 158–184 (2011). https://doi.org/10.1007/s00224-011-9341-8
9. Goswami, P., Bhattacharya, A., Das, R., Mandal, P.S.: Perpetual exploration of a ring in presence of byzantine black hole. In: 28th International Conference on Principles of Distributed Systems (OPODIS 2024), vol. 324, pp. 17:1–17:17 (2025). https://doi.org/10.4230/LIPIcs.OPODIS.2024.17
10. Markou, E., Paquette, M.: Black hole search and exploration in unoriented tori with synchronous scattered finite automata, pp. 239–253 (2012). https://doi.org/10.1007/978-3-642-35476-2_17

Brief Announcement: Consensus in Systems with Eventual Quiescence

Quentin Bramas$^{(\boxtimes)}$ [iD]

Strasbourg University, ICUBE, CNRS, Strasbourg, France
bramas@unistra.fr

Abstract. We define a simple and general condition (*eventually detectable quiescence*) under which consensus becomes solvable in asynchronous systems with crash failures. Unlike previous work that ties solvability to specific failure thresholds (e.g., contention levels), our approach isolates the essential property enabling agreement: the existence of an eventually quiescent period, i.e., a time where processes know that no more crash can occur. The key insight is that once quiescence is detected by a process, we can implement an Eventual Perfect failure detector, by monitoring processes activity, and execute the Lo & Hadzilacos consensus algorithm using this failure detector. Despite its simplicity, our algorithm improves the previous results that focused on contention-based failures.

1 Introduction

The impossibility of consensus in asynchronous systems prone to even a single crash failure is a fundamental result in distributed computing [5]. To circumvent this impossibility, researchers have explored various system assumptions and failure detector abstractions. The failure detector approach, pioneered by Chandra and Toueg [1], identifies the minimal information about failures needed to solve consensus. Recently, Durand et al. [4] introduced λ-constrained crash failures, where processes can crash only before the system contention (the number of process that have started their execution) reaches a threshold λ. They provide two algorithms. The first one solves consensus tolerating up to k λ-constrained crashes when $\lambda = n - k$ using objects with consensus number k (so in particular, their algorithm can tolerate one $(n-1)$-constrained failure using only atomic RW-registers). The second algorithm, assuming k divides n, solves the consensus tolerating up to $2k - 1$ λ-constrained crashes when $\lambda = n - 2k + 1$ using objects with consensus number k.

In this short paper, we take a step back and ask a more fundamental question: *What is the essential property that enables consensus in systems with restricted failure patterns?* We argue that the answer lies not in the specifics of contention thresholds, but in a more general property: the existence of an eventually quiescent period, *i.e.*, a time where the processes know that no more crash can occur.

This work was partially funded by the ANR project ref. ANR-22-CE25-0012.

S. Bonomi et al. (Eds.): SSS 2025, LNCS 16350, pp. 88–92, 2026.
https://doi.org/10.1007/978-3-032-11127-2_9

We present a simple algorithm solving consensus under this assumption. The algorithm may seem trivial, but it simplifies and improve the results of Durand et al. [4], while being more general, hence could be applied to a wider range of constrained failure (not only contention base).

In particular, we can derive from our general algorithm, a specific algorithm solving consensus tolerating k λ-constrained crashes when $\lambda = n - k$ for any $k \in [1, \ldots, n]$, **using only atomic RW-registers**. This bound is tight due to the impossibility result by Durand et al. [4].

Related Work. Consensus has been extensively studied under various system and failure assumptions. After the FLP impossibility result [5], a wealth of work explored weaker guarantees or stronger assumptions. Notably, failure detectors [1,8] and indulgent algorithms [6,7] enable progress under partial synchrony while preserving safety. Variants of the consensus problem, such as k-set agreement [2], and impossibility results under generalized or weak failure models [9,10], further characterize the limits of agreement. More recently, contention-aware failures [3,4] have been proposed to relax failure models without requiring synchrony. Our work follows this line, aiming to generalize and simplify the underlying principles that make consensus solvable under such constrained models.

2   System Model and Definitions

We consider a system of n asynchronous processes $p_1, \ldots, p_n$ that communicate through atomic read/write registers. Processes may crash (halt permanently) but do not exhibit Byzantine behavior. A process is *correct* in an execution if it does not crash, and *faulty* otherwise. A λ-constrained crash is a crash that occurs before λ processes started executing their algorithm.

The **consensus problem** requires a set of n processes to agree on a single value, satisfying the following properties: *Termination* (every correct process eventually decides), *Agreement* (no two processes decide differently), and *Validity* (the decided value must be one of the initial values proposed by the processes). Consensus is a fundamental coordination task in distributed systems, particularly challenging in asynchronous settings with crash failures.

We introduce two definitions to describe a system where no crash occurs after a given time.

Definition 1 (Eventually Quiescent System). *An execution is* eventually quiescent *if there exists a finite time T such that no process crashes after time T. We call T the* quiescence time.

This definition captures a broad class of failure patterns where, regardless of the specific conditions that determine when crashes stop occurring, the system eventually reaches a state where no further crashes happen.

Definition 2 (Eventually Detectable Quiescence). *A system has* eventual detectable quiescence *if there exists a predicate Q such that:*

1. *Q can be evaluated by correct processes using only local computation and shared memory reads*
2. *Once Q becomes true for a correct process, it remains true and no further crashes occur in the system.*
3. *Q eventually becomes true for every correct process.*

Failure Detector. Chandra and Toueg [1] formalized *failure detectors* as abstract modules that provide (possibly unreliable) information about process crashes. These are classified based on two properties: *completeness*, which ensures that crashed processes are eventually suspected, and *accuracy*, which bounds the incorrect suspicion of correct processes. Each property has both *strong* and *weak* variants, as well as *eventual* versions. For example, *Strong Completeness* requires that every crashed process is eventually permanently suspected by all correct processes, while *Eventual Weak Accuracy* ensures that some correct process is eventually never suspected. A failure detector satisfying Strong Completeness and Eventual Weak Accuracy is called an *Eventual Strong failure detector*. If, additionally, no correct process is ever suspected after some finite time (i.e., *Eventual Strong Accuracy*), the detector is known as an *Eventual Perfect failure detector*. More precisely, an Eventual Perfect failure detector satisfies:

- Strong Completness: there is a time after which every process that crashes is permanently suspected by every correct process;
- Eventual Strong Accuracy: there is a time after which correct processes are not suspected by any correct process

Lo & Hadzilacos [8] gave two consensus algorithms for asynchronous shared-memory systems using atomic read/write registers, one that uses a Strong failure detector, and one that uses an Eventual Strong failure detector, which can be implemented using an Eventual Perfect failure detector.

3   Consensus for Eventually Quiescent Systems

A General Algorithm. The key insight is that once quiescence is detected by the upper level, we can implement an Eventual Perfect failure detector ($\diamond$ P), by monitoring processes activity, and execute the Lo & Hadzilacos [8] consensus algorithm using it. The algorithm assumes that it starts after quiescence time.

We prove that the array **trusted** of Algorithm 1 is a $\diamond$P detector, *i.e.*, an Eventual Perfect failure detector.

Lemma 1. *Array **trusted** of Algorithm 1 is an Eventual Perfect failure detector ($\diamond P$) in post-quiescence systems.*

Proof. **Strong Completeness:** Since no process crashes after quiescence, any process that was crashed before quiescence will never update its entry in the **trusted** array.

Eventual Strong Accuracy: Since no crashes occur after quiescence, each correct process eventually writes in its entry in the **trusted** array.

Algorithm 1: Consensus Algorithm for Post-Quiescence Systems

Shared variables:;
trusted[1..n]: boolean array, initially [**false**, ..., **false**];

Operation *propose(in$_i$)*
> trusted[i] ← **true**;
> **Execute Lo & Hadzilacos consensus algorithm** with array trusted
> as a failure detector;

Theorem 1. *The algorithm that first wait until quiescence is reached using predicate Q and then executes Algorithm 1, solves consensus in any system with eventually detectable quiescence.*

Proof. By Lemma 1, we have an Eventual Perfect failure detector after quiescence is detected. The Lo & Hadzilacos algorithm solves consensus using such a failure detector.

Instantiation for Contention-Based Failures. For the specific case of contention-based failures, we define the simple following algorithm that first detects quiescence and then applies Algorithm 1.

Algorithm 2: Consensus Algorithm with Contention-Based Failures

---

Shared variables:;
PARTICIPATION[1..n]: boolean array, initially [**false**, ..., **false**];
Operation *propose* (*in$_i$*)
> // Signal participation
> PARTICIPATION[i] ← **true**;
> **Repeat**
> > *participated$_i$* ← asynchronous reading of PARTICIPATION[1..n] ;
> > **if** *participated$_i$*[1..n] *contains at most k entries with false* **then**
> > > break;
>
> **Run Algorithm 1;**

Theorem 2. *Let $n \geq k$ and $\lambda = n - k$. Algorithm 2 solves consensus in the presence of at most k λ-constrained crash failures.*

Proof. First, observe that a process cannot be stuck in the **Repeat** loop, because there are at most k failures. When algorithm 1 is started, we know that $\lambda = n - k$ processes have started their execution, so the quiescence period has started. The correctness then follows from Theorem 1.

Observe here that k can be any number between 1 and n and does not require the existence of a k-consensus object, like it is assumed in [4]

4   Conclusion and Possible Extensions

We have defined a simple and general condition (*eventually detectable quiescence*) under which consensus becomes solvable in asynchronous systems with crash failures. Unlike previous work that ties solvability to specific failure thresholds (e.g., contention levels), our approach isolates the essential property enabling agreement: the existence of a point in time after which no further failures can occur, and processes can deduce this stability. Our general framework applies beyond contention-based failures:

- Time-Based Quiescence: Systems where no crashes occur after a known duration T.
- Resource-Based Quiescence: Systems where no crashes occur once resource utilization reaches a threshold, indicating system stabilization.
- External Event-Based Quiescence: Systems where no crashes occur after some externally observable event.

An interesting direction for future work is to extend our work to message-passing systems, since the Chandra-Toueg consensus algorithm [1] using the Omega failure detector assumes that at least a majority of processes are correct, which imposes a limit on the number of faults we could tolerate using our scheme.

References

1. Chandra, T.D., Toueg, S.: Unreliable failure detectors for reliable distributed systems. J. ACM **43**(2), 225–267 (1996)
2. Chaudhuri, S.: More choices allow more faults: set consensus problems in totally asynchronous systems. Inf. Comput. **105**(1), 132–158 (1993)
3. Durand, A., Raynal, M., Taubenfeld, G.: Contention-related crash failures: definitions, agreement algorithms, and impossibility results. Theor. Comput. Sci. **909**, 76–86 (2022)
4. Durand, A., Raynal, M., Taubenfeld, G.: Reaching Consensus in the Presence of Contention-Related Crash Failures, pp. 193–205. Springer, Heidelberg (2022). https://doi.org/10.1007/978-3-031-21017-4_13
5. Fischer, M.J., Lynch, N.A., Paterson, M.S.: Impossibility of distributed consensus with one faulty process. J. ACM **32**(2), 374–382 (1985)
6. Gafni, E.: Round-by-round fault detectors (extended abstract): unifying synchrony and asynchrony. In: Proceedings of the Seventeenth Annual ACM Symposium on Principles of Distributed Computing - PODC '98, pp. 143–152. ACM Press (1998)
7. Guerraoui, R., Raynal, M.: The information structure of indulgent consensus. IEEE Trans. Comput. **53**(4), 453–466 (2004)
8. Lo, W.-K., Hadzilacos, V.: Using failure detectors to solve consensus in asynchronous shared-memory systems. In: Tel, G., Vitányi, P. (eds.) WDAG 1994. LNCS, vol. 857, pp. 280–295. Springer, Heidelberg (1994). https://doi.org/10.1007/BFb0020440
9. Taubenfeld, G.: A closer look at fault tolerance. Theory Comput. Syst. **62**(5), 1085–1108 (2017)
10. Taubenfeld, G.: Weak failures: definitions, algorithms and impossibility results. In: Podelski, A., Taïani, F. (eds.) NETYS 2018. LNCS, vol. 11028, pp. 51–66. Springer, Cham (2019). https://doi.org/10.1007/978-3-030-05529-5_4

Centroid Approximation
with Multidimensional Approximate Agreement Protocols

Mélanie Cambus[1] and Darya Melnyk[2(✉)]

[1] Aalto University, Espoo, Finland
[2] TU Berlin, Berlin, Germany
`melnyk@tu-berlin.de`

Abstract. In this paper, we present distributed fault-tolerant algorithms that approximate the centroid (i.e., the average) of a set of n data points in $\mathbb{R}^d$. Our work falls into the broader area of multidimensional Byzantine approximate agreement. We show that state-of-the-art algorithms, such as agreeing inside the convex hull of all non-faulty vectors, or minimum-diameter averaging (MDA), in the worst case either prevent us from agreeing on a vector close to the centroid (in terms of approximation quality), or allow Byzantine parties to influence the output considerably (in terms of validity).

To design better approximation algorithms, we propose a novel concept of defining an approximation ratio of the centroid by including the vectors of the Byzantine adversaries in the definition. We analyze synchronous algorithms in the public channel communication model. We show that the standard agreement algorithms based on agreeing inside the convex hull of all non-faulty vectors do not allow us to compute a better approximation than $2d$ of the centroid. On the other hand, MDA can be used to achieve constant approximation at the cost of only satisfying strong validity. As a trade-off, we develop an approach that reaches a $2\sqrt{d}$-approximation of the centroid, while satisfying box validity. Our approach provides optimal resilience, allowing up to $t < n/3$ faulty nodes.

Keywords: averaging agreement · byzantine fault tolerance · barycenter · vector consensus · collaborative learning

1 Introduction

Multidimensional Byzantine approximate agreement (MBAA) is an important subroutine that allows nodes in a network to approximately agree on vectors that are close to a convergence vector, in the presence of any type of node failures, from crashing to organized malicious behavior. The convergence vector should typically satisfy a so-called validity condition. This condition ensures that the convergence vector is not just a trivial predetermined vector, but depends on the inputs of the non-faulty nodes. MBAA is particularly interesting for practical applications where the fast running time of an agreement algorithm is more important than agreement on the same vector. In practical applications, there is

S. Bonomi et al. (Eds.): SSS 2025, LNCS 16350, pp. 93–110, 2026.
https://doi.org/10.1007/978-3-032-11127-2_10

often a desired convergence vector that the system should agree on, for example, the centroid, the weighted average, or the geometric median. This work considers the centroid as the desired convergence vector, as it is a well-known and broadly used representative vector in many applications such as vector quantization [4], collaborative learning [20], data anonymization [22], large-scale elections [31], community detection algorithms [25,26] and distributed voting [30].

Different approaches have been considered to evaluate the quality of the output of an MBAA algorithm: The traditional approach is to classify the output using validity conditions. However, even the most restrictive validity condition introduced for MBAA—the convex validity condition [30]—only guarantees that the convergence vector is in the convex hull of all input vectors of the non-faulty nodes. There are two reasons why this polytope does not represent the centroid well. If the centroid is inside the polytope, the polytope can be large such that the convergence vector is far from the centroid of the non-faulty input vectors. If the polytope is small, the polytope itself may lie far away from the centroid of the non-faulty input vectors. Another approach is to compare the absolute distance to the centroid to the maximum distance between any two non-faulty input vectors [20]. This measure only captures the worst-case scenario where Byzantine inputs are the only outliers in the data. If one single non-faulty outlier is present in the data, the measure cannot guarantee that the output is closer to the centroid than the distance to this outlier. The same holds when no Byzantine parties are present in the system.

In this paper, we provide a novel definition for centroid approximation (see Sect. 3.2). The approximation is defined relative to the input distribution: if the input distribution is unfavorable (e.g., there is one outlier among the non-faulty nodes), even for an optimal algorithm, it is impossible to differentiate between a non-faulty outlier and a Byzantine node. On the other hand, if the input distribution is favorable (e.g., the Byzantine party shares a similar vector to the non-faulty inputs), an optimal algorithm should be capable of agreeing on a vector close to the centroid. Our approximation definition compares an approximate agreement algorithm to what an optimal algorithm that cannot detect Byzantine behavior can achieve for a given input distribution. To ensure that an adversary cannot change the convergence vector arbitrarily, we make use of validity conditions. We thus focus on finding a trade-off between the quality of the approximation of the non-faulty nodes and the validity condition satisfied.

We focus on the following validity conditions in this work (fully defined in Sect. 3.1 and 3.3), presented from loose to more restrictive conditions: The **weak validity** condition [13,14,38], where if there are no failures and all input vectors are identical, then the output vector of each node is required to be the input vector; The **strong validity** condition [3,8,13,16], where if all non-faulty nodes have the same input vector then this input vector must be the output vector of all non-faulty nodes; The **box validity** condition [28,37], where the output vector of each non-faulty node must be in the smallest coordinate-parallel box containing all input vectors of non-faulty nodes; The **convex validity** condition [1,6,24,30,32,36], where the output vector of each non-faulty node must be in the convex hull of all input vectors of non-faulty nodes.

Our contributions are as follows:

- We prove that any algorithm guaranteeing the convex validity condition has an approximation ratio of at least $2d$ (Sect. 4),
- we give an algorithm guaranteeing box validity condition that has an approximation ratio of at most $2\sqrt{d}$, giving at least a quadratic improvement compared to the *SafeArea* algorithms from [30] in terms of approximation of the centroid while sacrificing very little on the validity condition (Sect. 5),
- we also analyze other agreement algorithms to show different trade-offs between validity conditions and approximation ratio (see Table 1).

Note that the resilience (i.e., maximum amount of tolerated Byzantine parties) of our algorithm is optimal, contrary to the other studied algorithms. Finally, our algorithm has only polynomial local computation steps, whereas the algorithms described and analyzed in Sect. 4 and in the full version of this paper [12] require exponential local computations. Our main result is shown in Theorem 1. We present a summary of the results of this paper in Table 1.

Remark 1. Observe that the proven lower bounds for different validity conditions also hold for exact Byzantine agreement, which is beyond the scope of this paper.

Table 1. Overview of the results of this paper, some of which only appear in the full version. We consider weak, strong, box and convex validity conditions.Please note that methods based on *MDA* and *SafeArea* rely on exponential local computation time, while $RB-TM$ and the Box Algorithm can be computed in polynomial time.

Method	Validity				Approximation
	weak	strong	box	convex	
SafeArea Algorithms [24,30] (full version [12])	✓	✓	✓	✓	∞
SafeArea Approaches[a] (Proposition 2)	✓	✓	✓	✓	$\Omega(d)$
Centroid & *SafeArea* (full version [12])	✓	✗	✗	✗	1
MDA [20] (full version [12])	✓	✓	✗	✗	3.8
MDA & *SafeArea* (full version [12])	✓	✓	✗	✗	2
Box Algorithm (Theorem 1)	✓	✓	✓	✗	$O(\sqrt{d})$
$RB-TM$ [20] (Corollary 1)	✓	✓	✓	✗	$O(\sqrt{d})$

[a] We refer to agreement inside the *SafeArea* (defined in Sect. 3.1) as the *SafeArea* approaches.

Technical Overview. The first technical challenge is to fix a meaningful definition of approximation ratio in the presence of Byzantine parties. A traditional way of defining the approximation ratio of an algorithm would be to compute the ratio between the algorithm's result and the result of an optimal algorithm. Note that with such an absolute measure, a Byzantine party can change the centroid arbitrarily. The best possible upper bound on the absolute distance is the maximum distance between any two input vectors (see [20]). Instead, we use the fact that even an optimal algorithm cannot always identify Byzantine parties to our advantage: the true centroid can be any centroid computed from at least $n - t$ received vectors (as up to t nodes can be Byzantine). An optimal algorithm that cannot identify Byzantine parties has to minimize the maximum distance to all of those possible centroids. This is a particular case of the Minimax problem, whose solution is to find the center of the smallest enclosing ball, i.e. the center of the smallest ball that contains all possible centroids. Our definition of approximation directly follows from this and is given in Sect. 3.2.

Another technical challenge is to get a significant improvement in the centroid approximation while satisfying a sound validity condition. Since we show in Sect. 4 that satisfying the convex validity condition makes the approximation at least $2d$, we aim for the next best validity condition: the box validity condition. We define the smallest box containing all non-faulty input vectors (in which the output vectors of non-faulty nodes need to be), but also the smallest box containing all possible centroids of $n - t$ vectors (Sect. 3.3 and 5). We call those boxes the trusted box and the centroid box respectively, and show that their intersection is non-empty. Agreeing in the intersection of those boxes guarantees the approximation ratio to be at most $2\sqrt{d}$, and also guarantees the box validity condition. However, those boxes are not computable locally due to the presence of Byzantine parties: nodes do not know which vectors come from Byzantine parties. We consider that Byzantine parties can send different vectors to different non-faulty nodes at each round, and can also not send any vector (see Sect. 3). To solve these issues, we define for each node a local trusted box and a local centroid box, which are guaranteed to be contained in their global counterpart. We also show these local boxes always intersect, making it possible for our algorithm to converge (see Definition 1). The details of this are given in the proof of Theorem 1.

2 Related Work

Approximate Byzantine agreement has been introduced for the one-dimensional case as a way to agree on arbitrary real values inside a bounded range [18]. Observe that any algorithm for solving synchronous Byzantine agreement exactly requires at least $t + 1$ communication rounds [23]. In contrast, a synchronous approximate agreement algorithm [2,18,21] converges to a solution inside an ε small ball in $O(\log(\delta(V)/\varepsilon))$ rounds, which gives an exponential speed-up over exact algorithms in the general case $t = \Theta(n)$.

The multidimensional version of approximate agreement was first introduced by Mendes and Herlihy [29] and Vaidya and Garg [35]. These results were later combined in [30]. The focus of this line of work is to establish agreement on a vector that is relatively "safe". In particular, the vector should lie inside the convex hull of all non-faulty vectors. This can be achieved through the Tverberg partition [34]. The authors provide synchronous and asynchronous algorithms to solve exact and approximate agreement in this setting. This approach can only tolerate up to $t < \min\{n/3, n/(d+1)\}$ Byzantine nodes. This algorithm idea has recently received much attention in different distributed models [1,6,19,24, 32,36,37].

Our work deviates from the previous work on approximate multidimensional agreement in the sense that we relax the strong validity property of agreeing inside the convex hull of all non-faulty vectors. To this end, we use the box validity condition that to date has been regarded as weak or impractical [1, 30]. A similar goal has been pursued by El-Mhamdi, Farhadkhani, Guerraoui, Guirguis, Hoang, and Rouault [20]. The authors focus on collaborative learning under Byzantine adversaries and reduce the corresponding problem to averaging agreement. In averaging agreement, the task is to agree on a vector close to the centroid. However, the computed solution is bounded using the distance between the two farthest vectors from non-faulty nodes, and no focus is put on the validity of the solution.

Our work studies validity conditions for multidimensional Byzantine agreement protocols, with the goal of finding a trade-off between the quality of the solution and the validity requirement. One of the first studies of the quality of Byzantine agreement protocols was presented by Stolz and Wattenhofer [33]. The authors consider the one-dimensional multivalued Byzantine agreement and provide a two-approximation of the median of the non-faulty nodes. Melnyk and Wattenhofer [28] improve the two-approximation result for the median computation to output an optimal approximation. The same paper also introduces the box validity property for multidimensional synchronous agreement used in this work. The first paper discussing validity conditions for multidimensional Byzantine agreement focused on the application of Byzantine agreement in voting protocols [27]. There, a validity condition was introduced that satisfies unanimity among non-faulty voters. This validity condition is also satisfied by the convex validity condition in multidimensional approximate agreement [30]. Allouah, Guerraoui, Hoang, and Villemaud [5] extend the unanimity condition to sparse unanimity, but they only focus on centralized Byzantine-resilient voting protocols. Civit, Gilbert, Guerraoui, Komatovic, and Vidigueira [15] initiate a first study on the impact of validity conditions on the solvability of Byzantine consensus in the partially synchronous communication model.

3   Model and Definitions

We consider a networked system of n nodes, each holding an input vector $v \in \mathbb{R}^d$. Note that, in order for nodes in the network to exchange vectors, we assume that a coordinate system is fixed, and none of the studied or presented algorithms proceed to changing this coordinate system. We assume the standard Euclidean structure on $\mathbb{R}^d$, and use the prevailing $\|\cdot\|_2$ norm. For two vectors $x = (x_1, \ldots, x_d)$ and $y = (y_1, \ldots, y_d)$, we measure the distance in terms of the Euclidean distance: $\mathrm{dist}(x, y) = \sqrt{\sum_{i=1}^{d}(x_i - y_i)^2} = \|x - y\|_2$.

The nodes of the network can communicate with each other in a fully connected peer-to-peer network via public channels. We assume that up to $t < n/3$ nodes can show Byzantine behavior, i.e. such nodes can arbitrarily deviate from the protocol and are assumed to be controlled by a single adversary. Non-faulty nodes we call correct. Similar to [30], we require that the communication between the nodes is reliable. Reliable broadcast has been introduced by Bracha [9] for the asynchronous communication and it guarantees the following two properties while tolerating up to $t < n/3$ Byzantine nodes:

- If a correct node broadcasts a message reliably, all correct nodes will accept this message.
- If a Byzantine node broadcasts a message reliably, all correct nodes will either accept the same message or not accept the Byzantine message at all.

The second condition of reliable broadcast is only satisfied eventually. Therefore, when using the reliable broadcast as a subroutine in agreement protocols, we can only guarantee the following condition:

- If two correct nodes accept a message from a Byzantine party, the accepted message must be the same.

The latter condition is also referred to as the consistency condition in the literature [11], and the corresponding broadcast routine as consistent broadcast. Observe that the reliable broadcast routine from [9] can also be applied in the synchronous communication model by increasing the number of communication rounds compared to simple broadcast.

We consider synchronous communication, in which the nodes communicate in discrete rounds. In each round, a node can send a message, receive messages, and perform some local computation. A message sent by a correct node in round i arrives at its destination in the same round.

3.1   Multidimensional Approximate Agreement

In this work, we design deterministic algorithms that solve multidimensional approximate agreement; every node thereby executes the same algorithm:

Definition 1 (Multidimensional approximate agreement). *Given n nodes, up to t of which can be Byzantine, the goal is to design a deterministic distributed algorithm that satisfies:*

ε**-Agreement:** *Every correct node decides on a vector s.t. any two vectors of correct nodes are at a Euclidean distance of at most ε from each other.*
Strong Validity: *If all correct nodes started with the same input vector, they should agree on this vector as their output.*
Termination: *Every correct node terminates after a finite number of rounds.*

The strong validity condition was originally introduced in the definition of the consensus problem [7,9,10] and it is widely used in the literature [3,8,13,16]. Sometimes, a weaker version of validity is considered in the literature [13,14,38], called the weak validity condition. This validity condition has been used for k-set agreement [17] and BFT protocols [14,38].

Weak Validity: If all nodes are correct and start with the same input vector, they should agree on this vector as their output.

In many state-of-the-art multidimensional approximate agreement protocols [1,6,24,30,32,36], the validity condition is instead replaced by the more restrictive convex validity condition:

Convex Validity: Each correct node decides on a vector that is inside the convex hull of all correct input vectors.

To guarantee convex validity in Byzantine agreement protocols, Mendes, Herlihy, Vaidya, and Garg [30] have the nodes terminate on a vector inside the so-called *SafeArea*:

Definition 2 (*SafeArea*). *The SafeArea of a set of vectors $\{v_i, i \in [n]\}$ that can contain up to t Byzantine vectors is defined as,*

$$SafeArea = \bigcap_{\substack{I \subseteq [n] \\ |I| = n-t}} \mathrm{Conv}\big(\{v_i, i \in I\}\big),$$

where $\mathrm{Conv}\big(\{v_i, i \in I\}\big)$ *is the convex hull of* $\{v_i, i \in I\}$.

The process of agreeing inside the *SafeArea* is costly. In [30], the authors showed that the upper bound on the number of Byzantine parties must be $t < \min(n/3, n/(d+1))$ in the synchronous case.

The goal of this paper is to establish approximate agreement on a representative vector: the centroid of the input vectors of all correct nodes. The centroid is defined as follows:

Definition 3 (Centroid). *The centroid of a finite set of n vectors $\{v_i, i \in [n]\}$ is $\frac{1}{n}\sum_{i=1}^{n} v_i$.*

We use $\mathrm{Cent}^\star$ to denote the centroid of the correct input vectors. We also refer to the set of correct input vectors as the *true vectors* and to $\mathrm{Cent}^\star$ as the *true centroid* respectively.

In Sect. 4, we show that the restriction of the strong validity condition to the convex validity condition cannot provide a good approximation of $\mathrm{Cent}^\star$. We therefore relax convex validity in order to improve both the approximation of the centroid and the resilience of the agreement protocols.

3.2   Approximation of the Centroid

The correct nodes cannot agree on the true centroid if Byzantine parties are present in the system. Let $f \leq t$ denote the actual number of Byzantine parties that are present. A Byzantine party can follow the protocol while choosing worst-case input vectors and thus be indistinguishable from a correct node. While it changes the outcome, such Byzantine behavior is not detectable, and even an optimal approach cannot find out whether all t nodes or only a subset of them are Byzantine. We therefore define the approximation ratio of a true centroid based on all n input vectors for the worst case when $f = t$.

Given all n input vectors, consider all subsets of $n - t$ vectors. At least one of these contains only true vectors. In the case $f < t$, there can be multiple such sets, while in the case $f = t$ there is only one. As mentioned earlier, we cannot determine which of these sets is a set of correct nodes. Thus, each subset of $n - t$ vectors could potentially be the (only) subset containing only true vectors. In the following, we define the set of all possible centroids of $n - t$ nodes, i.e., candidates for being $\text{Cent}^\star$ assuming the worst case where $f = t$ nodes are Byzantine.

Definition 4 (Set of possible centroids). *The set containing all possible centroids of $n - t$ nodes is denoted* S_Cent *and defined as*

$$\text{S}_\text{Cent} := \Big\{ \frac{1}{n - t} \sum_{i \in I} v_i \mid \forall I \subset [n] \ \ s.t. \ |I| = n - t \Big\}.$$

Observe that, in practice, $\text{Cent}^\star$ is not necessarily one of the elements of S_Cent, as the centroids in S_Cent might have been computed from fewer than $n - f$ vectors. We can however make the following observation, which allows us to use S_Cent also in the case $f < t$:

Observation 1. $\text{Cent}^\star \in \text{Conv}(\text{S}_\text{Cent})$.

For $f = t$, the observation simply follows from the fact that $\text{Cent}^\star \in \text{S}_\text{Cent}$. In other cases, let $\{v_i \mid i \in [n - f]\}$ denote the set of true vectors for a better overview. The true centroid is defined as $\text{Cent}^\star = \frac{1}{n-f} \sum_{i \in [n-f]} v_i$. Note that the sum of centroids from S_Cent that only contain true vectors is a multiple of $\text{Cent}^\star$ (indeed, we consider all possible sets of $n - t$ vectors, implying that all true vectors appear in exactly the same number of sets). Therefore, $\text{Cent}^\star$ can be written as a convex combination of elements of S_Cent and it is inside $\text{Conv}(\text{S}_\text{Cent})$.

In order to compute the best-possible approximation of the true centroid w.r.t. the Euclidean distance, we need to determine a vector that minimizes the maximum distance to all computed centroids, or as centrally as possible in $\text{Conv}(\text{S}_\text{Cent})$. Finding this point corresponds to solving a Minimax problem known as the 1-center problem for the set S_Cent, which requires finding the smallest enclosing ball of S_Cent and computing its center.

Proposition 1 (Best-possible centroid approximation vector). *Given n input vectors, f of which are Byzantine (where $f = t$ is the worst case). The best possible approximation vector of the true centroid* $\text{Cent}^\star$ *in the worst case is the center of the smallest enclosing ball* $\text{Ball}(\text{S}_\text{Cent})$ *of* S_Cent.

Given that the best-possible approximation of the true centroid is the center of $\text{Ball}(\text{S}_\text{Cent})$, and $\text{Cent}^\star$ could be on its boundary, the distance between the best-possible approximation and the true centroid can be up to $\text{Rad}(\text{Ball}(\text{S}_\text{Cent}))$. Using this idea, we can now define the optimal centroid approximation for any algorithm as follows:

Definition 5 (Optimal centroid approximation). *Let* $\mathrm{Rad}(\mathrm{Ball}(\mathrm{S}_{\mathrm{Cent}}))$ *be the radius of* $\mathrm{Ball}(\mathrm{S}_{\mathrm{Cent}})$. *All vectors at a distance of at most* $\mathrm{Rad}(\mathrm{Ball}(\mathrm{S}_{\mathrm{Cent}}))$ *from* $\mathrm{Cent}^\star$ *are defined to provide an optimal approximation of the true centroid.*

We can now define the approximation ratio of an algorithm for computing the centroid depending on the radius of the smallest enclosing ball $\mathrm{Rad}(\mathrm{Ball}(\mathrm{S}_{\mathrm{Cent}}))$:

Definition 6 (c-approximation of the centroid). *Let* $\mathcal{A}$ *be an algorithm computing an approximation vector of the true centroid. Let* $O_{\mathcal{A}}$ *be the output of the algorithm on some input* I. *The approximation ratio of an algorithm* $\mathcal{A}$ *on input* I *is defined as the smallest* c *for which holds*

$$\mathrm{dist}(O_{\mathcal{A}}, \mathrm{Cent}^\star) \le c \cdot \mathrm{Rad}(\mathrm{Ball}(\mathrm{S}_{\mathrm{Cent}})).$$

We further say that $\mathcal{A}$ *computes a c-approximation of the true centroid if, for any admissible input to the algorithm, the approximation ratio of the output of* $\mathcal{A}$ *is upper bounded by c.*

Note that admissible inputs consist in each of the n nodes having an input vector in $\mathbb{R}^d$, and at most t of these nodes being Byzantine.

3.3 Weakening the Convex Validity Condition

In addition to centroid approximation, we care about validity conditions. We strive for a more restrictive one than the strong validity from Definition 1. However, the *SafeArea* approaches do not give a good approximation of the centroid, which makes the convex validity condition too restrictive (Sect. 4). To achieve a better approximation ratio without sacrificing too much on the validity condition, we relax the convex validity condition to the "coordinate-parallel box" of true vectors:

Definition 7 (Trusted box). *Let* $f \le t$ *be the number of Byzantine nodes and let* v_i, $i \in [n - f]$ *denote the true vectors. Let* $v_i[k]$ *denote the kth coordinates of these vectors. The trusted box* TB *is the Cartesian product of* $[\min_{i \in [n-f]} v_i[k], \max_{i \in [n-f]} v_i[k]]$, *for* $k \in [d]$. *Denote* $TB[k]$ *the orthogonal projection of* TB *onto the kth coordinate.*

In other words, TB is the smallest box containing all true vectors. Given this definition, we can define a relaxed version of the convex validity condition, referred to as the box validity condition in the literature [28]:

Box validity: The output vectors of the correct nodes at termination should be inside the trusted box TB.

The trusted box cannot be locally computed by the correct nodes if the Byzantine parties submit their vectors. We hence define a locally trusted box that is computable locally by each correct node. We use M_i to denote the set of messages received by node i in a communication round. Observe that $m_i := |M_i| \ge n - t$. We also use the following notation that allows us to rearrange the values in each coordinate and reassign the indices:

Notation 1. (Coordinate-wise sorting). *For each coordinate $k \in [d]$, let $v_j[k]$ denote the kth coordinate of the vector received from node j. We order the values $v_j[k], \forall k \in [n]$ in increasing order and relabel the indices of the vectors accordingly. Thus, $v_j[k]$ now holds the jth smallest value in coordinate k. Note that after the renaming, for two different coordinates k and l, $v_j[k]$ and $v_j[l]$ may not hold coordinates received from the same node.*

With this notation, we can now define the locally trusted box:

Definition 8 (Locally trusted box). *For all received vectors $v_j \in M_i$ by node i, denote $v_j[k]$ the kth coordinate of the respective vector. For each coordinate, we reassign the indices as in Notation 1. The number of Byzantine values for each coordinate is at most $m_i - (n - t)$. The locally trusted box TB_i computed by node i is the Cartesian product of $\left[v_{m_i - (n-t)+1}[k], v_{n-t}[k] \right]$ for $k \in [d]$. Denote $TB_i[k]$ the orthogonal projection of TB_i onto the kth coordinate.*

In other words, since in the synchronous setting each node i is guaranteed to receive the $n - t$ true vectors, the number of received messages m_i allows node i to infer the number of Byzantine vectors it received and trim this exact number of values on each side of the interval for each coordinate. This ensures that $TB_i[k]$ always contains $n - 2t$ values, for every coordinate k.

Observe that since we remove the maximum number of potentially Byzantine values for each coordinate on each side when computing the locally trusted box TB_i, any locally trusted box is contained in the trusted box TB.

4   Lower Bound on the Approximation of the *SafeArea* Approach

In this section, we prove that satisfying convex validity comes at the cost of good centroid approximation.

Proposition 2. *The approximation ratio of the true centroid that can be achieved by any algorithm satisfying convex validity is at least $2d$.*

In the full version of this paper [12], we show that it is necessary to agree inside the *SafeArea* (Definition 2) to satisfy convex validity in the presence of Byzantine parties. We show a lower bound on the centroid approximation by using the fact that *SafeArea* can be reduced to a single point, at distance $2d$ from Cent*, forcing this lower bound on any algorithm agreeing in *SafeArea*, and therefore on any algorithm that satisfies the convex validity condition. The full proof is presented in [12]. In the following, we visualize the proof idea in 3 dimensions.

The idea of the proof in dimension $d = 3$ is illustrated by Fig. 1. In this example, the input vectors are divided into four groups of identical vectors: one group at $v_0 = (0, 0, 0)$ containing input vectors of $t + 1$ nodes, and the three groups close to v, each containing input vectors of t nodes. The key idea is that the three groups of t vectors are at infinitesimal distance from each other,

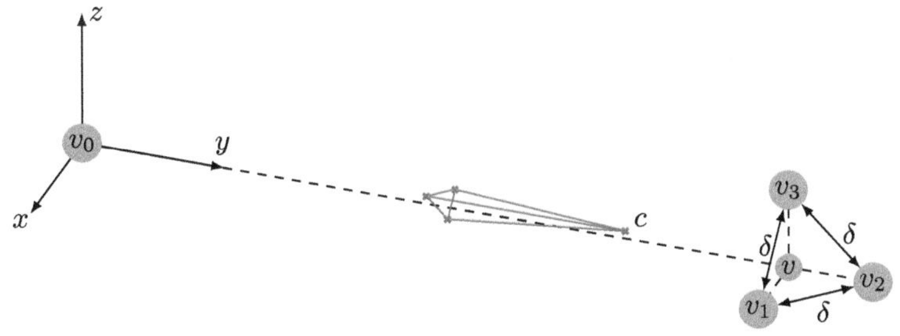

Fig. 1. Lower bound construction with $d = 3$ and $n = 4t + 1$. The input vectors are spread into $d + 1 = 4$ groups, one at v_0 ($t + 1$ of them), and the others at v_i, for $i = 1, 2, 3$ (groups of t vectors). Here $\mathrm{Conv}(v_0, v_1, v_2) \cap \mathrm{Conv}(v_0, v_1, v_3) \cap \mathrm{Conv}(v_0, v_2, v_3) = \{v_0\} = SafeArea$. The light blue area is $\mathrm{Conv}(S_{\mathrm{Cent}})$, where $c :- \mathrm{Cent}^\star$.

but are all linearly independent. Each of them will be ignored in computing one of the polytopes whose intersection gives the $SafeArea$, but the group at v_0 will be in all polytopes as it contains $t + 1$ vectors. Thus, the intersection of all polytopes has to be reduced to $\{v_0\}$. The lower bound comes from the fact that the t Byzantine input vectors, as well as the convergence vector since $SafeArea = \{v_0\}$, could be at v_0. In such a case, the true centroid $\mathrm{Cent}^\star$ is close to the three groups of t vectors, these vectors being all correct. More precisely, assimilating the infinitesimally closed vectors together at $v = (1, 0, 0)$, we get than $\mathrm{Cent}^\star = 3t/(3t + 1) \cdot v$, implying that $\mathrm{dist}(SafeArea, \mathrm{Cent}^\star) = 3t/(3t + 1)$. Then, $2\mathrm{Rad}(\mathrm{Ball}(S_{\mathrm{Cent}})) = 3t/(3t + 1) - 2t/(3t + 1) = t/(3t + 1)$ and $\mathrm{dist}(SafeArea, \mathrm{Cent}^\star)/\mathrm{Rad}(\mathrm{Ball}(S_{\mathrm{Cent}})) = (3t/(3t + 1)) \cdot (3t + 1/t) \cdot 2 = 6$.

5   A $2\sqrt{d}$-Approximation of the Centroid with Box Validity

In this section, we relax the convex validity from [30] to the box validity condition. Though this condition is looser than the convex validity, it is still way more restrictive than the strong validity condition. This allows us to improve the approximation ratio as well as the resilience of the $SafeArea$ approach from Sect. 4, but also allows us to satisfy a better validity condition than the ones satisfied by the algorithms based on MDA (see full version [12]). This is shown by Theorem 1.

In Fig. 2, we show an example where neither S_{Cent}, nor the smallest enclosing ball of S_{Cent} intersect TB_i. Thus, no node can locally compute an optimal approximation of $\mathrm{Cent}^\star$. Note that the smallest box around S_{Cent} would intersect TB_i in the same example. We formally define this box as follows:

Definition 9 (Centroid box). *The centroid box CB is the smallest box containing S_{Cent}, and the local centroid box CB_i of node $i \in [n]$ is the smallest box containing $S_{\mathrm{Cent}}(i)$.*

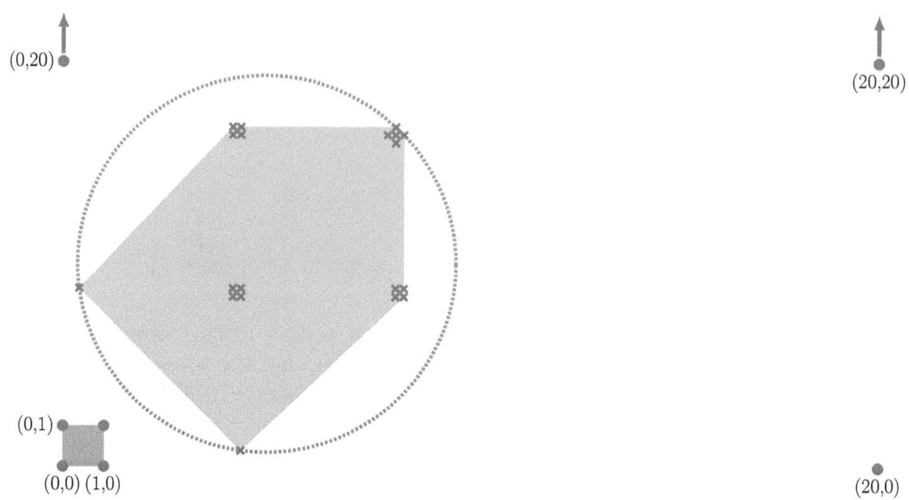

Fig. 2. An example of initial configuration for the 2-dimensional agreement problem. Solid circles represent the $n = 7$ input vectors, including up to $t = 2$ Byzantine vectors. The crosses represent all possible centroid positions, with the blue area showing their convex hull. The orange box will be contained in TB_i for all $i \in [n]$, and contains the *SafeArea*. The blue circle is the area where an optimal approximation of the centroid can be achieved. (Color figure online)

In the following, we make the observation that the local centroid box can be indeed computed in polynomial time locally:

Observation 2. *CB_i can be computed in polynomial time as follows: for each coordinate, compute the average of the $n - t$ smallest values and the average of the $n - t$ largest values. This yields the interval defining the box over this coordinate.*

Next, we define the midpoint function that is used in the algorithm.

Definition 10 (Midpoint). *The* Midpoint *of a box X is defined as*

$$\mathrm{Midpoint}(X) = \Big(\mathrm{Midpoint}\big(X[1]\big), \ldots, \mathrm{Midpoint}\big(X[d]\big)\Big),$$

where $X[k]$ is the set containing all k^{th} coordinates of vectors of the set X, and the one-dimensional Midpoint *function returns the midpoint of the interval spanned by a finite multiset of real values.*

Algorithm 1 solves multidimensional approximate agreement with box validity. The idea of this algorithm is to compute a box containing trusted vectors and a box containing all possible centroids. In every round, the new input vector is set to be the midpoint of the intersection of the two boxes. The presented algorithm has optimal resilience of $t < n/3$ and reaches a $2\sqrt{d}$ approximation of the true centroid, while using polynomial local computation time. Thus, the

Algorithm 1. Synchronous approximate agreement with box validity and resilience $t < n/3$

1: Let C be a large constant. Each node $i \in [n]$ with input vector v_i executes the following code in each round $r = 1, 2, \ldots, C \cdot \log_2\left(\frac{1}{\varepsilon} \cdot \mathcal{LE}\left(\text{TB}_i\right)\right)$:

2:      Broadcast v_i reliably to all nodes

3:      Reliably receive up to n messages $M_i = \{v_j, j \in [n]\}$

4:      Compute TB_i from M_i by excluding $|M_i| - (n - t)$ values on each side

5:      Compute CB_i from M_i

6:      Set v_i to Midpoint$(\text{TB}_i \cap \text{CB}_i)$.

---

algorithm outperforms the *SafeArea* approach in resilience, consensus quality, and computation time. The termination round is defined by each node locally in the first round of the algorithm. If some node terminates earlier than others, the other nodes can use its last sent vector for all following rounds. In order to analyze the convergence of the algorithm, we will define the convergence rate with respect to the longest edges of the boxes:

Definition 11 (Length of the longest edge ($\mathcal{LE}$)). *The length of the longest edge of a box B is defined as*

$$\mathcal{LE}(B) = \max_{\substack{k \in [d] \\ u,v \in B}} \left|u[k] - v[k]\right|.$$

Note that $\mathcal{LE}(\text{TB}) = \max_{k \in [d]; i,j \in I_{\text{correct}}} \left|v_i[k] - v_j[k]\right|$, where I_{correct} denotes the set of indices of the true vectors, and the longest edge thus corresponds to the diameter of the true vectors. The following theorem shows the main properties of the algorithm.

Theorem 1. *Algorithm 1 achieves a $2\sqrt{d}$-approximation of $\text{Cent}^\star$ in the synchronous setting. After $O\left(\log_2\left(\frac{1}{\varepsilon} \cdot \mathcal{LE}\left(TB\right)\right)\right)$ synchronous rounds, the vectors of the correct nodes satisfy multidimensional approximate agreement with box validity. The resilience of the algorithm is $t < n/3$.*

In the following, we prove that Algorithm 1 is well-defined, by showing that TB_i and CB_i intersect if $t < n/2$. This way, all correct nodes are able to choose a new input vector in line 6 of Algorithm 1. The proofs of correctness, convergence, box validity as well as resilience can be derived from this lemma and are presented in the full version of this paper [12].

Lemma 1. *For all correct nodes i, the local trusted box and the local centroid box have a non-empty intersection: $CB_i \cap TB_i \neq \emptyset$.*

Proof. Let $\text{TB}_i[k]$ and $\text{CB}_i[k]$ be the (coordinate-wise) projections of TB_i and CB_i on the kth coordinate, respectively. We prove the statement of the lemma for some correct node i (note that this proof works for every correct node).

For each coordinate $k \in [d]$, let $v_j[k]$ denote the kth coordinate of the vector received from node j. We sort the values $v_j[k], \forall k \in [n]$ as in Notation 1. Recall that node i receives $m_i \geq n - t$ vectors and that there are m_i values in each coordinate. Because we remove $m_i - (n - t)$ on each side of the interval for each coordinate k, each $\mathrm{TB}_i[k]$ contains $2(n - t) - m_i \geq n - 2t$ values. Our goal is to construct a point that is both in TB_i and CB_i.

Consider the averaged sum c_ℓ of the $n - t$ smallest received values in coordinate k: $c_\ell = \frac{1}{n-t} \sum_{j=1}^{n-t} v_j[k]$. Observe that there exists a subset of $n - t$ vectors with this averaged sum in their k^{th} coordinate. The corresponding vector is the centroid of the $n - t$ nodes with these input vectors. It is contained in CB_i and c_ℓ is its orthogonal projection on coordinate k. Similarly, there exists an averaged sum $c_u = \frac{1}{n-t} \sum_{j=m_i-(n-t)+1}^{m_i} v_j[k] \in \mathrm{S}_{\mathrm{Cent}}[k]$ formed by the $n - t$ largest values in coordinate k, that corresponds to the kth coordinate of some centroid in CB_i. This definition makes sure that $[c_\ell, c_u] \subseteq \mathrm{CB}_i[k]$.

Let $V_k = \frac{1}{2(n-t)-m_i} \sum_{j=m_i-(n-t)+1}^{n-t} v_j[k]$ be the average of values, where we truncate the $m_i - (n - t)$ smallest and largest values in this coordinate. This is possible since $t < n/3$ and $m_i \geq n - t$. Since the values $v_j[k]$ are ordered, $\frac{1}{2(n-t)-m_i} \sum_{j=m_i-(n-t)+1}^{n-t} v_j[k] \geq c_\ell$ and $\frac{1}{2(n-t)-m_i} \sum_{j=m_i-(n-t)+1}^{n-t} v_j[k] \leq c_u$. Therefore, $V_k \in [c_\ell, c_u] \subseteq \mathrm{CB}_i[k]$. Also, $V_k \in \left[v_{m_i-(n-t)+1}[k], v_{n-t}[k] \right] = \mathrm{TB}_i[k]$, since it is the mean of values $v_{m_i-(n-t)+1}[k]$ to $v_{n-t}[k]$.

Now, consider a vector V s.t. $\forall k \in [d]: V[k] = V_k$. By the definitions of TB_i and CB_i, since $V[k] \in \mathrm{TB}_i[k] \cap \mathrm{CB}_i[k], \forall k \in [d]$, this vector will be inside the intersection of the boxes, i.e. $V \in \mathrm{TB}_i \cap \mathrm{CB}_i$.

Observe that Algorithm 1 can be adjusted such that the nodes terminate after $\left\lceil \log_2\left(\frac{1}{\varepsilon} \cdot \mathcal{LE}(\mathrm{TB})\right) \right\rceil$ synchronous rounds. This can be achieved by letting the nodes mark their last sent message, such that other nodes can use this last vector for future computations. This way, the algorithm satisfies ε-agreement and termination. We show that Algorithm 1 with the stopping condition provides a $2\sqrt{d}$-approximation of the true centroid.

Lemma 2. *The approximation ratio of the true centroid by Algorithm 1 is upper bounded by* $2\sqrt{d}$.

Proof. Following Definition 6, we need to show that the ratio between the distance from any output of Algorithm 1 to $\mathrm{Cent}^\star$ and the optimal radius $\mathrm{Rad}\big(\mathrm{Ball}(\mathrm{S}_{\mathrm{Cent}})\big)$ is less than $2\sqrt{d}$. We first lower bound the radius of the smallest enclosing ball of $\mathrm{S}_{\mathrm{Cent}}$. Observe that the radius is always at least $\max_{x,y \in \mathrm{S}_{\mathrm{Cent}}} \big(\mathrm{dist}(x,y)\big)/2$.

Next, we focus on upper bounding the distance between $\mathrm{Cent}^\star$ and the furthest possible point from it inside CB. W.l.o.g., we assume that $\max_{x,y \in \mathrm{S}_{\mathrm{Cent}}} \big(\mathrm{dist}(x, y)\big) = 1$. We consider the relation between the smallest enclosing ball and CB. Observe that each face of the box CB has to contain at least one point of $\mathrm{S}_{\mathrm{Cent}}$. If CB is contained inside the ball, i.e. if the vertices of the box would lie on the ball surface, the computed approximation of Algorithm 1 would always be optimal.

The worst case is achieved if the ball is (partly) contained inside CB. Then, the optimal solution might lie inside CB and the ball, while the furthest node may lie on one of the vertices of CB outside of the ball. The distance of any node from $\text{Cent}^\star$ in this case is upper bounded by the diagonal of the box. Since the longest distance between any two points was assumed to be 1, the box is contained in a unit cube. Thus, the largest distance between two points of CB is at most $\sqrt{d}$. The approximation ratio of Algorithm 1 can be upper bounded by:

$$\frac{\max_{x \in \text{CB}}\big(\text{dist}(\text{Cent}^\star, x)\big)}{\text{Rad}\big(\text{Ball}(S_{\text{Cent}})\big)} \leq 2 \cdot \frac{\max_{x \in \text{CB}}\big(\text{dist}(\text{Cent}^\star, x)\big)}{\max_{x,y \in S_{\text{Cent}}}\big(\text{dist}(x, y)\big)} \leq 2 \cdot \frac{\sqrt{d}}{1} = 2\sqrt{d}.$$

This lemma concludes the proof of Theorem 1 and the analysis of the centroid approximation for the synchronous case.

Observe that Theorem 1 directly implies Corollary 1 for the $RB-TM$ algorithm from [20].

Corollary 1. *The $RB-TM$ algorithm [20] with a resilience of $t < n/3$ achieves a $2\sqrt{d}$-approximation of $\text{Cent}^\star$ in the synchronous setting.*

Proof. Observe that the trimmed mean computed in [20] corresponds to the vector V_k constructed in the proof of Lemma 1. Thus, the trimmed mean lies in the intersection of TB_i and CB_i for every node i.

6   Conclusion

In this work, we proposed a new definition of centroid approximation in the presence of Byzantine parties in the system. This approximation measure allowed us to analyze the quality of agreement algorithms not only in the worst case, but for any input distribution of the correct and Byzantine parties. Using this definition, we showed that the well-known algorithm for multi-dimensional approximate agreement [30] did not provide a bounded approximation of the centroid, whereas its variants may at best achieve a d-approximation. We hence proposed an algorithm that is based on one-dimensional approximate agreement that achieves a $2\sqrt{d}$-approximation of the centroid while providing optimal resilience and improving efficiency.

Note that all results in this work can be adapted to the asynchronous case by slight modifications of the definitions. In the asynchronous case, the lower and upper bounds only deviate by a small constant compared to the bounds in Table 1. While the results in Sect. 4 are only of theoretical interest, as the algorithms have small resilience and exponential running time, we believe that the algorithms in Sect. 5, and the algorithms based on MDA (full version [12]) are also of importance in practical applications. In future work, we want to investigate the performance of algorithms that have a good approximation in the context of collaborative learning [20]. In the full version of this paper [12], we also outline other interesting questions for future research.

Acknowledgments. We would like to thank Manish Kumar, Stefan Schmid, Jukka Suomela, and Jara Uitto for useful discussions. We would also like to thank the anonymous reviewers for the very helpful feedback they have provided for previous versions of this work. This research was supported by the Academy of Finland, Grant 334238, and by the Austrian Science Fund (FWF), Grant I 4800-N (ADVISE).

References

1. Abbas, W., Shabbir, M., Li, J., Koutsoukos, X.: Resilient distributed vector consensus using centerpoint. Automatica **136**, 110046 (2022)
2. Abraham, I., Amit, Y., Dolev, D.: Optimal resilience asynchronous approximate agreement. In: Higashino, T. (ed.) OPODIS 2004. LNCS, vol. 3544, pp. 229–239. Springer, Heidelberg (2005). https://doi.org/10.1007/11516798_17
3. Abraham, I., Devadas, S., Nayak, K., Ren, L.: Brief announcement: practical synchronous byzantine consensus. In: 31st International Symposium on Distributed Computing (DISC 2017) (2017). https://doi.org/10.4230/LIPIcs.DISC.2017.41
4. Ai, L., Yu, J., Wu, Z., He, Y., Guan, T.: Optimized residual vector quantization for efficient approximate nearest neighbor search. Multimedia Syst. **23** (2017)
5. Allouah, Y., Guerraoui, R., Hoang, L., Villemaud, O.: Robust sparse voting. CoRR arxiv:2202.08656 (2022)
6. Attiya, H., Ellen, F.: The step complexity of multidimensional approximate agreement. In: 26th International Conference on Principles of Distributed Systems (OPODIS 2022) (2023)
7. Bar-Noy, A., Dolev, D.: Families of consensus algorithms. In: Reif, J.H. (ed.) AWOC 1988. LNCS, vol. 319, pp. 380–390. Springer, New York (1988). https://doi.org/10.1007/BFb0040405
8. Berman, P., Garay, J., Perry, K.: Towards optimal distributed consensus. In: 30th Annual Symposium on Foundations of Computer Science (1989). https://doi.org/10.1109/SFCS.1989.63511
9. Bracha, G.: Asynchronous byzantine agreement protocols. Inf. Comput. **75**(2), 130–143 (1987)
10. Bracha, G., Toueg, S.: Resilient consensus protocols. In: Proceedings of the Second Annual ACM Symposium on Principles of Distributed Computing, PODC '83 (1983). https://doi.org/10.1145/800221.806706
11. Cachin, C., Guerraoui, R., Rodrigues, L.: Introduction to Reliable and Secure Distributed Programming, 2nd edn. Springer, Heidelberg (2014)
12. Cambus, M., Melnyk, D.: Centroid approximation with multidimensional approximate agreement protocols (2025). https://arxiv.org/abs/2306.12741
13. Civit, P., et al.: Byzantine consensus is θ (n^2): the Dolev-Reischuk bound is tight even in partial synchrony! In: 36th International Symposium on Distributed Computing (DISC 2022). Schloss Dagstuhl-Leibniz-Zentrum für Informatik (2022)
14. Civit, P., Gilbert, S., Gramoli, V.: Polygraph: accountable byzantine agreement. In: 2021 IEEE 41st International Conference on Distributed Computing Systems (ICDCS), pp. 403–413. IEEE (2021)
15. Civit, P., Gilbert, S., Guerraoui, R., Komatovic, J., Vidigueira, M.: On the validity of consensus. In: Proceedings of the 2023 ACM Symposium on Principles of Distributed Computing, PODC '23, pp. 332–343 (2023). https://doi.org/10.1145/3583668.3594567
16. Crain, T., Gramoli, V., Larrea, M., Raynal, M.: DBFT: efficient leaderless byzantine consensus and its application to blockchains. In: 2018 IEEE 17th International Symposium on Network Computing and Applications (NCA) (2018). https://doi.org/10.1109/NCA.2018.8548057

17. De Prisco, R., Malkhi, D., Reiter, M.K.: On k-set consensus problems in asynchronous systems. In: Proceedings of the Eighteenth Annual ACM Symposium on Principles of Distributed Computing, PODC '99 (1999). https://doi.org/10.1145/301308.301368

18. Dolev, D., Lynch, N.A., Pinter, S.S., Stark, E.W., Weihl, W.E.: Reaching approximate agreement in the presence of faults. J. ACM **33**(3), 499–516 (1986)

19. Dotan, M., Stern, G., Zohar, A.: Validated byzantine asynchronous multidimensional approximate agreement. CoRR arxiv:2211.02126 (2022). https://doi.org/10.48550/ARXIV.2211.02126

20. El-Mhamdi, E.M., Farhadkhani, S., Guerraoui, R., Guirguis, A., Hoang, L.N., Rouault, S.: Collaborative learning in the jungle (decentralized, byzantine, heterogeneous, asynchronous and nonconvex learning). In: Advances in Neural Information Processing Systems, vol. 34. Curran Associates, Inc. (2021)

21. Fekete, A.D.: Asymptotically optimal algorithms for approximate agreement. Distrib. Comput. **4**(1), 9–29 (1990)

22. Fiorina, M.P., Plott, C.R.: Committee decisions under majority rule: an experimental study. Am. Polit. Sci. Rev. **72**(2), 575–598 (1978)

23. Fischer, M.J., Lynch, N.A.: A lower bound for the time to assure interactive consistency. Inf. Process. Lett. **14**(4), 183–186 (1982)

24. Függer, M., Nowak, T.: Fast multidimensional asymptotic and approximate consensus. In: 32nd International Symposium on Distributed Computing (DISC 2018) (2018)

25. Li, B., Kamuhanda, D., He, K.: Centroid-based multiple local community detection. IEEE Trans. Comput. Soc. Syst. (2022)

26. Lloyd, S.: Least squares quantization in PCM. IEEE Trans. Inf. Theory **28**(2), 129–137 (1982)

27. Melnyk, D., Wang, Y., Wattenhofer, R.: Byzantine preferential voting. In: Web and Internet Economics (WINE), pp. 327–340 (2018)

28. Melnyk, D., Wattenhofer, R.: Byzantine agreement with interval validity. In: 2018 IEEE 37th Symposium on Reliable Distributed Systems (SRDS) (2018)

29. Mendes, H., Herlihy, M.: Multidimensional approximate agreement in byzantine asynchronous systems. In: Proceedings of the Forty-Fifth Annual ACM Symposium on Theory of Computing, STOC (2013)

30. Mendes, H., Herlihy, M., Vaidya, N., Garg, V.K.: Multidimensional agreement in byzantine systems. Distrib. Comput. **28**(6), 423–441 (2015)

31. Naz, A., Piranda, B., Goldstein, S.C., Bourgeois, J.: Approximate-centroid election in large-scale distributed embedded systems. In: 2016 IEEE 30th International Conference on Advanced Information Networking and Applications (AINA), pp. 548–556 (2016)

32. Park, H., Hutchinson, S.A.: Fault-tolerant rendezvous of multirobot systems. IEEE Trans. Rob. **33**(3), 565–582 (2017). https://doi.org/10.1109/TRO.2017.2658604

33. Stolz, D., Wattenhofer, R.: Byzantine agreement with median validity. In: 19th International Conference on Priniciples of Distributed Systems. OPODIS (2015)

34. Tverberg, H.: A generalization of radon's theorem. J. Lond. Math. Soc. **s1-41**(1), 123–128 (1966). https://doi.org/10.1112/jlms/s1-41.1.123

35. Vaidya, N.H., Garg, V.K.: Byzantine vector consensus in complete graphs. In: Proceedings of the 2013 ACM Symposium on Principles of Distributed Computing, PODC (2013)

36. Wang, X., Mou, S., Sundaram, S.: A resilient convex combination for consensus-based distributed algorithms. Numer. Algebra Control Optim. **9**(3), 269–281 (2019)

37. Xiang, Z., Vaidya, N.H.: Relaxed byzantine vector consensus. In: 20th International Conference on Principles of Distributed Systems (OPODIS 2016) (2017)
38. Yin, M., Malkhi, D., Reiter, M.K., Gueta, G.G., Abraham, I.: Hotstuff: BFT consensus with linearity and responsiveness. In: Proceedings of the 2019 ACM Symposium on Principles of Distributed Computing, pp. 347–356 (2019)

Computing Tree Structures in Anonymous Graphs via Mobile Agents

Prabhat Kumar Chand[1]([⊠]) , Manish Kumar[2] , and Anisur Rahaman Molla[1]

[1] Indian Statistical Institute, Kolkata, India
pchand744@gmail.com, molla@isical.ac.in
[2] Indian Institute of Technology, Madras, Chennai, India
ic39275@imail.iitm.ac.in

Abstract. Minimum Spanning Tree (MST) and Breadth-First Search (BFS) tree constructions are classical problems in distributed computing, typically studied in the message-passing model, where static nodes communicate via messages. This paper examines these problems in an agent-based network, where computational devices are modelled as mobile agents that explore a graph and perform computations. Each node serves as a container for agents, and communication occurs when agents meet at the same node. We consider the setting where n agents are dispersed (one per node) on an anonymous, arbitrary n-node, m-edge graph G. The goal is to construct tree structures such that each tree edge is recognized by at least one of its endpoint agents, minimizing both time and memory. We work in the synchronous model, measuring time in rounds, and assume agents have no prior knowledge of graph parameters such as n, m, D, Δ. A known solution constructs a BFS tree in $O(D\Delta)$ rounds with $O(\log n)$ memory per agent, assuming the root is known. We present a deterministic algorithm that constructs a BFS tree in $O(\min(D\Delta, m \log n) + n \log n + \Delta \log^2 n)$ rounds with $O(\log n)$ bits per agent, without any prior root knowledge. In discovering the root, we solve leader election and MST. Our leader election and MST algorithms run in $O(n \log n + \Delta \log^2 n)$ rounds using $O(\log n)$ memory. Previous results require $O(m)$ rounds and $O(\log^2 n)$ memory for leader election, and $O(m + n \log n)$ rounds and $O(\max(\Delta, \log n) \log n)$ memory for MST. Our results improve over this prior work. We assume each agent knows λ, the maximum identifier, bounded by a polynomial in n (i.e., $\lambda \leq n^c$ for some constant $c \geq 1$).

Keywords: Distributed Graph Algorithms · Leader Election · Minimum Spanning Tree · Breadth First Search Tree · Mobile Agents · Mobile Robots · Autonomous Agents

1 Introduction

Tree structures are fundamental in computer science, enabling efficient organization and retrieval across domains such as databases, networking, and AI. Among them,

The work of Manish Kumar is supported by CyStar at IIT Madras.
A. R. Molla is supported, in part, by ANRF-SERB Core Research Grant, file no. CRG/2023/009048, and R. C. Bose Centre's internal research grant.

S. Bonomi et al. (Eds.): SSS 2025, LNCS 16350, pp. 111–127, 2026.
https://doi.org/10.1007/978-3-032-11127-2_11

Breadth-First Search (BFS) Trees and Minimum Spanning Trees (MST) are widely used in both classical and distributed systems. BFS trees support shortest-path discovery in unweighted graphs [34], facilitating applications in networking, pathfinding, and distributed computing. MSTs and BFS trees are particularly useful in decentralized systems where mobile agents perform local computations, minimizing communication overhead—a key advantage in scenarios like disaster response, urban management, and military operations.

Agent-based models have proven effective for memory-efficient graph exploration [34], subgraph enumeration [26], and autonomous dispersion [17,22]. Mobile agents are widely applied in network exploration, from underwater navigation [11] to network-centric warfare [25], social network analysis [27,41], and search-and-rescue missions [33,38–40]. In Wireless Sensor Networks (WSNs), drone-based agents enhance sensor deployment in remote areas [1,12,15,16]. Recent studies explore agent-based solutions for maximal independent sets (MIS) [29,32], small dominating sets [10], triangle counting [7], BFS tree construction [8], and message-passing simulations [18]. This paper uses a mechanism for agents to meet their neighbours by exploiting their ID bits without relying on graph parameters, enabling improved MST construction over [18] and enhancing BFS tree formation from [8]. Techniques that leverage agent identifiers for coordination and traversal have indeed been explored in the literature, particularly in deterministic rendezvous and treasure hunt problems using strongly universal exploration sequences (UXS) [37]. While our proposed protocol shares the high-level idea of employing agent labels, it differs in both the problem setting and the manner in which labels are used. Specifically, our protocol focuses on enabling agents to meet their neighbors within a designated round window in an anonymous graph, using labels to coordinate local interactions rather than global exploration. A fully extended version of this paper with all details, proofs and pseudocodes is available at [9].

1.1 Our Contribution

In this paper, we study the complexity of constructing two fundamental tree structures—BFS and MST—without assuming any prior knowledge of graph parameters. Each agent only knows the value of a parameter λ, which denotes the highest identifier among all agents and is independent of the underlying graph structure. As part of the MST construction, our algorithm also performs leader election. The main results of this paper are summarized below.

Theorem 1 (MST). *Given a dispersed configuration of n agents with unique identifiers positioned one agent at every node of n nodes, m edges, maximum degree Δ, and a graph G with no node identifier. Then, a deterministic algorithm constructs an MST in $O(\Delta \log^2 n + n \log n)$ rounds and $O(\log n)$ bits per agent. The algorithm requires no prior knowledge of any graph parameters and assumes that each agent knows only λ, the maximum among all agent identifiers. (Sect. 4)*

The memory complexity is optimal since any algorithm designed in the agent-based model with n agents needs $\Theta(\log n)$ bits per agent [2,21]. Besides its merits regarding improved time and memory complexities for an important problem, the root of the MST provides an immediate result for leader election, as stated in the following theorem.

Theorem 2 (Leader Election). *Given a dispersed configuration of n agents with unique identifiers positioned one agent at every node of n nodes m edges, maximum degree Δ graph G with no node identifier. Then, there is a deterministic algorithm that elects a leader in $O(\Delta \log^2 n + n \log n)$ rounds and $O(\log n)$ bits per agent. The algorithm requires no prior knowledge of any graph parameters and assumes that each agent knows only λ, the maximum among all agent identifiers. (Sect. 4)*

The elected leader, with the help of the MST algorithm, eventually works as the root of the BFS tree and we have the following theorem.

Theorem 3 (BFS Tree). *Given a dispersed configuration of n agents with unique identifiers positioned one agent at every node of n nodes m edges, maximum degree Δ, and diameter D graph G with no node identifier. Then, there is a deterministic algorithm that constructs a BFS tree in $O(\min(D\Delta, m \log n) + n \log n + \Delta \log^2 n)$ rounds and $O(\log n)$ bits per agent. The algorithm requires no prior knowledge of any graph parameters and assumes that each agent knows only λ, the maximum among all agent identifiers. (Sect. 5)*

Theorem 3 improves results for graphs with large diameter (D) and high maximum degree (Δ), which balance global reach with local connectivity. Such graphs frequently appear in network design, social networks, and distributed computing, where efficient traversal is essential.

To establish the above theorems, we introduced a protocol, termed `Meeting with Neighbor()` (details in Sect. 4.1), to systematically enable agents to meet each of their neighbors within $O(\log n)$ time per neighbor. This protocol operates without requiring any graph-specific parameters and only needs knowledge of the ID of the highest ID agent and therefore allows computation between neighboring agents without using any knowledge of graph parameters, unlike previous approaches [7,8,10]. Using `Meeting with Neighbor()`, we construct a BFS tree in an arbitrary anonymous graph. To facilitate this, agents first identify a root and gather essential graph parameters such as m and Δ by constructing a Minimum Spanning Tree, after which the Breadth First Search Tree is constructed. Table 1 lists and compares all the established results with the previous results.

While our graph G is unweighted, we adapt our approach from the GHS algorithm due to its inherently parallel structure, which offers a distinct advantage over the existing agent-based MST approach in [18] (See Table 1). Although [18] also utilizes a variant of the GHS algorithm, it operates sequentially. In addition to an independent MST construction study using agents, the GHS algorithm allows for the construction of a spanning tree with a root. This is therefore advantageous than the spanning tree construction method used in [8].

1.2   Challenges, Techniques and High-Level Ideas

In most agent-based graph algorithms, particularly in [7, 10, 29], agents need to know Δ, the maximum degree of the graph, to coordinate meetings effectively between adjacent agents, especially when operating autonomously. This knowledge of Δ allows agents to create a window in some multiples of Δ rounds, ensuring that agents meet with all their neighbors within this timeframe. However, our approach overcomes this requirement by employing a new ID system, where agents concatenate a copy of their complemented padded ID bits. This mechanism guarantees that an agent r_i seeking to meet its neighbor r_j can do so without needing agents to know Δ. This methodology allows an agent r_u to meet any neighboring agent within $O(\log n)$ rounds. The agents only require prior knowledge of the highest ID among the agents to get a bound on the ID length of the n agents. The maximum ID across the n agents is denoted as λ, so the IDs of the agents fit within a $\log \lambda$ length.

To construct a spanning tree, we adapt the distributed GHS Minimum Spanning Tree (MST) algorithm for our mobile agent model [13]. The first challenge is assigning distinct weights to the graph's edges to ensure the existence of a unique minimum spanning tree (a requirement for the GHS algorithm). To assign a distinct weight to an adjacent edge, the agents use their ID and port numbering of the outgoing edge. Each agent computes the weight of its adjacent edges deterministically whenever it is required (without storing these weights) based on the IDs of the agents across the edge and their port numbers. The spanning tree construction works similarly to the GHS algorithm, where initially there are n separate tree fragments. In each phase, these fragments compute the Minimum Weight Outgoing Edge (MWOE) and use it to merge or absorb other fragments. The final spanning tree is achieved when no further MWOE exists, and all nodes belong to a single fragment. A critical challenge arises during the MWOE selection. An agent who discovers a new outgoing edge must send this information to its fragment leader via the tree so that the leader can calculate the minimum weight. The leader can then gather all the edge weights and send the MWOE throughout the tree. However, this process requires the agents to move sequentially to their children to propagate this information. To manage this, agents keep moving to and fro between their home nodes and parent nodes periodically, allowing those at the same level to collect the MWOE from their parents in a single round. Once the MWOE is identified, the agents use these edges to merge different fragments. Upon completing the spanning tree construction, the leader of the final fragment, designated as $r^\star$, becomes the leader of all the n agents. Through the MST, agents also gather the necessary graph parameters. The BFS construction now begins from the leader as the root. The BFS construction can be completed in $O(D\Delta)$ rounds as demonstrated in [8]. However, we introduce another approach that completes the BFS construction in $O(m \log n)$ rounds by utilizing the MST built in the previous step. This MST enables the agents to gather important global parameters, such as m and Δ. By integrating both BFS approaches, our algorithm achieves an overall round complexity of $O(\min(D\Delta, m \log n))$ for BFS construction, offering improved time complexity depending on the graph's structure.

2 Model and Problem Statements

Graph: The underlying graph $G(V, E)$ that is connected, undirected, unweighted and anonymous with $|V| = n$ nodes and $|E| = m$ edges. Nodes of G do not have any distinguishing identifiers or labels. These nodes do not possess any memory and hence cannot store any information. The degree of a node $v \in V$ is denoted by $\delta(v)$ and the maximum degree of G is Δ. Edges incident on v are locally labeled using port numbers in the range $[0, \delta(v) - 1]$. We denote a port number that connects the nodes u to v by p_{uv}. p_{uv} represents the edge with the outgoing port number of node u and the incoming port number of node v. In general, $p_{uv} \neq p_{vu}$. The edges of the graph serve as *routes* through which the agents can commute. Any number of agents can travel through an edge at any given time. We also consider a weighted graph for the MST problem, where the edge weights are positive numbers bounded by $O(n^c)$ for some constant $c \geq 1$. The edge weights are associated with the port numbers of the respective edges. An agent at a node v can see the weights of the edges adjacent to v.

Mobile Agents: n agents enumerated as $\mathcal{R} = \{r_1, r_2, \dots, r_n\}$ residing on the nodes of the graph with each having a unique ID $\in [0, n^{O(1)}]$ and has $O(\log n)$ bits of memory to store information. The highest ID among the n agents, denoted by λ, is known to all agents[1]. Two or more agents can be present (*co-located*) at a node or pass through an edge in G but cannot stay on an edge. An agent can recognize the port number through which it has entered and exited a node, but they do not have any visibility beyond their (current) location at a node. An agent at a node v can only realize its adjacent ports (connecting to edges) at v. Only the collocated agents at a node can sense each other and exchange information. An agent can exchange all the information stored in its memory instantaneously. We consider that n agents start from a *dispersed initial configuration*[2] such that each node of the graph G has exactly one agent.

Communication Model: We consider a synchronous system where the agents are synchronized to a common clock and the *local communication* model, where, only co-located agents (i.e., agents at the same node) can communicate among themselves. In each round, an agent r_i performs the $Communicate - Compute - Move\ (CCM)$ task-cycle as follows: (i) *Communicate:* r_i may communicate with other agents at the same node, (ii) *Compute:* Based on the gathered information and subsequent computations, r_i may perform all manner of computations within the bounds of its memory, and (iii) *Move:* r_i may move to a neighboring node using the computed exit port. We measure the complexity in two metrics, namely, time/round and memory. The *time complexity* of an algorithm is the number of rounds required to execute the algorithm. The *memory complexity* is measured w.r.t. the amount of memory (in bits) required by each agent for computation.

---

[1] We use $O(\log n)$ instead of $O(\log \lambda)$ throughout the paper in theorem and result formation to ensure the results are expressed using graph parameters. This also helps to show the direct comparison with the existing results.

[2] For other starting configurations of agents, dispersion is first achieved in $O(n \log^2 n)$ rounds using [36].

Table 1. Comparing previous and some recently developed results for three problems in a graph G in the agent-based model. '$-$' means no a priori knowledge of root and other parameters. λ is the highest ID agent in an n-node graph with m edges, diameter D, and maximum degree Δ.

Algorithm	Knowledge	Time	Memory/Agent	Initial Config.
Leader Election				
Section 4	λ	$O(n \log n + \Delta \log^2 n)$	$O(\log n)$	Dispersed
Kshemkalyani *et al.* [18]	$-$	$O(m)$	$O(\log^2 n)$	Arbitrary
Kshemkalyani *et al.* [19]	n, Δ	$O(n \log^2 n + D\Delta \log n)$	$O(\log n)$	Arbitrary
Kshemkalyani *et al.* [20]	$-$	$O(n \log^2 n)$	$O(\log n)$	Arbitrary
MST				
Section 4	λ	$O(n \log n + \Delta \log^2 n)$	$O(\log n)$	Dispersed
Kshemkalyani *et al.* [18]	$-$	$O(m + n \log n)$	$O(\max(\Delta, \log n) \log n)$	Arbitrary
Kshemkalyani *et al.* [19]	n, Δ	$O(m + n \log^2 n + D\Delta \log n)$	$O(\Delta \log n)$	Arbitrary
Kshemkalyani *et al.* [20]	$-$	$O(m + n \log^2 n)$	$O(\Delta \log n)$	Arbitrary
BFS				
Section 5	λ	$O(\min(m \log n, D\Delta) + n \log n + \Delta \log^2 n)$	$O(\log n)$	Dispersed
Section 5	root	$O(\min(m \log n, D\Delta))$	$O(\log n)$	Dispersed
Chand *et al.* [8]	root	$O(D\Delta)$	$O(\log n)$	Dispersed

2.1 Problem Statements

Consider an n-node simple, arbitrary, connected, and anonymous graph $G(V, E)$. Assume that n autonomous agents are initially placed on the n nodes of the graph, with exactly one agent per node (*dispersed initial configuration*). The objective is to solve the following problems and analyze their computational complexity:

1. *Construction of a Minimum Spanning Tree (MST):* To construct an MST of G such that each edge belonging to the MST is known to at least one of its endpoint agents.
2. *Leader Election:* To designate a unique leader among the n mobile agents.
3. *Construction of a Breadth-First Search (BFS) Tree:* To construct a BFS tree of G such that each edge in the BFS tree is known to at least one of its endpoint agents.

3 Related Work

The leader election problem was first stated by Le Lann [24] in the context of token ring networks, and since then it has been central to the theory of distributed computing. Awerbuch [5] provided a deterministic algorithm with time complexity $O(n)$ and message complexity $O(m + n \log n)$. Peleg [31] provided a deterministic algorithm with optimal time complexity $O(D)$ and message complexity $O(mD)$. Recently, an algorithm is given in [23] with message complexity $O(m)$ but no bound on time complexity, and another algorithm with $O(D \log n)$ time complexity and $O(m \log n)$ message complexity. Additionally, it was shown in [23] that the message complexity lower bound is $\Omega(m)$ and time complexity lower bound is $\Omega(D)$ for deterministic leader election in graphs. Leader election was studied in the agent-based model in [18–20].

One of the earliest works on distributed MST construction was by Spira [35], which analyzed the communication complexity of the problem. This work laid the foundation for the classical GHS algorithm [13], a cornerstone in distributed network algorithms. In the paper, the authors presented algorithms for constructing an MST in $O(n \log n)$ rounds using $O(n \log n + m)$ messages, where each message contained at most one weight value—a non-negative integer less than $\log n$, along with three additional bits. Later, the time complexity was reduced to $O(n)$ rounds in subsequent work [3] and further to $O(\sqrt{n} \log^\star n + D)$ in [14]. Additionally, [30] established a lower bound of $O(\frac{\sqrt{n}}{\log n} + D)$ rounds for the time complexity of distributed spanning tree construction. More recently, MST construction for an agent-based model was introduced in [18]. This study demonstrated an algorithm to build an MST with n agents, achieving a time complexity of $O(m + n \log n)$ rounds. For agents beginning in a dispersed initial configuration, the algorithm requires $O(\max(\Delta, \log n) \log n)$ bits of memory per agent; however, if agents start from arbitrary positions, the memory requirement per agent increases to $O(n \log n)$ bits. Further studies were done in [19, 20].

The earliest agent-based approach to BFS tree construction is found in the work of Awerbuch [4, 6], where a single agent explores the graph via a BFS traversal. To manage resource limitations, this agent periodically returns to its starting node, a technique known as *Piecemeal Exploration* (for instance, to recharge). In more recent work, Palanisamy et al. [28] proposed a BFS traversal method for a multi-agent distributed system by dividing the graph into clusters, assigning an agent to each cluster to construct parts of the BFS tree in parallel. BFS traversal has also been applied to solve the dispersion problem. For example, Kshemkalyani et al. [22] used a BFS-based strategy to disperse agents across the graph, ensuring nearly one agent per node. Under the local communication model—where agents can only communicate when co-located—their BFS-based algorithm achieves a time complexity of $O(D\Delta(D + \Delta))$ rounds when all agents start from a single node. In the case where agents start from arbitrary nodes, their BFS-based dispersion requires $O((D + k)\Delta(D + \Delta))$ rounds, but this assumes a global communication model, allowing agents to communicate at any time, regardless of location. Both algorithms require each agent to have $O(\log D + \Delta \log k)$ bits of memory. Most recently, in [8], n agents were employed to construct a BFS tree in an arbitrary, anonymous n-node graph. Starting with a root node, the agents first built a spanning tree to determine Δ, which was then used to complete the BFS tree. This algorithm completes BFS construction in $O(D\Delta)$ rounds when the root node is initially known. Additional related work on distributed BFS construction is available in [8]. In [19], the authors addressed leader election and MST construction in a similar model, with the main focus on minimizing time and memory when the graph parameters Δ and n are known to the agents a priori. They gave a deterministic leader election algorithm running in $O(n \log^2 n + D\Delta \log n)$ rounds with $O(\log n)$ bits per agent, and an MST construction in $O(m + n \log n)$ rounds using $O(\Delta \log n)$ bits per agent.

4   Agent-Based MST Construction

We construct a Breadth-First Search (BFS) tree starting from an initial configuration where n agents are dispersed across the n-node graph G. Since a BFS tree requires a root, we begin by electing a leader using the Minimum Spanning Tree (MST) construction. This MST construction, based on the parallel GHS algorithm [13], serves both to elect the leader and to collect key graph parameters needed for BFS. Our approach improves upon the existing agent-based MST method [18], which follows a more sequential variant of GHS.

Although G is unweighted, we assign virtual weights to edges to enable MST construction (details in Sect. 4.2). For MST construction, a meeting of two adjacent agents is an essential condition. The challenging part is synchronizing the movement of the agents. As a solution, we develop an approach that requires $O(\log n)$ rounds to meet two adjacent agents across the same edge (in Sect. 4.1).

4.1 Meeting with Neighboring Agents

The agents operate autonomously without any centralized control, so, ensuring that an agent successfully meets its intended neighbor is challenging. This difficulty arises because the targeted neighbor might simultaneously move to meet another of its own neighbors in the same round. Therefore, a mechanism is needed to guarantee that two adjacent agents, intending to meet, can do so within a specified time frame. One such mechanism, known as `Protocol MYN`, has been employed in several studies [7,8,10], where agents use the knowledge of Δ (the maximum degree) along with their ID bits to ensure that each agent meets all its neighbors at least once within $O(\Delta \log \lambda)$ rounds. However, this approach has two key drawbacks. First, agents must have prior knowledge of Δ, the highest degree in the graph. Secondly, the order in which neighbors are met is dictated by the ID bits rather than the node's port numbers. In some algorithms, agents may be required to meet only a specific neighbor or visit the neighbors in a sequence of port numbers. Our proposed approach overcomes both of these limitations.

Our proposed algorithm `Meeting with Neighbor()` (Algorithm 1) is as follows. Consider two agents, r_u and r_v, initially located at adjacent nodes u and v, respectively. Let p_u and p_v denote the ports connecting these nodes. Each agent r_u first pads its ID $r_u.ID$ with leading 0s to ensure a binary length of exactly $\log \lambda$ bits. Let us call this padded ID as $r_u.b$ (Line 1). It then computes a new identifier, new_ID, by appending the bitwise complement of its ID to the left (Line 2), resulting in a $2 \log \lambda$-bit string. The protocol runs for $4 \log \lambda$ rounds, with two rounds corresponding to a bit position. Agents scan their new_ID from the least significant bit (LSB) to the most significant bit (MSB). In each of these rounds, if the current bit of $r_u.new\_ID$ (new_ID of agent r_u) is 1, r_u moves to neighboring node v; otherwise, it stays at node u. Since the agents have distinct IDs, there must exist a bit position in $r_u.new\_ID$ and $r_v.new\_ID$ where the former has a 1 and the latter has a 0. This guarantees that r_u moves to v and encounters r_v.

For example, let r_u have ID 100 and $\lambda = 1100$. Padding gives 0100, and computing new_ID results in 10110100. If r_v has ID 110, then $r_v.new\_ID$ is 10010110. In this case, r_u meets r_v in the $2 \cdot 6 = 12$th round.

Observation 4. *Let r_u be an agent located at node u with degree δ_u. Let v be a neighboring node of u hosting an agent r_v. Then r_u can meet r_v within $4 \log \lambda$ rounds. Furthermore, r_u can meet all its neighbors in at most $\delta_u \cdot 4 \log \lambda$ rounds.*

Algorithm 1. Meeting with Neighbor()

Input: Two agents r_u and r_v with unique IDs, located at adjacent nodes u and v.
Output: The agent r_u meets r_v at v within $4 \log \lambda$ rounds.
 1: Pad each agent's ID to $\log \lambda$ bits by adding leading zeros and call it b.
 2: $r_u.new\_ID \leftarrow (\sim r_u.b) || (r_u.b)$.
 3: Let us suppose $round\_no$ represents the round number initialized to 0.
 4: **for** $4 \log \lambda$ rounds **do**
 5:      **if** $round\_no \bmod 2 = 0$ **then**
 6:          r_u scans $r_u.new\_ID$ from LSB to MSB.
 7:          **if** $(round\_no/2 + 1)^{th}$ bit of $r_u.new\_ID$ is 1 **then**
 8:              r_u moves to node v via port p_u and return to node u
 9:          **else**
10:              r_u remains at node u for 2 consecutive rounds
11:          **end if**
12:      **end if**
13: **end for**

Observe that concatenating the IDs with their bitwise complement is crucial; otherwise, the two agents may not meet. For instance, consider agents r_u and r_v with IDs 0010 and 0110, respectively. Suppose r_u aims to meet r_v, but r_v is simultaneously engaged in another meeting with a neighbor $r_w \neq r_u$. In the second round, both r_u and r_v move, preventing their meeting. In the subsequent rounds, r_u remains stationary, failing to meet r_v. However, using new_ID, this issue is avoided. The new IDs for r_u and r_v are 11010010 and 10010110, respectively. In this case, r_u is guaranteed to meet r_v in the 7th round, where r_u has a 1 and r_v has a 0.

Having established the meeting with neighboring agents, we now move to the next section, where we construct a minimum spanning tree of a graph using agents.

4.2 Constructing a Spanning Tree on an Unweighted Graph

Now, we present an algorithm that enables agents to construct a spanning tree for a given graph $G(V, E)$. Our approach is an adaptation of the GHS algorithm [13], a classical algorithm for constructing a Minimum Spanning Tree (MST) in the message-passing model of distributed computing. To adapt the GHS algorithm to our agent-based model, we first address the requirement of unique edge weights, as the algorithm assumes distinct weights for all m edges. Since we consider a graph with unweighted edges, we must assign unique weights to meet this requirement. While this could be achieved by assigning random weights to the edges, our algorithm is deterministic, necessitating a deterministic method for weight assignment. To this end, we leverage the agents' unique identifiers (IDs) and the port numbering at each node to systematically assign distinct edge weights. Furthermore, this weight assignment approach introduces additional memory requirements for storing and managing the assigned weights. For an agent r_u, the memory required would be $O(\delta_u \log \lambda)$ bits, where δ_u denotes the degree of node u and λ represents the maximum ID value of the agents. To eliminate this overhead, we do not store edge weights in the agents' memory. Instead, these weights are computed dynamically when needed. To obtain a distinct edge weight for the edges, an agent r_u does the following:

- r_u meets its neighbors one by one.
- Once r_u meets with its neighbor r_v, the weight of the edge (r_u, r_v) is evaluated as $r.ID + \frac{1}{p_{uv}+2}$, where $r.ID = \min\{r_u.ID, r_v.ID\}$ and p_{uv} is the port number at agent r leading to the other agent.
- Both r_u and r_v assign the same weight $r.ID + \frac{1}{p_{uv}+2}$ to the edge (r_u, r_v).

Lemma 1 (Distinct Edge Weights). *The edge weights assigned to each edge are distinct.*

The proof the the above lemma can be found in the extended version of the paper [9]. Now, to compute the weight of an edge, an agent must interact with its neighbouring agent, which can be achieved using the `Meeting with Neighbor()` protocol. Through this process, each agent determines the weights of its outgoing edges one by one. Once distinct edge weights have been assigned, the algorithm proceeds to construct a spanning tree. We divide our algorithm into the following three main stages:

1. **Minimum Weight Outgoing Edge (MWOE) Calculation:** Each fragment's leader determines the minimum weight outgoing edge (MWOE) among all its outgoing edges.
2. **MWOE Identification and Selection:** In this stage, the agent gets to know whether its outgoing edge is MWOE. For this, the fragment's leader broadcasts the minimum weight value to all the agents in its fragment. Once an agent realizes that one of its incident edges is the MWOE, it proceeds to the next stage.
3. **Fragment Merging:** In this stage, as the agent holding the MWOE (say, r_u) meets with its adjacent agent (say, r_v), r_u shares its fragment level (*treelevel*) and its intention to merge with or absorb the fragment of r_v. At this point, agent r_u executes either `Merge()` or `Absorb()`, depending on the respective *treelevel* values of agents r_u and r_v and whether the edge (r_u, r_v) is the MWOE for both fragments.

The process continues until all agents become part of a single fragment with no remaining outgoing edges, at which point the algorithm terminates. We now describe the three stages. Some protocols (for example `Merge()`, `Absorb()`, etc.) are briefly outlined in this section, while full details and pseudocodes are available in the extended version [9].

Stage 1 - Minimum Weight Outgoing Edge (MWOE) Calculation: Each edge is assigned a unique virtual weight, as derived from the previous section. Initially, each of the n agents individually represents a distinct tree fragment. As the algorithm progresses, these fragments merge via the Minimum Weight Outgoing Edge (MWOE). Each fragment's leader[3] computes the MWOE for its fragment. The algorithm begins

---

[3] During MST construction, the leader is assumed to be the fragment leader unless specified otherwise. Once all fragments merge into a single one, the fragment's leader becomes the leader of all agents.

with each agent exploring its neighbors in parallel. For each neighbor, the agent checks whether the edge connecting to it is internal or outgoing by comparing their respective *treelabel* values. If the *treelabel* values match, the edge is internal; otherwise, it is considered outgoing. Then each agent sequentially examines its neighbors to identify the minimum-weight outgoing edge. During this process, it maintains a variable r_u.min_weight to store the smallest weight encountered among all outgoing edges. Although each agent searches its local neighborhood sequentially, these searches occur in parallel across the agents. The agent then aggregates the minimum weights received from its children, storing the local minimum in the variable *min_local*. This value is subsequently propagated up in the fragment's hierarchy with the parent node, ensuring that the fragment leader obtains the global minimum. The algorithm terminates when only a single fragment remains, which occurs when there are no outgoing edges. Each agent is also equipped with the variable *treelevel*, which tracks the level of a given fragment. Specifically, when two fragments with the same *treelevel* merge, the newly formed fragment's *treelevel* is incremented. The variables *treelabel* and *treelevel* serve distinct roles: *treelabel* identifies agents within the same fragment (which serves like a tree ID), while *treelevel* facilitates the merging of different fragments. Initially, $treelevel = 0$, and *treelabel* is set to the fragment leader's ID. Since each agent starts as an independent fragment, its *treelabel* is initially assigned as its own ID. As the fragments merge, the *treelabel* of the agents in the merged fragment is determined by the leader of the new fragment.

Before proceeding to the selection of MWOE, we highlight two important remarks. The first deals with memory complexity, which enables the agents to construct the MST using $O(\log \lambda)$ bits. The second introduces a simple synchronization technique that allows the leader of a fragment to efficiently communicate with all agents in its subtree.

Remark 1 (Memory Complexity). The primary memory overhead of the algorithm arises from maintaining child pointers. An agent with up to Δ children would typically require $O(\Delta \log \lambda)$ bits for storage. However, this complexity can be reduced using a `sibling` variable. When an agent discovers its first child, it sets its `child` pointer to the corresponding outgoing port. For each subsequent child, instead of storing multiple child pointers, the agent updates the previously added child's `sibling` variable with the new child's outgoing port. This cascading approach eliminates the need for Δ child pointers, allowing the `child` pointers to be stored efficiently in $O(\log \lambda)$ bits. The other variables `parent`, `child`, and `treelabel` require $O(\log \lambda) + O(\log \Delta) = O(\log \lambda)$ bits, while other variables require only $O(1)$ bits. Thus, the overall memory complexity is $O(\log \lambda)$ or $O(\log n)$ bits, since $\lambda \leq n^{O(1)}$.

Remark 2 (`parent-child` communication). To efficiently exchange information between a parent and its children in a single round, we introduce the following strategy: After every $4 \log \lambda$ rounds of neighbor interactions, an additional round is reserved in which each agent (except the leader) moves to its parent for updates. Thus, after every $4 \log \lambda$ round, each child oscillates once between its node and its parent node. When the leader initiates a broadcast, its immediate children receive the value in the first oscillation, followed by the grandchildren in the next, continuing until all agents are updated after D oscillations, where D is the tree diameter. To ensure updates propagate correctly, once all children at a given level have received the update, they stay at their node for $4 \log \lambda$ rounds, allowing deeper levels to update sequentially. This ensures a smooth transfer of information. Conversely, during the same reserved round, children can send information to their parents simultaneously. Thus, transmitting any value requires an additional $O(n)$ rounds in the worst case.

Stage 2 - MWOE Identification and Selection: To determine the MWOE of a fragment, an agent must verify whether any of its adjacent edges has been selected by the leader as the MWOE for connecting to another fragment. This requires the leader to broadcast the minimum computed weight value to all agents. Each agent checks if any of its adjacent edges is the MWOE by comparing its variable $r_u.min\_weight$ with the value sent by the leader; if they match, one of its adjacent edges is identified as the MWOE. Since agents do not store edge weights in memory, they must reevaluate the weights to identify the edge corresponding to the MWOE weight received from the leader, introducing an additional computational factor of $\Delta \log n$. With the MWOE of the fragment determined, we now proceed to describe the merging of fragments.

Stage 3 - Fragment Merging: Let F_1 and F_2 be two fragments connected via a MWOE e. Then the fragments F_1 and F_2 merge based on the following rule:

1. **Rule 1** - If $F_1.treelevel > F_2.treelevel$, then F_1 joins the fragment F_2 and the $treelabel$ of the new fragment will be $F_2.treelabel$. Moreover, the leader of the fragment will be the leader of fragment F_2.
2. **Rule 2** - If $F_1.treelevel = F_2.treelevel$, then F_1 and F_2 will merge if and only if, e is the common MWOE of F_1 and F_2. In such a case, the leader of the fragment will be the leader of F_1 and F_2 that has the minimum ID.
3. **Rule 3** - In other cases, fragments wait till one of the rules 1 or 2 applies to them.

Now, let us consider two adjacent agents r_u and r_k connected via the MWOE e of r_u. Now, r_u does the following

1. If $r_u.treelevel > r_k.treelevel$, r_u does nothing.
2. If $r_u.treelevel = r_k.treelevel$ and e is the MWOE of the fragment belonging to r_k as well, r_u executes `Merge()`.
3. If $r_u.treelevel = r_k.treelevel$ and e is **not** the MWOE of the fragment belonging to r_k, r_u does nothing.
4. If $r_u.treelevel < r_k.treelevel$, r_u executes `Absorb()`.

Specifically, `Merge()` is used when two fragments have the same level, requiring an increment in their $treelevel$ value. In contrast, `Absorb()` is typically applied when two fragments have different $treelevel$ values and one fragment gets subsumed by the other. During the merging process, the merging fragment terminates all ongoing operations and updates its child and parent pointers, as well as $treelabel$ and $treelevel$ accordingly.

Lemma 2. *The respective time complexities of the 3 stages of the MST construction are respectively* $O(\Delta \log \lambda + n)$, $(\Delta \log \lambda + n)$ *and* $O(n)$ *rounds.*

Proof. For *Stage 1*, each agent takes $O(\Delta \log \lambda)$ rounds in the worst case to search its neighbors, which is done in parallel by the agents of a fragment. To report the minimum weight to the leader, the agents must send this value via their parent, and the leader computes the minimum of all the received weights. Since this transmission occurs via the tree, therefore in the worst case it can take $O(n)$ time. So, calculating the MWOE by the leader takes $O(\Delta \log \lambda + n)$ rounds. To distribute the MWOE to the whole fragment, the agents can take a maximum of $O(n)$ rounds as the agents that are at the same distance from the root parallelly collect this MWOE from their parents. Now the agents once again evaluate the edge weights to compute the matching edge with the selected edge weight. So, *Stage 2* also takes $O(n + \Delta \log \lambda)$ rounds. Finally, for *Stage 3*, the agent that connects its fragment via the MWOE needs to visit the chain of parents to inform them about the merging and changing of their child-parent pointers, respectively. This modification takes $O(n)$ rounds in the worst case.   □

According to the GHS algorithm, the number of fragments while constructing the spanning tree reduces by at least half in each phase, so it takes $O(\log n)$ phases for the fragments to merge into a single fragment, giving out the final spanning tree. Now since, during each phase, weight is evaluated (at most twice) followed by merging of fragments, the time complexity of constructing the spanning tree is given by $O(\log n) \cdot (O(\Delta \log \lambda) + O(\Delta \log \lambda + n) + O(n + \Delta \log \lambda) + O(n)) = O(\Delta \log n \log \lambda + n \log n)$ rounds. Considering that $O(\log \lambda) = O(\log n)$, we can rewrite the complexity as $O(\Delta \log^2 n + n \log n)$. The maximum memory overhead for the agents is used to store the IDs and pointers. Since the IDs are bound by n^c for some constant $c \geq 1$, the memory requirement for each agent is $O(\log n)$ bits. Combining other details from the discussion in Remark 1, the correctness of the original GHS algorithm, and the preceding lemma, we have the following theorem.

Theorem 5. *Given an arbitrary simple connected graph G with n nodes m edges, maximum degree Δ and diameter D. Then, an MST can be constructed in $O(\Delta \log^2 n + n \log n)$ rounds, when the n agents begin in a dispersed initial configuration with $O(\log n)$ bits of memory per agent. The agents do not require any prior knowledge of graph parameters but need to know an upper bound on the ID of the agents, λ.*

4.3   From the Unweighted-MST to Leader Election and Weighted-MST

Following the construction of the spanning tree, all fragments successfully merge into a single component, resulting in a unique leader election that is part of a single fragment having n agents. Consequently, we establish Theorem 2, which guarantees the existence of a single leader within the graph. In the case of a weighted graph with distinct edge weights, the algorithm proceeds analogously to MST construction algorithm, with the primary distinction being that edge weights are predefined and do not require explicit computation. Consequently, all analytical results, including time and memory complexity, remain consistent with those derived for the MST construction algorithm. This leads to the formulation of Theorem 1. With this knowledge, we proceed to the construction of the BFS tree of the graph using agents (discussed in the next section). The BFS tree construction requires that all agents be aware of the values of n, m, and Δ beforehand.

Utilising the spanning tree structure, agents compute fundamental graph parameters—number of nodes (n), number of edges (m), and maximum degree (Δ). This process follows a methodology similar to the Minimum Weight Outgoing Edge (MWOE) computation and dissemination, as described in the MST construction algorithm. Each agent transmits its locally computed values to its parent, accumulating the required information at the MST leader within $O(n)$ rounds. The leader then consolidates these values and propagates them back across the spanning tree, ensuring all agents acquire a global view of the network parameters.

5   Agent-Based BFS Tree Construction

The paper [8] presents a BFS construction algorithm that requires $O(D\Delta)$ rounds when a root is specified. In this section, we introduce a new BFS construction algorithm that completes in $O(m \log n)$ rounds, given a known root node and the number of edges m. By combining these two algorithms, we achieve a round complexity of $O(\min(D\Delta, m \log n))$ for the BFS construction. The assumptions of a known root and m can be removed using the leader election protocol as in the MST construction algorithm. In fact, the values of m, and any other necessary graph parameters can also be computed. The BFS algorithm heavily uses the `Meeting with Neighbor()` as a subroutine, which requires λ to be known to the agents. Below, we discuss an overview of the algorithm. All details are available in the extended version [9].

High-Level Idea: The algorithm proceeds by propagating levels and marking tree edges in the BFS tree, similar to the flooding-based algorithm in the message-passing model. We use the `Meeting with Neighbor()` (Algorithm 1) to send level information from one agent to its neighboring agents, one by one. Suppose an agent $r^\star$ is located at the specified root node, which is at level 0 in the BFS tree. We denote an agent's level as the level of its corresponding node in the BFS tree. Thus, the root agent $r^\star$ sets its level as $r^\star.level = 0$, and every other agent r_u sets its level as $r_u.level = \perp$ initially. Whenever an agent r_u updates its level, it informs its neighboring agents of its new level. A neighboring agent r_v updates its level such that $r_v.level = r_u.level + 1$ if, $r_v.level > r_u.level + 1$ or $r_v.level = \perp$. r_v considers r_u as its parent and the corresponding edge is considered as the edge of the BFS. In case more than one neighbor sends their level simultaneously (in the same round) such that they can be considered as the parent, then the minimum level neighbor is considered as the parent. In case, there is more than one such minimum considerable level send their level then one of them is arbitrarily considered as the parent. The agent informs its neighbors one by one based on port numbering (in increasing order) using the `Meeting with Neighbor()` that takes $O(4 \log \lambda)$ rounds. Notice that to meet any neighbor `Meeting with Neighbor()` start the procedure in the round number that is divisible by $4 \log \lambda$ and this procedure runs for $2m$ times, after which each agent learns its actual level in the BFS tree.

Let us now briefly explain the $O(D\Delta)$ rounds BFS tree construction algorithm. Chand *et al.* constructed the BFS tree in $O(D\Delta)$ rounds and $O(\log n)$ bits of memory per agent in [8]. For the BFS construction, they assumed the dispersed initial configuration and prior knowledge of the root node. For that, first, the algorithm computed the Δ

in $O(D\Delta)$ rounds, and then each level took $O(\Delta)$ rounds to complete a level during the BFS tree construction. Thus, their algorithm required $O(D\Delta)$ rounds and $O(\log n)$ bits of memory per agent. After constructing BFS, with the help of a parent port, each agent informed its parent, and eventually, the root got to know about the BFS formation, and the algorithm terminated. For the detailed analysis, refer to [8].

Now, we have two algorithms; one with round complexity $O(m \log n)$ and another with round complexity $O(D\Delta)$ (from [8]). We consider the best of these two algorithms for BFS construction. For that, first, we run the $O(D\Delta)$ round algorithm; if that algorithm terminates within $O(m \log n)$ rounds, then we have the BFS tree. Otherwise, we run this new algorithm that terminates in $O(m \log n)$ rounds. Thus, we get the best round complexity of these two algorithms and consequently have Theorem 3.

6   Conclusion and Future Works

In this paper, we analyze the complexity of constructing a BFS tree using mobile agents, improving upon [8] in two key ways: (1) eliminating the need for a pre-designated root and (2) reducing the round complexity from $O(D\Delta)$ to $O(\min(D\Delta, m \log n) + n \log n + \Delta \log^2 n)$. We also develop a Minimum Spanning Tree (MST) protocol for gathering graph parameters and electing a leader for BFS root selection, improving over previous works. Both constructions use a new neighbor-meeting protocol, allowing adjacent agents to meet in $O(\log n)$ time without global graph parameter knowledge. This eliminates the need for the Δ assumption, and this may further reduce time complexity in existing solutions. Future work includes establishing round lower bounds for agent-based BFS, and analyzing fault tolerance under crash and Byzantine failures, given their real-world relevance.

References

1. Akyildiz, I.F., Su, W., Sankarasubramaniam, Y., Cayirci, E.: A survey on sensor networks. IEEE Commun. Mag. (2002)
2. Augustine, J., Moses Jr, W.K.: Dispersion of mobile robots: a study of memory-time trade-offs. In: ICDCN, pp. 1:1–1:10 (2018)
3. Awerbuch, B.: Optimal distributed algorithms for minimum weight spanning tree, counting, leader election, and related problems. In: STOC (1987)
4. Awerbuch, B., Betke, M., Rivest, R., Singh, M.: Piecemeal graph exploration by a mobile robot. Massachusetts Institute of Technology (1995)
5. Awerbuch, B., Gallager, R.: A new distributed algorithm to find breadth first search trees. IEEE Trans. Inf. Theory (1987)
6. Awerbuch, B., Betke, M., Rivest, R.L., Singh, M.: Piecemeal graph exploration by a mobile robot. Inf. Comput. (1999)
7. Chand, P.K., Das, A., Molla, A.R.: Agent-based triangle counting and its applications in anonymous graphs. In: AAMAS (2024)
8. Chand, P.K., Kumar, M., Molla, A.R.: Agent-driven BFS tree in anonymous graphs with applications. In: NETYS (2024)
9. Chand, P.K., Kumar, M., Molla, A.R.: Computing tree structures in anonymous graphs via mobile agents. arXiv:2506.19365 (2025)

10. Chand, P.K., Molla, A.R., Sivasubramaniam, S.: Run for cover: dominating set via mobile agents. In: ALGOWIN (2023)
11. Cong, Y., Gu, C., Zhang, T., Gao, Y.: Underwater robot sensing technology: a survey. Fundam. Res. (2021)
12. Dunbabin, M., Marques, L.: Robots for environmental monitoring: significant advancements and applications. IEEE Robot. Autom. Mag. (2012)
13. Gallager, R.G., Humblet, P.A., Spira, P.M.: A distributed algorithm for minimum-weight spanning trees. ACM Trans. Program. Lang. Syst. (1983)
14. Garay, J., Kutten, S., Peleg, D.: A sublinear time distributed algorithm for minimum-weight spanning trees. SIAM J. Comput. (1998)
15. Huang, H., Savkin, A.V., Ding, M., Huang, C.: Mobile robots in wireless sensor networks: a survey on tasks. Comput. Netw. (2019)
16. Isler, V.: Robotic sensor networks for environmental monitoring. In: AAAI Spring Symposium: AI, The Fundamental Social Aggregation Challenge (2012)
17. Kshemkalyani, A.D., Ali, F.: Efficient dispersion of mobile robots on graphs. In: ICDCN (2019)
18. Kshemkalyani, A.D., Kumar, M., Molla, A.R., Sharma, G.: Brief announcement: agent-based leader election, MST, and beyond. In: DISC (2024)
19. Kshemkalyani, A.D., Kumar, M., Molla, A.R., Sharma, G.: Faster leader election an its applications for mobile agents with parameter advice. In: ICDCIT (2025)
20. Kshemkalyani, A.D., Kumar, M., Molla, A.R., Sharma, G.: Near-linear time leader election in multiagent networks. In: AAMAS (2025)
21. Kshemkalyani, A.D., Molla, A.R., Sharma, G.: Fast dispersion of mobile robots on arbitrary graphs. In: ALGOSENSORS, pp. 23–40 (2019)
22. Kshemkalyani, A.D., Molla, A.R., Sharma, G.: Dispersion of mobile robots using global communication. J. Parallel Distrib. Comput. (2022)
23. Kutten, S., Pandurangan, G., Peleg, D., Robinson, P., Trehan, A.: On the complexity of universal leader election. J. ACM $62(1)$, 7:1–7:27 (2015)
24. Le Lann, G.: Distributed systems - towards a formal approach. In: IFIP (1977)
25. Lee, J., Shin, S., Park, M., Kim, C.: Agent-based simulation and its application to analyze combat effectiveness in network-centric warfare considering communication failure environments. Math. Probl. Eng. (2018)
26. Levinas, I., Scherz, R., Louzoun, Y.: BFS-based distributed algorithm for parallel local-directed subgraph enumeration. J. Complex Netw. (2022)
27. Meivel, S., et al.: Mask detection and social distance identification using internet of things and faster R-CNN algorithm. Comput. Intell. Neurosci. (2022)
28. Palanisamy, V., Vijayanathan, S.: Cluster based multi agent system for breadth first search. In: ICTER (2020)
29. Pattanayak, D., Bhagat, S., Gan Chaudhuri, S., Molla, A.R.: Maximal independent set via mobile agents. In: ICDCN (2024)
30. Peleg, D., Rubinovich, V.: A near-tight lower bound on the time complexity of distributed MST construction. In: FOCS (1999)
31. Peleg, D.: Time-optimal leader election in general networks. J. Parallel Distrib. Comput. $8(1)$, 96–99 (1990)
32. Pramanick, S., Samala, S.V., Pattanayak, D., Mandal, P.S.: Filling MIS vertices of a graph by myopic luminous robots. In: ICDCIT (2023)
33. Rihan, M., Selim, M.M., Xu, C., Huang, L.: D2D communication underlaying UAV on multiple bands in disaster area: stochastic geometry analysis. IEEE Access (2019)
34. Ryu, H., Chung, W.K.: Local map-based exploration using a breadth-first search algorithm for mobile robots. Int. J. Precis. Eng. Manuf. $16(10)$, 2073–2080 (2015). https://doi.org/10.1007/s12541-015-0269-9

35. Spira, P.: Communication complexity of distributed minimum spanning tree algorithms. In: Proceedings of the Second Berkeley Conference on Distributed Data Management and Computer Networks (1977)
36. Sudo, Y., Shibata, M., Nakamura, J., Kim, Y., Masuzawa, T.: Near-linear time dispersion of mobile agents. In: DISC (2024)
37. Ta-Shma, A., Zwick, U.: Deterministic rendezvous, treasure hunts, and strongly universal exploration sequences. ACM Trans. Algorithms (2014)
38. Wilk-Jakubowski, G., Harabin, R., Ivanov, S.: Robotics in crisis management: a review. Technol. Soc. (2022)
39. Wu, M.: Robotics applications in natural hazards. Highlights in Science, Engineering and Technology (2023)
40. Zhao, N., et al.: UAV-assisted emergency networks in disasters. IEEE Wirel. Commun. (2019)
41. Zhuge, C., Shao, C., Wei, B.: An agent-based spatial urban social network generator: a case study of Beijing, China. J. Comput. Sci. (2018)

Knowledge-Guided Machine Learning for Stabilizing Near-Shortest Path Routing

Yung-Fu Chen[1(✉)] [ID], Sen Lin[2], and Anish Arora[1] [ID]

[1] The Ohio State University, Columbus, OH 43210, USA
chen.6655@osu.edu, arora.9@osu.edu
[2] University of Houston, Houston, TX 77204, USA
slin50@central.uh.edu

Abstract. We propose a simple algorithm that requires only a few data samples from a single graph for learning local routing policies that generalize across a rich class of geometric random graphs in Euclidean metric spaces. We thus address the all-pairs near-shortest path problem by training deep neural networks (DNNs) that let each graph node efficiently and scalably routes packets by considering only the node's state and the state of the neighboring nodes. Our algorithm design exploits network domain knowledge in the selection of input features and design of the policy function for learning an approximately optimal policy. Domain knowledge also provides theoretical assurance that the choice of a "seed graph" and its node data sampling is sufficient for generalizable learning. Remarkably, one of the DNNs we train —using distance-to-destination as the only input feature— learns a policy that replicates the well-known Greedy Forwarding behavior, which forwards packets to the neighbor with the shortest distance to the destination. We also learn a new policy, which we call *Greedy Tensile* routing —using both distance-to-destination and node stretch as the input features— that almost always outperforms greedy forwarding. We demonstrate the explainability and ultra-low latency run-time operation of Greedy Tensile routing by symbolically interpreting its DNN in low-complexity terms of two linear actions.

Keywords: zero-shot learning · local routing · reinforcement learning · all-pairs shortest path routing · geographic routing · self-stabilization

1 Introduction

There has been considerable interest in machine learning to mimic the human ability of learning new concepts from just a few instances. While formal frameworks for machine learning, such as the language-identification-in-the-limit framework [7] and probably approximately correct (PAC) learning [26], have yet to match the high data efficiency of human learning, few-shot (or one-shot) learning methods [18, 19, 28] are attempting to bridge the gap by using prior

© The Author(s), under exclusive license to Springer Nature Switzerland AG 2026
S. Bonomi et al. (Eds.): SSS 2025, LNCS 16350, pp. 128–145, 2026.
https://doi.org/10.1007/978-3-032-11127-2_12

knowledge to rapidly generalize to new instances given training on only a small number of samples with supervised information.

In this paper, we explore the learnability of policies that generalize in the domain of graph routing, where designing a one-size-fits-all solution based on local or global network states remains challenging. Routing using global network states tends to be inefficient in terms of time and communication when the class of graphs has strict scalability limits for network capacity [31] or has significant graph dynamics [8,10]. Manually designed algorithms and heuristics using local or regional network states often cater to particular network conditions and come with tradeoffs of complexity, scalability, and generalizability across diverse graphs. Many machine learned algorithms in this space incur relatively high computational complexity during training [23], have high overhead at run-time that limits their use in graphs with high dynamics, or are applicable only for small-scale graphs or graphs of limited types (e.g., high-density graphs). Our work focuses attention on answering the following question:

Can we design a sample-efficient machine learning algorithm for graph routing based on local information that addresses scalability, generalizability, and complexity issues all at once?

We answer this question in the affirmative for the all-pairs near-shortest path (APNSP) problem over the class of uniform random graphs in Euclidean metric spaces. It is well known that uniform random graphs in Euclidean metric spaces can represent the topologies inherent in wireless networks [13]; the policies we learn are thus applicable to real-world wireless networks. Our key insight is that —in contrast to pure black-box approaches— *exploiting domain knowledge in various ways—input feature selection, policy design, sample selection—enables learning of near-optimal, low-complexity routing from only a few samples that generalizes to (almost) all graphs in this class and adapts quickly to network dynamics.*

Near-Optimal, Low-Complexity Routing. To motivate our focus on routing based solely on local information, we recall that approaches to solving the APNSP problem fall into two categories: global information and local information. Policies using global information encode the entire network state into graph embeddings [20] and find optimal paths, whereas policies using local information need only node embeddings [9] or the coordinates of the neighbors and the destination to predict the next forwarder on the optimal path. The complexity (in time and space) resulting from the latter is inherently better than that of the former. However, typically, the latter comes with the penalty of sub-optimality of the chosen path relative to the optimal (i.e., shortest) path. While this holds for both manual designs and machine learned algorithms, we show that by exploiting domain knowledge with respect to bounded forwarding, it is possible to learn a routing policy that achieves near-optimality in (almost) all graphs in the chosen class.

Generalizability. We also develop a domain theory for achieving efficient learning that generalizes over the class of graphs. The theory is based on a condition under which the ranking of neighboring forwarders based on local information matches that based on globally optimal information. We empirically validate the condition by demonstrating a strong correlation between local ranking and global ranking metrics for the class of graphs. The theory then implies the APNSP objective can be realized with high probability by training a deep neural network (DNN) that characterizes the local ranking metric of each neighbor as a potential forwarder. Moreover, it implies the DNN policy can generalize even if it is trained from only a few data samples chosen from a single "seed" graph. The theory also guides our selection of seed graphs as well as corresponding training data and is corroborated by empirical validation of our learned routing solutions.

Quick Adaptation to Network Dynamics. By virtue of its generalizability to all graphs in the class, the same DNN can be used without any retraining when the network changes. Moreover, since the DNN uses only local information, route adaptation to dynamics is quick. In fact, the policy is inherently self-stabilizing.

In this paper, we develop knowledge-guided learning as follows: We model the APNSP problem as a Markov decision process (MDP) and use a DNN architecture to learn a "single-copy" local routing policy based on the features of distance-to-destination and node stretch. At each routing node and at each time, the DNN only considers the states from that node and one of its neighbors to predict a local metric (a Q-value) for routing. Routing selects the neighbor with the highest Q-value.

The approach yields a routing policy that we call *Greedy Tensile* routing. *Greedy Tensile* is a light-weight routing policy for the chosen class of graphs, in following ways: (a) Low complexity: It is rapidly learned from a small dataset that is easily collected from a single "seed" graph; (b) Generalizability: It can be used on all nodes of a graph, and is able to generalize across almost all graphs in the class without additional training on the target networks; and (c) Scalability: Its routing decision only depends on the local network state, i.e., the state of the node and its one-hop neighbor nodes, and thus is much more efficient than routing using global states and can be used even when the topology changes.

The main contributions and findings of this paper are summarized as follows:

1. Generalization from few-shot learning from a single graph is feasible for APNSP and theoretically assured by domain knowledge.
2. Domain knowledge also guides the selection of input features and training samples to increase the training efficiency and testing accuracy.
3. Learning from a single graph using only a distance-to-destination metric matches the well-known greedy forwarding routing.
4. Learning from a single graph using both distance-to-destination and node stretch relative to a given origin-destination node pair yields a new policy, *Greedy Tensile* routing, that achieves even better generalized APNSP routing over Euclidean random graphs.
5. The *Greedy Tensile* DNN can be symbolically interpreted in a low-complexity fashion—it is approximated by a policy with two linear actions.
6. Reinforcement learning from a single graph achieves comparable generalization performance for APNSP.

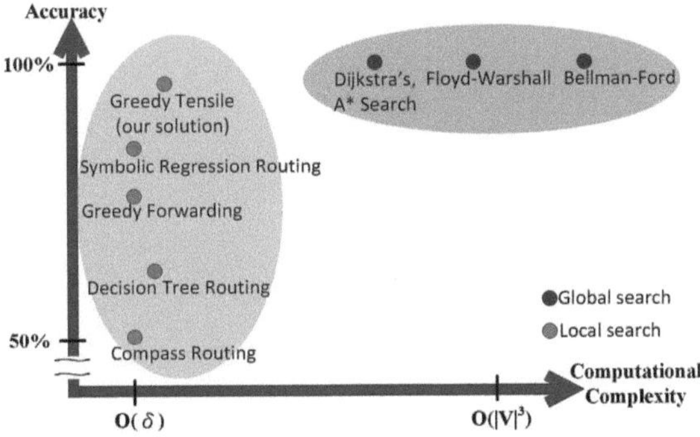

Fig. 1. A comparison of different (machine learned) routing policies for the per-hop best forwarder prediction for all-pair near-shortest path problem (APNSP).

2   Related Work

2.1   Feature Selection for Local Routing

A classic feature for local routing comes from Greedy Forwarding [6], in which the distance to the destination node is used to optimize forwarder selection. It has been proven that this feature achieves nearly optimal routing for certain network classes, including scale-free networks [14,22]. Notable other features for forwarder selection include Most Forward within Radius (MFR) [24], Nearest with Forwarding Progress (NFP) [11], the minimum angle from neighbor u to the line between origin O and destination D, $\angle uOD$, (aka Compass Routing) [15], and Random Progress Forwarding (RPF) [21].

Moreover, a recent study [5] shows that searching for shortest paths in uniform random graphs can be restricted to an elliptical search region with high probability. Its geographic routing protocol, *QF-Geo*, uses Node Stretch $\frac{\overline{Ou+uD}}{\overline{OD}}$, the stretch factor of node u from the origin-destination-line $\overline{OD}$, as an input feature to determine whether a node's neighbor u lies in the search region and to forward packets only within the predicted elliptical region. We refer to this domain knowledge as *Elliptical Region for Bounded Search*.

2.2   Local Versus Global Routing Policies

Figure 1 compares the accuracy and computational complexity[1] of *Greedy Tensile* with other routing policies based on local states or global states. With respect

---

[1] The computational complexity resulting from the global states is related to the network size $|V|$ and the number of edges $|E|$. FloydâĂŞWarshall algorithm solves all-pairs (near) shortest paths with time complexity $O(|V|^3)$. Dijkstra's, A* search, and BellmanâĂŞFord algorithms, which find single-source shortest paths, can be extended to solve APNSP by iterating $|V|$ sources and then incur the time complexity: $O(|V|^2 \log |V|)$, $O(|V|^2 \log |V|)$, $O(|V|^2 |E|)$, respectively.

to the accuracy (in terms of near-shortest paths) for predicting the best forwarder, local routing policies yield diverse performances across machine-learned models or geographic routing policies. We explored symbolic regression [2] routing and decision tree routing to select a subset of all candidate features[2] that are significant to predict the optimal routing metric: both tended to choose node stretch as their primary feature, but did not yield a model whose accuracy across diverse network configurations (of sizes and densities) was ideal. We recall that extant geographic routing techniques that use a simple linear function over only one feature, such as Greedy Forwarding (using distance) and Compass Routing (based on angle), do not achieve generalizability across various network topologies either. *Greedy Tensile*, which considers both distance-to-destination and node stretch as the input features, allows for searching a larger decision space and thus learns a model that is more sophisticated than a linear regression function and that achieves a near-optimal prediction.

2.3   Generalizability of Machine Learned Routing

Only recently has machine learning research started to address generalizability in routing contexts. For instance, generalizability to multiple graph layout distributions, using knowledge distillation, has been studied for a capacitated vehicle routing problem [1]. Some explorations have considered local states: i.e., wireless network routing strategies using local states based on deep reinforcement learning [16,17] have been shown to generalize to other networks of up to 100 nodes, in the presence of diverse dynamics including node mobility, traffic pattern, congestion, and network connectivity. Deep learning has also been leveraged for selecting an edge set for a well-known heuristic, Lin-Kernighan-Helsgaun (LKH), to solve the Traveling Salesman Problem (TSP) [30]. The learned model generalizes well for larger (albeit still modest) sized graphs and is useful for other network problems, including routing. Likewise, graph neural networks and learning for guided local search to select relevant edges have been shown to yield improved solutions to the TSP [12]. In related work, deep reinforcement learning has been used to iteratively guide the selection of the next solution for routing problems based on neighborhood search [29].

3   Problem Formulation for Generalized Routing

Consider the class $\mathbb{G}$ of all graphs $G = (V, E)$ whose nodes are uniformly randomly distributed over a 2-dimensional Euclidean geometric space. Each node $v \in V$ knows its global coordinates. For each pair of nodes $v, u \in V$, edge $(v, u) \in E$ holds if and only if the distance between v and u is at most the communication radius R, a user-defined constant.

---

[2] For a given node u and its origin-destination node pair (O, D), the feature set includes: distance to the destination $\overline{uD}$, node stretch $\frac{\overline{Ou} + \overline{uD}}{\overline{OD}}$, node degree, and node angle $\angle uOD$.

Let ρ denote the network density, where network density is defined to be the average number of nodes per R^2 area, and n the number of nodes in V. It follows that if all nodes in V are distributed in a square area, its side must be of length $\sqrt{\frac{n \times R^2}{\rho}}$.

3.1 All-Pairs Near-Shortest Path Problem (APNSP)

The objective of APNSP routing problem is to locally compute for all node pairs of any graph $G \in \mathbb{G}$ their near-shortest path. Here, near-shortest path is defined as one whose length is within a user-specified factor (≥ 1) of the shortest path length.

Formally, let $d_e(O, D)$ denote the Euclidean distance between two endpoints O and D, and $d_{sp}(O, D)$ denote the length of the shortest path between these endpoints. Further, let $\zeta(O, D)$ denote the path stretch of the endpoints, i.e., the ratio $\frac{d_{sp}(O,D)}{d_e(O,D)}$.

The APNSP Problem: Learn a routing policy π such that, for any graph $G = (V, E) \in \mathbb{G}$ and any origin-destination pair (O, D) where $O, D \in V$, $\pi(O, D, v) = u$ finds v's next forwarder u and in turn yields the routing path $p(O, D)$ with path length $d_p(O, D)$ that with high probability is a near-shortest path. In other words, π optimizes the accuracy of $p(O, D)$ as follows:

$$\max \quad Accuracy_{G,\pi} = \frac{\sum_{O,D \in V} \eta(O, D)}{|V|^2}, \tag{1}$$

$$s.t. \quad \eta(O, D) = \begin{cases} 1, & if \ \frac{d_p(O,D)}{d_{sp}(O,D)} \leq \zeta(O, D)(1 + \epsilon) \\ 0, & otherwise \end{cases} \tag{2}$$

Note that the user-specified factor for APNSP is $\zeta(O, D)(1 + \epsilon)$, where $\epsilon \geq 0$. The use of path stretch $\zeta(O, D)$ in Eq. 2 is to normalize the margin for shortest paths prediction (given ϵ is constant) across networks with different densities since the variance of path stretch becomes more significant as the network density ρ decreases[3].

3.2 MDP Formulation for the APNSP Problem

To solve the APNSP problem, we first formulate it as a Markov decision process (MDP), which learns to choose actions that maximize the expected future reward. In general, an MDP consists of a set of states S, a set of actions A for any state $s \in S$, an instantaneous reward $r(s, a)$, indicating the immediate reward

---

[3] The result in [5] shows that, in sparse graphs (e.g., ρ=1.4, 2), the path stretch can vary in $[1.0, 10.0]$ since some of the shortest paths are prone to be found around the holes. However, the path stretch in dense graphs (e.g., ρ=4, 5) is likely to vary only within the range of $[1.0, 3.5]$. Thus, we exploit $\zeta(O, D)$ to mitigate the gap between sparse and dense networks for APNSP prediction.

for taking action $a \in A$ at state s, and a state transition function $P(s'|s, a)$ that characterizes the probability that the system transits to the next state $s' \in S$ after taking action a at the current state $s \in S$. To simply the routing behavior in the problem, the state transition is assumed to be deterministic. Specifically, each state s represents the features of a node v holding a packet associated with an origin-destination pair (O, D), and an action $a \in A = nbr(v)$ indicates the routing behavior to forward the packet from node v to one of its neighbors $u \in nbr(v)$, where $nbr(v)$ denotes the set of v's one-hop neighbors. Given the current state s and an action $a \in nbr(v)$ which selects one neighbor as the next forwarder, the next state s' is determined as the features of the selected neighbor u such that the probability $P(s'|s, a)$ is always one. The tuple (s, a, r, s') is observed whenever a packet is forwarded.

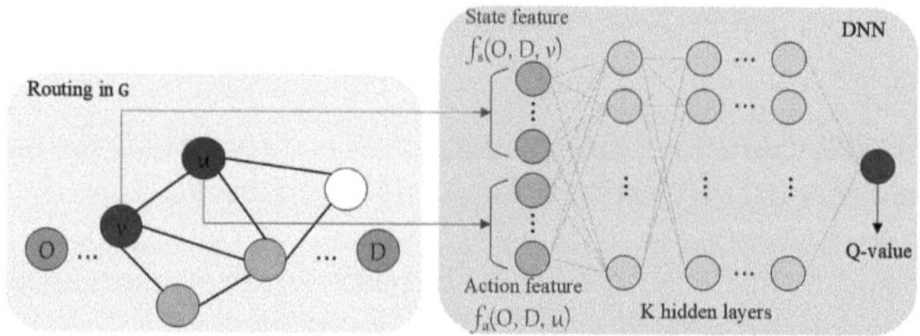

Fig. 2. Schema for solution using DNN to predict Q-value for selecting the routing forwarder.

Design of Q : In general, the Q-value is defined to specify the cumulative future reward from state s and taking action a:

$$Q(s_t = s, a_t = a) = \sum_{i=t}^{L} \gamma^{i-t} r(s_i, a_i)$$

where $\gamma, 0 \leq \gamma \leq 1$, is the discount factor. When $\gamma = 0$, the instantaneous reward is considered exclusively, whereas the future rewards are treated as equally important as the instantaneous reward in the Q-value if $\gamma = 1$.

To design the reward function r, we leverage the domain knowledge from the *Elliptical Region for Bounded Search* to prioritize forwarding that yields a routing path within an ellipse. In the APNSP problem, the user-specified factor, $\zeta(O, D)(1+\epsilon)$, implies the size of the ellipse for search. A routing path p between O and D with the length $d_p(O, D)$, not larger than $d_{sp}(O, D)\zeta(O, D)(1 + \epsilon)$, is considered a near-optimal path. For any action that will make the routing path stretch outside the elliptical region, a negative penalty is added to the reward r. Note that, in supervised learning, the path stretch $\zeta(O, D)$ for each source-destination pair is known for calculating the optimal Q values. In Reinforcement Learning (RL), the probabilistic bound of the path stretch (i.e., the size of the ellipse for search) can be modeled as a function of network density and the Euclidean distance between the two endpoints [5].

Therefore, we define the instantaneous reward as follows:

$$r(s,a) = \begin{cases} -d_e(s,s'), & if\ \frac{d_e(s,s')+d_{sp}(s',D)}{d_{sp}(s,D)} \leq \zeta(O,D)(1+\epsilon) \\ -d_e(s,s') - C\Delta(s,s',O,D), & otherwise \end{cases},$$

where $\Delta(s,s',O,D) \geq 0$ is the penalty to the reward and $C > 0$ is a constant. $\Delta(s,s',O,D) = (d_e(s,s') + d_{sp}(s',D)) - d_{sp}(s,D)\zeta(O,D)(1+\epsilon)$ calculates the part that exceeds the margin. We set $\gamma = 1$, and, therefore, the optimal Q-value $Q^*(s,a)$ is equal to the cumulative negative length of the shortest path from s to the destination through s', if the path is near-optimal.

To solve the APNSP problem, we seek to learn the optimal Q-value through a data-driven approach with DNNs. As depicted in Fig. 2, each state s and action a will be embedded into a set of input features denoted by $f_s(s)$ and $f_a(a)^4$, respectively. A DNN will be learned to approximate the optimal Q-value given the input features, based on which a near-shortest path routing policy can be obtained by taking actions with largest Q-values. We begin with theory that guides the learnability of a generalizable DNN for APNSP.

4 Explainable Learnability and Generalizability of Routing Policy

In this section, we identify two sufficient conditions, **Pointwise Monotonicity** and **RankPres** that, respectively, imply learnability and generalizability of an optimal routing policy. We also identify domain knowledge on the satisfiability of these two conditions in the context of the APNSP routing problem. Together, these explain the feasibility of learning a routing policy from limited data on a few nodes in a single graph that suffices for all graphs $G \in \mathbb{G}$. (For reasons of space, all proofs of the theory in this section are relegated to the appendix in [4].)

The basic concept in the APNSP context is of a ranking metric function for each node $v \in V$ with respect to each $u \in nbr(v)$. Let $f_s : V \to \mathbb{R}^I$ be a map of $v \in V$ to its state features (of cardinality I) and let $f_a : V \to \mathbb{R}^J$ be a map of $u \in nbr(v)$ to its action features (of cardinality J). We define a ranking metric $m : f_s \times f_a \to \mathbb{R}$ to be a *linear* or *non-linear* function over the input features associated with a node v and its neighbor u.

4.1 Learnability of Ranking Metric Function m

Definition 1 (Pointwise Vector Ordering $\leq$). *A vector ordering $\leq$ is pointwise if for any two vectors, $A = (a_1, a_2, ..., a_n) \in \mathbb{R}^n$ and $B = (b_1, b_2, ..., b_n) \in \mathbb{R}^n$, $A \leq B$ iff $(\forall k : 1 \leq k \leq n : a_k \leq b_k)$.*

---

[4] Since O and D will remain the same in all forwarder prediction associated with one (O,D) pair, for notational convenience, we will omit O and D in the DNN input and use $f_s(v)$ and $f_a(u)$ to represent the state and action features.

Definition 2 (Pointwise Monotonicity of m wrt $\leq$). *For any node v in V and ranking metric m, m is* **Pointwise Monotonic** *with respect to a pointwise vector ordering $\leq$ iff*

For any ordering $\langle u_1, ..., u_d \rangle$ of all nodes in $nbr(v)$, if $(f_s(v), f_a(u_1)) \leq ... \leq (f_s(v), f_a(u_d))$ then $m(f_s(v), f_a(u_1)) \leq ... \leq m(f_s(v), f_a(u_d)$

Theorem 1 (Learnability). *Assume the metric m is Pointwise Monotonic with respect to $\leq$ for a node v in V. Then there exists a DNN H, $H : \mathbb{R}^n \to \mathbb{R}$, with training samples $\langle X_v, Y_v \rangle$ that learns m for v.*

4.2 Generalizability of Ranking Metric Function m

We next define property **RankPres** that relates the local ranking metric m and the corresponding Q-values set Y_v, namely, the global ranking metric. RankPres suffices to apply m for ranking the neighbors $u \in nbr(v)$ using local node states to achieve the same ranking using Q-values, which allows routing to be performed locally as well as is achieved by using Q.

Definition 3 (RankPres). *A metric function satisfies RankPres for a node v in V iff*

For any ordering $\langle u_1, ..., u_d \rangle$ of all nodes in $nbr(v)$, if $\langle m(f_s(v), f_a(u_1)), ..., m(f_s(v), f_a(u_d)) \rangle$ is monotonically non-decreasing, then $\langle Q(v, u_1), ..., Q(v, u_d) \rangle$ is monotonically non-decreasing.

Next, we lift the sufficient condition to provide a general basis for ranking the neighbors of all nodes in a graph to predict the best forwarder, by training with only the samples derived from one (or a few) of its nodes. Subsequently, we further lift the sufficient condition for similarly ranking the neighbors of all graphs in $\mathbb{G}$.

For notational convenience, given an ordering $\langle u_1, ..., u_d \rangle$ of all nodes in $nbr(v)$, let $X_v = \{\langle f_s(v), f_a(u_1) \rangle, ..., \langle f_s(v), f_a(u_d) \rangle\}$, denote the set of vectors for each corresponding neighbor $u_k \in nbr(v), 1 \leq k \leq d$. Also, let $Y_v = \{Q(v, u_1), ..., Q(v, u_d)\}$ denote the corresponding set of Q-values.

Lemma 1 (Cross-Node Generalizability). *For any graph G, $G = (V, E)$, if ranking metric $m(f_s(v), f_a(u))$ satisfies RankPres for all $v \in V$ and is learnable with a DNN H trained with only a subset of training samples $\langle X_{V'}, Y_{V'} \rangle$, where $V' \subseteq V$, $X_{V'} = \bigcup_{v \in V'} X_v$, and $Y_{V'} = \bigcup_{v \in V'} Y_v$, then H approximates an optimal ranking policy for all $v \in V$.*

Note that if the $Q(v, u)$ value corresponds to the optimal (shortest) path Q-value for each (v, u) pair, then m indicated by Lemma 1 achieves an optimal routing policy for all nodes in V. Note also that if m satisfies RankPres not for all nodes but for almost all nodes, a policy learned from samples from one or more nodes v that satisfy RankPres may not achieve optimal routing for all nodes. Nevertheless, if the proportion of nodes that do not satisfy RankPres to the number of nodes that do satisfy RankPres is small then, with high probability, the policy achieves near-optimal routing.

Lemma 2 (Cross-Graph Generalizability). *If metric $m(f_s(v), f_a(u))$ satisfies RankPres for the nodes in all graphs $G \in \mathbb{G}$ and is learnable with a DNN H trained with samples from one or more nodes in one or more chosen seed graph(s) $G^* \in \mathbb{G}$, then H approximates an optimal ranking policy for all graphs G.*

Again, if Lemma 2 is considered in the context of Q-values corresponding to optimal shortest paths, the learned routing policy H generalizes to achieving optimal routing over all graphs $G \in \mathbb{G}$. And if we relax the requirement that **RankPres** holds for all nodes of all graphs in $\mathbb{G}$ to only require that for almost all graphs $G \in \mathbb{G}$, there is a high similarity between the ranking orders of neighbors using m and the ones using optimal Q-value over all nodes v, then with high probability the policy achieves near-optimal routing.

4.3   Learnability and Generalizability for APNSP

Theorem 2 (Metric Function for Optimal Policy). *If a metric function m satisfies RankPres and Pointwise Monotonicity, it yields an optimal generalizable and learnable ranking policy.*

We now instantiate the theory for provable learnability and generalizability for APNSP.

Proposition 1. *For APNSP, there exists a ranking metric $m_1(f_s(v), f_a(u))$ of the form $w_1.d(v, D) + w_2.d(u, D)$ based on the* **distance-to-destination** *input feature that satisfies the property of RankPres and Pointwise Monotonicity for almost all nodes in almost all graphs G.*

Also, there exists a ranking metric $m_2(f_s(v), f_a(u))$ of the form $w_1.d(v, D) + w_2.ns(O, D, v) + w_3.d(u, D) + w_4.ns(O, D, u)$ based on both **distance-to-destination** *and* **node stretch** *input features that satisfies the property of RankPres and Pointwise Monotonicity for almost all nodes in almost all graphs G.*

In [4], we present empirical validation of Proposition 1 for graphs in Euclidean space. We respectively choose two linear functions, m_1 and m_2, that satisfy Pointwise Monotonicity. And then RankPres is quantified in terms of Ranking Similarity; high ranking similarity for both m_1 and m_2 implies that RankPres holds with high probability. It follows that an efficient generalizable policy for APNSP is learnable for each of the two chosen input feature sets.

5   Single Graph Learning

5.1   Design of Input Features

To learn a routing protocol that generalizes across graphs with different scales, densities, and topologies, the input features of the DNN should be designed to be independent of global network configurations, including the identity of nodes and of packets. Recall that each node knows its own coordinates and the coordinates of the origin and the destination.

For feature selection, we first apply the decision tree and symbolic regression to filter the crucial features from a set of candidate features widely used in geographic routing protocols: distance from v to destination D, perpendicular distance from node v to the origin-destination-line $\overleftrightarrow{OD}$, angle between $\overline{vO}$ and $\overline{OD}$, and node stretch relative to (O, D) pair. Note that symbolic regression finds a mathematical function (using a combination of an operator set $\{+, -, *, /, cos, sin, log, min, max, ...\}$) that best describes the relationship between the input features and the output on a given dataset. In addition to the distance from v to destination D that is used to achieve near-optimal routing in scale-free graphs, node stretch is found to be the dominant feature for predicting the Q-values using both decision tree and symbolic regression. These results motivate us to adopt these two features as the node states for the best forwarder prediction for APNSP. Accordingly, we design the input features, including state and action features, as follows:

- State feature, $f_s(v)$. For a packet with its specified origin O and destination D at node v, the state features are the vectors with the elements below.
  1. **Distance to destination, $d(v, D)$:** the distance between v and D.
  2. **Node Stretch, $ns(O, D, v) = \frac{d(O,v)+d(v,D)}{d(O,D)}$:** the stretch of the indirect distance between O and D that is via v with respect to the direct distance between O and D.
- Action feature, $f_a(a) = f_s(u)$. The feature for the action that forwards a packet from v to $u \in nbr(v)$, $f_a(a)$ is chosen to be the same as the state feature of u, $f_s(u)$.

Henceforth, we consider learning with two different combinations of input features, one with only $d(v, D), d(u, D)$ and the other with both $d(v, D), d(u, D)$ and $ns(O, D, v), ns(O, D, u)$.

5.2   Selection of Seed Graph and Graph Subsamples

To achieve both cross-graph generalizability and cross-node generalizability, we develop a knowledge-guided mechanism with the following two selection components: (1) seed graph selection and (2) graph subsample selection.

Seed Graph Selection. The choice of seed graph depends primarily on the analysis of cross-node generalizability (Lemma 1) across a sufficient set of uniform random graphs with diverse sizes and densities/average node degrees. In [4], we empirically show that, with the use of distance-to-destination and node stretch, a good seed graph (with high $SIM_G(m, Q^*)$) is likely to exist in a set of graphs with modest size (e.g., 50) and high density (e.g., 5) in the Euclidean space.

There may be applications where analysis of large (or full) graphs is not always possible. In such situations, given a graph G, an alternative choice of seed graph can be from a small subgraph $G' = (V', E'), V' \subset V, E' \subset E$ with relatively high cross-node generalizability. Note that, in Lemma 1, for a graph $G = (v, E)$ satisfying the **RankPres** property for all $v \in V$, **RankPres** still holds for $v' \in V'$ in a subgraph $G' = (V', E')$ of G. This is because the learnable function m still preserves the optimal routing policy for all nodes in a subset of $nbr(v)$.

Graph Subsamples Selection. We provide the following scheme of subsample selection for a given graph $G = (V, E)$ to choose a set of ϕ nodes for generating $\phi\delta$ training samples, where δ is the average node degree (i.e., number of neighbors).

1. Select an origin and destination pair $(O, D), O, D \in V$.
2. Select ϕ nodes, $v_1, ..., v_\phi \in V \setminus D$.
3. For each chosen node $v_1, ..., v_\phi$, respectively, collect the subsamples $\langle X, Y \rangle$, where
   $X = \bigcup_{v \in \{v_1,...,v_\phi\} \wedge u \in nbr(v)} \{\langle f_s(v), f_a(u) \rangle\}$ and
   $Y = \bigcup_{v \in \{v_1,...,v_\phi\} \wedge u \in nbr(v)} \{Q^*(v, u)\}$.

5.3 Supervised Learning for APNSP with Optimal Q-Values

Given the dataset $\langle X, Y \rangle$ collected by using the subsampling policy for the seed graph, we train a DNN based on supervised learning to capture the optimal ranking policy. Specifically, suppose the DNN H is parameterized by θ. We seek to minimize the following loss function:

$$\min_\theta \quad \sum_{\langle X,Y \rangle} \|H_\theta(f_s(v), f_a(u)) - Q^*(v, u)\|^2. \tag{3}$$

Note that we assume that the optimal Q-values are known for the seed graph in the supervised learning above, which can be obtained based on the shortest path routing policies of the seed graph. By leveraging these optimal Q-values and supervised learning on the seed graph, a generalized routing policy is learned for APNSP routing over almost all uniform random graphs, as we validate later in the experiments.

5.4 Reinforcement Learning for APNSP

For the case where the optimal Q-values of graphs are unknown, we develop an approach for solving the APNSP problem based on RL. Using the same input features and the same seed graph selection procedure, the RL algorithm continuously improves the quality of Q-value estimations by interacting with the seed graph.

In contrast to the supervised learning algorithm, where we collect only a single copy of data samples from a set of chosen (shortest path) nodes once before training, in RL new training data samples from nodes in a shortest path, predicted by most recent training episode (i.e., based on the current Q-value estimation), are collected at the beginning of each training episode. Remarkably, the generalizability of the resulting RL routing policy across almost all graphs in U is preserved. The details of the RL algorithm, named as RL-APNSP-ALGO, are shown in the appendix in [4].

6   Routing Policy Performance for Scalability and Zero-Shot Generalization

In this section, we discuss implementation of our machine learned routing policies and evaluate their performance in predicting all-pair near-shortest paths for graphs across different sizes and densities in Python3. We use PyTorch 2.6.0 [25] on the CUDA 12.4 compute platform to implement DNNs as shown in Fig. 2. The appendix in [4] shows our simulation parameters for training and testing the routing policies.

6.1   Comparative Evaluation of Routing Policies

We compare the performance of the different versions of *Greedy Tensile* policies obtained using the following approaches:

1. **GreedyTensile-S**: Supervised learning from appropriately chosen seed graph G^* using graph subsamples selection from $\phi=3$.
2. **GreedyTensile-RL**: RL from appropriately chosen seed graph G^* using graph subsamples selection from $\phi=3$.
3. **GF**: Greedy forwarding that forwards packets to the one-hop neighbor with the minimum distance to the destination.
4. **SR-NS**: Symbolic Regression routing that learns a function, $ns(O, D, u) +$ 0.64, to forward packets to the one-hop neighbor with the minimum node stretch.

Note that for a given set of input features, both supervised learning and reinforcement learning schemes use the same DNN configuration to learn the routing policies. By using the subsampling mechanism, not only the sample complexity but also the training time will be significantly reduced in GreedyTensile-S and GreedyTensile-RL since they only require the data samples derived from $\phi=3$ nodes rather than all nodes in a seed graph.

6.2 Zero-Shot Generalization over Diverse Graphs

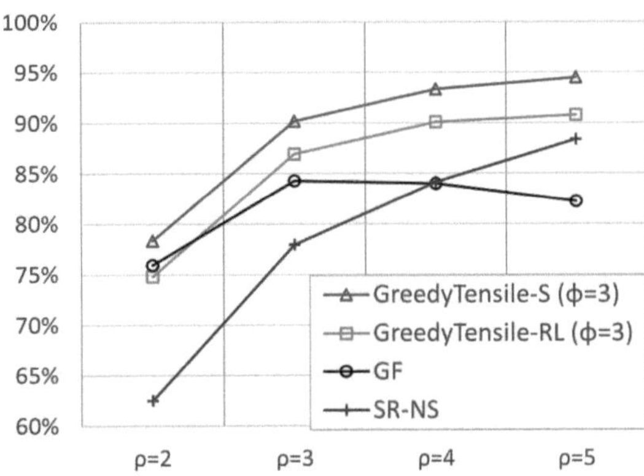

Fig. 3. Average APNSP prediction accuracy across graph sizes with various ρ in Euclidean space for *Greedy Tensile* policies.

To evaluate the scalability and generalizability of the policies, we directly (i.e., without any retraining or fine-tuning) test the policies learned from the seed graph G^* on new uniform random graphs with different combinations of (N_{test}, ρ_{test}) in Euclidean spaces. We select 80 random graphs for each pair and calculate the average prediction accuracy over these $80N_{test}^2$ shortest paths.

For the DNNs with input $d(v, D)$ and $d(u, D)$, the tests confirm that the performance of all the learned policies exactly match the prediction accuracy of GF.

For the *Greedy Tensile* DNNs with input $\langle d(v, D), ns(O, D, v), d(u, D), ns(O, D, u)\rangle$, we plot in Fig. 3 the respective average prediction accuracies across graph sizes in $\{27, 64, 125, 216\}$ with density in $\{2, 3, 4, 5\}$.

The Supervised approach (GreedyTensile-S) achieves the best performance among all the approaches. In particular, compared to GF, GreedyTensile-S improves the accuracy by up to 12. 22% over GF, while the RL approach (GreedyTensile-RL) shows at least comparable performance in low-density graphs ($\rho = 2$) and achieves an improvement of up to 8.55% in graphs with $\rho \geq 3$. The performance gap between DNNs and GF increases as the network density increases to a high level (e.g., $\rho = 5$), where GF was believed to work close to optimal routing. SR-NS, merely generalizing to a specific network configuration, yields an accuracy gap of up to 12. 42% over GF in low-density graphs ($\rho \leq 2$) and starts to outperform GF in high-density graphs ($\rho \geq 4$). The performance of SR-NS increases to come close to the performance of the GreedyTensile-RL policy as the density increases to $\rho = 5$.

The results imply that, comparing the features of distance-to-destination and node stretch, routing using only distance-to-destination (i.e., GF) or routing using only node stretch (i.e., SR-NS) tend to generalize only in sparse or dense wireless networks, respectively. Our DNNs take advantage of these two features and produce more sophisticated policies based on the current node and (O, D) information by searching for a larger input space, and then show comprehensive generaliability across diverse network configurations.

7   Discussion

Distributed Routing. Greedy Tensile adopts a fully distributed strategy where each node independently makes forwarding decisions using its own and one-hop neighbors' coordinates, and the source and destination coordinates embedded in the packet header. This design parallels Greedy Forwarding, eliminating the need for global topology information while preserving low complexity and scalability.

Stabilization. The learned policy is self-stabilizing in a straightforward manner: the routing decision using *Greedy Tensile* depends only on the local network states, $\langle f_s(v), f_a(u) \rangle$, $u \in nbr(v)$. *Greedy Tensile* thus recovers from arbitrarily inconsistent or stale states (e.g., due to topology changes or crashed nodes) by locally retrieving the coordinates of the current neighbors.

Handling of Inaccurate Routes. For handling of route failures, i.e., when the route leads to a dead-end, the policy can be extended by taking advantage of the knowledge that searching within the elliptical region will, with high probability, yield a path between the source and destination. As is typical, depth-first search (DFS) can be used to traverse all nodes within the ellipse. Let $G' = (V', E') \subseteq G$ be the subgraph including all nodes $v' \in V'$ and edges $e' \in E'$ within a given elliptical region. The time complexity of rediscovering a routing path using DFS is $O(|V'| + |E'|)$.

Generalizable Routing for Random Geometric Hyperbolic Space Graphs. We extend our experimental findings to show the zero-shot generalizability of *Greedy Tensile* for the APNSP problem in hyperbolic metric space. This space is known to represent the scale topologies of the Internet and social networks, where node degree follows a power-law distribution [3,27]. As with the Euclidean space, *Greedy Tensile* generalizes to all random hyperbolic graphs with different average node degrees. To the best of our knowledge, we are the first to provide a routing policy that outperforms GF in almost all random graphs; recall that GF has been shown to find almost optimal shortest path in scale-free topologies [22]. The detailed results are shown in [4].

Symbolic Interpretation of *Greedy Tensile.* We symbolically interpret the learned DNN with input of $d(v, D)$, $d(u, d)$, $ns(O, D, v)$, and $ns(O, D, v)$ to show that it can be approximated by a low-complexity two-linear action policy, which is visualized and elaborated upon in [4]. One action uses both distance-to-destination and node stretch features to select the next forwarder for those neighbors providing forward progress, and the other considers only distance-to-destination features in the absence of forward progress.

8   Conclusions and Future Work

We have shown that guiding machine learning with domain knowledge can lead (somewhat surprisingly) to the rediscovery of a well-known routing policy, in addition to discovering a new routing policy that performs well in terms of complexity, scalability, and generalizability. The theory we have presented in the paper is readily extended to other classes of graphs (such as non uniform cluster distributions) and MDP actions that span multiple neighbors. Thus, albeit our illustration intentionally uses relatively familiar input features and local routing architectures, richer domain theory will be useful to guide machine learning of novel routing algorithms. Moreover, the routing policies are likely to be competitive for richer classes of graphs than the class of uniform random graphs on which we have focused our validation.

While samples from nodes of a single seed graph suffice for generalizable learning, in practice, learning from multiple seed graphs may be of interest. For instance, if an ideal seed graph is not known a priori, online learning from better or multiple candidate seed graphs as they are encountered may be of interest for some applications. Along these lines, we recall that the set of ideal (and near-ideal) seed graphs is relatively large in the problem we considered. One way to relax the knowledge of ideal seed graphs is to leverage online meta-learning, for learning a good model initialization and continuing to improve the initialization based on better seed graphs as they are encountered. Towards this end, we have also been studying the merits of efficiently fine tuning the model for the target graph as an alternative to zero-shot generalization.

References

1. Bi, J., et al.: Learning generalizable models for vehicle routing problems via knowledge distillation. In: Advances in Neural Information Processing Systems 35: Annual Conference on Neural Information Processing Systems (NeurIPS) (2022)
2. Billard, L., Diday, E.: Symbolic regression analysis. In: Classification, Clustering, and Data Analysis: Recent Advances and Applications, pp. 281–288. Springer (2002)
3. Boguná, M., Papadopoulos, F., Krioukov, D.: Sustaining the internet with hyperbolic mapping. Nat. Commun. **1**(1), 62 (2010)
4. Chen, Y.F., Lin, S., Arora, A.: Knowledge-guided machine learning for stabilizing near-shortest path routing. arXiv preprint arXiv:2509.06640 (2025)
5. Chen, Y.F., Parker, K.W., Arora, A.: QF-Geo: Capacity aware geographic routing using bounded regions of wireless meshes. arXiv preprint arXiv:2305.05718 (2023)
6. Finn, G.G.: Routing and addressing problems in large metropolitan-scale internetworks. University of Southern California Marina Del Rey Information Sciences Inst, Technical report (1987)
7. Gold, E.M.: Language identification in the limit. Inf. Control **10**(5), 447–474 (1967)
8. Grossglauser, M., Tse, D.N.: Mobility increases the capacity of ad hoc wireless networks. IEEE/ACM Trans. Networking **10**(4), 477–486 (2002)
9. Grover, A., Leskovec, J.: node2vec: scalable feature learning for networks. In: Proceedings of the 22nd ACM SIGKDD International Conference on Knowledge Discovery and Data Mining, pp. 855–864 (2016)

10. Hekmat, R., Van Mieghem, P.: Interference in wireless multi-hop ad-hoc networks and its effect on network capacity. Wireless Netw. **10**, 389–399 (2004)
11. Hou, T.C., Li, V.: Transmission range control in multihop packet radio networks. IEEE Trans. Commun. **34**(1), 38–44 (1986)
12. Hudson, B., Li, Q., Malencia, M., Prorok, A.: Graph neural network guided local search for the traveling salesperson problem. arXiv preprint arXiv:2110.05291 (2021)
13. Iyer, S.K., Manjunath, D.: Topological properties of random wireless networks. Sadhana **31**(2), 117–139 (2006)
14. Kleinberg, J.M.: Navigation in a small world. Nature **406**(6798), 845–845 (2000)
15. Kranakis, E.: Compass routing on geometric networks. In: Proceedings of the 11th Canadian Conference on Computational Geometry (CCCG 1999), Vancouver (1999)
16. Manfredi, V., Wolfe, A., Zhang, X., Wang, B.: Learning an adaptive forwarding strategy for mobile wireless networks: resource usage vs. latency. In: Reinforcement Learning for Real Life (RL4RealLife) Workshop in the 36th Conference on Neural Information Processing Systems (NeurIPS) (2022)
17. Manfredi, V., Wolfe, A.P., Wang, B., Zhang, X.: Relational deep reinforcement learning for routing in wireless networks. In: 2021 IEEE 22nd International Symposium on a World of Wireless, Mobile and Multimedia Networks (WoWMoM), pp. 159–168 (2021)
18. Muggleton, S.: Hypothesizing an algorithm from one example: the role of specificity. Phil. Trans. R. Soc. A **381**(2251), 20220046 (2023)
19. Muggleton, S.: Learning from positive data. In: International Conference on Inductive Logic Programming, pp. 358–376. Springer (1996)
20. Narayanan, A., Chandramohan, M., Venkatesan, R., Chen, L., Liu, Y., Jaiswal, S.: graph2vec: learning distributed representations of graphs. arXiv preprint arXiv:1707.05005 (2017)
21. Nelson, R., Kleinrock, L.: The spatial capacity of a slotted aloha multihop packet radio network with capture. IEEE Trans. Commun. **32**(6), 684–694 (1984)
22. Papadopoulos, F., Krioukov, D., Boguná, M., Vahdat, A.: Greedy forwarding in dynamic scale-free networks embedded in hyperbolic metric spaces. In: IEEE INFOCOM, pp. 1–9 (2010)
23. Reis, J., Rocha, M., Phan, T.K., Griffin, D., Le, F., Rio, M.: Deep neural networks for network routing. In: 2019 International Joint Conference on Neural Networks (IJCNN), pp. 1–8. IEEE (2019)
24. Takagi, H., Kleinrock, L.: Optimal transmission ranges for randomly distributed packet radio terminals. IEEE Trans. Commun. **32**(3), 246–257 (1984)
25. The Linux Foundation: PyTorch 2.6.0 (2025). https://pytorch.org/get-started/pytorch-2.x/
26. Valiant, L.G.: A theory of the learnable. Commun. ACM **27**(11), 1134–1142 (1984)
27. Verbeek, K., Suri, S.: Metric embedding, hyperbolic space, and social networks. In: Proceedings of the Thirtieth Annual Symposium on Computational Geometry, pp. 501–510 (2014)
28. Wang, Y., Yao, Q., Kwok, J.T., Ni, L.M.: Generalizing from a few examples: a survey on few-shot learning. ACM Comput. Surv. (CSUR) **53**(3), 1–34 (2020)
29. Wu, Y., Song, W., Cao, Z., Zhang, J., Lim, A.: Learning improvement heuristics for solving routing problems. IEEE Trans. Neural Networks Learn. Syst. **33**(9), 5057–5069 (2021)

30. Xin, L., Song, W., Cao, Z., Zhang, J.: NeuroLKH: combining deep learning model with Lin-Kernighan-Helsgaun heuristic for solving the traveling salesman problem. In: Advances in Neural Information Processing Systems 34: Annual Conference on Neural Information Processing Systems (NeurIPS), pp. 7472–7483 (2021)
31. Xue, F., Kumar, P.R.: Scaling Laws for Ad Hoc Wireless Networks: An Information Theoretic Approach. Now Publishers Inc, Norwell (2006)

Brief Announcement: On the Impact of Unlimited Computational Power in $\mathcal{OBLOT}$: Consequences for Synchronous Robots on Graphs

Serafino Cicerone[1], Alessia Di Fonso[1(✉)], Gabriele Di Stefano[1], and Alfredo Navarra[2]

[1] Università degli Studi dell'Aquila, L'Aquila, Italy
{serafino.cicerone,alessia.difonso,gabriele.distefano}@univaq.it
[2] Università degli Studi di Perugia, Perugia, Italy
alfredo.navarra@unipg.it

Abstract. The $\mathcal{OBLOT}$ model in swarm robotics assumes robots are anonymous, disoriented, oblivious, and silent. Their only means of (implicit) communication is transferred to their positioning. These constraints make distributed algorithm design difficult, and prior research has mainly focused on task feasibility, measuring cost in movements or rounds while neglecting computational power. This paper shows that for synchronous robots on finite graphs, unlimited computational power (within finite time) is impactful, enabling a definitive resolution algorithm that solves a broad class of problems with minimal moves and rounds.

1 Introduction

The $\mathcal{OBLOT}$ model [14] has been extensively studied in theoretical swarm robotics. Over the past two decades, many research papers have focused on determining whether a specific problem can be solved under the $\mathcal{OBLOT}$ model and under which assumptions. Emphasis has been placed on feasibility issues rather than the computational power required by robots to accomplish a task. In this model, a set of distributed and mobile robots has limited capabilities. They are anonymous, disoriented, no memory of past events (oblivious), and cannot communicate directly. They interact by observing the environment and the positions of all the other robots.

The time is measured in *rounds* for synchronous robots. In one round, any active robot performs a so-called Look-Compute-Move (LCM) cycle, whereas inactive robots are just idle. The subset of robots activated within one round either coincides with the whole set of robots (FSync case), or is chosen by an ideal adversarial scheduler that provides some fairness, i.e., every robot is activated within finite time, infinitely often (SSync case). In the Look phase, a robot acquires a snapshot of the environment with information on the relative positioning of the other robots with respect to its own. During the Compute phase, a robot makes computations – without restrictions except for finite time – to decide its next destination. Computations are based only on the snapshot acquired during the current

© The Author(s), under exclusive license to Springer Nature Switzerland AG 2026
S. Bonomi et al. (Eds.): SSS 2025, LNCS 16350, pp. 146–152, 2026.
https://doi.org/10.1007/978-3-032-11127-2_13

LCM cycle and on the designed algorithm, the same for all the robots. During the `Move` phase, the robot moves toward the computed destination.

The robots need to coordinate to reach a shared goal. However, the limited robots' capabilities make the design of distributed algorithms a challenging task. The possibility of reaching the goal may also depend on other abilities, like the *strong multiplicity detection*: perceiving the exact number of robots forming a multiplicity (a vertex occupied by more than one robot) during the `Look` phase. Many researchers have studied which tasks robots can perform in the $\mathcal{OBLOT}$ model according to assumed robots' abilities. They characterized various problems based on the initial configurations of robots, the type of time scheduler, and its variants. The most relevant ones solved within the $\mathcal{OBLOT}$ model concern *Pattern Formation* [3,6,7,11,19,20], that is, robots must reach a final configuration (in terms of a particular disposal or of some property) within a finite time. Standard tasks within this model concern, for instance, *Gathering* [8–10,13,16], *Mutual Visibility* [12,18], and *Geodesic Mutual Visibility* [2,4].

One implicit assumption underlying the $\mathcal{OBLOT}$ model, which the research community accepts, is that robots have unbounded computational power. However, this assumption does not imply that all problems are solvable in the $\mathcal{OBLOT}$ model as the other constraints (i.e. anonymity, obliviousness...) as well as the adopted time scheduler (i.e. synchronous, asynchronous) pose significant algorithmic challenges. In addition, in many distributed computing environments, which differ from standard algorithmic theory, cost metrics focus on the exchanged information rather than on the computational complexity. For example, in *message-passing* models like $\mathcal{LOCAL}$ and $\mathcal{CONGEST}$ [15,17], the emphasis is on the number (and the size) of the exchanged messages; in $\mathcal{OBLOT}$, on the number of robot movements. As a result, the robots' computational power is usually considered irrelevant.

The Result. We show that, for robots moving on graphs, assuming unbounded computational power has a substantial impact. Exploiting this assumption, we design a definitive resolution algorithm for synchronous robots with strong multiplicity detection operating on finite graphs. Our algorithm applies to a broad class $\mathcal{P}$ of problems and, at the same time, optimizes the number of required moves and rounds. The generality of this result highlights the importance of restricting the robots' computational power to polynomial-time algorithms when addressing problems involving robots moving on graphs. We study feasibility issues in the $\mathcal{OBLOT}$ model for robots moving on the vertices of finite graphs, under the FSync scheduler. As in $\mathcal{OBLOT}$, we assume that robots face no restrictions on their computational power during the `Compute` phase, and we exploit this by allowing unbounded computational memory. Using this assumption, we design an algorithmic framework that applies to a broad class of problems, enabling us to determine whether a given problem is solvable from a specific initial configuration. Moreover, whenever the answer is positive, the framework also yields the minimum number of moves that robots must perform from each reachable configuration.

The importance of our findings relies on the theoretical implications. In the future, any attempts to solve problems under the $\mathcal{OBLOT}$ model on graphs should consider limited computational power for the `Compute` phase and polynomial in the size of the instance. The full version of this work is available at [5].

2   Solving General Formation Problem

A *configuration* C is modeled as $C = (G, \lambda)$, where G is an undirected finite graph and $\lambda : V(G) \to \mathbb{N}$ provides the number of robots on each vertex of G assumed that $k = \sum_{v \in V(G)} \lambda(v)$ is bounded.

Let Π be a generic problem defined in the $\mathcal{OBLOT}$ model. Although Π can be defined in different ways, for the sake of simplicity we first focus on *formation* problems where a solution can be specified by a set F of *final configurations*.[1] We call this class of problems $\mathcal{P}$. Notice that classical problems like *Gathering*, *Pattern Formation*, *Mutual visibility* fit in $\mathcal{P}$. Any problem $\Pi \in \mathcal{P}$ depends on a finite graph G, on an integer $k \geq 1$ representing the number of robots located on G, and on a set F of configurations $C = (G, \lambda)$, with $\sum_{v \in V(G)} \lambda(v) = k$. Note that the elements of F are final for Π, that is, the displacement of robots in $C \in F$ is considered a solution for Π. In summary, Π can be denoted by the triple (G, k, F) and formalized as follows:

Problem 1. $\Pi = (G, k, F) \in \mathcal{P}$
Input: Any configuration $C = (G, \lambda)$ with $\sum_{v \in V(G)} \lambda(v) = k$.
Goal: Design a distributed algorithm working under the $\mathcal{OBLOT}$ model
              that, starting from C, brings all robots to a configuration $C' \in F$.

We devised an algorithm called RobotMove that solves $\Pi = (G, k, F)$ as follows: if Π can be solved starting from a configuration C, RobotMove guides all the robots to move so that, with the minimum number of moves, a configuration $C' = (G, \lambda') \in F$ is created. If $C \in F$ or Π cannot be solved from C, RobotMove makes robots aware of this (and they do not move). According to this notation, the main result of this work can be stated as follows.

Theorem 1. *Let $\Pi = (G, k, F) \in \mathcal{P}$. Starting from a configuration C, by repeatedly applying Algorithm RobotMove, the synchronous robots reach a configuration in F within the minimum number of rounds if and only if C is solvable.*

3   The Algorithm

In what follows, we first provide notions, tools and data structures needed to define RobotMove, and then we give a short description of the algorithm.

Configurations' Automorphisms. Two undirected graphs $G = (V, E)$ and $G' = (V', E')$ are *isomorphic* if there is a bijection φ from V to V' such that

---

[1] Note that F could, alternatively, be provided as input in the form of some property that the robots can verify on any configuration with k robots located on G.

$\{u, v\} \in E$ if and only if $\{\varphi(u), \varphi(v)\} \in E'$. An *automorphism* of G is an isomorphism from G onto itself. The set of automorphisms of a given graph, under the composition operation, forms the automorphism group of the graph denoted by $\mathrm{Aut}(G)$. Two vertices u and v of G are *equivalent vertices* if there exists an automorphism $\varphi \in \mathrm{Aut}(G)$ such that $\varphi(u) = v$. The equivalence classes of the vertices of G under the action of the automorphisms are called vertex *orbits*.

Isomorphisms can be also defined for configurations: (G, λ) and (G', λ') are isomorphic if there exists $\varphi \in \mathrm{Aut}(G)$ such that $\lambda(v) = \lambda'(\varphi(v))$ for each vertex v. As for graphs, $\mathrm{Aut}(C)$ is the automorphism group of $C = (G, \lambda)$. In C, two robots r and r' are *equivalent robots* if they reside on vertices that are equivalent according to some automorphism of $\mathrm{Aut}(C)$. Symbols $\mathcal{O}_G$ and $\mathcal{O}_C$ represent the set of all the orbits of G and C, respectively.

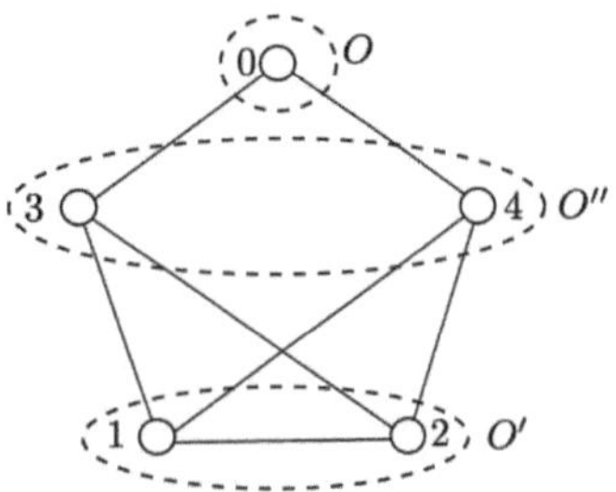

Fig. 1. An example of graph canonization, where dashed ovals enclose orbits. By the canonization, $O < O' < O''$.

Graph Canonization and Orbits Ordering. In graph theory, graph canonization is the problem of finding a canonical form of a given graph G. A canonical form is a labeled graph $\mathrm{Canon}(G)$ that is isomorphic to G, and such that every graph that is isomorphic to G has the same canonical form as G. Thus, from a solution to the graph canonization problem, one could also solve the problem of graph isomorphism: to test whether two graphs G' and G'' are isomorphic, compute their canonical forms $\mathrm{Canon}(G')$ and $\mathrm{Canon}(G'')$, and test whether these two canonical forms are identical. The formal computational study of graph canonization began in the 1970s. Quite recently, in [1], László Babai announced a quasi-polynomial-time algorithm for graph canonization, that is, one with running time $2^{O((\log n)^c)}$ for some fixed $c > 0$ and n being the number of vertices. Available algorithms can compute not only the canonical labeling but also the orbits[2].

Consider the situation where two robots, during the same **Look** phase, take a snapshot of the graph G. Since they are disoriented, each robot may represent G differently (say as G' and G''). However, if they both apply the same canonization algorithm, they obtain $\mathrm{Canon}(G')$ and $\mathrm{Canon}(G'')$, which must be identical since both derive from the same underlying graph G. In particular, if the canonization assigns integer labels from $[0, n-1]$ to the vertices of any n-vertex graph, then the vertex labeled i in $\mathrm{Canon}(G')$ is isomorphic to the vertex labeled i in $\mathrm{Canon}(G'')$. Consequently, the corresponding orbits in $\mathcal{O}_{G'}$ and $\mathcal{O}_{G''}$ receive the same set of labels. This ensures that the two robots can agree on a total ordering of the orbits of G: $O' < O''$ if the smallest label assigned to O' is smaller than the smallest label assigned to O'' (cf. Fig. 1). This approach extends to configurations, so robots also agree on a total ordering of the orbits of a configuration.

---

[2] e.g. *bliss*, see http://www.tcs.tkk.fi/Software/bliss/index.html.

Modeling Robots' Moves. Let $\bar{\mathcal{O}}_C \subseteq \mathcal{O}_C$ be the set of all the occupied orbits of C. A move in C is a function $m_C : \bar{\mathcal{O}}_C \to \mathcal{O}_C \cup \{nil\}$ that specifies the adjacent orbit that the equivalent robots lying on O have to reach, for each orbit $O \in \bar{\mathcal{O}}_C$. If $m_C(O) = nil$, for some $O \in \bar{\mathcal{O}}_C$, the robots lying on O do not move. Since the vertices in the destination orbit are equivalent, as well as the moving robots, the adversary decides the specific vertex that each robot reaches within the specified orbit. A move m_C can be seen as an ordered set of pairs $(O, m_C(O))$ for each orbit $O \in \bar{\mathcal{O}}_C$. Hence, moves can be ordered since, given two moves m and m', we can say that $m < m'$ if the ordered sequence of pairs in m is lexicographically less than the ordered sequence of pairs in m'. If $(O, m_C(O))$ is such that $m_C(O)$ is nil, we assume that the label associated with nil is less than any other label used in the canonization of the configuration.

Configuration Hypergraph. Let $C = (G, \lambda)$ be a configuration with k robots. From G and k, it is possible to define a directed hypergraph $H_{G,k} = (V, A)$ with vertex set $V(H_{G,k})$ containing all the configurations of k robots on G, that is $V(H_{G,k}) = \{C \mid C = (G, \lambda_C), \sum_{v \in V(G)} \lambda_C(v) = k\}$, and with $(C, \Delta) \in A(H_{G,k})$ (the hyperarc set of $H_{G,k}$) whenever $C \in V(H_{G,k})$ and $\Delta \subseteq V(H_{G,k})$ is the set of all the configurations (up to isomorphisms) reachable from C using a single move. We call $H_{G,k}$ the *configuration hypergraph* of G with k robots. Figure 2 shows the configuration hypergraph $H_{G,k}$ of the complete bipartite graph $G = K_{2,3}$ containing $k = 2$ robots. If (C, Δ) is a hyperarc, each element of Δ can be

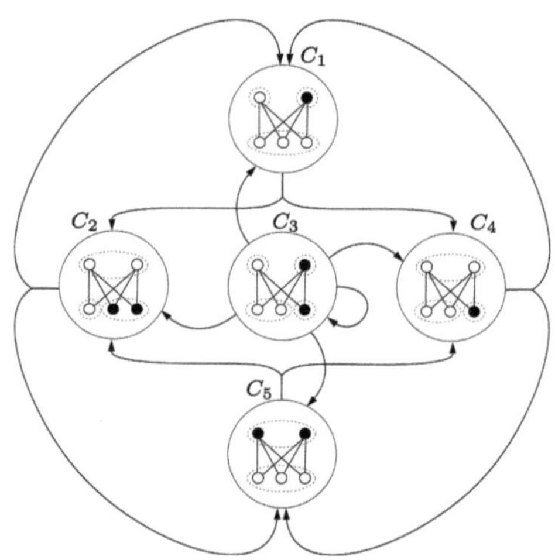

Fig. 2. The configuration hypergraph of graph $G = K_{2,3}$ with $k = 2$ robots. Dashed ovals enclose orbits.

reached from C by applying different moves. Hence, all such moves are contained in a uniquely determined set $\mathcal{M}((C, \Delta))$.

Description of the Algorithm. Consider a problem $\Pi = (G, k, F)$ defined on a graph G with k robots and concerning a set F of final configurations. Given $C = (G, \lambda)$ with k robots, RobotMove computes the canonical form of C and the canonical form of each configuration in F. Then, the algorithm builds the configuration hypergraph $H_{G,k}$ and the set $\mathcal{M}$ of all the moves associated to each hyperarc. Once all such data are available, RobotMove performs a graph traversal of the hypergraph to find a shortest path for reaching a configuration if F. Finding such a shortest path corresponds to providing each robot the knowledge of a minimum number of moves (rounds) to solve the problem in case the problem Π is feasible in C.

References

1. Babai, L.: Canonical form for graphs in quasipolynomial time: preliminary report. In: Charikar, M., Cohen, E. (eds.) Proceedings of the 51st Annual ACM SIGACT Symposium on Theory of Computing, STOC 2019, pp. 1237–1246. ACM (2019). https://doi.org/10.1145/3313276.3316356
2. Badri, S., Cicerone, S., Di Fonso, A., Di Stefano, G.: An optimal algorithm for geodesic mutual visibility on hexagonal grids. In: Masuzawa, T., Katayama, Y., Kakugawa, H., Nakamura, J., Kim, Y. (eds.) Stabilization, Safety, and Security of Distributed Systems - 26th International Symposium, SSS 2024, Nagoya, Japan, October 20–22, 2024, Proceedings. Lecture Notes in Computer Science, vol. 14931, pp. 161–176. Springer (2024). https://doi.org/10.1007/978-3-031-74498-3_12
3. Cicerone, S., Di Fonso, A., Di Stefano, G., Navarra, A.: Arbitrary pattern formation on infinite regular tessellation graphs. Theor. Comput. Sci. **942**, 1–20 (2023). https://doi.org/10.1016/j.tcs.2022.11.021
4. Cicerone, S., Di Fonso, A., Di Stefano, G., Navarra, A.: The geodesic mutual visibility problem: oblivious robots on grids and trees. Pervasive Mob. Comput. **95**, 101842 (2023). https://doi.org/10.1016/j.pmcj.2023.101842
5. Cicerone, S., Di Fonso, A., Di Stefano, G., Navarra, A.: On the impact of unlimited computational power in OBLOT: consequences for synchronous robots on graphs (2025). https://arxiv.org/abs/2509.04383
6. Cicerone, S., Di Stefano, G., Navarra, A.: Asynchronous arbitrary pattern formation: the effects of a rigorous approach. Distributed Comput. **32**(2), 91–132 (2019). https://doi.org/10.1007/S00446-018-0325-7
7. Cicerone, S., Di Stefano, G., Navarra, A.: Embedded pattern formation by asynchronous robots without chirality. Distributed Comput. **32**(4), 291–315 (2019). https://doi.org/10.1007/S00446-018-0333-7
8. Cicerone, S., Di Stefano, G., Navarra, A.: Gathering robots in graphs: the central role of synchronicity. Theor. Comput. Sci. **849**, 99–120 (2021). https://doi.org/10.1016/j.tcs.2020.10.011
9. Cieliebak, M., Flocchini, P., Prencipe, G., Santoro, N.: Distributed computing by mobile robots: gathering. SIAM J. Comput. **41**(4), 829–879 (2012)
10. D'Angelo, G., Di Stefano, G., Navarra, A.: Gathering on rings under the look-compute-move model. Distrib. Comput. **27**(4), 255–285 (2014)
11. Das, S., Flocchini, P., Santoro, N., Yamashita, M.: Forming sequences of geometric patterns with oblivious mobile robots. Distrib. Comput. **28**(2), 131–145 (2015)
12. Di Luna, G.A., Flocchini, P., Chaudhuri, S.G., Poloni, F., Santoro, N., Viglietta, G.: Mutual visibility by luminous robots without collisions. Inf. Comput. **254**, 392–418 (2017). https://doi.org/10.1016/j.ic.2016.09.005
13. Di Stefano, G., Navarra, A.: Optimal gathering of oblivious robots in anonymous graphs and its application on trees and rings. Distrib. Comput. **30**(2), 75–86 (2017)
14. Flocchini, P., Prencipe, G., Santoro, N. (eds.): Distributed Computing by Mobile Entities, Current Research in Moving and Computing, Lecture Notes in Computer Science, vol. 11340. Springer (2019). https://doi.org/10.1007/978-3-030-11072-7
15. Hirvonen, J., Suomela, J.: Distributed Algorithms. Aalto University, Finland (2025). https://jukkasuomela.fi/da2020/da2020.pdf
16. Klasing, R., Kosowski, A., Navarra, A.: Taking advantage of symmetries: gathering of many asynchronous oblivious robots on a ring. Theor. Comput. Sci. **411**, 3235–3246 (2010)

17. Linial, N.: Locality in distributed graph algorithms. SIAM J. Comput. **21**(1), 193–201 (1992). https://doi.org/10.1137/0221015
18. Poudel, P., Aljohani, A., Sharma, G.: Fault-tolerant complete visibility for asynchronous robots with lights under one-axis agreement. Theor. Comput. Sci. **850**, 116–134 (2021). https://doi.org/10.1016/j.tcs.2020.10.033
19. Suzuki, I., Yamashita, M.: Distributed anonymous mobile robots: formation of geometric patterns. SIAM J. Comput. **28**(4), 1347–1363 (1999)
20. Yamashita, M., Suzuki, I.: Characterizing geometric patterns formable by oblivious anonymous mobile robots. Theor. Comput. Sci. **411**(26–28), 2433–2453 (2010)

Gathering in Non-vertex-Transitive Graphs Under Round Robin

Serafino Cicerone[1], Alessia Di Fonso[1(✉)], Gabriele Di Stefano[1],
and Alfredo Navarra[2]

[1] Dipartimento di Ingegneria e Scienze dell'Informazione e Matematica, Università
degli Studi dell'Aquila, Via Vetoio, 67100 L'Aquila, Italy
{serafino.cicerone,alessia.difonso,gabriele.distefano}@univaq.it
[2] Dipartimento di Matematica e Informatica, Università degli Studi di Perugia, Via
Vanvitelli, 06123 Perugia, Italy
alfredo.navarra@unipg.it

Abstract. The GATHERING problem for a swarm of robots asks for a
distributed algorithm that brings such entities to a common place, not
known in advance. We consider the well-known $\mathcal{OBLOT}$ model with
robots constrained to move along the edges of a graph, hence gathering in one vertex, eventually. Despite the classical setting under which
the problem has been usually approached, we consider the 'hostile' case
where: i) the initial configuration may contain *multiplicities*, i.e. more
than one robot may occupy the same vertex; ii) robots cannot detect
multiplicities. As a scheduler for robots activation, we consider the 'favorable' *round-robin* case, where robots are activated one at a time.

Our objective is to achieve a complete characterization of the problem in the broad context of *non-vertex-transitive* graphs, i.e., graphs
where the vertices are partitioned into at least two different classes of
equivalence. We provide a resolution algorithm for any configuration of
robots moving on such graphs along with its correctness. Furthermore,
we analyze its time complexity.

Keywords: mobile robots · synchrony · gathering · graphs · time
complexity

1 Introduction

A particularly prominent model for swarm robotics from a theoretical perspective is the $\mathcal{OBLOT}$ model [18,19], which characterizes robots with weak
computational power. In this framework, robots operate according to repeated
Look-Compute-Move cycles: during each cycle, a robot observes its surroundings
(Look), computes its next move based on a deterministic algorithm (Compute),
and then proceeds to the selected destination (Move).

One of the most studied problems in this setting is the so-called GATHERING
problem. Robots are required to converge to a common, unspecified place where

S. Bonomi et al. (Eds.): SSS 2025, LNCS 16350, pp. 153–170, 2026.
https://doi.org/10.1007/978-3-032-11127-2_14

they eventually stop moving. For robots moving in the Euclidean plane, the GATHERING problem has been fully characterized. In fact, it is always solvable for synchronous robots, whereas in the asynchronous case it is unsolvable when just two robots are considered.

When robots are constrained to move along the edges of a graph, the situation is less clear. Apart from some impossibility results or basic conditions that guarantee the resolution of the GATHERING problem provided in [5,7,17], most of the literature usually focuses on specific topologies. Studied topologies are: Trees [9,10], Regular Bipartite graphs [21], Finite Grids [9], Infinite Grids [16], Tori [23], Oriented Hypercubes [3], Complete graphs [6,7], Complete Bipartite graphs [6,7], Butterflies [4], and Rings [2,11–15,17,22,24–27].

The most of such topologies are very symmetric, i.e., the vertices can be partitioned into a few classes of equivalence. Another crucial aspect of the cited works, applicable to both Euclidean and graphs, is the time scheduling under which the robots operate. Notably, the *asynchronous* scheduler, where robots are activated independently of one another, often presents the greatest challenges. However, there are cases where asynchrony makes the problem unsolvable, whereas the *synchronous* setting allows the development of non-trivial strategies, see, e.g. [7]. In any case, there are two very common assumptions in the literature:

A_1: robots are endowed with some kind of *multiplicity detection*. With this, robots are able to recognize whether a vertex contains a *multiplicity*, i.e., if two or more robots are located at the same vertex (not necessarily the exact number);

A_2: the *initial* configuration, i.e., the first configuration ever seen by any robot, does not contain multiplicities.

In a recent work [20], the version of the GATHERING problem that assumes A_2 is denoted as DISTINCT GATHERING, while GATHERING refers to the case without that assumption. From now on, we keep this distinction. As a time scheduler, we consider the *Round Robin* (RR). This is a specific type of synchronous scheduler, where k robots are activated one at a time in a fair sequence. That is, in the first k Look-Compute-Move cycles, all robots are activated. This constitutes the first *epoch*. Then, another epoch starts with the same order of activations fixed by an ideal adversary and unknown to the robots and repeating forever. Certainly, dealing with RR may seem much easier compared to other schedulers, especially the asynchronous one. Indeed, all the symmetries admitted by a given configuration are inherently broken each time by the single activated robot. However, the GATHERING problem becomes considerably more complicated without assumptions A_1 and A_2. In [20], the GATHERING problem has been fully characterized when robots freely move on the Euclidean plane: it is impossible for $k = 3$ robots but possible for any $k \geq 4$.

Our Results. In this paper, we continue the investigation of the GATHERING problem under the newly proposed RR scheduler, without relying on assumptions A_1 and A_2, and considering general graphs. In particular, our goal is to achieve a complete characterization of the problem in the broad context of non-vertex-transitive graphs. For these graphs, we demonstrate that the concepts of orbits[1] and graph canonization[2] serve as powerful analytical tools. In particular, we present a simple, time-optimal algorithm, $\mathcal{A}_{\mathbf{T}}$, designed for graphs that contain a terminal orbit—a novel graph-theoretic concept introduced in this work and linked to other graph-theoretic parameters. For graphs that lack terminal orbits, we identify structural properties that require the development of a more intricate gathering algorithm, $\mathcal{A}_{\neg\mathbf{T}}$. Nevertheless, it is a linear factor away, in the graph's size, from being time-optimal in terms of epochs.

2 Model

We consider the standard $\mathcal{OBLOT}$ model of distributed systems of autonomous mobile robots. In $\mathcal{OBLOT}$, the system is composed of a set $\mathcal{R} = \{r_1, r_2, \ldots, r_k\}$ of $k \geq 2$ computational robots that operate on a graph G. Each vertex of G is initially empty, occupied by one robot, or occupied by more than one robot (i.e., a *multiplicity*; recall that we are not using assumptions A_1 and A_2 as described in the Introduction). Robots can be characterized according to many different settings. In particular, they have the following basic properties:

- **Anonymous:** they have no unique identifiers;
- **Autonomous:** they operate without a centralized control;
- **Dimensionless:** they are viewed as points, i.e., they have no volume nor occupancy restraints;
- **Disoriented:** they have no common sense of orientation;
- **Oblivious:** they have no memory of past events;
- **Homogeneous:** they all execute the same deterministic algorithm with no type of randomization admitted;
- **Silent:** they have no means of direct communication.

Each robot in the system has sensory capabilities, allowing it to determine the location of other robots in the graph, relative to its location. Each robot refers to a `Local Reference System` (LRS) that might differ from robot to robot. Each robot has a specific behavior described according to the sequence of the following four states: `Wait`, `Look`, `Compute`, and `Move`. Such a sequence defines the computational activation cycle (or simply a cycle) of a robot. More in detail:

1. `Wait`: the robot is in an idle state and cannot remain as such indefinitely;
2. `Look`: the robot obtains a global snapshot of the system containing the positions of the other robots with respect to its `LRS`, by activating its sensors. Each robot is seen as a point in the graph occupying a vertex;

---

[1] Roughly speaking, an orbit is a maximal equivalence class of vertices, i.e., a maximal set of indistinguishable vertices.

[2] A method for finding a labeling of a graph G such that every graph isomorphic to G has the same labeling as G.

3. **Compute**: the robot executes a local computation according to a deterministic algorithm $\mathcal{A}$ (we also say that the robot executes $\mathcal{A}$). This algorithm is the same for all the robots, and its result is the destination of the movement of the robot. Such a destination is either the vertex where the robot is already located, or a neighboring vertex at one hop distance (i.e., only one edge per move can be traversed);
4. **Move**: if the computed destination is a neighboring vertex v, the robot moves to v; otherwise, it executes a *nil* movement (i.e., it does not move).

In the literature, the computational cycle is simply referred to as Look-Compute-Move (LCM) cycle, because when a robot is in the Wait state, we say that it is *inactive*. Thus, the LCM cycle only refers to the *active* states of a robot. It is also important to notice that since the robots are oblivious, without memory of past events, every decision they make during the Compute phase is based on what they can determine during the current LCM cycle. In particular, during the Look phase, the robots take a global snapshot of the system and they use it to elaborate the information, building what is called the *view* of the robot. Regarding the Move phase of the robots, the movements executed are always considered to be instantaneous. Thus, the robots are only able to perceive the other robots positioned on the vertices of the graph, never while moving. Regarding the position of a robot on a vertex, two or more robots may be located on the same vertex, i.e., they constitute a multiplicity.

A relevant feature that greatly affects the computational power of the robots is the *time scheduler*. In this work, we consider the standard Round Robin (RR):

- Robots are activated one at time. The time during which a robot is active is called a *round*. If there are k robots, then from round 1 to round k, all the robots are activated. In the subsequent k rounds, all the robots are again activated in the same order. Each of those intervals of k rounds is called an *epoch*. The order in which robots are activated is decided at the beginning by an adversarial scheduler.[3]

3   Problem Formulation and Preliminary Observations

The topology where robots are placed on is represented by a simple and connected graph $G = (V, E)$, with finite vertex set $V(G) = V$ and finite edge set $E(G) = E$. The cardinality of V is represented as $|V|$ or n. $G[X]$ denotes the subgraph of G induced by a subset of vertices $X \subset V$. We denote by $diam(G)$ the diameter of G, that is, the maximum distance between any pair of vertices of the graph. For each vertex $v \in V$, $N(v)$ is the set of neighboring vertices of v and $N[v] = N(v) \cup \{v\}$. Two vertices u and v are *false twins* if $N(u) = N(v)$ and *true twins* if $N[u] = N[v]$. A vertex v is *pendant* in G if $|N(v)| = 1$.

---

[3] Actually, our strategies also work if the order of the activations is changed by the adversary at each epoch.

A function $\lambda : V \to \{0,1\}$ indicates to the robots whether a vertex of G is empty or occupied. Note that more than one robot may occupy the same vertex, but robots cannot perceive such information. We call $C = (G, \lambda)$ a *configuration* – from the robots' perspective, whenever the actual number of robots is bounded and greater than zero. A subset $V' \subseteq V$ is said *occupied* if at least one of its elements is occupied, *unoccupied* otherwise. We denote by $\Delta(C)$ the maximum distance among any pair of vertices that are occupied in C, and by $occ(C) = \sum_{v \in G} \lambda(v)$ the number of vertices that are occupied in C.

A configuration C is *final* if all the robots are on a single vertex (i.e., $\exists u \in V : \lambda(u) = 1$ and $\lambda(v) = 0,\ \forall v \in V \setminus \{u\}$), i.e. $\Delta(C) = 0$. Any configuration C that is not final can be *initial*. The gathering problem can be formally defined as the problem of transforming an initial configuration into a final one. Throughout the paper, we assume that each initial configuration is composed of at least two robots occupying at least two vertices (otherwise, the problem is trivially solved). A *gathering algorithm* for this problem is a deterministic distributed algorithm that brings the robots in the system to a final configuration in a finite number of LCM-cycles from any given initial configuration, regardless of the adversary. Formally, an algorithm $\mathcal{A}$ solves the gathering problem for an *initial configuration* C if, for any execution $\mathbb{E} : C = C(0), C(1), C(2), \ldots$ of $\mathcal{A}$, there exists a time instant $t > 0$ such that $C(t)^4$ is final and no robots move after t, i.e., $C(t') = C(t)$ holds for all $t' \geq t$. Given an initial configuration $C = (G, \lambda)$, it is worth remarking that many different placements of robots correspond to C due to possible multiplicities. If there exists a gathering algorithm for all the placements of robots corresponding to C, we say that C is *gatherable*, otherwise we say that C is *ungatherable*. With respect to the number of epochs required to accomplish the gathering, we can state the following:

Lemma 1. *Given an initial configuration $C = (G, \lambda)$, any resolution algorithm for* GATHERING *on C under* RR *requires $\Omega(\Delta(C))$ epochs.*

Proof. Let r and r' be two robots occupying two vertices that determine $\Delta(C)$. In order to solve the GATHERING problem, r and r' should meet, eventually. The fastest way to do it is that they move toward each other along a shortest path. This requires $\Delta(C)/2$ moves for each robot and hence, due to the RR scheduler, $\Delta(C)/2$ epochs. $\qquad\square$

Note that, as we are dealing with finite graphs, the above lemma can be stated also with respect to $\Omega(diam(G))$ as it is easy to provide a configuration C with $\Delta(C) = diam(G)$.

We now introduce a tool that will play a fundamental role in describing a solution for the problem at hand. In particular, we recall from [7] and [17] the notions of automorphism and orbit for general graphs.

Graph Automorphisms and Orbits. Two undirected graphs $G = (V, E)$ and $G' = (V', E')$ are *isomorphic* if there is a bijection φ from V to V' such that $\{u, v\} \in E$ if and only if $\{\varphi(u), \varphi(v)\} \in E'$. An *automorphism* on a graph G is an

---

^{4} $C(t)$ represents the configuration C as observed at time instant t.

isomorphism from G to itself. The set of automorphisms of a given graph, under the composition operation, forms the automorphism group of the graph, denoted by $\mathrm{Aut}(G)$. Two vertices u and v of G are *equivalent vertices* if there exists an automorphism $\varphi \in \mathrm{Aut}(G)$ such that $\varphi(u) = v$. The equivalence classes of the vertices of G under the action of the automorphisms are called vertex *orbits*. The partition of the vertex set of G consisting of the orbits by G is called *orbit partition* of G and is denoted by $\mathcal{O}_G$. It is easy to observe that when G contains only one orbit, then it is *vertex-transitive*, otherwise it is *non-vertex-transitive*. We say that two orbits O and O' are *adjacent* if and only if there is an edge (x, y) in G such that $x \in O$ and $y \in O'$.

Graph Canonization and Orbits Ordering. In graph theory, graph canonization is the problem of finding a canonical form of a given graph G. A canonical form is a labeled graph $\mathrm{Canon}(G)$ that is isomorphic to G, and such that every graph that is isomorphic to G has the same canonical form as G. Thus, from a solution to the graph canonization problem, one could also solve the problem of graph isomorphism: to test whether two graphs G' and G'' are isomorphic, compute their canonical forms $\mathrm{Canon}(G')$ and $\mathrm{Canon}(G'')$, and test whether these two canonical forms are identical. The formal computational study of graph canonization began in the 1970s. Quite recently, in [1], László Babai announced a quasi-polynomial-time algorithm for graph canonization, that is, one with running time $2^{O((\log n)^c)}$ for some fixed $c > 0$ and n being the number of vertices. Available algorithms can compute not only the canonical labeling but also the orbits.[5]

Consider the situation where two robots, during the same **Look** phase, take a snapshot of the graph G. Since they are disoriented, each robot may represent G differently (say as G' and G''). However, if they both apply the same canonization algorithm, they obtain $\mathrm{Canon}(G')$ and $\mathrm{Canon}(G'')$, which must be identical since both derive from the same underlying graph G. In particular, if the canonization assigns integer labels from $[0, n-1]$ to the vertices of any n-vertex graph, then the vertex labeled i in $\mathrm{Canon}(G')$ is isomorphic to the vertex labeled i in $\mathrm{Canon}(G'')$. Consequently, the corresponding orbits in $\mathcal{O}_{G'}$ and $\mathcal{O}_{G''}$ receive the same set of labels. This ensures that the two robots can agree on a total ordering of the orbits of G: $O' < O''$ if the smallest label assigned to O' is smaller than the smallest label assigned to O'' (cf. Fig. 1).

Our Approach and Methodology.. The algorithms introduced in this paper to solve the GATHERING problem on general non-vertex-transitive graphs follow the methodology proposed in [8]. We briefly outline how a generic algorithm $\mathcal{A}$, intended to solve a problem $\mathcal{P}$, can be designed according to this approach.

Since robots have extremely limited capabilities, it is beneficial to decompose $\mathcal{P}$ into simpler tasks denoted as $T_1, T_2, \ldots, T_q$. Among these tasks, at least one is designated as the *terminal* one.

Since robots operate according to the LCM cycle, they must identify the correct task to execute based on the configuration they observe during the **Look** phase.

---

[5] e.g. *bliss*, see http://www.tcs.tkk.fi/Software/bliss/index.html.

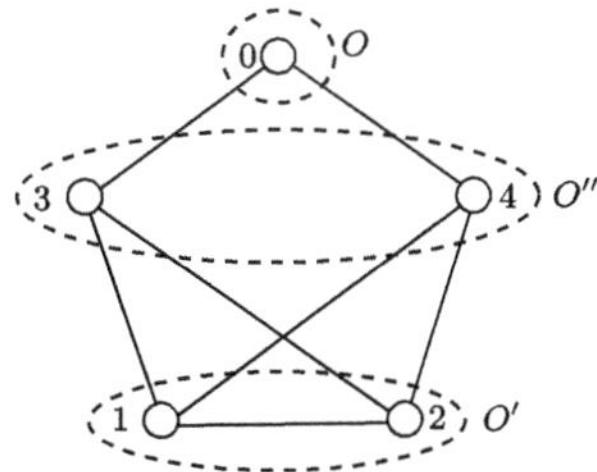

Fig. 1. An example of graph canonization. Note that $O < O' < O''$, according to the vertex labeling provided by the canonization.

This task recognition is triggered by associating a predicate P_i with each task T_i. More concretely, if the predicates are well-defined, a robot r observing that predicate P_i is true, can executes the corresponding move m_i associated with task T_i. To ensure the correctness of this method, each predicate must satisfy the following properties:

- $Prop_1$: each predicate P_i must be computable based on the configuration C observed during the **Look** phase;
- $Prop_2$: the predicates must be mutually exclusive, i.e., $P_i \wedge P_j = \mathtt{false}$ for all $i \neq j$, ensuring that each robot unambiguously selects a single task;
- $Prop_3$: for every possible configuration C, there must be at least one predicate P_i that is evaluated as true.

To define each predicate P_i, we first identify a set of basic variables that capture relevant properties of the configuration C, e.g., metric, numerical, ordinal, or topological features, that can be computed by each robot using only its local observation. Then, such variables are used to assemble a pre-condition $\mathtt{pre}_i$ that must be satisfied for task T_i to be applicable. Finally, predicate P_i can be defined as:

$$P_i = \mathtt{pre}_i \wedge \neg(\mathtt{pre}_{i+1} \vee \mathtt{pre}_{i+2} \vee \cdots \vee \mathtt{pre}_q).$$

This definition guarantees that $Prop_2$ is always satisfied and imposes a specific order on the task evaluation. Specifically, predicates are evaluated in reverse order: the robot first checks $P_q = \mathtt{pre}_q$, then $P_{q-1} = \mathtt{pre}_{q-1} \wedge \neg\mathtt{pre}_q$, and so on. If all predicates from P_q down to P_2 evaluate to false, then P_1 must be true and task T_1 executed. When a robot performs a generic task T_i in a configuration C, the algorithm may lead to a new configuration C' where another task T_j must be performed. In such a case, we say that the algorihtm induces a transition from T_i to T_j. Collectively, all such transitions form a directed graph called the *transition graph*. The terminal task, which marks the successful resolution of problem $\mathcal{P}$, must correspond to a sink vertex in this graph. As shown in [8], the correctness of an algorithm designed in this way is guaranteed if the following conditions hold:

H_1: the transition graph is correct, i.e., for each task T_i, the reachable tasks are exactly those depicted in the transition graph;

H_2: all the loops in the transition graph, including self-loops not involving sink vertices, must be executed a finite number of times;

H_3: with respect to the studied problem $\mathcal{P}$, no a-priori proved unsolvable configuration is generated by $\mathcal{A}$.

4   Configurations on Non-Vertex-Transitive Graphs

In this section, we consider configurations defined on connected and non-vertex-transitive graphs. In other words, any graph G considered here to define a configuration admits at least two orbits.

Definition 1. *Let G be a non-vertex-transitive graph and O be an orbit of G. We say that O is* terminal *if the following property holds:*

$$\forall u \in V(G) \setminus O, \; \forall v \in O, \; \exists u, v\text{-path } P \text{ such that } P \cap O = \{v\}. \tag{1}$$

Informally, when an orbit O is terminal in G, it is possible to move from any vertex u not in O to any vertex v in O without traversing vertices in O except when reaching v. This is equivalent to saying that removing all but one vertex of O produces a connected induced subgraph of G. Figure 2 shows graphs admitting or not terminal orbits. In the first case, the orbit containing all the pendant vertices is terminal (but the second orbit containing all the remaining vertices is not terminal). In the second example, each side of the complete bipartite graph is a terminal orbit. In the latter example, there are two orbits (one of them formed by the vertices with degree 4), and it is easy to check that none of them is terminal.

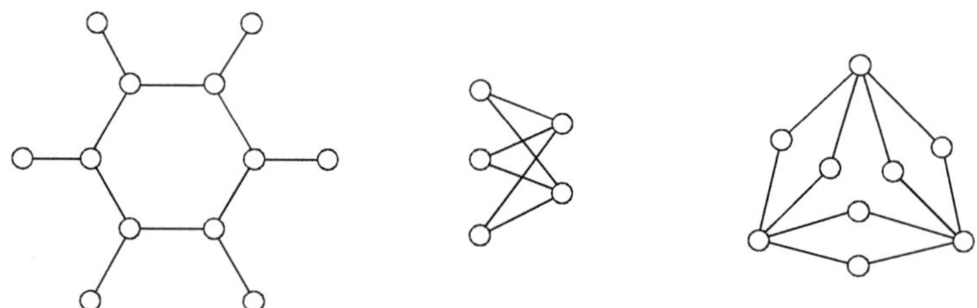

Fig. 2. Three graphs with two orbits each. The first one admits one terminal orbit constituted by the pendant vertices. In the second one, that is a complete bipartite graph $K_{3,2}$, the two orbits are constituted by the two partitions, resp., and they are both terminal. In the last graph, one orbit is constituted by the vertices of degree four and the other by the vertices of degree two, and they are both not terminal (cf. Definition 1).

Let G be a non-vertex-transitive graph, O be its smallest terminal orbit, and $C = (G, \lambda)$ an initial configuration. We define a gathering algorithm $\mathcal{A}_{\mathbf{T}}$ based on just three tasks:

T_1: O has two or more occupied vertices. If the active robot r is in O, then r moves out of O;

T_2: O has one occupied vertex v. If the active robot r is in O, then r does not move. If the active robot r is outside O, then r moves toward v without passing through O before reaching v;

T_3: O has no occupied vertices. The active robot r moves toward an arbitrary vertex v in O.

Theorem 1. *Given a non-vertex-transitive graph G and an initial configuration $C = (G, \lambda)$, if G has terminal orbits, algorithm $\mathcal{A}_\mathbf{T}$ is correct under* RR, *requiring $\Theta(diam(G))$ epochs.*

Proof. We prove that this algorithm solves the gathering problem in $O(diam(G))$ epochs. If at time $t_1 = 0$ task T_1 is executed in C, since the smallest orbit O is always computable by all robots, the algorithm proceeds by keeping T_1 under execution until time $t_2 > t_1$ when the condition that activates T_2 holds. This happens after at most $k-1$ robots move out of O. The whole task then requires at most one epoch. Each robot takes one LCM cycle to do so because G is connected and every vertex in O must admit an edge leading outside O. From time t_2, the configuration evolves according to one or more executions of T_2 until time $t_3 > t_2$, when the gathering will be achieved. This holds because O is terminal, so during the robots' movements toward orbit O, this orbit always remains with only one occupied vertex. In this case, each robot may perform at most $O(diam(G))$ moves, hence the whole process may require at most $O(diam(G))$ epochs. The same analysis holds if T_2 (instead of T_1) is performed at time $t_1 = 0$. Conversely, if T_3 is performed at time $t_1 = 0$, as soon as the first robot reaches O within one epoch, T_2 is performed in the obtained configuration, and therefore gathering is assured. Overall, $\mathcal{A}_\mathbf{T}$ requires $O(diam(G))$ epochs to solve GATHERING. According to Lemma 1, $\mathcal{A}_\mathbf{T}$ is basically time optimal as the initial configuration C might admit $\Delta(C) = diam(G)$. $\qquad\square$

The following statement provides a property that can be used to characterize graphs not admitting terminal orbits.

Theorem 2. *Given a non-vertex-transitive graph G, if it admits a proper subset of orbits whose vertices induce a connected subgraph of G, then G has a terminal orbit.*

Proof. Let $\mathcal{O}_G$ be the set containing all the orbits of G, and let $\mathcal{O}'$ be any proper subset of $\mathcal{O}_G$ whose vertices induce a connected subgraph G' of G. Let $O \in \mathcal{O}_G \setminus \mathcal{O}'$ an orbit with vertices adjacent to vertices in G'. By definition of orbit, any vertex in O is adjacent to a vertex in G'. Since G' is connected, we have that any vertex $v \in O$ can be reached from any vertex in $u \in G' \setminus O$ through a path P such that $P \cap O = \{v\}$, i.e. O is terminal with respect to the graph G'' induced by the vertices in $\mathcal{O}' \cup \{O\}$. Furthermore, we notice that G'' is a connected subgraph of G larger than G'. Hence, any orbit O' not belonging to $\mathcal{O}' \cup \{O\}$ but adjacent to them is terminal with respect to the subgraph induced by the vertices in $\mathcal{O}' \cup \{O, O'\}$. If we keep proceeding in this way, since G is connected, then we conclude that there must exist at least one orbit that is terminal with respect to the entire graph G.   □

Notice that the opposite is not true; in fact, in the complete bipartite graph $K_{m,n}$ with $m \neq n$, there are two orbits, each of which is terminal, but neither orbit induces a connected subgraph. Moreover, the previous theorem can be alternatively expressed as follows: if G is a non-vertex-transitive graph that does not contain any terminal orbit, then every proper subset of orbits of G induces a disconnected subgraph. In particular, *each orbit of G must induce a disconnected subgraph.*

4.1   Algorithm for Graphs Without Terminal Orbits

In this section, we describe $\mathcal{A}_{\neg \mathbf{T}}$, an algorithm able to solve the GATHERING problem in each configuration $C = (G, \lambda)$ defined on a connected non-vertex-transitive graph G *without terminal orbits*. It is worth remarking that, according to Theorem 2, each orbit O of G induces a disconnected subgraph $G[O]$. Moreover, each connected component of $G[O]$ must be vertex-transitive, with vertices adjacent to the vertices of another orbit, since the whole graph G is connected.

Algorithm.   The strategy of $\mathcal{A}_{\neg \mathbf{T}}$ is to confine all robots within a subgraph G' of G that admits a terminal orbit. Hence, applying $\mathcal{A}_{\mathbf{T}}$ for graphs admitting terminal orbits on the configuration restricted to G' would solve the gathering (Task T_4). Subgraph G' is constructed as follows. At each activation, during the Compute phase, a robot applies the graph canonization method in order to obtain an ordered sequence of the orbits of G. Let O_1 be the smallest orbit and O_2 be the first orbit in the sequence that is adjacent to O_1. The subgraph G' is composed of one connected component CC of $G[O_2]$, along with all vertices in O_1 that are adjacent to any vertex in CC. Ideally, in a first phase (Task T_1), all robots are moved to the vertices of O_1. In the second phase (Task $T_{2.i}$), one robot moves from a vertex in O_1 to a vertex in O_2, defining the connected component CC. In the third phase (Tasks $T_{2.ii}$, $T_{2.iii}$), all the remaining robots should move from O_1 to CC.

Naturally, the paths of these robots are not always direct from O_1 to O_2; they may pass through other orbits (even all orbits may be involved). To manage this, our algorithm allows at most one connected component of O_2 to contain occupied vertices, and permits exactly one additional robot to move through the graph to reach CC. However, if the moving robot r is part of a multiplicity (i.e., multiple robots sharing a vertex), all robots in that multiplicity are moved one by one together with r (Task $T_{2.iv}$).

Since there are cases where CC contains just one occupied vertex and the moving robot steps over O_2 as well, there might be some 'confusion' in detecting CC, hence we have a fourth phase (Task $T_{3.i}$) that deals with such special occurrences. Also in this phase, if the moving robot r is on a multiplicity, all the robots in the multiplicity are moved with r one by one (Tasks $T_{3.ii}$ and $T_{3.iii}$).

In what follows, we formalize each of these tasks. Note that, according to the methodology sketched in Sect. 2, we assume that the proposed algorithm $\mathcal{A}_{\neg \mathbf{T}}$ evaluates the preconditions that must hold for each task in the reverse order from pre_4 to pre_1, and performs the first task for which the precondition is true.

Task 1: If none of the preconditions pre_4, pre_3, and pre_2 hold for a given configuration, algorithm $\mathcal{A}_{\neg \mathbf{T}}$ makes all the robots that are not on a vertex of O_1 move toward O_1.
Precondition pre_1: *true*.
Move m_1: any robot not in O_1 moves toward O_1.

Task 2: This task is the core of algorithm $\mathcal{A}_{\neg \mathbf{T}}$. If all robots are in O_1, any robot in O_1 moves toward O_2. If all robots are in O_1 or in a single connected component of $G[O_2]$, called CC, any robot on a vertex of O_1 closest to CC moves toward CC. If there is a single occupied vertex u except those in CC and O_1, which is also closer to CC than any other occupied vertex in O_1, any robot on u moves toward CC. Unfortunately, it could be the case that on u there is a multiplicity, then, after a move from u to say v, both vertices u and v are occupied and the algorithm deals with this case by moving all the remaining robots on u toward v before continuing.
Precondition pre_2: the set of occupied vertices V' is the union of three sets F, U and M such that:
- F is the set of all occupied vertices in O_1, $|F| \geq 0$;
- U is the set of all occupied vertices in a connected component CC in $G[O_2]$;
- M consists of an occupied vertex u, if any, closer to CC than any occupied vertex in F and an occupied vertex u, if present, such that $uv \in E$ and v is on a shortest path from u to CC. In any case, $M \cup F$ and U form a partition of V'.

Furthermore, one of the following conditions holds:
i) $|M| = 0$, $|U| = 0$, $|F| > 0$;
ii) $|M| = 0$, $|U| > 0$, $|F| > 0$;
iii) $M = \{u\}$, $|U| > 0$, $|F| \geq 0$;
iv) $M = \{u, v\}$, $|U| > 0$, $|F| \geq 0$.

Move m_2:
if i), any robot in O_1 moves toward O_2;
if ii), any robot on a vertex of F and closest to CC moves toward CC;
if iii), any robot on u moves toward CC;
if iv), any robot on u moves toward v.

Task 3: This task deals with all the configurations in which the connected component CC of $G[O_2]$, where the robots are supposed to go, could be confused with another component of $G[O_2]$. If this happens, it means there are exactly two connected component in $G[O_2]$ containing robots. We call CC_1 and CC_2 these two components.
Precondition pre_3: the set of occupied vertices is the union of three sets F, U_1, and U_2 such that:

- F is the set of all occupied vertices in O_1, $|F| \geq 0$;
- U_1 (U_2, resp.) consists of an occupied vertex v in CC_1 (CC_2, resp.) closer to CC_2 (CC_1, resp.) than any occupied vertex in F and an occupied vertex u, if present, such that $uv \in E$ and v is on a shortest path from u to CC_2 (CC_1, resp.).

Furthermore, one of the following conditions holds:

i) $|U_1| = 1$, $|U_2| = 1$;

ii) $|U_1| = 1$, $|U_2| = 2$;

iii) $|U_1| = 2$, $|U_2| = 2$.

Move m_3:

if i), any active robot in U_1 (U_2, resp.) moves toward CC_2 (CC_1, resp.);

if ii), any active robot on u moves toward vertex v;

if iii), any active robot on u in U_1 (U_2, resp.) moves toward vertex v.

Task 4: This task deals with a configuration where all the robots are on a subgraph G' of G having a terminal orbit, that is gatherable by applying $\mathcal{A}_\mathbf{T}$, cf. Theorem 1.

Precondition pre_4: all robots are on vertices of a subgraph G' composed by one connected component CC belonging to $G[O_2]$ and all the vertices in O_1 adjacent to vertices in CC.

Move m_4: apply the algorithm designed for terminal configurations confined to G'.

Correctness. With the next lemmas, we prove that condition H_1 holds, that is, the correctness of the transition graph described by Table 1. Then, with Theorem 3, we also prove that condition H_2 holds. With respect to condition H_3, this is satisfied as there are no forbidden configurations.

Table 1. A tabular representation of the transition graph of the algorithm for non-vertex-transitive graphs without terminal orbits.

T_1	T_2, T_3, T_4
$T_{2.i}$	$T_{2.ii}$
$T_{2.ii}$	$T_{2.ii}, T_{2.iii}, T_{2.iv}, T_3$
$T_{2.iii}$	$T_{2.ii}, T_{2.iii}, T_{2.iv}, T_3, T_4$
$T_{2.iv}$	$T_{2.iii}, T_{2.iv}, T_{3.i}, T_{3.ii}, T_4$
$T_{3.i}$	$T_{2.iii}, T_{2.iv}, T_{3.i}, T_{3.ii}, T_4$
$T_{3.ii}$	$T_{3.i}, T_{3.ii}$
$T_{3.iii}$	$T_{3.ii}, T_{3.iii}$
T_4	GATHERING

Lemma 2. *Let C be a configuration in T_1. From C, $\mathcal{A}_{\neg\mathbf{T}}$ leads to a configuration belonging to T_2, T_3, or T_4.*

Proof. According to move m_1, any robot that awakes outside O_1 moves toward O_1. As long as the configuration remains in T_1, robots have to move at most $diam(G)$ hops, hence requiring at most $diam(G)$ epochs. If at some point the obtained configuration does not belong to T_1 anymore, it may concern any other task among T_2, T_3, or T_4. Notice that we are guaranteed that such a transition occurs because if not earlier, it happens once all the robots have reached O_1 and pre_2 with condition i) holds. $\qquad\square$

Lemma 3. *Given a configuration C in T_2, $\mathcal{A}_{\neg\mathbf{T}}$ from C leads to a configuration belonging to $T_{2.ii}$, $T_{2.iii}$, $T_{2.iv}$, T_3 or T_4.*

Proof. According to move m_2, there are four possible cases. When $|U| > 0$, let G' be the subgraph of G consisting of the connected component CC belonging to O_2 and all vertices in O_1 adjacent to vertices in CC.

i) When $|M| = 0$, $|U| = 0$ and $|F| > 0$, then the first robot that awakes moves toward O_2, and the configuration is necessarily in $T_{2.ii}$, as $|M| = 0$, $|U| = 1$ and $|F| > 0$. The configuration achieved cannot be in T_4 since that would imply that C was already in T_4. The configuration achieved cannot be in T_3 since there is just one robot out of O_1.

ii) When $|M| = 0$, $|U| > 0$ and $|F| > 0$ then any robot r in O_1, closest to the only connected component CC occupied by robots in O_2, moves toward it. The move may lead to a configuration C' still in $T_{2.ii}$, if r moves on a vertex of CC. C' might be in $T_{2.iii}$ or $T_{2.iv}$ (if r moves from a multiplicity) or, similarly, in T_3 if the robot moves on a vertex of O_2.
C' cannot be in T_4 since there is at least another robot at the same original distance of r from CC, otherwise $|M| > 0$.

iii) When $M = \{u\}$, $|U| > 0$ $|F| \geq 0$, then any robot on u moves toward CC. If there is only one robot on u, the resulting configuration could remain in $T_{2.iii}$ or lead to a configuration in $T_{2.ii}$, if the robot reaches a vertex of CC. The resulting configuration could be in T_3 if the robot reaches a vertex in O_2, or even in T_4, if the moving robot reaches a vertex in G'. If, instead, there is a multiplicity in u, then the movement may lead to $T_{2.iv}$ with $|M| = 2$, $|U| > 0$ and $|F| \geq 0$.

iv) When $|M| = \{u, v\}$, $|U| > 0$ $|F| \geq 0$, any robot on u moves on v toward CC. If there is a multiplicity on u, the resulting configuration clearly remains in $T_{3.iv}$. If there is a single robot on u, the resulting configuration could be in $T_{2.iii}$, in T_4 (if v is in O_1 and adjacent to a vertex in CC) or even in $T_{3.i}$ or $T_{3.ii}$ (if v is in O_2 and u on a different orbit). $\qquad\square$

Lemma 4. *Given a configuration C in T_3, $\mathcal{A}_{\neg\mathbf{T}}$ from C leads to a configuration belonging to $T_{2.ii}$, $T_{2.iii}$, $T_{2.iv}$, T_3 or T_4.*

Proof. In case C is in $T_{3.i}$, according to move $m_{3.i}$ one robot on a vertex in $U_1 \cup U_2$ moves. Without loss of generality, assume that the active robot is on u and moves toward CC_2. Assume first no multiplicity is on u. Of course, in one step a moving robot cannot reach its target since by definition there are no edges between CC_1 and CC_2. If the reached vertex w is in O_2 the obtained configuration still belongs to $T_{3.i}$, otherwise the configuration is in $T_{2.iii}$, or even in T_4 if u is in G'. If there is a multiplicity in u, the reached configuration still has occupied vertices in two connected components of $G[O_2]$. Then the configuration is in $T_{3.ii}$, or in $T_{2.iv}$ with $M = \{u, w\}$ and $|U| = 1$.

In case C is in $T_{3.ii}$, assume v be the vertex closest to CC_1 and $U_1 = \{u, v\}$. Any active robot on u moves to v. If there is a multiplicity in u, the resulting configuration is clearly still in $T_{3.ii}$. Otherwise, the configuration is in $T_{3.i}$.

In case C is in $T_{3.iii}$, any active robot on u of either U_1 or U_2 moves to v. If there is a multiplicity in u, the resulting configuration is clearly still in $T_{3.iii}$, otherwise it is in $T_{3.ii}$. $\qquad\square$

Lemma 5. *Given a configuration $C = (G, \lambda)$ in T_4, $\mathcal{A}_{\neg \mathbf{T}}$ from C solves* GATH-ERING *in at most $O(|V(G)|)$ epochs.*

Proof. Let G' be the subgraph of G induced by the unique connected component CC occupied by robots in O_2 along with each neighboring vertex in O_1. It is easy to observe that G' can be considered non-vertex-transitive by exploiting the distinction among vertices in O_1 and O_2. Moreover, there are vertices in O_1 that belong to a terminal orbit of G'. By considering the sub-configuration $C' = (G', \lambda)$, according to Theorem 2 we can use algorithm $\mathcal{A}_{\neg \mathbf{T}}$ to solve the GATHERING problem in G' (all robots are gathered in a vertex belonging to O_1). By Theorem 1, the algorithm requires $O(diam(G'))$ epochs, which in turn is bounded by $O(|V(G)|)$. $\qquad\square$

Theorem 3. *Given a non-vertex-transitive graph G with no terminal orbits and an initial configuration $C = (G, \lambda)$, the* GATHERING *problem can be solved form C under the* RR *scheduler in $O(occ(C) \cdot \Delta(C) + |V(G)|)$ epochs.*

Proof. By Table 1, clearly Task T_1 can be performed only from an initial configuration since no transition leads to it. By move m_1, each robot is moved on a vertex of O_1 tracing a path of length at most $\Delta(C)$. Then, the task is performed within $\Delta(C)$ epochs. Similarly, Task $T_{2.i}$, where all robots are in O_1 and one of them moves to O_2, is performed at most once. This task requires only one move. Furthermore, also Task T_4 is performed at most once. By Lemma 5, it requires at most $|V(G)|$ epochs. The overall number of epochs required by the above tasks is then bounded by $O(|V(G)|)$.

It remains to analyze the other cases, that is, configurations in $T_{2.ii}$, $T_{2.iii}$, $T_{2.iv}$ or T_3. Note that, if from one of the tasks $T_{2.ii}$, $T_{2.iii}$ and $T_{3.i}$, a robot r moves from a multiplicity and the achieved configuration is in one of the tasks among $T_{2.iv}$, $T_{3.ii}$ and $T_{3.iii}$, then, by the respective moves $m_{3.ii}$, $m_{3.iii}$ and $m_{2.iv}$, all the robots in the multiplicity reach r within one epoch. So, in what follows, we consider only configurations in $T_{2.ii}$, $T_{2.iii}$ and $T_{3.i}$, without considering the transient configurations in $T_{2.iv}$, $T_{3.ii}$ and $T_{3.iii}$. On the one hand, it allows us to analyze the movement of a multiplicity like a move of a single robot. On the other hand, we assume that moving a robot to an adjacent vertex requires one epoch, since it could be the case it moved from a multiplicity.

We now show that the total number B of robots in at most two components of $G[O_2]$ always increases.

Let us analyze the case in which no configuration in T_3 is generated, that is, a connected component CC of $G[O_2]$, possibly generated after tasks T_1 or $T_{2.i}$, or present in the initial configuration, is always recognizable.

Assume that the configuration is $T_{2.ii}$, that is, all occupied vertices are in O_1 or in CC. Then, a robot r on a vertex v closest to CC moves toward CC. From now on, the generated configuration is in $T_{2.iii}$, until a configuration in T_3 or T_4 is generated or the moving robot reaches CC. In conclusion, in a configuration of tasks $T_{2.ii}$ and $T_{2.iii}$, a moving robot r and all robots in the same multiplicity of r, reach a configuration in $T_{2.ii}$, T_3 or T_4 in $diam(G)$ epochs. Meanwhile, the number B of robots in CC (or in CC_1 and CC_2 for a new configuration in T_3) increases by at least one.

It remains to discuss the case in which the configuration is in T_3 that is, it is not possible to distinguish a single connected component in $G[O_2]$. Recall that, if C is in $T_{3.ii}$ or in $T_{3.iii}$, the repeated application of the corresponding moves eventually generates a configuration in $T_{3.i}$ within one epoch. Meanwhile, the connected components CC_1 and CC_2 do not change.

Then, consider a configuration C in $T_{3.i}$ and let B be the number of robots in CC_1 and CC_2. Without loss of generality, let r_1 be the activated robot in CC_1 that moves toward CC_2. The obtained configuration is then in $T_{2.iii}$ as soon as r_1 moves outside CC_1. Now, from the above discussion, a configuration in T_4 is generated, or again a configuration in $T_{3.i}$ with the occupied vertex in CC_1 closer to CC_2 then in C. Symmetrically, let r_2 be the robot in CC_2 eventually activated and moving toward CC_1. Then, either the two robots meet in a single component CC or a configuration in T_4 is generated. This requires at most $\Delta(C)$ epochs. The total number of robots B in CC_1 and CC_2 is not increased, but the obtained configuration, if not in T_4, is in now in $T_{2.ii}$ or $T_{2.iii}$ and from there, B will increase again.

In summary, moving the robots from a vertex occupied by $t \geq 1$ robots from a configuration in T_2 or T_3 to a new configuration in T_2 or T_4 requires $O(\Delta(C))$ epochs, and if the resulting configuration is in T_2, value B increases by t. Since value B cannot be greater than k (i.e., it cannot increase more than k), the algorithm converges to a configuration in T_4, eventually. The total number of epochs required for moving all robots from the initial occupied vertices of configurations in T_2 and T_3 is then bounded by $O(occ(C) \cdot \Delta(C))$.

In conclusion, the overall number of epochs to achieve the gathering is $O(occ(C) \cdot \Delta(C) + |V(G)|)$. $\qquad\qquad\square$

5   Concluding Remarks and Future Work

We have approached the GATHERING problem for a swarm of robots moving on non-vertex-transitive graphs under a Round Robin scheduler.

We have designed two general resolution algorithms dealing with configurations admitting or not terminal orbits, respectively. In particular, we have proposed a time-optimal algorithm, $\mathcal{A}_{\mathbf{T}}$, designed for graphs that contain a terminal orbit. For graphs that lack terminal orbits, we have designed an algorithm, $\mathcal{A}_{\neg\mathbf{T}}$, that for any initial configuration $C = (G, \lambda)$ guarantees the GATHERING within $O(occ(C) \cdot \Delta(C) + |V(G)|)$ epochs. It remains open whether it is possible to provide a time optimal resolution algorithm with respect to the basic lower bound of $\Omega(\Delta(C))$ provided by Lemma 1.

As a research direction for future work, we aim to investigate the GATHERING problem in vertex-transitive graphs.

References

1. Babai, L.: Canonical form for graphs in quasipolynomial time: preliminary report. In: Charikar, M., Cohen, E. (eds.) Proceedings of the 51st Annual ACM SIGACT Symposium on Theory of Computing, STOC 2019, pp. 1237–1246. ACM (2019). https://doi.org/10.1145/3313276.3316356
2. Bonnet, F., Potop-Butucaru, M., Tixeuil, S.: Asynchronous gathering in rings with 4 robots. In: Mitton, N., Loscri, V., Mouradian, A. (eds.) ADHOC-NOW 2016. LNCS, vol. 9724, pp. 311–324. Springer, Cham (2016). https://doi.org/10.1007/978-3-319-40509-4_22
3. Bose, K., Kundu, M.K., Adhikary, R., Sau, B.: Optimal gathering by asynchronous oblivious robots in hypercubes. In: Gilbert, S., Hughes, D., Krishnamachari, B. (eds.) ALGOSENSORS 2018. LNCS, vol. 11410, pp. 102–117. Springer, Cham (2019). https://doi.org/10.1007/978-3-030-14094-6_7
4. Cicerone, S., Di Fonso, A., Di Stefano, G., Navarra, A.: Gathering of robots in butterfly networks. In: Masuzawa, T., Katayama, Y., Kakugawa, H., Nakamura, J., Kim, Y. (eds) Stabilization, Safety, and Security of Distributed Systems. SSS 2024. LNCS, vol. 14931, pp. 106–120. Springer (2024). https://doi.org/10.1007/978-3-031-74498-3_7
5. Cicerone, S., Di Stefano, G., Navarra, A.: Asynchronous robots on graphs: gathering. In: Flocchini, P., Prencipe, G., Santoro, N. (eds.) Distributed Computing by Mobile Entities, Current Research in Moving and Computing, LNCS, vol. 11340, pp. 184–217. Springer (2019). https://doi.org/10.1007/978-3-030-11072-7_8
6. Cicerone, S., Di Stefano, G., Navarra, A.: On gathering of semi-synchronous robots in graphs. In: Ghaffari, M., Nesterenko, M., Tixeuil, S., Tucci, S., Yamauchi, Y. (eds.) SSS 2019. LNCS, vol. 11914, pp. 84–98. Springer, Cham (2019). https://doi.org/10.1007/978-3-030-34992-9_7
7. Cicerone, S., Di Stefano, G., Navarra, A.: Gathering robots in graphs: the central role of synchronicity. Theor. Comput. Sci. **849**, 99–120 (2021). https://doi.org/10.1016/j.tcs.2020.10.011
8. Cicerone, S., Di Stefano, G., Navarra, A.: A structured methodology for designing distributed algorithms for mobile entities. Inf. Sci. **574**, 111–132 (2021). https://doi.org/10.1016/j.ins.2021.05.043

9. D'Angelo, G., Di Stefano, G., Klasing, R., Navarra, A.: Gathering of robots on anonymous grids and trees without multiplicity detection. Theor. Comput. Sci. **610**, 158–168 (2016)
10. D'Angelo, G., Di Stefano, G., Navarra, A.: Gathering asynchronous and oblivious robots on basic graph topologies under the look-compute-move model. In: Alpern, S., Fokkink, R., Gąsieniec, L., Lindelauf, R., Subrahmanian, V. (eds.) Search Theory, pp. 197–222. Springer, New York (2013). https://doi.org/10.1007/978-1-4614-6825-7_13
11. D'Angelo, G., Di Stefano, G., Navarra, A.: Gathering on rings under the look-compute-move model. Distrib. Comput. **27**(4), 255–285 (2014)
12. D'Angelo, G., Di Stefano, G., Navarra, A.: Gathering six oblivious robots on anonymous symmetric rings. J. Discrete Algorithms **26**, 16–27 (2014)
13. D'Angelo, G., Di Stefano, G., Navarra, A., Nisse, N., Suchan, K.: Computing on rings by oblivious robots: a unified approach for different tasks. Algorithmica **72**(4), 1055–1096 (2015)
14. D'Angelo, G., Navarra, A., Nisse, N.: A unified approach for gathering and exclusive searching on rings under weak assumptions. Distrib. Comput. **30**(1), 17–48 (2017)
15. D'Emidio, M., Di Stefano, G., Frigioni, D., Navarra, A.: Characterizing the computational power of mobile robots on graphs and implications for the Euclidean plane. Inf. Comput. **263**, 57–74 (2018)
16. Di Stefano, G., Navarra, A.: Gathering of oblivious robots on infinite grids with minimum traveled distance. Inf. Comput. **254**, 377–391 (2017)
17. Di Stefano, G., Navarra, A.: Optimal gathering of oblivious robots in anonymous graphs and its application on trees and rings. Distrib. Comput. **30**(2), 75–86 (2017)
18. Flocchini, P., Prencipe, G., Santoro, N. (eds.): Distributed Computing by Mobile Entities, Current Research in Moving and Computing. Lecture Notes in Computer Science, vol. 11340. Springer, Heidelberg (2019). https://doi.org/10.1007/978-3-030-11072-7
19. Flocchini, P., Prencipe, G., Santoro, N. (eds.): Distributed Computing by Oblivious Mobile Robots. Synthesis Lectures on Distributed Computing Theory. Morgan & Claypool Publishers (2012)
20. Frei, F., Wada, K.: Brief announcement: distinct gathering under round robin. In: 38th International Symposium on Distributed Computing (DISC 2024). LIPIcs, vol. 319, pp. 48:1–48:8. Schloss Dagstuhl – Leibniz-Zentrum für Informatik (2024). https://doi.org/10.4230/LIPIcs.DISC.2024.48
21. Guilbault, S., Pelc, A.: Gathering asynchronous oblivious agents with local vision in regular bipartite graphs. Theor. Comput. Sci. **509**, 86–96 (2013)
22. Izumi, T., Izumi, T., Kamei, S., Ooshita, F.: Time-optimal gathering algorithm of mobile robots with local weak multiplicity detection in rings. IEICE Transactions **96-A**(6), 1072–1080 (2013)
23. Kamei, S., Lamani, A., Ooshita, F., Tixeuil, S., Wada, K.: Asynchronous gathering in a torus. In: 25th International Conference on Principles of Distributed Systems (OPODIS). LIPIcs, vol. 217, pp. 9:1–9:17. Schloss Dagstuhl - Leibniz-Zentrum für Informatik (2021)
24. Klasing, R., Kosowski, A., Navarra, A.: Taking advantage of symmetries: gathering of many asynchronous oblivious robots on a ring. Theor. Comput. Sci. **411**, 3235–3246 (2010)
25. Klasing, R., Markou, E., Pelc, A.: Gathering asynchronous oblivious mobile robots in a ring. Theor. Comput. Sci. **390**, 27–39 (2008)

26. Navarra, A., Piselli, F.: Oblivious robots under round robin: gathering on rings. In: Chau, V., Dürr, C., Li, M., Lu, P. (eds.) IJTCS-FAW 2025. LNCS, vol. 15828, pp. 166–180. Springer, Singapore (2025). https://doi.org/10.1007/978-981-96-8312-3_13
27. Ooshita, F., Tixeuil, S.: On the self-stabilization of mobile oblivious robots in uniform rings. In: Richa, A.W., Scheideler, C. (eds.) SSS 2012. LNCS, vol. 7596, pp. 49–63. Springer, Heidelberg (2012). https://doi.org/10.1007/978-3-642-33536-5_6

Optimal Dispersion of Silent Robots in a Ring

Bibhuti Das[1]([✉]), Barun Gorain[1], Kaushik Mondal[2],
Krishnendu Mukhopadhyaya[3], and Supantha Pandit[4]

[1] Indian Institute of Technology Bhilai, Bhilai, India
dasbibhuti905@gmail.com, barun@iitbhilai.ac.in
[2] Indian Institute of Technology Ropar, Rupnagar, India
kaushik.mondal@iitrpr.ac.in
[3] Indian Statistical Institute, Kolkata, India
krishnendu.mukhopadhyaya@gmail.com
[4] Dhirubhai Ambani Institute of Information and Communication Technology,
Gandhinagar, India

Abstract. Given a set of co-located mobile robots in an unknown anonymous graph, the robots must relocate themselves in distinct graph nodes to solve the *dispersion* problem. In this paper, we consider the *dispersion* problem for *silent* robots, i.e., no direct, explicit communication between any two robots. The robots are deployed on the nodes of an oriented n node ring network. They operate in synchronous rounds. The *dispersion* problem for *silent* mobile robots has been studied in arbitrary graphs where the robots start from a single source. In this paper, we focus on the *dispersion* problem for *silent* mobile robots where robots can start from multiple sources. The robots have unique labels from a range $[0, L]$ for some positive integer L. Any two co-located robots do not have the information about the label of the other robot. The robots have *weak multiplicity detection* capability, which means they can determine if it is alone on a node. The robots are assumed to be able to identify an increase or decrease in the number of robots present on a node in a particular round. However, the robots can not get the exact number of increase or decrease in the number of robots. We have proposed a deterministic distributed algorithm that solves the *dispersion* of k robots in an oriented ring in $O(\log L + k)$ synchronous rounds with $O(\log L)$ bits of memory for each robot. A lower bound $\Omega(\log L + k)$ on time for the dispersion of k robots on a ring network is presented to establish the optimality of the proposed algorithm.

Keywords: Dispersion · Ring Network · Mobile Robots · Multiple Sources · Lower Bound · Deterministic Algorithm

1 Introduction

In distributed computing, the *dispersion* problem for mobile robots in a graph has recently become very popular. The goal of the *dispersion* problem is starting

© The Author(s), under exclusive license to Springer Nature Switzerland AG 2026
S. Bonomi et al. (Eds.): SSS 2025, LNCS 16350, pp. 171–191, 2026.
https://doi.org/10.1007/978-3-032-11127-2_15

from one or multiple source nodes; a set of mobile robots must be placed in the graph nodes by ensuring that no two robots are co-located. The *dispersion* problem was first introduced by Augustine and Moses Jr. [2] for a set of mobile robots. They investigated the problem by focusing on minimizing the memory required by each robot and the number of rounds needed to achieve dispersion in an arbitrary graph. Over the years, the *dispersion* problem has been well-studied in the literature over various models [1,9,14–16,18,19,24]. As the *dispersion* problem aims at distributing a group of robots it has got the potential for several practical applications. One such application is related to electric vehicles and charging stations. To reduce the waiting time of a vehicle in a charging station, the cars can be spread over the charging stations by ensuring that each charging station is occupied by one car.

The *dispersion* problem is closely related to several other well-studied problems on a graph network such as exploration [6,7,10], scattering [3,11,20,21,23], load balancing [4,8,22,25,27] etc. Dispersion of robots is very similar to the robot scattering or uniform-deployment problem on graphs [3,11,20,21,23], both of which require the spreading of robots uniformly in the graph. The dispersion problem is closely related to the exploration problem where n robots start at a given node and must explore the graph in as few rounds as possible [6,7,10]. For the dispersion problem, if the number of robots is equal to the number of nodes, then every possible solution to the dispersion problem would also solve the n robot exploration under the same set of conditions. The dispersion problem can be viewed as a variant of load balancing on graphs [4,8,22,25,27], wherein the nodes usually begin with an arbitrary amount of load and are required to transfer load using edges until each node has nearly the same amount.

1.1   Motivations

Our work is motivated by the result of gathering by Bouchard et al. [5]. The authors solved the problem of gathering robots without communication capability. Gorain et al. [12] investigated the dispersion problem on arbitrary graphs considering the same model described in the paper [5]. They assumed that a robot could identify the following: (1) whether it is alone on a node, (2) whether the number of robots changes at the node compared to the previous round. They proposed a deterministic algorithm that achieves dispersion on any arbitrary graph in time $O(k \log L + k^2 \log \Delta)$ where Δ is the maximum degree of a node in the graph. Each robot uses $O(\log L + \log \Delta)$ additional memory, starting from a single source. The memory $O(\log L + \log \Delta)$ required in [12] optimal. However, the authors did not comment on the optimality of the dispersion time in this model. We investigate the dispersion problem with silent robots on the class of cycles and proposed an optimal solution even if the mobile robots were initially placed in multiple source nodes.

1.2 Model and Problem Definition

We consider a set of k robots located on the nodes of an anonymous $n \geq k$ node ring denoted by $\mathscr{R} = (V,\ E)$. The nodes of the graphs are anonymous. However, the edges incident to a node v are labeled 0 and 1. The ring is oriented, i.e., the robots have a global notion of clockwise and counter-clockwise directions. As a consequence, the labeling of the ports has the same orientation. For a node v, the adjacent node connected through port 0 is said to be a predecessor of v and denoted by $pre(v)$. Similarly, the adjacent node connected through port 1 is said to be a successor of v and denoted by $succ(v)$. In an initial rooted configuration, all the robots are co-located at a single node. In this paper, multiple nodes in $\mathscr{R}$ can initially contain more than one robot (a multiplicity), known as an un-rooted configuration. There is no restriction on the initial placement of such multiplicity nodes.

The robots move in synchronous rounds, and a robot can traverse at most one edge in every round. Each round is comprised of two different stages. In the first stage, the robot at a node v completes all the local computations. In the second stage, based on the computations done in the previous stage, a robot either moves along one of the edges incident to v or stays at v. When two robots choose to move in the same round along the same edge from opposite endpoints, neither robot is capable of detecting the other's movement. In other words, a robot can detect the presence of other robots only at nodes in the graph, not at edges.

Every mobile robot has a unique ID in the $[0,\ L]$ range, where $L \geq k$. A mobile robot is aware of its ID and the value of L but is uninformed of the identities of other robots. The robots are silent, i.e., no two robots can directly communicate. However, at any round, a robot is capable of detecting the following local activities:

1. The robots have *weak multiplicity detection* capability. If the robot is present at a node v in a round and no other robot is co-located with it, then it can detect that it is alone in this round. A local variable $alone = true$ represents the robot's detection of this activity.
2. If the robot stays at a node v in a round t, and the number of robots co-located at v increases at the end of round t, then the robot can detect in round $t+1$ that the number of co-located robots increased in the previous round. A local variable $increase = true$ represents the detection of this activity by the robot in round $t+1$. If no such activity is detected, then $increase = false$.
3. If the robot stays at a node v in a round t, and the number of robots co-located at v decreases at the end of round t, then the robot can detect in round $t+1$ that the number of co-located robots increased in the previous round. A local variable $decrease = true$ represents the detection of this activity by the robot in round $t+1$. If no such activity is detected, then $decrease = false$.

Note that the robots can not identify the exact number of co-located robots that increased or decreased. Also, if there is an equal number of arriving and leaving robots at node v, then the robot M can not detect anything, and both the variables *increase* and *decrease* will be set as false in the next round.

Let $MaxSize$ denote the length of the binary representation of the maximum ID of a robot, which is L. As L is known to all the robots, $MaxSize = \lfloor \log L \rfloor + 1$ is a global knowledge. Without loss of generality, we may assume that the IDs of each robot are of size $MaxSize$. Otherwise, every robot can append the necessary numbers of zeros in the front of its binary label to make the size of the label equal to $MaxSize$.

1.3   Our Contributions

This work considers the dispersion problem starting from multiple sources for silent robots. Our proposed algorithm works without direct, explicit communication among the robots other than identifying the local variables *alone*, *increase*, and *decrease* as explained in Sect. 1.2.

1. Our proposed algorithm solves the dispersion problem within $O(\log L + k)$ synchronous rounds with $O(\log L)$ bits of memory for each robot.
2. We have shown that dispersion of silent mobile robots requires at least $\Omega(\log L + k)$ synchronous rounds on a ring.

The main idea behind our algorithm is to separate a group of co-located robots based on the bits in the binary representation of their labels [5,12]. Since each robot has a distinct ID, at least one bit of two labels differs, and hence, if a robot moves clockwise for bit 1 and stays for bit 0, the robots with different bits will be separated. Once two robots get separated, it is not very difficult to ensure that they remain separated in the subsequent rounds. However, in a multi-source scenario, the above approach imposes various challenges. One of the main challenges is while getting separated from its source group by moving clockwise, a group of robots may end up merging with a group of robots with different sources, and it may be possible that none of these two groups identify this merging due to lack of explicit communication. Our main contribution to this paper is resolving this difficulty using very limited local information.

2   Related Work

Augustine and Moses Jr. [2] were the first to introduce the *dispersion* problem. The authors investigated the problem with the assumption that the number of robots (say k) is equal to the number of vertices (say n). The problem has been investigated in paths, rings, trees, and arbitrary graphs. They proved that, for any graph G of diameter D, any deterministic algorithm must take $\Omega(\log k)$ bits of memory by each robot and $\Omega(\log D)$ number of rounds. For arbitrary graphs, they provided an algorithm that requires $O(m)$ rounds where each robot requires $O(n \log n)$ bits of memory. For paths, rings, and trees, they proposed algorithms for each kind of graph such that each robot requires $O(\log n)$ bits of memory and takes $O(n)$ rounds. Furthermore, their proposed algorithm for rooted trees takes $O(D^2)$ rounds, and each robot requires $O(\Delta + \log n)$ bits of memory.

In the paper, [13], Kshemkalyani and Ali proposed five different *dispersion* problems on graphs. Their first three algorithms require $O(k \log \Delta)$ bits for each robot and $O(m)$ steps running time, where m is the number of edges and Δ is the degree of the graph. These three algorithms differ in whether they use the synchronous or the asynchronous system model and in what, where, and how data structures are maintained. Their fourth algorithm considers the asynchronous model and uses $O(D \log \Delta)$ bits of memory at each robot and $O(\Delta^D)$ rounds, where D is the graph diameter. Their fifth algorithm under the asynchronous model uses $O(\max(\log k, \log \Delta))$ bits memory at each robot and $O((m - n)k)$ rounds. In [14], Kshemkalyani et al. provided a deterministic algorithm in arbitrary graphs in a synchronous model that requires $O(\min(m, k\Delta) \log k)$ rounds and $O(\log n)$ bits of memory by each robot.

The *dispersion* problem was first studied on a grid graph by Kshemkalyani et al. [16]. They provided two deterministic algorithms: (i) for the local communication model (where robots can only communicate with other robots that are present at the same node.) and (ii) for the global communication model (where a robot can communicate with any other robot in the graph possibly at different nodes (but the graph structure is not known to robots)). For the local communication model, they proposed an algorithm that requires $O(\min(k, \sqrt{n}))$ rounds, and each robot uses $O(\log k)$ bits of memory. For the local communication model, their algorithm requires $O(\sqrt{k}))$ rounds, and each robot uses $O(\log k)$ bits of memory. An extension of this work on general graphs is considered by Kshemkalyani et al. [15]. They proposed an algorithm for arbitrary graphs using DFS traversal that requires $O(\min(m, k\Delta))$ rounds and uses $\Theta(\log(\max(k, \Delta)))$ bits memory at each robot. They also proposed a BFS traversal-based algorithm for arbitrary graphs that requires $O(\max(D, k)\Delta(D + \Delta))$ rounds and uses $O(\max(D, \Delta \log k))$ bits memory at each robot. Further, they proposed an algorithm for arbitrary trees using BFS traversal that requires $O(D \max(D, k))$ rounds and uses $O(\max(D, \Delta \log k))$ bits memory at each robot. Kshemkalyani et al. [17] presented a multi-source DFS traversal algorithm solving dispersion on arbitrary anonymous graphs in $O(\min(m, k\Delta))$ rounds and using $O(\log(k + \Delta))$ bits at each robot. Sudo et al. [26], presented significantly faster algorithms for dispersion in both the rooted and the general settings than the known fastest algorithm in the literature. First, an algorithm for the rooted setting that solves the dispersion problem in $O(k \log \min(k, \Delta)) = O(k \log k)$ time using $O(\log(k + \Delta))$ bits of memory per agent is proposed. Next, an algorithm for the general setting that achieves dispersion in $O(k \log k . \log \min(k, \Delta)) = O(k \log^2 k)$ time using $O(log(k + \Delta))$ bits is proposed. Finally, for the rooted setting, a time-optimal (i.e., $O(k)$-time) algorithm with $O(\Delta + \log k)$ bits of space per agent is presented.

Agarwalla et al. [1] explored the *dispersion* problem on dynamic rings. In the paper [19], Molla et al. introduced fault-tolerant in *dispersion* problem in a ring in the presence of Byzantine robots.

Molla et al. [18] used randomness in the *dispersion* problem and proposed an algorithm assuming $O(\log \Delta)$ bits of memory for each of the robots. For any randomized algorithm for the *dispersion* problem, a matching lower bound of $\Omega(\log \Delta)$ bits is presented. The authors also investigated a generalized version of dispersion, namely k-dispersion problem where $k > n$ robots must be dispersed over n nodes by ensuring that at most $\dfrac{k}{n}$ robots are placed at each node. Das et al. [9] provided an optimal algorithm such that each robot uses $O(\log \Delta)$ bits of memory.

3   Dispersion from Multiple Sources

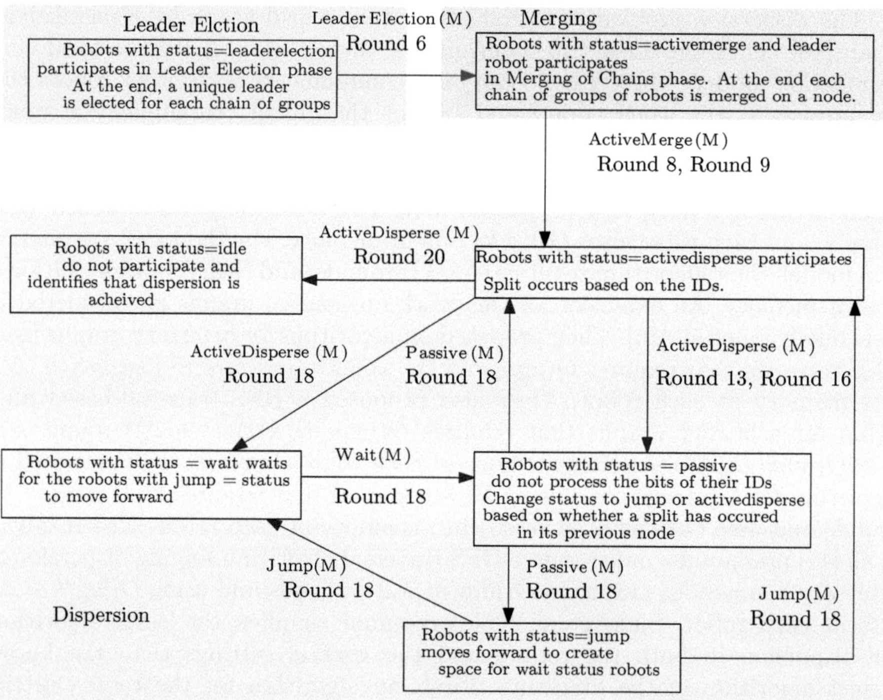

Fig. 1.   Flowchart showing the major activities during execution of *AlgorithmMultiStart*.

3.1   High Level Idea

Before going to the details of the proposed algorithm, as a warm-up, we present an informal description of dispersing mobile robots from a single source in an oriented ring. Initially, all the robots are located at the source node s. Since the robots have distinct IDs, in the binary representation of the IDs of two robots, from right to left, at least one bit must be different. Based on this fact, the robots process the bits of the binary representations of the IDs one by one from right to left, and for each processed bit by a robot with a value of 1, the robot moves clockwise one step. If the bit is 0, the robot does not move. Hence, two robots, having different j^{th} bits, will be separated while processing the jth bit of their IDs. Therefore, the set of robots will be separated into two groups: the group of robots with j^{th} bit 0 stays in the previous node, and the group of robots with the jth bit 1 moves clockwise 1 step. We call such an event a *split*. There may be an integer j such that all the robots have the same j^{th} bit. Hence, split only happens if at least two robots have different j^{th} bit. Thus, if the length of the binary representation of the maximum ID is ℓ when ℓ bits are processed, every robot will be separated from each other.

In order to implement the above idea, we must ensure that if two groups of robots are separated once, they must not be co-located again. If this happens, there may be two robots, one from each of these two separated groups, for which the bits in the rest of the IDs are the same. Hence, they will not be separated using the above process.

In order to achieve this, we 'activate' alternate groups in a particular phase. Specifically, among two consecutive groups, exactly one group of robots processes their bits in a particular phase, and the next group remains inactive. As a consequence, it would ensure that the split happens only in one group, and the other group of robots can detect this split. If a split occurs, the next group, which was inactive, moves one step forward in the next phase to create a space for this new group that has been created due to the split in the previous phase. In this way, no two groups, which have been separated before due to a split, will ever be merged again. Hence, a natural approach is to activate all the groups at the odd distance from the source node in one phase and, in the next phase, activate all the groups that are at an even distance from the source node.

There is another important reason why alternate groups are active in a particular phase. If we allow the robots to be active in two adjacent nodes simultaneously, the robots may not identify a split. For example, consider two groups of robots G_1 and G_2 present on adjacent nodes v_1 and v_2, respectively. Suppose the robots on both nodes become active at the same round, processing the jth bit. Next, consider the following two different scenarios:

1. No split happened at the nodes v_1 and v_2. For both groups, the jth is 0 for all the robots.
2. A split happened at the nodes v_1 and v_2. Consider the case when a split happened for both groups such that the number of incoming robots is exactly equal to the outgoing robots at the node v_2.

However, the robots of G_2 can not possibly distinguish between the above two cases. In particular, the robots of G_2 can not identify the split for the second case. To proceed, the robots must check the $(j+1)^{th}$ bit for a possible split for the first case. In doing so, robots from G_1 and G_2 co-located on v_2 would start processing the $(j+1)^{th}$ bit without realizing they are from different groups. As discussed earlier, there may be two robots that might not be separated in the future.

The main difficulty in implementing the above idea for the un-rooted case is determining which group will be active or inactive in a particular phase (A phase is a collection of rounds. The details will be defined later). In the rooted case, in the odd phases, groups positioned at odd distances from the source are active, and in the even phases, even distanced groups are active. In case of arbitrary start, having multiple sources, two consecutive groups (from two different sources) may become active in the same phase, which creates a problem in identifying a split. To avoid such a scenario, we follow the steps described below.

- We ensure that no two groups are consecutive initially before they start processing the bits of their labels. To do so, first, we merge consecutive groups into a single group. This is done in the following way. Every group tries to elect a leader robot by processing the bits of their labels. After every split, the robots in each group will move to the predecessor node to check whether a robot becomes alone or not due to the split. Since only the first group of robots of a chain of consecutive groups have an empty neighbor in the backward direction, only this group can elect the leader successfully. (The robots execute subroutine $LeaderElection(M)$ to achieve this task)
- Once the leader is elected, the leader moves forward to visit all the groups of robots until it finds an empty node. When visited by a robot in a specific round, the other groups understand this as a signal to merge and merge with the last group in the chain. (The robots execute subroutine $ActiveMerge(M)$ to achieve this task)
- Once the merging is done, the algorithm executes as explained in the rooted case. However, the leader, who was previously elected, moves forward in each phase to ensure that an empty node is present to expand the chain, which is now created due to splits. Whenever the leader finds another group of agents ahead, the group of agents synchronizes their relative positions (odd or even) with respect to the previous group (The robots execute subroutine $ActiveDisperse(M)$ to achieve this task).

As discussed in the above three cases, the robot's movements are used to 'communicate' different activities that are needed to achieve dispersion. For example, a robot may identify $increase = true$ in a particular round while the leader election algorithm is executed. On the other hand, a group of robots in a chain may find $increase = true$ in the merging phase after the leader election is done. These two events are different for the robots. If robots cannot differentiate between these two events, there can be serious chaos, and the algorithm may not be correctly executed as desired. In order to avoid such ambiguity, we allocate unique 'slots' for different activities. To be specific, we divide time into phases of 20 consecutive rounds. In each phase, each of these 20 rounds has a dedicated task. A robot finding $increase = true$ in the ith round in a particular slot means different than a robot finding $increase = true$ in the jth round of the same slot for $i \neq j \in [1, 20]$. The role of each round in a particular phase is explained in Sect. 3.2.

The details of the above process are explained later while describing the algorithm. A flowchart showing the major activities during execution of $AlgorithmMultiStart$ is presented in Fig. 1.

Chain of Groups: Let G_i denote the group of robots on the node v_i. A set of group of robots $C_j = \{G_1, G_2, \ldots, G_p\}$ for some $j, p > 0$ is said to form a chain of groups if $\forall i \in \{2, \ldots, p-1\}$, $pred(v_i) = v_{i-1}$ and $succ(v_i) = v_{i+1}$.

A chain of groups refers to consecutive groups of robots without any empty nodes. Initially, there can be multiple chains of groups. While executing our proposed algorithm $AlgorithmMultiStart$, all the robots belonging to the same chain of groups are merged on a single node. Later on, the robots would form a chain of groups of robots due to a split of robots by the execution of $AlgorithmMultiStart$.

Table 1. Description of Variables

Variables	Values	Descriptions
$move$	0, 1, 2	The variable $move$ is reset to 0 before the start of each phase. During the execution of $DispersionMultiStart$, the variable can be updated to its permitted values from the set $\{0, 1, 2\}$ based on the robot movement
$proceed$	0, 1, 2	The variable $proceed$ is not reset to 0 like the variable $move$ before the start of each phase. During the execution of $DispersionMultiStart$, the variable can be updated to its permitted values from the set $\{0, 1, 2\}$ based on the robot movement
$start$	0, 1	Let a robot M be at the node v, and w be the predecessor of v. If all the robots up to w are $idle$ and M is a possible candidate that may become $idle$ at the beginning of its next phase, then M would update the variable $start$ to 1. Note that the $start$ variable is not reset to 0 at the beginning of a phase
$settle$	0, 1	Let a robot M be at the node v, and w be the predecessor of v. If all the robots up to w are $idle$, and M is alone at the node v, then the $settle$ is updated to 1
$alone$	true, false	If a robot M is the only robot that lies on a node, it updates its $alone$ variable to $true$. Otherwise, the $alone$ variable is set to $false$
$increase$	true, false	Initially, $increase$ is set to $false$ for all the robots. If the number of robots co-located at node u increases in round t, then for all the robots present at node u, the variable $increase$ is set to $true$ at the beginning of round $t + 1$. Otherwise, $increase = false$
$decrease$	true, false	Initially, $decrease$ is set to $false$. If the number of robots co-located at the node u decreases in round t, the variable $decrease$ is set to $true$ at the beginning of round $t + 1$. Otherwise, $decrease = false$
$leader$	true, false	Initially, the variable $leader$ is set to $false$. If a robot M identifies as a leader, the variable $leader$ is set to $true$
$election$	true, false	Initially, the variable $election$ is set to $false$. If a leader is elected, the variable $election$ is set to $true$
$advance$	true, false	Initially, the variable $advance$ is set to 0 for all the robots. If the variable $leader = true$, then at $MaxSize + 1$ rounds, a leader M updates the variable $advance$ to 1

During the execution of $DispersionMultiStart$, a robot checks its status and variables to decide its next action. All the robots have the following variables:

move, *advance*, *leader*, *election*, *start*, *settle*, *increase*, *decrease*, and *alone* stored in its memory (Table 1).

Each robot have a variable called *status* which can be either *leaderelection*, *activemerge*, *activedisperse*, *passive*, *wait*, *jump*, *or idle*. Initially, all the robots have the status *leaderelection*. A robot M with status $X \in \{activemerge, activedisperse, passive, wait, jump, idle\}$ calls the procedure $X(M)$ at the start of any phase. During the execution of a subroutine $X(M)$, if a robot with status X updates its status to Y, then it continues the execution of $X(M)$ until the current phase completes and in the next phase it starts executing $Y(M)$. The proposed algorithm runs in phases where each phase consists of 20 synchronous rounds. Based on the status an robot M executes either $LeaderElection(M)$ $ActiveMerge(M)$, $ActiveDisperse(M)$, $Passive(M)$, $Wait(M)$, and $Jump(M)$. A robot with status *idle* never participates in any algorithm.

3.2   Description of *AlgorithmMultiStart*

In this section, we present a round-wise description of a particular phase of our proposed algorithm *AlgorithmMultiStart*. The time is divided into consecutive phases, each of which contains 20 synchronous rounds. By round j for $1 \leq j \leq 20$, we mean the jth round of a particular phase.

Rounds Designated for Leader Election: In a particular phase, rounds 1–6 are designated for the leader election step. Initially, all the robots have *status = leaderelection* and $i = 1$. A leader is elected for each chain of groups by the execution of *LeaderElection*. A robot with *status = leaderelection* participates in rounds 1–6. For the rest of the rounds, a robot with *status = leaderelection* does not participate. Once a leader is elected, all the robots in a chain update *election = true*. However, the robots wait until $MaxSize$ phases to update *status = activemerge* to ensure that all the robots start merging chains synchronously.

- **Round 1:** In this round, only the robots with **status = leaderelection** participates. The robots process the ith bit of $l(M)$. If the ith bit of $l(M)$ is 1 and *election = false*, the robot M updates the variable *proceed = 1* (By checking the variable *proceed*, a robot with *status = leaderelection* can identify that it has moved one step forward in round 1), and moves through port 1. Otherwise, the robot M stays at the current node. A split will occur if at least two robots have different i^{th} bit. If *alone = true* and *proceed = 0*, then the robot updates *leader = true*.
- **Round 2:** In this round, only the robots with **status = leaderelection** participates. If a split happened in round 1, then the robots that did not move (By checking whether the variable *proceed = 0*) can identify *decrease = true* in this round. As a consequence, they become aware of the split in round 1. However, the robots that moved forward in round 1 (*proceed = 1*) can not possibly identify the split because they identify neither decrease nor increase. To inform them about the split, the robots with *status = leaderelection*,

$proceed = 0$, and $decrease = true$ update $proceed = 2$ and move through port 1 in this current round.

- **Round 3:** In this round, only the robots with $status = leaderelection$ participates. Due to the movement of robots with $status = leaderelection$, $proceed = 0$, and $decrease = true$ in round 2, the robots with $proceed = 1$ (moved in round 1) identify $increase = true$. As a consequence, they become aware of the split in round 1. The robots with $proceed = 1$ and $increase = true$, together with the robots with $proceed = 2$, move through port 0 (they return to the node where the split happened). If a robot with $proceed = 1$ identifies $increase = false$, then it becomes aware that a split did not happen in round 1 as the ith bit was 1 for all the robots. In such a scenario ($status = leaderelection$, $proceed = 0$ and $decrease = false$), the robots move through port 0 and update $proceed = 0$.
- **Round 4:** In this round, only the robots with $status = leaderelection$ participates. If $proceed = 1$, then the robots move through port 0. Note that according to the definition of a chain of groups, there exists at least one node (say v_1) whose predecessor node is empty. As all the robots have distinct IDs, for the robots in v_1, there exists one phase in which only a single robot with $proceed = 1$ would move to the empty node in round 4. The robots with $proceed = 2$ would remain at the current node till $MaxSize$ rounds (They can be viewed as a disqualified candidate for being a leader).
- **Round 5:** In this round, only the robots with $status = leaderelection$ participates. If a robot with $proceed = 1$ identifies $alone = true$, it updates $leader = true$, $election = true$ and moves through port 1.
- **Round 6:** In this round, only the robots with $status = leaderelection$ participates. If a robot identifies $alone = false$, $increase = true$ and $proceed = 0$, it updates $election = true$. In case a robot identifies $alone = false$ and $proceed = 1$, they move through port 1 and update $proceed = 0$. If $i = MaxSize$, the robots terminate the leader election by updating $status = activemerge$. Else, they update $i = i + 1$.

Rounds Designated for Merging of Chains before Dispersion: Rounds 7–9 are designated for merging the groups in a chain. Each chain of groups starts the merging process independently by the execution of $ActiveMerge$. Robots with $status = activemerge$ participate in rounds 7–9. For the rest of the rounds, robots with $status = activemerge$ do not participate. Once the merging is done, all the robots, including the leader, update $status = activedisperse$ and $i = 1$.

- **Round 7:** In this round, only the robots with $status = activemerge$ participates. If a robot identifies $leader = true$ and $status = activemerge$, it moves through port 1 to start the merging. It continues to move forward in this current round until it finds an empty node.
- **Round 8:** In this round, only the robots with $status = activemerge$ participates. In this round, the leader checks whether it has moved to an empty node. If $alone = true$ and $leader = true$, the leader identifies that the merging is complete. It moves through port 0 and terminates the merging by updating $status = activedisperse$ and $i = 1$.

- **Round 9:** In this round, only the robots with $status = activemerge$ participates. In this round, a non-leader robot discovers whether the leader has returned or not in round 8. If a robot identifies $leader = false$ and $increase = false$, it becomes aware that the merging is incomplete. It moves through port 1 to continue merging. In case $leader = false$ and $increase = true$, a robot identifies that the merging is complete and the leader has returned. It updates $status = activedisperse$ and $i = 1$ and terminates merging.

Rounds Designated for Dispersion: Rounds 10–20 are designated for the dispersion of robots. The robots with $status = activedisperse$, $status = passive$, $status = wait$, and $status = jump$ participate. Each robot with $status = X$ executes subroutine $X(M)$.

- **Round 10:** In this round, only a leader with status **activedisperse** or **passive** participates. If the leader identifies $advance = 0$ and $alone = false$, it moves through port 1 (to check whether the successive node is empty) and updates $advance = 1$.
- **Round 11:** In this round, only a leader with status **activedisperse** or **passive** participates. If the leader identifies $advance = 1$ and $alone = true$, it moves through port 1 (to check whether the successive node is empty).
- **Round 12:** In this round, only a leader with status **activedisperse** or **passive** participates. If the leader identifies $advance = 1$ and $alone = true$, it identifies that it has moved to an empty node. It moves through port 0 and updates $advance = 0$. The leader becomes aware of the fact that the successive node is empty.
- **Round 13:** In this round, only a non-leader with status **activedisperse** or **passive** participates. If a non-leader robot identifies $increase = true$ due to the movement of a leader in round 12, it moves through port 0 and updates $status = passive$.
- **Round 14:** In this round, only a robot with status **activedisperse** participates. If it identifies $alone = true$ and $start = 0$, then it updates $start = 1$. In case $alone = true$ and $start = 0$, then a robot updates $settle = 1$ (To remember that it might become $idle$). If the last bit of $l(M)$ is 1, then M updates $move = 1$ (to identify that it has moved in round 14) and moves through port 1. If at least two robots have different last bits, a split would occur.
- **Round 15:** In this round, only a non-leader with status **activedisperse** or **jump** participates. If a robot with status $activedisperse$ identifies $decrease = true$, it becomes aware of the split in round 14. To inform about the split, the robot updates $move = 2$ (to remember that it moved in round 15 to inform about the split) and moves through port 1. If the robot identifies $status = jump$, it moves through port 1 to create space for the robots that arrived in the previous phase.
- **Round 16:** In this round, only a non-leader with status **activedisperse** or **passive** participates. In case a robot with $status = activedisperse$ identifies

$move = 0$, it realizes that no split happened in round 14 as all the robots had the last bit 0. It updates $status = passive$. Otherwise, if $move = 1$ and $increase = false$, the robot becomes aware that no split happened in round 14 as all the robots had the last bit 1. In case $move = 2$, it identifies that a split has occurred, and it moves to inform about the split. It moves through port 0 and updates $status = passive$. If a robot with $status = passive$ identifies $increase = true$ due to a split in round 14, it sets $move = 1$ and becomes aware that it must vacate the current node.

- **Round 17:** In this round, only a non-leader with status **passive** participates. If a robot with $status = passive$ identifies $move = 1$, it realizes that it must vacate the current node by informing the incoming group that it must wait. To inform the incoming group, the robot moves through port 0.

- **Round 18:** In this round, only a non-leader with status **activedisperse** or **passive** or **jump** or **wait** participates. If a robot with $status = activedisperse$ identifies $move = 1$ and $decrease = true$, it realizes that it has moved to an occupied node. It updates $status = wait$. If a robot with $status = activedisperse$ identifies $move = 1$ and $decrease = false$, it realizes it has moved to an empty node. It updates $status = activedisperse$. If a robot with $status = passive$ identifies $move = 0$, then it $updates = activedisperse$. Consider the case when a robot with $status = passive$ identifies $move = 1$. It learns that a split happened, and it has moved in round 17 to inform about the split. It updates $status = jump$ and moves through port 1. If a robot with $status = jump$ identifies $decrease = false$, it learns that there were no robots in the current node when it jumped from the predecessor node. It updates $status = activedisperse$. Consider the case when a robot with $status = jump$ identifies $decrease = true$. It becomes aware of the fact that it jumped to an occupied node. It updates $status = wait$. if a robot has $status = wait$, it updates $status = passive$.

- **Round 19:** In this round, only a robot with status **activedisperse** participates. If a robot with $status = activedisperse$ identifies $settle = 1$, then it moves through port 1 (to inform its successive robots it will become $idle$).

- **Round 20:** In this round, only a robot with status **activedisperse** and **passive** participates. If a robot with $status = activedisperse$ identifies $settle = 1$, then it moves through port 0 and updates $status = idle$. If a robot with $status = passive$ identifies $increase = true$, then it learns that the robot in the predecessor node will become $idle$, and the robot has arrived to inform the same. It updates $set = 1$ (To remember that the predecessor robot has become $idle$).

4   Correctness of *DispersionMultiStart*

The following two lemmas prove that exactly one leader will be elected for every chain after executing the procedure *LeaderElection*.

Lemma 1. *For some $1 \leq j \leq MaxSize$, consider the execution of $LeaderElection(M)$. Let A_j be the set of robots at a node v for which proceed $= 0$ and B_j be the set of robots for which proceed $= 2$ at the beginning of Phase j. Also, let $A_j(0)$ (similarly $B_j(0)$) and $A_j(1)$ (similarly $B_j(1)$) be the set of robots with j-th bit 0 and 1, respectively, in A_j (similarly in B_j). If $|A_j| > 1$, then*

1. $A_{j+1} = A_j$ and $B_{j+1} = B_j$, when $A_j(1) = A_j$ or $A_j(0) = A_j$.
2. $A_{j+1} = A_j(1)$ and $B_{j+1} = B_j \cup A_j(0)$, otherwise.

Lemma 2. *Let $C_j = \{G_1, G_2, \ldots, G_p\}$ for some $j, p > 0$ be a chain of groups in $\mathcal{R}$ where G_i is a group of robots at node v_i. If there are $m > 1$ number of robots at the node v_1, then after $MaxSize$ phases, there would be a unique leader for C_j.*

Proof. Proof by contradiction. We claim that $A_{MaxSize} = 1$. To prove this, suppose that $|A_{MaxSize}| > 1$. This implies that there exist two robots M_1 and M_2 such that after $MaxSize$ rounds, $M_1, M_2 \in A_{MaxSize}$. By Lemma 1, they have the exact same bits for all $j \in \{1, 2, \ldots, MaxSize\}$. This is a contradiction to the unique labels of the robots.

Next, we show that only a robot M from the group G_1 will find $alone = true$ in round 5 (steps 15–16) of the execution of $LeaderElection(M)$ for some j and no robot from any other group will find $alone = true$ in the execution of $LeaderElection(M)$ for any j. To prove this for the robots in G_1, let $|A_j| = 1$ and $M \in A_j$. During the execution of $LeaderElection(M)$, this robot, in round 1, moves to $succ(v_1)$ and updates $proceed = 1$. In the round 2, the robots of $A_j(0)$ would move to $succ(v_1)$ and update $proceed = 2$. In the round 3, the robots of $A_j(0) \cup A_j(1)$ move back to the node v. In the round 4, the robot M move to $pred(v_1)$ and the robots in $A_j(0)$ stay at the node v_1. In round 5, the robot M identifies $alone = true$ and updates $leader = true$.

Since a robot of a group G_i can set $leader = true$ in the 5th round by identifying $alone = true$ at the node $pred(v_i)$ in the execution of $LeaderElection$, it is enough to show that for any robot $M \in G_t$, for $t > 1$, the node v_{t-1} contains at least one robot of the group G_{t-1} in round 4 for every execution of $LeaderElection(M)$, $j = 1, \ldots, MaxSize$. If this is true, then any robot of G_t which is present at v_{t-1} identifies $alone = false$ and hence never sets $leader = true$.

For some $M \in G_{t-1}$, consider the execution of $LeaderElection(M)$ for some $k > 1$. If $\exists M$ at v_{t-1} for which $proceed = 2$, then this robot will remain at v_{t-1} for all the successive phases till $MaxSize$ phases. As a consequence, any robot of G_t present at v_{t-1} will not be able to identify $alone = true$ in round 5. Otherwise, we have the following three possible cases:

Case 1. $\forall M \in G_{t-1}$, the kth bit of $l(M)$ is 0: We have $A_k(0) = A_k$ and $A_k(1) = \emptyset$. All the robots would stay at node v_{t-1} in round 1. The robots would identify $decrease = false$ and stay at node v_{t-1} in round 2. In round 3, the robots would identify $increase = false$, but as $proceed = 0$, they would stay at the current node. The robots would stay at the current node as $proceed = 0$ in rounds 4–6. The robots would stay at the current node for the rest of the rounds.

Case 2. $\forall M \in G_{t-1}$, the kth bit of $l(M)$ is 1: We have $A_t(1) = A_t$ and $A_t(0) = \emptyset$. All the robots would move to $succ(v)$ in round 1. The robots would stay at $succ(v_{t-1})$ in round 2. In round 3, the robots would identify $proceed = 1$ and $increase = false$. The robots would update $proceed = 0$ and move to v_{t-1}. For the rest of the rounds, the robots will stay at v_{t-1}.

Case 3. $\exists M_1, M_2 \in G_{t-1}$, the kth bit of $l(M_1)$ is 0 and the kth bit of $l(M_2)$ is 1: We have $A_k = A_k(0)$ and $B_k = A_k(1)$. In the round 1, the robots of $A_k(1)$ would move to $succ(v)$ and update $proceed = 1$. In the round 2, the robots of $A_k(0)$ would move to $succ(v)$ and update $proceed = 2$. In the round 3, the robots of $A_k(0) \cup A_k(1)$ would move back to the node v. In round 4, the robots of $A_k(0)$ would move to $pred(v_{t-1})$, and the robots of $A_k(1)$ would stay at the node v_{t-1}. In the round 5, the robots of $A_k(0)$ would return to v_{t-1} and update $proceed = 0$. The robots will stay at the current node for the rest of the rounds.

The above three cases show that at the beginning of round 5, during the execution of $LeaderElection(M, k)$, the node v_{t-1} will contain at least one robot from G_{t-1}. Thus, any robot of G_t present at v_{t-1} will not be able to identify $alone = true$ in round 5.

The following lemma proves that every chain merges to a single group in the chain.

Lemma 3. *Let $C_j = \{G_1, G_2, \ldots, G_p\}$ for some $j, p > 0$ be a chain of groups in $\mathscr{R}$ where G_i is a group of robots at node v_i. After the execution of $ActiveMerge$, the chain would be merged into a single group of robots (say G) within $O(p)$ synchronous rounds at the node v_p.*

The following lemma shows that the robots from two different chains remain separated and merged into different groups before they start executing $ActiveDisperse$.

Lemma 4. *Let $C_i = \{H_1, H_2, \ldots, H_q\}$ and $C_j = \{G_1, G_2, \ldots, G_p\}$ be two chain of groups such that $u_q = pred(w_1), w_1 = pred(w_2), \ldots w_l = pred(v_1)$ where $n - 2 \geq l \geq 1$. During the execution of subroutines $LeaderElection$ and $ActiveMerge$, no two robots from different chains will be co-located.*

Observation 1. *If a robot M updates $status = ActiveDisperse$ from $status = ActiveMerge$ at the end of phase t, then for all the successive phases, M would never update its status to $ActiveMerge$.*

Observation 2. *If a group of robots G_i located at node v_i updates $status = ActiveDisperse$ in phase t, the robots will occupy at most one vacant successive node in phase $t' > t$ when a split occurs.*

According to our proposed algorithm $ActiveMerge$, for two different chains, the merging of the groups in the respective chains will start in the same phase. However, since the length of the chains may be different, the merging may be completed in different rounds. Hence, one may think that when a group (after merging) starts *ActiveDisperse* and expands by splitting into smaller groups, this expansion may meet the next group that is still merging. The next lemma shows that this event can not happen. To be specific, when a group of robots, while executing *ActiveDisperse*, meet the next group, by that time, this next group must have started executing *ActiveDisperse*.

Lemma 5. *Let $C_i = \{H_1, H_2, \ldots, H_q\}$ and $C_j = \{G_1, G_2, \ldots, G_p\}$ be two chain of groups such that $u_q = pred(w_1), w_1 = pred(w_2), \ldots, w_l = pred(v_1)$ where $n - 2 \geq l \geq 1$. If two robots, one from C_i and one from C_j, ever become adjacent in some round i, then none of them can have status $= activemerge$ in round i.*

Remark 1. Let $M \in G_i$ be a robot of some chain C_j that is present at a node v_i at the start of any phase t. Consider the case when M with $status = activedisperse$ or $status = passive$ identifies $increase = true$ due to the incoming of a leader from a previous chain (say C_i). In round 13, M would move to $pred(v_i)$ and update $status = passive$ in phase t. By moving to $pred(v_i)$ it becomes a part of the chain of groups C_i. For the incoming of a leader for each of the previous chain of groups, a robot M would update $status = passive$. Thus, there can be at most k number of such updates of $status = passive$. Otherwise, M would either stay at v_i or move to $succ(v_i)$ as discussed below:

1. M with $status = activedisperse$ would execute *ActiveDisperse*. During any execution of*ActiveDisperse*, there are two possible cases: a split happens, or no split happens. In both cases, the robots would only move to $succ(v_i)$.
2. M with $status = passive$ would remain at v_i by the execution of *Passive*.
3. M with $status = jump$ would execute *Jump*. The robots would move to $succ(v_i)$.
4. M with $status = waiting$ would remain at v_i and wait for other robots to move forward.

The correctness of our algorithm depends on the fact that two consecutive groups of robots will not simultaneously split into smaller groups. This fact is proved in the next lemma.

Lemma 6. *During the execution of the algorithm ActiveDisperse, at the beginning of some phase t, let w_1 and w_2 be two consecutive occupied nodes. Let M_1 be a robot at w_1 and M_2 be a robot at w_2. Then, exactly one of the following statements is true.*

a. *The status of M_1 is in $\{activedisperse, wait, jump\}$ and the status of M_2 is passive.*
b. *The status of M_2 is in $\{activedisperse, wait, jump\}$ and the status of M_1 is passive.*

Definition 1. *Two robots are called distinguished in a phase; if they are either not co-located at the beginning of a phase or if co-located, then their status is not the same.*

The following lemma says that once two robots become distinguished, they remain distinguished in the subsequent phases.

Lemma 7. *If two robots are distinguished in some phase t, they remain distinguished in each phase $t' > t$.*

Proof. Let M_1 and M_2 be two robots co-located at the node v_0 with j^{th} bit 0 and 1, respectively. Suppose M_1 and M_2 are separated by processing the jth bit in phase t by the execution of *activedisperse*. In the phase $t+1$, M_1 is at v_o and M_2 is at $succ(v_0)$. As M_1 and M_2 are not co-located at the same node, they are distinguished at phase $t+1$. Next, consider the case when M_1 and M_2 become co-located for some phase $t' > t+1$ at a node (say v_i). As the non-leader robots move only forward by one step in each phase, the robot M_1 must have moved from the node $pred(v_i)$ to v_i in the phase $t'-1$. Thus, the robots at $pred(v_i)$ have either *status = activedisperse* or a subset containing M_1 have *status = jump* in the phase $t' - 1$. By Lemma 6, the robots at v_i have *status = passive* in the phase $t' - 1$. In the phase t', all the group of robots which moved to v_i would have *status = wait* (including M_1). The robots which were at the node v_i would have *status = jump*. Thus, M_1 and M_2 are distinguished in the phase t'. Hence, if two robots are distinguished in some phase t, they remain distinguished in each phase $t' > t$.

The dispersion is eventually achieved by splitting larger groups into smaller groups of robots, which is done by processing the bits of the labels of the robots during subroutine *ActiveDisperse*. Hence, we have to make sure that sufficient numbers of *Activedisperse* must be called during the execution of our algorithm. The following lemma proves the fact that the robots with a status other than *activedisperse* eventually change their status to *activedisperse* and hence guarantees the splitting as mentioned above.

Lemma 8. *If in some phase, a robot has status $\in \{passive, jump, wait\}$, then within a finite number of phases, it changes its status to activedisperse.*

Lemma 9. *A robot, once changes its status to activedisperse for the first time, can have status wait in at most k phases and status jump in at most $2k$ phases.*

Now onwards, our technical results will converge in proving the final result that proves that dispersion is achieved and it is achieved in $O(MaxSize + k)$ rounds.

Lemma 10. *If a robot $M_1 \in G$ in some chain C has updated status $=$ activedisperse for MaxSize number of phases, then for any other robot $M_2 \in G$, M_1 and M_2 must be distinguished.*

Proof. If possible, let M_1 and M_2 remain co-located, and both have the same status after $MaxSize$ number of phases. As all the robots have unique IDs, $\exists j$, $1 \leq j \leq MaxSize$ such that the robots M_1 and M_2 have distinct j^{th} bit. Consequently, they separated by processing the jth bit at some phase, say t.

In phase $t + 1$, M_1 and M_2 would not be co-located. Thus, they would become distinguished in phase $t + 1$. By Lemma 7, they would remain distinguished for all phases $t' > t$. A contradiction. Hence, if a robot $M_1 \in G$ in some chain C has updated $status = activedisperse$ for $MaxSize$ number of phases, then for any other robot $M_2 \in G$, M_1 and M_2 must be distinguished.

Lemma 11. *A robot can have a passive status for at most $MaxSize + 2k$ phases.*

The final result is stated in the following theorem that proves that dispersion is achieved.

Theorem 1. *Within $O(MaxSize + k)$ rounds after the start of Algorithm MultiStart, no two robots on the ring are co-located.*

Proof. Initially, all the robots have $status = leaderelection$. From Lemma 2, it follows that within $MaxSize$ rounds, a unique leader would be selected for each chain of groups. Lemma 3 ensures that a chain of length p would be merged within $O(p)$ rounds. Lemma 4 ensures that no two robots from different chains would be co-located during the execution of $ActiveMerge$ and $LeaderElection$. By Observation 6, it follows that if a robot updates $status = activedisperse$ from $status = activemerge$, then for all the successive phases, the robot would never update its status to $activemerge$. From Lemma 5, it follows that if two robots from different chains become adjacent, none of them can have $status = activemerge$. From Lemma 6, robots with $status = passive$ would be guaranteed to be present in alternate nodes. Lemma 7 ensures that once two robots become distinguished, they will remain distinguished for all the successive nodes. Lemma 8 guarantees that if in some phase, a robot has status $\in \{passive, jump, wait\}$, then within a finite number of phases, it changes its status to $activedisperse$. It is guaranteed by Lemma 10 that if a robot $M_1 \in G$ in some chain C has updated $status = activedisperse$ for $MaxSize$ number of phases, then for any other robot $M_2 \in G$, M_1 and M_2 must be distinguished. By the Lemma 9, it follows that if a robot once changes its status to $activedisperse$ for the first time, it can have $status = wait$ in at most k phases and $status = jump$ in at most $2k$ phases. Lemma 11 shows that a robot can have $status = passive$ for at most $MaxSize + 2k$ phases. Hence, within $O(MaxSize + k)$ rounds after the start of $Algorithm MultiStart$, no two robots on the ring are co-located. Since $MaxSize = \lfloor \log L \rfloor + 1$, within $O(\log L + k)$ rounds after the start of $Algorithm MultiStart$, no two robots on the ring will be co-located.

5   Lower Bound

In this section, we show a lower bound $\Omega(k + \log L)$ on time for the dispersion of k robots starting from a single node of a ring. Since there are k robots, at least one must reach a node at a distance at least $\lceil \frac{k}{2} \rceil$ from the starting node. Since a robot can travel at most one edge in a round, at least $\lceil \frac{k}{2} \rceil$ many rounds are needed before dispersion is achieved. This proves the lower bound $\Omega(k)$ on time of dispersion. Therefore, showing a lower bound $\Omega(\log L)$ is enough. The following lemma shows a lower bound $\Omega(\log L)$ on time of dispersion.

Lemma 12. *Any algorithm that solves dispersion on a ring must take $\Omega(\log L)$ rounds.*

Proof. We prove this lemma by contradiction. Suppose that there exists an algorithm $\mathcal{A}$ that solves the problem of dispersion in at most y where $y \leq \frac{\log L}{2}$ rounds. At any time $t > 0$, a robot has three possible choices according to $\mathcal{A}$: (i) move clockwise from the current node, (ii) move counter-clockwise from the current node, and (iii) stay at the current node. Define a vector $X_i = (x_i^1, x_i^2, \ldots, x_i^y)$ for the robot M_i such that,

$$
x_i^j = \begin{cases} -1, & \text{if the robot } M_i \text{ moves clockwise at round } j \\ 1, & \text{if the robot } M_i \text{ moves counter-clockwise at round } j \\ 0, & \text{if the robot } M_i \text{ does not move at round } j \end{cases}
$$

As all the robots are initially placed at the same node of the ring, if for any two robots M_i and M_j, $X_i = X_j$, then these robots stay co-located even after y rounds. There are at most $3^{\frac{\log L}{2}} < L$ different possible vectors of length y. Also, there are at most L different possible labels the set of robots may receive initially. Therefore, by Pigeon hole principle, there exist two integers $L_1, L_2 \leq L$ such that if the two robots have labels L_1 and L_2, the respective movement vectors are the same. This shows that even after round y, the robots with labels L_1 and L_2 remain co-located. This contradicts the fact that dispersion can be solved within y rounds.

From Lemma 12, along with the lower bound $\Omega(k)$, we summarize the desired result in the following theorem.

Theorem 2. *Any algorithm that solves dispersion on a ring must take $\Omega(\log L + k)$ rounds.*

6   Conclusions

We have presented a deterministic solution for the dispersion problem in an oriented ring, allowing multiple initial points. Our proposed algorithm solves the dispersion problem within $O(\log L + k)$ rounds. We have presented a lower bound $\Omega(\log L + k)$ on time for the dispersion of k robots on a ring. The assumption of a common orientation has a significant impact on the results. It would be interesting to study the problem in an un-oriented ring. Another assumption is about the knowledge of L, which has helped design our proposed algorithm. An algorithm without the assumption of knowledge of L can be explored in the future. The dispersion problem with multiple sources can be investigated for other kinds of graphs, namely cactus, grids, torus, and general graphs.

Acknowledgement. The first author acknowledges the support by IIT Bhilai Innovation and Technology Foundation under the PostDoc fellowship scheme.

References

1. Agarwalla, A., Augustine, J., Moses Jr., W.K., Madhav, S.K., Sridhar, A.K.: Deterministic dispersion of mobile robots in dynamic rings. In: ICDCN, pp. 19:1–19:4 (2018)
2. Augustine, J., Moses Jr., W.K.: Dispersion of mobile robots: a study of memory-time trade-offs. In: ICDCN, pp. 1:1–1:10 (2018)
3. Barrière, L., Flocchini, P., Barrameda, E.M., Santoro, N.: Uniform scattering of autonomous mobile robots in a grid. Int. J. Found. Comput. Sci. **22**(3), 679–697 (2011)
4. Berenbrink, P., Friedetzky, T., Zengjian, H.: A new analytical method for parallel, diffusion-type load balancing. J. Parallel Distrib. Comput. **69**(1), 54–61 (2009)
5. Bouchard, S., Y., Pelc, A.: Want to gather? No need to chatter! In: PODC, pp. 253–262 (2020)
6. Brass, P., Cabrera-Mora, F., Gasparri, A., Xiao, J.: Multirobot tree and graph exploration. IEEE Trans. Robot. **27**(4), 707–717 (2011)
7. Brass, P., Vigan, I., Xu, N.: Improved analysis of a multirobot graph exploration strategy. In: ICARCV, pp. 1906–1910 (2014)
8. Cybenko, G.: Dynamic load balancing for distributed memory multiprocessors. J. Parallel Distrib. Comput. **7**(2), 279–301 (1989)
9. Das, A., Bose, K., Sau, B.: Memory optimal dispersion by anonymous mobile robots. In: Mudgal, A., Subramanian, C.R. (eds.) CALDAM 2021. LNCS, vol. 12601, pp. 426–439. Springer, Cham (2021). https://doi.org/10.1007/978-3-030-67899-9_34
10. Dereniowski, D., Disser, Y., Kosowski, A., Pajak, D., Uznanski, P.: Fast collaborative graph exploration. Inf. Comput. **243**, 37–49 (2015)
11. Elor, Y., Bruckstein, A.M.: Uniform multi-agent deployment on a ring. Theor. Comput. Sci. **412**(8–10), 783–795 (2011)
12. Gorain, B., Mandal, P.S., Mondal, K., Pandit, S.: Collaborative dispersion by silent robots. J. Parallel Distrib. Comput. 104852 (2024)
13. Kshemkalyani, A.D., Ali, F.: Efficient dispersion of mobile robots on graphs. In: ICDCN, pp. 218–227 (2019)
14. Kshemkalyani, A.D., Molla, A.R., Sharma, G.: Fast dispersion of mobile robots on arbitrary graphs. In: ALGOSENSORS, pp. 23–40 (2019)
15. Kshemkalyani, A.D., Molla, A.R., Sharma, G.: Dispersion of mobile robots in the global communication model. In: ICDCN, pp. 12:1–12:10 (2020)
16. Kshemkalyani, A.D., Molla, A.R., Sharma, G.: Dispersion of mobile robots on grids. In: WALCOM, pp. 183–197 (2020)
17. Kshemkalyani, A.D., Sharma, G.: Near-optimal dispersion on arbitrary anonymous graphs. J. Comput. Syst. Sci. **152**, 103656 (2025)
18. Molla, A.R., Moses Jr., W.K.: Dispersion of mobile robots: the power of randomness. In: TAMC, pp. 481–500 (2019)
19. Molla, A.R., Mondal, K., Moses Jr., W.K.: Efficient dispersion on an anonymous ring in the presence of weak byzantine robots. In: ALGOSENSORS, pp. 154–169 (2020)

20. Mondal, M., Chaudhuri, S.G.: Uniform scattering of robots on alternate nodes of a grid. In: Proceedings of the 23rd International Conference on Distributed Computing and Networking, pp. 254–259 (2022)
21. Poudel, P., Sharma, G.: Fast uniform scattering on a grid for asynchronous oblivious robots. In: Devismes, S., Mittal, N. (eds.) SSS 2020. LNCS, vol. 12514, pp. 211–228. Springer, Cham (2020). https://doi.org/10.1007/978-3-030-64348-5_17
22. Sauerwald, T., Sun, H.: Tight bounds for randomized load balancing on arbitrary network topologies. In: 2012 IEEE 53rd Annual Symposium on Foundations of Computer Science, pp. 341–350. IEEE (2012)
23. Shibata, M., Mega, T., Ooshita, F., Kakugawa, H., Masuzawa, T.: Uniform deployment of mobile agents in asynchronous rings. In: PODC, pp. 415–424 (2016)
24. Shintaku, T., Sudo, Y., Kakugawa, H., Masuzawa, T.: Efficient dispersion of mobile agents without global knowledge. In: SSS, pp. 280–294 (2020)
25. Subramanian, R., Scherson, I.D.: An analysis of diffusive load-balancing. In: Proceedings of the Sixth Annual ACM Symposium on Parallel Algorithms and Architectures, pp. 220–225 (1994)
26. Sudo, Y., Shibata, M., Nakamura, J., Kim, Y., Masuzawa, T.: Near-linear time dispersion of mobile agents. In: 38th International Symposium on Distributed Computing (2024)
27. Xu, C.-Z., Lau, F.C.M.: Analysis of the generalized dimension exchange method for dynamic load balancing. J. Parallel Distrib. Comput. 16(4), 385–393 (1992)

Self-stabilizing Mutual Exclusion in Dynamic Networks with Bounded Temporal Diameter

Stéphane Devismes[1], Swan Dubois[2], François Malenfer[2], Franck Petit[2(✉)], and Mouna Safir[3]

[1] Université de Picardie Jules Verne, MIS, Amiens, France
[2] Sorbonne Université, CNRS, LIP6, DELYS, Paris, France
`franck.petit@sorbonne-universite.fr`
[3] ENS de Lyon, CNRS, Inria, Université Claude Bernard Lyon 1, LIP, Lyon, France

Abstract. We consider distributed systems subject to frequent topological changes. Specifically, we assume the network topology evolves as a dynamic graph in which, at any point in time, the temporal distance between any two processes is at most Δ. Under a synchronous message-passing model where processes have unique identifiers and know both Δ and an upper bound N on the number of processes n, we provide a distributed self-stabilizing mutual exclusion algorithm working in that class of dynamic graphs. Our solution stabilizes in $\mathcal{O}(\Delta.N)$ rounds using bounded local memories. Moreover, it achieves optimal waiting time: once stabilized, the maximum delay before a process enters its critical section is at most $n-1$ rounds. Our algorithm is actually a composition of several self-stabilizing building blocks that respectively achieve Leader Election, Unison, and Ranking. We also provide original self-stabilizing solutions for the latter two problems; for the self-stabilizing leader election, we use a solution given by Altisen *et al.* (Theoretical Computer Science, 2023).

1 Introduction

Usually, distributed systems are modeled as static, undirected, connected graphs, where vertices map to processes and edges represent bidirectional communication links between pairs of processes. Clearly, such modeling is no longer suitable for modern networks such as Mobile Ad-Hoc Networks, Vehicular Ad-Hoc Networks, Disruption-Tolerant Networks, Wireless Sensor Networks, the Internet of Things, . . . Indeed, these networks are subject to frequent topological changes over time due to their mobility and instability.

Various models have been introduced to describe how topological changes occur over time. These models are designed to reflect the inherently dynamic

This work has been partially supported by the ANR project SkyData (`ANR-22-CE25-0008`).

nature of real-world systems by explicitly considering temporal features such as the appearance and disappearance of links, node mobility, and communication delays; refer to [9–11] for more details. In [7], the authors introduced the *evolving graphs* with the aim of unifying many of the previous approaches. They proposed modeling the time as a sequence of discrete instants and the system dynamics by a sequence of static graphs sharing the same set of nodes, one for each instant. Another graph formalism, called *Time-Varying Graphs* (TVG), has been provided in [12]. Evolving and Time-Varying Graphs are closely related. However, TVGs allow continuous time, while time is necessarily discrete in evolving graphs. In [12], TVGs/evolving graphs are gathered into classes that are partially ordered according to two main features: the quality of connectivity among the participating nodes and the possibility/impossibility to solve given tasks.

In this paper, we consider *long-lived highly* dynamic networks, *i.e.*, networks with unbounded lifetime and recurrent connectivity, meaning that every process can communicate by routing with each other, infinitely often, despite the high frequency of edge adding and removal. These networks fall into the class denoted $\mathcal{R}$, one of the most general class of infinite TVGs introduced so far [11,12]. However, as shown in [6], providing algorithms for Class $\mathcal{R}$ presents significant challenges and, in many cases, is actually provably impossible. Intuitively, this impossibility stems from the inherent uncertainty in such networks: without additional assumptions, no node is able to determine whether an adjacent edge, whose appearance and disappearance may vary arbitrarily over time, will reappear or is gone forever. To overcome this issue, several assumptions are commonly used, such as (1) T-interval connectivity which ensures a connected spanning subgraph is always present during every period of T times [29], (2) the recurrence of edges which states that an edge either appears infinitely often or never [12], or (3) the knowledge of an upper bound Δ on the temporal diameter of the network [24]. Here, we will focus on this latter assumption. The temporal diameter is based on the notion of temporal distance. Intuitively, the temporal distance between processes u and v at a given moment represents the minimum time required for u to reach v through the dynamic network starting from that moment. Then, the temporal diameter is at most Δ if at any point in time, the temporal distance between any two processes is at most Δ.

Unsurprisingly, handling highly dynamic environments becomes much harder in fault-prone systems [6]. Self-stabilization [17] is a well-known general paradigm that captures the ability of a distributed algorithm to withstand transient faults, *i.e.*, faults such as memory corruption that occur unpredictably, but do not cause permanent hardware damage and happen infrequently. Indeed, starting from an arbitrary configuration (potentially caused by such faults), a self-stabilizing algorithm makes a distributed system reach within a finite time a configuration from which its behavior is correct. It is worth noting that self-stabilizing algorithms designed for arbitrary network topologies can tolerate, up to a certain extent, some topological changes (*i.e.*, the addition or the removal of communication links or nodes). Specifically, if such changes are eventually detected locally by the involved processes and occur infrequently, they can be treated as transient faults. This perspective is supported by several works, *e.g.*, [15,18,21]. Furthermore, several approaches, like *superstabilization* [20] and *gradual stabilization* [3], have been proposed to complement self-stabilization by efficiently handling topological changes, provided these changes are sparse in both time and space. However, all these aforementioned approaches [3,15,18,20,21] become totally ineffective when the frequency of topological changes drastically increases.

Several self-stabilizing works [1,2,8,13,19,23] deal with a highly dynamic context. In [2], the authors study conditions under which leader election can be solved and provide three self-stabilizing leader election algorithms for Classes $\mathcal{B}(\Delta)$, $\mathcal{Q}(\Delta)$, and $\mathcal{R}$, respectively. Class $\mathcal{B}(\Delta)$ includes the dynamic networks where the temporal diameter is at most Δ, *i.e.*, at *any point in time*, the temporal distance between any two processes is at most Δ. Class $\mathcal{Q}(\Delta)$ contains the dynamic networks where the temporal distance between any two processes is at most Δ, *infinitely often*. Notice that $\mathcal{B}(\Delta) \subsetneq \mathcal{Q}(\Delta) \subsetneq \mathcal{R}$ [1]. In [1], the same authors consider self-stabilization and its weakened form called *pseudo-stabilization*. They establish conditions under which (pseudo- and self-) stabilizing leader election is solvable in various refined classes of highly dynamic networks. In [13], the authors present a self-stabilizing unison algorithm for $\mathcal{B}(\Delta)$ that stabilizes within at most 3Δ rounds, using finite but unbounded local memory. However, notice that, in this work, Δ is unknown to the processes. The works in [8,19,23] deal with models widely different than the one we consider here. In [8,23], self-stabilizing leader election is addressed in highly dynamic systems through the population protocol model [4]. In this model, communications are achieved by atomic rendezvous between pairs of anonymous processes, where ties are non-deterministically broken. In [19], Dolev *et al.* assume that the system is endowed with a routing algorithm which allows any two processes to communicate, provided that the sender knows the identifier of the receiver. This black-box protocol abstracts the dynamics of the system: the network dynamics make it fair lossy, non-FIFO, and duplication-prone. Moreover, the channel capacity is assumed to be bounded. Based on this weak routing algorithm, they build a stronger routing protocol which is reliable, FIFO, and which prevents duplication.

In this paper, we address the self-stabilizing *mutual exclusion* problem in highly dynamic networks. Mutual Exclusion is a benchmark problem that has been widely studied in many distributed contexts. It consists in managing the concurrent access to a shared resource by multiple processes while ensuring that no two processes access it concurrently, and that every process eventually succeeds. In [27], superstabilizing algorithms that solve (token-based) mutual exclusion are proposed. Specifically, Herman established necessary conditions for implementing a superstabilizing token circulation and provides algorithms. However, his study is restricted to ring-shape networks, and no more than one topological change can occur once the system is stabilized. In [28], the authors study one of the efficiency measures introduced in [27], the *latency* (the maximal time to recover after a topological change occur in a legitimate configuration). Still for ring-shaped networks, they provide an optimal algorithm for this measure. To the best of our knowledge, [27,28] are the only work addressing the self-stabilizing mutual exclusion in non-static networks.

Contribution. We consider a synchronous message-passing model where processes have unique identifiers and communicated using local broadcast primitive. Our focus is on the class $\mathcal{B}(\Delta)$. In this context, any process can communicate with any other process through routing within at most Δ rounds. Assuming processes know both Δ and an upper bound N on the number of processes n, we provide a distributed self-stabilizing mutual exclusion algorithm working in such dynamic systems. Our solution stabilizes in $\mathcal{O}(\Delta.N)$ rounds using messages of size $\mathcal{O}(\log N + \log \Delta + \log |IDSET|)$ bits and requiring $\mathcal{O}(N(\log N + \log \Delta + \log |IDSET|))$ bits of memory per process, where $IDSET$ denotes the domain of process IDs. Moreover, our solution achieves optimal waiting time: once stabilized, the maximum delay before a process executes its critical section is at most $n-1$ rounds. We adopt a modular approach which combines self-stabilizing solutions to three fundamental problems: leader election (we use the algorithm provided in [2]), unison, and ranking. In addition to mutual exclusion, we also provide a novel self-stabilizing algorithms for unison and ranking, which may be of independent interest.

Roadmap. The next section introduces the computational model and basic definitions. Sections 3, 4, and 5 present our self-stabilizing solutions for unison, ranking, and mutual exclusion, respectively. We conclude with some perspectives in Sect. 6.

2   Preliminaries

In this section, we adopt the dynamic graph formalism and computational model from [2,14]. Also, we define self-stabilization in the context of dynamic networks, and present the algorithm composition used in the design of our solutions. Finally, we present the problems addressed in this work.

2.1   Dynamic Graphs

For any directed graph G, we denote its vertex set by $V(G)$ and its set of directed edges by $E(G)$. A *dynamic graph* $\mathcal{G}$ *with vertex set* V is an infinite sequence of loopless directed graphs $(G_i)_{i \geq 1} = G_1, G_2, \ldots$ such that $V(G_i) = V$, for every $i \in \mathbb{N}^*$. We denote by $\mathcal{G}_{i\triangleright}$ the dynamic graph $G_i, G_{i+1}, \ldots$ with vertex set V, *i.e.*, the suffix of $\mathcal{G}$ starting from position $i \in \mathbb{N}^*$.

A *journey* $\mathcal{J}$ can be thought as a path over time from a starting vertex u_1 to a destination vertex v_k, *i.e.*, given a dynamic graph $\mathcal{G} = G_1, G_2, \ldots$, a *journey* $\mathcal{J}$ is a finite non-empty sequence of pairs $(e_1, t_1), (e_2, t_2), \ldots, (e_k, t_k)$ where $\forall i \in \{1, \ldots, k\}$, $t_i \in \mathbb{N}^*$, $e_i = (u_i, v_i) \in E(G_{t_i})$, and $i < k \Rightarrow v_i = u_{i+1} \wedge t_i < t_{i+1}$. Vertices u_1 and v_k are respectively called the *initial* and *final extremities* of $\mathcal{J}$. We respectively denote by $departure(\mathcal{J})$ and $arrival(\mathcal{J})$ the *starting time* t_1 and the *arrival time* t_k of $\mathcal{J}$. The *temporal length* of $\mathcal{J}$ is equal to $arrival(\mathcal{J}) - departure(\mathcal{J}) + 1$. A *journey from u to v* is a journey whose initial and final extremities are u and v, respectively. We denote by $\mathcal{J}_{\mathcal{G}}(u,v)$ the set of journeys from u to v in $\mathcal{G}$. Let $\overset{\mathcal{G}}{\rightsquigarrow}$ be the binary relation over V such that $u \overset{\mathcal{G}}{\rightsquigarrow} v$ iff $u = v$ or there exists a journey from u to v in $\mathcal{G}$.

The *temporal distance* from u to v in $\mathcal{G}$, $\hat{d}_{\mathcal{G}}(u,v)$, is defined as follows: $\hat{d}_{\mathcal{G}}(u,v) = 0$ if $u = v$, $\hat{d}_{\mathcal{G}}(u,v) = \min\{arrival(\mathcal{J}) : \mathcal{J} \in \mathcal{J}_{\mathcal{G}}(u,v)\}$ otherwise (by convention, we let $\min \emptyset = +\infty$). Roughly speaking, the temporal distance from u to v in $\mathcal{G}$ gives the minimum timespan for u to reach v in $\mathcal{G}$ through a journey.

A class of dynamic graphs is defined as a particular subset of dynamic graphs. Here, we consider the class $\mathcal{B}(\Delta)$ where every vertex is *always* at temporal distance at most Δ from any other vertex. Formally, $\mathcal{G} \in \mathcal{B}(\Delta)$ iff $\forall u, v \in V, \forall i \in \mathbb{N}^*, \hat{d}_{\mathcal{G}_{i\triangleright}}(u,v) \leq \Delta$.

2.2   Computational Model

We assume a *distributed system* made of a set V of n *processes*. We assume that each process p has a unique identifier (ID for short). The ID of p is denoted by $id(p)$ and taken in an arbitrary domain $IDSET$ totally ordered by $<$. We also assume that processes know a common value $N \geq n$. $|IDSET|$ and N are not related, $|IDSET|$ may be arbitrarily large compared to N. We denote by ℓ the process holding the smallest identifier. We let $p_0, \ldots, p_j, \ldots, p_{n-1}$ be the processes of V ordered by increasing identifiers, *i.e.*, $p_0 = \ell$ and for every $i \in [1..n-1]$, $id(p_{i-1}) < id(p_i)$.

Processes are assumed to communicate by message passing through an interconnected network that evolves over time. The dynamic topology of the network is then conveniently modeled by a dynamic graph $\mathcal{G} = (G_i)_{i \geq 1}$ with vertex set V (the set of processes). In the remainder of this paper, we use the terms *process*, *vertex*, and *node* interchangeably. Likewise, *edge* and *(communication) link* as used as synonyms. Processes execute their local algorithms in *synchronous rounds*. For every $i \in \mathbb{N}^*$, the communication network at *round i* is defined by G_i, *i.e.*, the graph at position i in $\mathcal{G}$. $\forall p \in V, \forall i \in \mathbb{N}^*$, we denote by $\mathcal{IN}(p)^i = \{q \in V : (q,p) \in E(G_i)\}$ the set of p's incoming neighbors at round i. $\mathcal{IN}(p)^i$ is assumed to be (*a priori*) unknown to p, whatever be the value of i.

A *distributed algorithm* $\mathcal{A}$ is a collection of n local algorithms $\mathcal{A}(p)$, one per process $p \in V$. At each time instant, the *state* of each process $p \in V$ in $\mathcal{A}$ is defined by the values of its variables in $\mathcal{A}(p)$. Some variables may be constant in which case their values are predefined. A *configuration* of $\mathcal{A}$ for V is a vector of n components $\gamma = (s_1, s_2, \ldots, s_n)$, where s_1 to s_n represent the states of the processes in V. Let γ_1 be an initial configuration of $\mathcal{A}$ for V. For any (synchronous) round $i \geq 1$, the system moves from the current configuration γ_i to some configuration γ_{i+1}, where γ_i (resp. γ_{i+1}) is referred to as the configuration *at the beginning (resp. end) of round i*. Such a move is atomically performed by every process $p \in V$ according to the following three steps, defined in its local algorithm $\mathcal{A}(p)$:

S: p sends a message consisting of all or a part of its state in γ_i using the primitive SEND(),
R: using Primitive RECEIVE(), p receives all messages sent by processes in $\mathcal{IN}(p)^i$, and
C: p computes its state in γ_{i+1}.

An *execution* of a distributed algorithm $\mathcal{A}$ in the dynamic graph $\mathcal{G} = (G_i)_{i \geq 1}$ is an infinite sequence of configurations $(\gamma_i)_{i \geq 1} = \gamma_1, \gamma_2, \ldots$ of $\mathcal{A}$ for V such that γ_1 is arbitrary and $\forall i \geq 1$, γ_{i+1} is obtained by executing a synchronous round of $\mathcal{A}$ on γ_i based on the communication network at round i, *i.e.*, the graph G_i. In the following, we denote by $x(p)^i$ the value of variable $x(p)$ at the beginning of Round i.

2.3   Self-Stabilization in $\mathcal{B}(\Delta)$

We recall the definition of self-stabilization from [2], which is an adaptation of the well-known definition in [17] to accommodate dynamic context.

Let $\mathcal{A}$ be a distributed algorithm, SP be a specification (*i.e.*, a predicate over configuration sequences).

Definition 1 (Self-stabilization). *An algorithm $\mathcal{A}$ is* self-stabilizing *for SP on $\mathcal{B}(\Delta)$ iff for every set of processes V, there exists a subset of configurations $\mathcal{L}$ of $\mathcal{A}$ for V, called* legitimate configurations, *such that for every $\mathcal{G} \in \mathcal{B}(\Delta)$ with set of processes V, (i) every execution of $\mathcal{A}$ in $\mathcal{G}$ contains a configuration of $\mathcal{L}$* (Convergence)*, and (ii) for every execution e in $\mathcal{G}$ starting from a configuration of $\mathcal{L}$, $SP(e)$ holds* (Correctness)*.*

The *length of the stabilization phase* of an execution e is the length of its longest prefix containing no legitimate configuration. The *stabilization time* in rounds is the maximum length of a stabilization phase over all possible executions.

2.4   Composition

A classic way of simplifying the design and proof of distributed algorithms is to compose several algorithms which, together, solve a given problem. Depending on the model and the problem being addressed, various composition techniques have been proposed, *e.g.*, [5,16,26]. In this paper, we use the synchronous composition.

Definition 2 (Synchronous Composition). *Let $\mathcal{A}_1$ and $\mathcal{A}_2$ be two distributed algorithms. The synchronous composition of $\mathcal{A}_1$ and $\mathcal{A}_2$, denoted by $\mathcal{A}_1 \circ \mathcal{A}_2$, is the distributed algorithm where for each process p, the local algorithm $(\mathcal{A}_1 \circ \mathcal{A}_2)(p)$ is defined as follows.*

- *$(\mathcal{A}_1 \circ \mathcal{A}_2)(p)$ contains all variables of $\mathcal{A}_1(p)$ and $\mathcal{A}_2(p)$;*
- *Each (synchronous) round of $\mathcal{A}_1 \circ \mathcal{A}_2(p)$ comprises in the following three steps, executed in sequence:*

1. *the **S** step of $(\mathcal{A}_1 \circ \mathcal{A}_2)(p)$ consists in sending a message $\langle m_{\mathcal{A}_1}, m_{\mathcal{A}_2} \rangle$, where $m_{\mathcal{A}_1}$ (resp., $m_{\mathcal{A}_2}$) is the data sent during the **S** step of $\mathcal{A}_1(p)$ (resp. of $\mathcal{A}_2(p)$);*
2. *the **R** step of $(\mathcal{A}_1 \circ \mathcal{A}_2)(p)$ consists in receiving all messages of the form $\langle m_{\mathcal{A}_1}, m_{\mathcal{A}_2} \rangle$ sent by the incoming neighbors of the current round.*
3. *the **C** step of $(\mathcal{A}_1 \circ \mathcal{A}_2)(p)$ consists of executing a **C** step of $\mathcal{A}_1(p)$ based on all received $m_{\mathcal{A}_1}$ data, followed by a **C** step of $\mathcal{A}_2(p)$ based on all received $m_{\mathcal{A}_2}$ data.*

When we compose more than two algorithms, we assume $\circ$ is left-associative.

Property 1 Let $\mathcal{A}_1$ and $\mathcal{A}_2$ be two distributed algorithms and SP be a specification. Let $x, \overline{x} \in \{1, 2\}$ with $x \neq \overline{x}$. If no variable written by $\mathcal{A}_{\overline{x}}$ appears in $\mathcal{A}_x$ and $\mathcal{A}_x$ is self-stabilizing for SP on $\mathcal{B}(\Delta)$, then $\mathcal{A}_1 \circ \mathcal{A}_2$ is self-stabilizing for SP on $\mathcal{B}(\Delta)$. Moreover, the stabilization time of $\mathcal{A}_1 \circ \mathcal{A}_2$ for SP is at most the one $\mathcal{A}_x$ for SP.

2.5 Studied Problems

Leader Election. In identified networks, the *leader election* problem consists in making all processes agree on one of the identifiers held by processes. The identifier of the elected process is stored at each process p as an output variable, denoted here by $lid(p)$.

Definition 3 (SP_{LE}). *A sequence of configurations $(\gamma_i)_{i \geq 1}$ satisfies SP_{LE} if and only if $\exists p \in V$ such that $\forall i \geq 1$, $\forall q \in V$, $lid(q) = id(p)$ in Configuration γ_i.*

An algorithm is said to be a *self-stabilizing leader election algorithm* for $\mathcal{B}(\Delta)$ if it is self-stabilizing for SP_{LE} on $\mathcal{B}(\Delta)$. In this work, we will use the self-stabilizing leader election algorithm for $\mathcal{B}(\Delta)$ introduced in [2], hereafter referred to as $LE_{\mathcal{B}(\Delta)}$. Below, we provide a theorem that summarizes results about $LE_{\mathcal{B}(\Delta)}$.

Theorem 1 ([2]). *Algorithm $LE_{\mathcal{B}(\Delta)}$ is a self-stabilizing leader election algorithm for $\mathcal{B}(\Delta)$ that elects the process with the smallest identifier. Its stabilization time is at most 3Δ rounds. It requires $\mathcal{O}(N(\log |IDSET| + \log \Delta))$ bits per process and messages of size $\mathcal{O}(\log |IDSET| + \log \Delta)$ bits.*

Synchronous Unison. In the synchronous *unison* problem [22, 25], each process p maintains a local clock $C(p)$ in such a way that all clocks of the network are always equal while incrementing infinitely often. In this paper, we focus on clocks that are periodic, *i.e.*, the clock domain is $[0..K-1]$ where $K \geq 2$ is called the period and clocks increment modulo K.

Definition 4 (SP_U). *A sequence of configurations $(\gamma_i)_{i \geq 1}$ satisfies SP_U iff $\forall i \geq 1$,*

- $\forall p, q \in V$, $C(p) = C(q)$ in γ_i *(Safety); and*
- $\forall p \in V$, $\exists j \geq i$, $C(p)$ *is incremented (by one modulo K) at round j (Liveness).*

A distributed algorithm is said to be a *self-stabilizing (synchronous) unison algorithm* for $\mathcal{B}(\Delta)$ if it is self-stabilizing for SP_U on $\mathcal{B}(\Delta)$.

Ranking. Recall that $p_0, \ldots, p_j, \ldots, p_{n-1}$ denotes the n processes sorted in increasing order of identifiers. The *ranking* problem [30] consists in making each process p_j compute its index j. This index is called the *rank* of p_j and is stored in its output variable $Rank(p_j)$.

Definition 5 (SP_R). *A sequence of configurations $(\gamma_i)_{i\geq 1}$ satisfies SP_R if and only if $\forall i \geq 1$, $\forall j \in [0..n-1]$, $Rank(p_j) = j$ in γ_i.*

A distributed algorithm is said to be a *self-stabilizing ranking algorithm* for $\mathcal{B}(\Delta)$ if it is self-stabilizing for SP_R on $\mathcal{B}(\Delta)$.

Mutual Exclusion. In the *Mutual Exclusion* problem, each process has a special section of code called *critical section*, denoted CS. Two properties should be then achieved: no two processes execute their critical section concurrently (*Safety*) , and each process executes its critical section infinitely often (*Liveness*).

Definition 6 (SP_{ME}). *A sequence of configurations $(\gamma_i)_{i\geq 1}$ satisfies SP_{ME} iff $\forall p \in V$, $\forall i \geq 1$, if p is executing CS at Round i, then $\forall q \neq p \in V$, q does not execute CS at Round i (Safety); and $\forall p \in V, \forall i \geq 1, \exists j \geq i$ such that p enters CS at Round j (Liveness).*

A distributed algorithm is said to be a *self-stabilizing mutual exclusion algorithm* for $\mathcal{B}(\Delta)$ if it is self-stabilizing for SP_{ME} on $\mathcal{B}(\Delta)$. In our context, the *waiting time* of a self-stabilizing mutual exclusion algorithm is the maximum time a process has to wait before executing CS, once $\mathcal{A}$ has stabilized.

3   Synchronous Unison in $\mathcal{B}(\Delta)$

Our self-stabilizing unison algorithm is called Algorithm $Sync_{\mathcal{B}(\Delta)}$. It ensures that eventually the clock of all processes increment (modulo K) at each step while remaining synchronized. To ensure this synchronization, the basic idea is to gradually make all processes adopting the clock value of a particular process. We designate this latter using the self-stabilizing leader election for Class $\mathcal{B}(\Delta)$ proposed in [2] and denoted by $LE_{\mathcal{B}(\Delta)}$. After at most 3Δ rounds, Algorithm $LE_{\mathcal{B}(\Delta)}$ constantly outputs the smallest identifier of the system, $id(\ell)$, in the variable $lid(p)$ of each process p; see Theorem 1, page 6. The clock synchronization is actually achieved by Algorithm $Unison_{\mathcal{B}(\Delta)}$ where each process p takes $lid(p)$ as input. The code of $Unison_{\mathcal{B}(\Delta)}$ is given in Algorithm 1. Hence, $Sync_{\mathcal{B}(\Delta)}$ is defined, using the synchronous composition, as follows:

$$Sync_{\mathcal{B}(\Delta)} = LE_{\mathcal{B}(\Delta)} \circ Unison_{\mathcal{B}(\Delta)}.$$

In Algorithm $Unison_{\mathcal{B}(\Delta)}$, each process p has three inputs: its identifier $id(p)$, the clock period K, and $lid(p)$, the latter being the output variable of $LE_{\mathcal{B}(\Delta)}$. Each process p maintains three other variables: $C(p)$, $BestTTL(p)$, and $MSGs(p)$. $C(p)$ is the local clock of p. $BestTTL(p) \in [0..\Delta + 1]$ is a *freshness indicator* of the value stored in $C(p)$. $MSGs(p)$ is a multiset of at most N messages used as a buffer to store the messages received during the round.

The basic principle of the algorithm is to implement a contamination broadcast. At each round, every process p sends its pair $(BestTTL(p), C(p))$. Next, p adopts, among the receiving pairs and its own, the one with the smallest $BestTTL$-value; see Lines 13–15. (If multiple pairs in $MSGs(p)$ share the smallest $BestTTL$-value, then p selects the pair with the smallest ID by choosing the lexicographically smallest among them.) Then, p increments $C(p)$ modulo K (see Line 16), and adjusts $BestTTL(p)$ as follows. If p believes to be the leader, $BestTTL(p)$ is reset to 0; see Lines 18 and 19. Otherwise, $BestTTL(p)$ is incremented, except if its value is already $\Delta + 1$; see Line 17.

At most one round after the stabilization of $LE_{\mathcal{B}(\Delta)}$, the freshness indicator of the leader ℓ, $BestTTL(\ell)$, is forever equal to 0 and all other freshness indicators are forever greater than 0. From that point on, the clock value of ℓ gradually contaminates all other processes. Indeed, at each round, ℓ initiates a global broadcast in the network. During the broadcast, the clock value stored in a message initiated by ℓ will be maintained synchronized with its own clock and the freshness indicator inside the message will be incremented at each hop. Assume some process q receives a message initiated by ℓ. Consider the one with the smallest freshness indicator I and let C be the clock value stored in the message. If q is still not synchronized with ℓ, then I is necessarily smaller than its own freshness indicator $BestTTL(q)$ and consequently, q becomes synchronized with ℓ by adopting both C and I.

Correctness of Algorithm $Sync_{\mathcal{B}(\Delta)}$. We define a *legitimate configuration* as any configuration of Algorithm $Sync_{\mathcal{B}(\Delta)}$ where $\forall p \in V$, $C(p) = C(\ell)$.

Recall that Algorithm $LE_{\mathcal{B}(\Delta)}$ stabilizes to a configuration from which $lid(p) = id(\ell)$ forever, for every process p, within at most 3Δ rounds; see Theorem 1, page 6. So, from the beginning of Round $3\Delta + 1$, $lid(p) = id(\ell)$ forever, for every process p. Consequently, we can deduce the following remark from the code of Algorithm 1 (Lines 18 and 19).

Remark 1. During any round $t \geq 3\Delta + 1$, the test of Line 18 is true for process p if and only if $p = \ell$. Consequently, at the beginning of any round $t \geq 3\Delta + 2$, we have $id(p) = id(\ell) \Leftrightarrow BestTTL(p) = 0$, for every process p.

For every $t \geq 3\Delta + 2$, $BestTTL(\ell) = 0$ at the beginning of t, by Remark 1. So, ℓ does not execute Line 15 during Round t. Consequently, we have the following observation.

Remark 2. Let $t \geq 3\Delta + 2$ and c be the value of $C(\ell)$ at the beginning of Round t. $\forall t' \geq t$, $C(\ell) = (c + t - t') \mod K$ at the beginning of Round t'.

Algorithm 1: $Unison_{\mathcal{B}(\Delta)}$ for any process p

---

```
 1  Inputs:
 2      id(p) ∈ IDSET // identifier of p
 3      K ∈ ℕ* // period of the clocks
 4      Δ ∈ ℕ* // upper bound on the temporal diameter
 5      lid(p) ∈ IDSET // input from LE_{B(Δ)}
 6  Local Variables:
 7      C(p) ∈ {0, ..., K − 1}; BestTTL(p) ∈ {0, ..., Δ + 1}; MSGs(p): Multiset of at most N
        messages
 8  foreach round do
 9        SEND(BestTTL(p), C(p));
10        MSGs(p) ← RECEIVE() ;
11        if MSGs(p) ≠ ∅ then
12            Let (MinT, MinC) the smallest couple in MSGs in the lexicographical order;
13            if MinT < BestTTL(p) then
14                BestTTL(p) ← MinT;
15                C(p) ← MinC;

16        C(p) ← (C(p) + 1) mod K ;
17        BestTTL(p) ← min(BestTTL(p) + 1, Δ + 1)
18        if id(p) = lid(p) then
19            BestTTL(p) = 0;
```

In the next lemma we show that, from the beginning of Round $3\Delta + 2$, $BestTTL(p)$ is strictly increasing with rounds for each process p such that $C(p) \neq C(\ell)$ until reaching $\Delta + 1$.

Lemma 1. *At the beginning of any round $3\Delta + t$ with $t \geq 2$, $C(p) \neq C(\ell) \Rightarrow BestTTL(p) \geq \min(t - 1, \Delta + 1)$, for every process p.*

Let $i \in [0..\Delta]$. We let $Stab_i = \{p \in V \mid \hat{d}_{\mathcal{G}_{3\Delta+2\triangleright}}(\ell, p) \leq i\}$. Thanks to the previous lemma as well as Remarks 1 and 2, we can establish by induction on i, that starting from configuration $\gamma_{3\Delta+2}$, when a process $p \in Stab_i$ is hit for the first time by a broadcast initiated by ℓ, if p is still not synchronized with ℓ, then $BestTTL(p)$ is strictly greater than the freshness indicator in the message, and consequently becomes synchronized with ℓ by adopting the clock value stored in the message. Hence, the next lemma follows.

Lemma 2. $\forall i \in [0..\Delta], \forall p \in Stab_i$, $C(p) = C(\ell) \wedge BestTTL(p) \leq i$ *at the beginning of Round $3\Delta + i + 2$.*

By the definition of $\mathcal{B}(\Delta)$, $\forall p \in V, \hat{d}_{\mathcal{G}_{3\Delta+2\triangleright}}(\ell, p) \leq \Delta$. Hence $Stab_\Delta = V$ and, by applying the previous lemma with $i = \Delta$, we can deduce that the configuration at the beginning of Round $4\Delta + 2$ is legitimate.

Then, from the code of the algorithm, we can observe that

1. the set of legitimate configurations of Algorithm $Sync_{\mathcal{B}(\Delta)}$ is trivially closed, and
2. if the configuration is legitimate at the beginning of some round, then every process increments its clock (modulo K) during the round.

Indeed, let c be the clock value common to all processes at the beginning of a round starting from a legitimate configuration. Then, every message received by any process during the round (if any) contains the clock value c. Consequently, the clock value of any process at the end of the round is equal to $(c+1) \bmod K$. Hence, from Lemma 2 and the code of the algorithm, we deduce the following theorem:

Theorem 2. *Algorithm $Sync_{\mathcal{B}(\Delta)}$ is a self-stabilizing (synchronous) unison algorithm for $\mathcal{B}(\Delta)$. Its stabilization time is at most $4\Delta + 1$ rounds. It uses messages of size $\mathcal{O}(\log|IDSET| + \log K + \log \Delta)$ bits and can receive up to N messages. Thus, it needs $\mathcal{O}(N(\log|IDSET| + \log K + \log \Delta))$ bits of memory per process.*

4   Ranking in $\mathcal{B}(\Delta)$

Our ranking algorithm is called Algorithm $Rank_{\mathcal{B}(\Delta)}$. Recall that $p_0, \ldots, p_j, \ldots, p_{n-1}$ denotes the n processes sorted in increasing order of identifiers and the index j of any process p_j is referred to as the *rank* of p_j. Actually, using Algorithm $Rank_{\mathcal{B}(\Delta)}$, each process computes both its own rank and n, the actual number of processes in the system. Again, Algorithm $Rank_{\mathcal{B}(\Delta)}$ is a synchronous composition of several subalgorithms. Precisely,

$$Rank_{\mathcal{B}(\Delta)} = SetRk_{\mathcal{B}(\Delta)} \circ LE^+_{\mathcal{B}(\Delta)} \circ Sync^R_{\mathcal{B}(\Delta)}.$$

Algorithm $Sync^R_{\mathcal{B}(\Delta)}$ is an instance of Algorithm $Sync_{\mathcal{B}(\Delta)}$ (presented in the previous section) where the clock period K is set to $3(\Delta+1)(N+1)+1$. Thus, $Sync^R_{\mathcal{B}(\Delta)}$ stabilizes within at most $4\Delta+1$ rounds (Theorem 2) and maintains the clocks within the range $0..3(\Delta+1)(N+1)$. These clocks are used as inputs to Algorithm $SetRk_{\mathcal{B}(\Delta)}$.

Algorithm $LE^+_{\mathcal{B}(\Delta)}$ is a slightly derived version of Algorithm $LE_{\mathcal{B}(\Delta)}$ [2] that makes use of *augmented identifiers*. Precisely, the identifier domain $IDSET$ is extended with an additional bit. We let $IDSET^+ = \{0,1\} \times IDSET$ be this extended version of the domain $IDSET$. Since $IDSET$ is an order set, $IDSET^+$ can be ordered using the lexicographic order, denoted here by $\prec$: $(a,b) \prec (c,d) \equiv (a < c \vee (a = c \wedge b < d))$. To obtain $LE^+_{\mathcal{B}(\Delta)}$, we modify $LE_{\mathcal{B}(\Delta)}$ as follows. The input of $LE_{\mathcal{B}(\Delta)}$ at each process p (*i.e.*, $id(p)$) is replaced by the pair $(S(p), id(p))$, where $S(p) \in \{0,1\}$ is an input from Algorithm $SetRk_{\mathcal{B}(\Delta)}$. Moreover, the order used by $LE^+_{\mathcal{B}(\Delta)}$ is $\prec$, instead of $<$. Hereafter, we refer to the pair $(S(p), id(p))$ as $id^+(p)$, and we use the terms *pending* process when $S(p) = 0$, and *settled* process when $S(p) = 1$. The pair $id^+(p) = (S(p), id(p))$ of any process p is always unique since $id(p)$ is unique, by definition. Recall also that, assuming constant identifiers picked in an arbitrary ordered set, $LE_{\mathcal{B}(\Delta)}$ stabilizes in $\mathcal{B}(\Delta)$ within at most 3Δ rounds, and once stabilized, we have $lid(p)$ that is equal to the minimum identifier used in the system, for every process p; see Theorem 1. Hence, using the augmented identifier $id^+(p)$ of each process p

as input, $LE^+_{\mathcal{B}(\Delta)}$ outputs in lid-variables a value of type $IDSET^+$ and ensures the following property:

Property 2. $\forall t \in \mathbb{N}^*$, if for every process p, $S(p)$ is constant from the beginning of Round t to the beginning of Round $t+3\Delta$, then for every process p, $lid(p)^{t+3\Delta} = \min_\prec \{(S(q)^t, id(q)) \mid q \in V\}$.

In the following, $lid(p).S$ and $lid(p).id$ will respectively denote the left and right elements of the augmented identifier stored into $lid(p)$.

Algorithm 2: $SetRk_{\mathcal{B}(\Delta)}$ for any process p

```
 1  Inputs:
 2      id(p) ∈ IDSET // identifier of p
 3      N ∈ ℕ* // upper bound on the number of nodes
 4      Δ ∈ ℕ* // upper bound on the temporal diameter
 5      C(p) ∈ {0,...,(3Δ+1)(N+1)} // input from Sync^R_{B(Δ)}
 6      lid(p) ∈ IDSET⁺ // input from LE⁺_{B(Δ)}
 7  Local Variables:
 8      Rank(p) ∈ {0,...,N−1}; NbP(p) ∈ {1,...,N}; PNok(p), S(p) ∈ {0,1}
 9  foreach round do
10      if C(p) = (3Δ+1)(N+1) then
11          S(p) ← 0;
12          PNok(p) ← 0;
13      else
14          if C(p) mod (3Δ+1) = 3Δ then
15              if lid(p).S = 0 then
16                  if lid(p).id = id(p) then
17                      Rank(p) ← ⌊C(p)/(3Δ+1)⌋ ;
18                      S(p) ← 1
19              else
20                  if PNok(p) = 0 then
21                      NbP(p) ← ⌊C(p)/(3Δ+1)⌋ ;
22                      PNok(p) = 1
```

We now focus on the last building block of $Rank_{\mathcal{B}(\Delta)}$: Algorithm $SetRk_{\mathcal{B}(\Delta)}$. The code of this algorithm is given in Algorithm 2. Algorithm $SetRk_{\mathcal{B}(\Delta)}$ has several inputs. First, recall that each process uses $id(p)$ as a part of its augmented identifier, $id^+(p)$. Then, the upper bounds Δ and N are necessary, in particular to bound variable domains. Finally, p reads its clock in $Sync^R_{\mathcal{B}(\Delta)}$, $C(p)$, and the output of the leader election $LE^+_{\mathcal{B}(\Delta)}$, $lid(p)$. Using these inputs, p maintains four variables: $S(p)$, $\mathrm{PNok}(p)$, $Rank(p)$, and $NbP(p)$. Actually, $S(p)$ and the flag $\mathrm{PNok}(p)$ are used to compute the $Rank_{\mathcal{B}(\Delta)}$'s outputs at each process p: $Rank(p)$ and $NbP(p)$, which should ultimately be set to the rank of p and the total number of processes n, respectively.

The main idea behind Algorithm $SetRk_{\mathcal{B}(\Delta)}$ is to assign ranks to processes using leader elections. Ideally, n elections would be enough. However, processes do not *a priori* know n. Instead, we perform $N + 1$ elections. Each election is followed by a decision round where output variables may be updated. An election together with its decision round form we call an *epoch*. Epochs are numbered from 0 to N. Among them, only the $n + 1$ first ones are really useful, but again processes do not *a priori* know n. Precisely, at the end of an epoch j with $0 \le j < n$, p_j correctly sets its rank $Rank(p_j)$ to j. Then, thanks to the flag $\mathtt{PNok}(p)$, every process p detects the end of epoch n and sets $NbP(p)$ accordingly. These $N + 1$ epochs are preceded by a reset round and together form what we call a *cycle*. Each epoch is made of $3\Delta + 1$ rounds: 3Δ rounds for the election (Property 2) followed by the decision round. This justify why we choose the period $K = 3(\Delta + 1)(N + 1) + 1$.

According to Theorem 2, after $4\Delta + 1$ rounds, we have the guarantee that all clocks are synchronized. However, at that point, the common clock value is arbitrary. Now, we impose a cycle to start (with its reset round) from a configuration where all clocks have their maximum value, $K - 1$. Such configurations are termed *almost legitimate*.

Definition 7 (Almost Legitimate). *A configuration of $Rank_{\mathcal{B}(\Delta)}$ is almost legitimate if $C(p) = (3\Delta + 1)(N + 1)$ $(= K - 1)$, for every process p.*

From Property 1, Theorem 2 and the code of Algorithm $Unison_{\mathcal{B}(\Delta)}$, we have:

Remark 3. From any almost legitimate configuration, all processes increment their clock by one at each round.

Definition 8 (Cycle). *A cycle is any interval $[t..t']$ such that:*

- *the configuration at Round t is almost legitimate and*
- *$t' = t + (3\Delta + 1)(N + 1) + 1$ $(= t + K)$.*

From Remark 3, follows.

Property 3. Let $[t..t']$ be a cycle. For every $i \in [t + 1..t']$, for every process p, we have $C(p) = i - (t + 1)$ at the beginning of Round i.

Remark that, for any cycle $[t..t']$, the first almost legitimate configuration after the configuration at the beginning of Round t is the one at the beginning of Round t'.

Remark 4. The first almost legitimate configuration appears at most $K - 1$ rounds after the stabilization of $Sync^R_{\mathcal{B}(\Delta)}$ and then, the execution suffix is only made of cycles.

The overall structure and execution flow of our solution are shown in Fig. 1. From the previous remark, we have a finite period, we call *setup phase*, between the stabilization of $Sync^R_{\mathcal{B}(\Delta)}$ and the first cycle. After this setup phase, the execution is partitioned into infinitely many cycles. The first step of any cycle is a reset round where every process p resets both $S(p)$ and $\mathrm{PNok}(p)$ to zero (Lines 11–12). Then, the first epoch starts: starting from the common clock value 0, all processes are pending during the 3Δ first rounds of the epoch. Consequently, after these rounds, $lid(p) = (0, id(p_0))$ for each process p, by Property 2. During the following round (*i.e.*, the decision round of epoch 0), p_0 sets its rank $Rank(p_0)$ to $\left\lfloor \frac{C(p)}{3\Delta+1} \right\rfloor = 0$ since $C(p_0) = 3\Delta$ (Line 17) and becomes settled by setting its flag $S(p_0)$ to 1; see Line 18. From now on, $id^+(p_0)$ remains constant until the end of the cycle. Moreover, until the end of the next epoch, $id^+(p_0) = (1, id(p_0))$ becomes the largest ID (*w.r.t.* $\prec$) in $IDSET^+$ among the n processes. In addition, the output of $LE^+_{\mathcal{B}(\Delta)}$, $(0, id(p_0))$, no more corresponds to any input $id^+(p)$ of $LE^+_{\mathcal{B}(\Delta)}$. So, $LE^+_{\mathcal{B}(\Delta)}$ re-stabilizes and within 3Δ rounds "elects" $(0, id(p_1))$. Thus, p_1 becomes settled during the decision round of epoch 1, thereby excluding itself as a future leader in the next epochs, and so on so forth.

Following this process, after $3n\Delta$ rounds, all processes are settled and participate to the leader election of epoch n with their S flag set to 1. Then, after an additional 3Δ rounds, p_0 is elected again, but with $lid(p_0).S$ equal to 1. In the following round, each process p assigns $NbP(p)$ to $\left\lfloor \frac{C(p)}{3\Delta+1} \right\rfloor = n$, the total number of processes in the system; see Line 21. The flag $\mathrm{PNok}(p)$ is also set to 1 to indicate that the computation of $NbP(p)$ is now complete; Line 22. Thanks to $\mathrm{PNok}(p)$, the remaining epochs, if any, does not modify the outputs of $SetRk_{\mathcal{B}(\Delta)}$. Subsequently, the same cycle repeats indefinitely. Nevertheless, as shown below, the variables $Rank(p)$ and $NbP(p)$ remain unchanged, since their values are merely overwritten with their correct values.

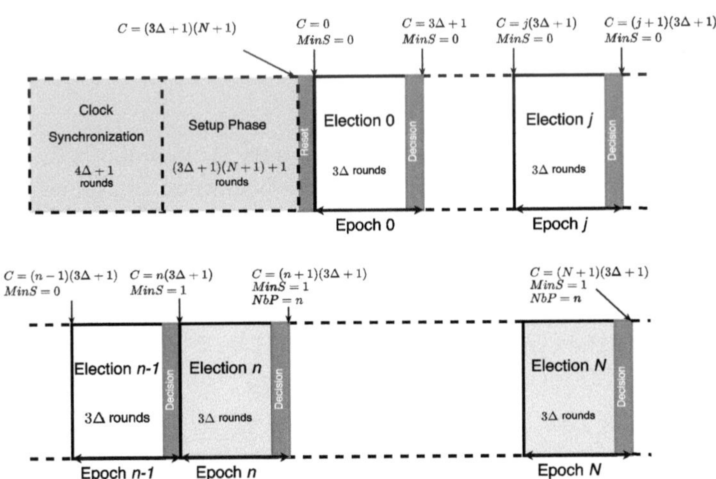

Fig. 1. Structure and execution flow of $Rank_{\mathcal{B}(\Delta)}$, where $MinS = \min\{S(p) \mid p \in V\}$.

Correctness of Algorithm $Rank_{\mathcal{B}(\Delta)}$.

Definition 9 (Legitimate). *A configuration γ of $Rank_{\mathcal{B}(\Delta)}$ is legitimate if γ is almost legitimate, and $\forall i \in [0..n-1]$, $Rank(p_i) = i$ and $NbP(p_i) = n$.*

Theorem 3 establishes the self-stabilization and the complexity of Algorithm $Rank_{\mathcal{B}(\Delta)}$. Theorem 3 is immediate from Corollary 1. This latter is obtained with two technical lemmas: Lemmas 3 and 4. Lemma 3 below is shown by induction on i. Basically, it establishes that, thanks to $S(p_j)$, the rank of any process p_j is set exactly once during a cycle: in the last round of epoch j, *i.e.*, the decision round of epoch j.

Lemma 3. *Let $[t..t']$ be a cycle. $\forall i \in [t+1..t']$, $\forall j \in [0..n-1]$,*

1. – $Rank(p_j)^i = Rank(p_j)^t$ **if** $i - (t+1) < (3\Delta + 1).(j+1)$,
    – $Rank(p_j)^i = j$ **otherwise**;
2. *for every process p, $lid(p)^i = (0, id(p_j))$* **if** $i - (t+1) = (3\Delta + 1).j + 3\Delta$;
3. $S(p_j)^i = 0$ **if and only if** $i - (t+1) < (3\Delta + 1).(j+1)$.

Lemma 4 below is also shown by induction on i. It is quite similar to Lemma 3, but focusing on NbP-variables. It shows that, thanks to the flag $\texttt{PNok}(p)$, all processes p simultaneously set $NbP(p)$ to n exactly once during a cycle: in the last round of epoch n, *i.e.*, the decision round of epoch n.

Lemma 4. *Let $[t..t']$ be a cycle. $\forall i \in [t+1..t']$, for every process p,*

– $NbP(p)^i = NbP(p)^t$ *and* $\texttt{PNok}(p)^i = 0$ **if** $i - (t+1) < (3\Delta + 1).(n+1)$,
– $NbP(p)^i = n$ *and* $\texttt{PNok}(p)^i = 1$ **otherwise**.

Corollary 1. *For every cycle $[t..t']$, the configuration at Round t' is legitimate. Moreover, if the configuration at Round t is legitimate, all variables $Rank$ and NbP are constant all along the cycle.*

Proof. The corollary is immediate from Property 3, Lemmas 3 and 4. Indeed, first, by Property 3, the configuration at the beginning of Round t' is almost legitimate. Then, by applying Lemmas 3 and 4 with $i = t' = t + (3\Delta + 1)(N + 1)+1$, and owing the fact that $N \geq n$, we can deduce that the configuration at the beginning of Round t' is legitimate. Finally, if the configuration at the beginning of Round t is legitimate, then we can deduce that all variables $Rank$ and NbP are constant all along the cycle: indeed, we have: (1) $\forall j \in [0..n-1]$, $Rank(p_j)^t = j$ in Lemma 3; and (2) for every process p, $NbP(p)^t = n$ in Lemma 4. $\square$

From the previous corollary, we know that from any legitimate configuration, the specification of the ranking holds and NbP-variables are constantly equal to n. Moreover, by Remark 4, the system reaches an almost legitimate configuration at most $K - 1$ $(= (3\Delta + 1)(N + 1))$ rounds after the unison has stabilized. From that point on, the system reaches a legitimate configuration within the next cycle (Corollary 1), *i.e.*, after at most K additional rounds. Hence, by Property 1 and Theorem 2, we have the following theorem:

Theorem 3. *Algorithm $Rank_{\mathcal{B}(\Delta)}$ is a self-stabilizing ranking algorithm for $\mathcal{B}(\Delta)$ that also self-stabilizingly computes the number of processes n in the network. It stabilizes in at most $(6\Delta + 2)(N + 1) + (4\Delta + 2)$ rounds. Moreover, it requires messages of size $\mathcal{O}(\log N + \log \Delta + \log |IDSET|)$ bits and $\mathcal{O}(N(\log N + \log \Delta + \log |IDSET|))$ bits of memory per process.*

5   Mutual Exclusion in $\mathcal{B}(\Delta)$

In this section, we propose the algorithm that served as the main motivation for this work. This algorithm, denoted by $Mutex_{\mathcal{B}(\Delta)}$, provides a self-stabilizing solution to the mutual exclusion problem in $\mathcal{B}(\Delta)$ whose waiting time is optimal. It is based on the unison and ranking algorithms introduced in the previous two sections.

The overall approach is to use Algorithm $Rank_{\mathcal{B}(\Delta)}$ to assign a unique name from the interval $[0..n-1]$ into the $Rank$-variable of each process. Then, we use an instance of Algorithm $Sync_{\mathcal{B}(\Delta)}$, referred to as $Sync^{ME}_{\mathcal{B}(\Delta)}$, where each process instantiates the period K with its NbP-variable from $Rank_{\mathcal{B}(\Delta)}$. Once $Rank_{\mathcal{B}(\Delta)}$ has stabilized (after at most $(6\Delta + 2)(N + 1) + (4\Delta + 2)$ rounds, according to Theorem 3), the period remains fixed at n forever. Note that since Variable NbP is an output of Algorithm $Rank_{\mathcal{B}(\Delta)}$, there is no guarantee that the process clocks belong to the interval $[0..n-1]$ before $Rank_{\mathcal{B}(\Delta)}$ has stabilized. Consequently, the clocks variables are only guaranteed to fall within $[0..n-1]$ only after at most $(6\Delta + 2)(N + 1) + (4\Delta + 2) + 1$ rounds, *i.e.*, the time necessary for $Rank_{\mathcal{B}(\Delta)}$ to stabilize, plus one additional round for $Sync^{ME}_{\mathcal{B}(\Delta)}$ to ensure that clocks have taken a value in $[0..n-1]$ thanks to the incrementation modulo $K(= n)$. Then, the clocks remain permanently within $[0..n-1]$ and become synchronized within at most $4\Delta + 1$ additional rounds (Theorem 2). From that point on, the mutual exclusion is simply achieved as follows: for every $i \in [0..n-1]$, process p_i executes its critical section (*i.e.*, CS) if $C(p_i) = Rank(p_i)$; see the code of Algorithm $ME_{\mathcal{B}(\Delta)}$ (Algorithm 3). Once stabilized, any process enters its critical section at most every $n-1$ rounds, which is optimal, since clocks increment at each round.

Hence, still using the synchronous composition, Algorithm $Mutex_{\mathcal{B}(\Delta)}$ is formally defined as follows: $Mutex_{\mathcal{B}(\Delta)} = ME_{\mathcal{B}(\Delta)} \circ Rank_{\mathcal{B}(\Delta)} \circ Sync^{ME}_{\mathcal{B}(\Delta)}$.

Algorithm 3: $ME_{\mathcal{B}(\Delta)}$ any process p

```
1  Inputs:
2      N ∈ ℕ* // upper bound on the number of nodes
3      rank(p) ∈ {0, ..., N − 1} // input from Rank_B(Δ)
4      C(p) ∈ {0, ..., N − 1} // input from Sync^ME_B(Δ)

5  foreach round do
6      if  C(p) = rank(p) then
7          CS;
```

One can define the set of legitimate configuration of $Mutex_{\mathcal{B}(\Delta)}$ as any configuration γ of $Mutex_{\mathcal{B}(\Delta)}$ such that: (1) the projection of γ onto the variables of $Rank_{\mathcal{B}(\Delta)}$ is a legitimate configuration of $Rank_{\mathcal{B}(\Delta)}$, (2) the projection of γ onto the variables of $Sync^{ME}_{\mathcal{B}(\Delta)}$ is a legitimate configuration of $Sync_{\mathcal{B}(\Delta)}$, and (3) all clocks have a value in $[0..n-1]$ in γ.

First, observe that the specification of the mutual exclusion is straightforwardly achieved from any legitimate configuration of $Mutex_{\mathcal{B}(\Delta)}$. Then, from Property 1 and Theorem 3, after at most $(6\Delta+2)(N+1)+(4\Delta+2)$ rounds the processes have constant unique names in $[0..n-1]$ and the period of $Sync^{ME}_{\mathcal{B}(\Delta)}$ is forever equal to n. Moreover, from that point on, after at most $4\Delta+2$ rounds, all reached configurations satisfy (2) and (3) and a legitimate configuration of $Rank_{\mathcal{B}(\Delta)}$ is periodically reached every $(3\Delta+1)(N+1)+1$ rounds; see Corollary 1. Indeed, a cycle terminates every $(3\Delta+1)(N+1)+1$ rounds, by definition. Now, $4\Delta+2 < (3\Delta+1)(N+1)+1$. So, after at most $\mathcal{O}(\Delta N)$ rounds, the systems reaches a legitimate configuration of $Mutex_{\mathcal{B}(\Delta)}$ and we can conclude with the following theorem:

Theorem 4. *Algorithm* $Mutex_{\mathcal{B}(\Delta)}$ *is a self-stabilizing mutual exclusion algorithm for* $\mathcal{B}(\Delta)$*, it stabilizes in* $\mathcal{O}(\Delta.N)$ *rounds. It requires messages of size* $\mathcal{O}(\log N + \log \Delta + \log |IDSET|)$ *bits and needs* $\mathcal{O}(N(\log N + \log \Delta + \log |IDSET|))$ *bits of memory per process. Its waiting time is optimal, i.e., at most* $n-1$ *rounds.*

6   Conclusion

We considered dynamic networks where, at any instant, each process can communicate by routing with any other process within at most Δ rounds. In this context, we presented a self-stabilizing mutual exclusion algorithm that uses bounded local memories and achieves a stabilization time in $\mathcal{O}(N\Delta)$ rounds, where N is an upper bound on the number of processes. Optimal in waiting time, our approach relies on self-stabilizing solutions to leader election, unison, and ranking problems. Self-stabilizing solutions for these two latter problems are also provided in this paper.

Several directions emerge from this work, among which two merit special attention. The first concerns reducing the memory footprint of our ranking and mutual exclusion algorithms (it seems very challenging to significantly improve the memory footprint of our unison algorithm). Then, following an approach similar to [2], the second involves investigating the feasibility of designing self-stabilizing mutual exclusion algorithms for more general classes of dynamic graphs, including Classes $\mathcal{Q}(\Delta)$ and $\mathcal{R}$.

References

1. Altisen, K., Devismes, S., Durand, A., Johnen, C., Petit, F.: On implementing stabilizing leader election with weak assumptions on network dynamics. In: Miller, A., Censor-Hillel, K., Korhonen, J.H. (eds) PODC '21: ACM Symposium on Principles of Distributed Computing, pp. 21–31, Virtual Event, Italy, July 26-30 2021. ACM

2. Altisen, K., Devismes, S., Durand, A., Johnen, C., Petit, F.: Self-stabilizing systems in spite of high dynamics. Theor. Comput. Sci. **964**, 113966 (2023)
3. Altisen, K., Devismes, S., Durand, A., Petit, F.: Gradual stabilization. J. Parallel Distributed Comput. **123**, 26–45 (2019)
4. Angluin, D., Aspnes, J., Diamadi, Z., Fischer, M.J., Peralta, R.: Computation in networks of passively mobile finite-state sensors. Distrib. Comput., 235–253 (2006)
5. Beauquier, J., Gradinariu, M., Johnen, C.: Cross-over composition - enforcement of fairness under unfair adversary. In: Datta, A.K., Herman, T. (eds.) WSS 2001. LNCS, vol. 2194, pp. 19–34. Springer, Heidelberg (2001). https://doi.org/10.1007/3-540-45438-1_2
6. Braud-Santoni, N., Dubois, S., Kaaouachi, M.H., Petit, F.: The next 700 impossibility results in time-varying graphs. Int. J. Netw. Comput. **6**(1), 27–41 (2016)
7. Bui-Xuan, B.M., Ferreira, A., Jarry, A.: Computing shortest, fastest, and foremost journeys in dynamic networks. Int. J. Found. Comput. Sci. **14**(2), 267–285 (2003)
8. Cai, S., Izumi, T., Wada, K.: How to prove impossibility under global fairness: On space complexity of self-stabilizing leader election on a population protocol model. Theory Comput. Syst. **50**(3), 433–445 (2012)
9. Casteigts, A., Flocchini, P.: Deterministic algorithms in dynamic networks: formal models and metrics. Technical report, Defence Research and Development Canada, 2013-020 (2013)
10. Casteigts, A., Flocchini, P.: Deterministic algorithms in dynamic networks: problems, analysis, and algorithmic tools. Technical report, Defence Research and Development Canada, 2013-020 (2013)
11. Casteigts, A.: A Journey through Dynamic Networks (with Excursions) (2018)
12. Casteigts, A., Flocchini, P., Quattrociocchi, W., Santoro, N.: Time-varying graphs and dynamic networks. Int. J. Parallel Emergent Distrib. Syst. **27**(5), 387–408 (2012)
13. Charron-Bost, B., de Monterno, L.P.: Self-stabilizing clock synchronization in dynamic networks. In Hillel, E., Palmieri, R., Rivière, E., (eds) 26th International Conference on Principles of Distributed Systems, OPODIS 2022, December 13-15, 2022, Brussels, Belgium, vol. 253 of LIPIcs, pp. 28:1–28:17. Schloss Dagstuhl - Leibniz-Zentrum für Informatik (2022)
14. Charron-Bost, B., Moran, S.: The firing squad problem revisited. Theor. Comput. Sci. **793**, 100–112 (2019)
15. Datta, A.K., Larmore, L.L.: Self-stabilizing leader election in dynamic networks. Theory Comput. Syst. **62**(5), 977–1047 (2018)
16. Datta, A.K., Larmore, L.L., Devismes, S., Heurtefeux, K., Rivierre, Y.: Self-stabilizing small k-dominating sets. Int. J. Netw. Comput. **3**(1), 116–136 (2013)
17. Dijkstra, E.W.: Self-stabilizing systems in spite of distributed control. Commun. ACM **17**(11), 643–644 (1974)
18. Dolev, S.: Optimal time self-stabilization in uniform dynamic systems. Parallel Process. Lett. **8**(1), 7–18 (1998)
19. Dolev, S., Hanemann, A., Schiller, E.M., Sharma, S.: Self-stabilizing End-to-End Communication in (Bounded Capacity, Omitting, Duplicating and non-FIFO) Dynamic Networks. In: Richa, A.W., Scheideler, C. (eds.) SSS 2012. LNCS, vol. 7596, pp. 133–147. Springer, Heidelberg (2012). https://doi.org/10.1007/978-3-642-33536-5_14
20. Dolev, S., Herman, T.: Superstabilizing protocols for dynamic distributed systems. Chic. J. Theor. Comput. Sci., (1997)
21. Dolev, S., Israeli, A., Moran, S.: Self-stabilization of dynamic systems assuming only read/write atomicity. Distrib. Comput. **7**(1), 3–16 (1993)

22. Even, S., Rajsbaum, S.: Unison in distributed networks. In: Capocelli, R.M. (ed) Sequences, pp. 479–487, New York, NY (1990). Springer New York
23. Fischer, M., Jiang, H.; Self-stabilizing leader election in networks of finite-state anonymous agents. In: Shvartsman, M.M.A.A. (ed) Principles of Distributed Systems, pp. 395–409, Berlin, Heidelberg. Springer Berlin Heidelberg (2006)
24. Gómez-Calzado, C., Casteigts, A., Lafuente, A., Larrea, M.: A connectivity model for agreement in dynamic systems. In: Träff, J.L., Hunold, S., Versaci, F. (eds.) Euro-Par 2015. LNCS, vol. 9233, pp. 333–345. Springer, Heidelberg (2015). https://doi.org/10.1007/978-3-662-48096-0_26
25. Gouda, M.G., Herman, T.: Stabilizing unison. Inf. Process. Lett. **35**, 171–175 (1990)
26. Gouda, M.G., Herman, T.: Adaptive programming. IEEE Trans. Software Eng. **17**(9), 911–921 (1991)
27. Herman, T.: Superstabilizing mutual exclusion. Distributed Comput. **13**(1), 1–17 (2000)
28. Katayama, Y., Ueda, E., Fujiwara, H., Masuzawa, T.: A latency optimal superstabilizing mutual exclusion protocol in unidirectional rings. J. Parallel Distrib. Comput. **62**(5), 865–884 (2002)
29. Kuhn, F., Lynch, N.A., Oshman, R.: Distributed computation in dynamic networks. In: Schulman, L.J. (ed.) Proceedings of the 42nd ACM Symposium on Theory of Computing, STOC 2010, Cambridge, Massachusetts, USA, 5-8 June 2010, pp. 513–522. ACM (2010)
30. Zaks, S.: Optimal distributed algorithms for sorting and ranking. IEEE Trans. Comput. **34**(4), 376–379 (1985)

Brief Announcement: PQ-STAR
Post-Quantum Stateless Auditable Rekeying

Shlomi Dolev[1], Avraham Yagudaev[1(✉)], and Moti Yung[2]

[1] Department of Computer Science, Ben-Gurion University of the Negev Beer Sheva,
Beersheba, Israel
`avraamy@post.bgu.ac.il`
[2] Google, New York, NY, USA

Abstract. Rekeying is an effective technique for protecting symmetric ciphers against side-channel and key-search attacks. Since its introduction, numerous rekeying schemes have been developed [16]. We introduce *Post-Quantum Stateless Auditable Rekeying (PQ-STAR)*, a novel post-quantum secure [4] stateless rekeying scheme with *audit* support. PQ-STAR is presented in three variants of increasing security guarantees: (*i*) *Plain PQ-STAR* lets an authorized auditor decrypt and verify selected ciphertexts; (*ii*) *Commitment-based PQ-STAR* with the additional binding guarantee from the commitments, preventing a malicious sender from potentially claiming a random or wrong session key. (*iii*) *Zero-knowledge PQ-STAR* equips each session key with a signature-based zero-knowledge proof (ZKP), which proves that the session key was derived honestly, without ever revealing the secret preimage. We provide informal arguments that all variants achieve *key-uniqueness*, *index-hiding*, and *forward-secrecy*, even if a probabilistic polynomial-time (PPT) adversary arbitrarily learns many past session keys. PQ-STAR provides a verified, stateless, and audit-capable rekeying primitive that can be seamlessly integrated as a post-quantum upgrade for existing symmetric-key infrastructures.

Keywords: Post-Quantum Cryptography · Stateless Rekeying · Side Channel Resistance · Symmetric-Key Updates · Zero-Knowledge Proof · Auditability

1 Introduction

Side-channel attacks [13] endanger practical cryptographic implementations by exploiting physical leakages, such as timing, power consumption, or electromagnetic emissions. A well-established countermeasure is rekeying, i.e., updating

Partially supported by the Google Research Grant, the Rita Altura Trust Chair in Computer Science, the BGU Data Center, the Frankel Center for Computer Science, and the Israeli Science Foundation (Grant No. 465/22).

S. Bonomi et al. (Eds.): SSS 2025, LNCS 16350, pp. 211–215, 2026.
https://doi.org/10.1007/978-3-032-11127-2_17

cryptographic keys at sufficiently short intervals so that an adversary observes only negligible leakage for any single key.

Rekeying [2] can be categorized as either *stateless (parallel)* or *stateful (sequential)*. Stateless rekeying—also called *parallel rekeying* [1,2]—derives every session key directly from a long-term master key, which must withstand both Simple Power Analysis (SPA) and Differential Power Analysis (DPA) [10,11]. In contrast, stateful rekeying maintains a mutable state for generating subsequent keys and was popularized by Kocher's index-based scheme [12].

In a related direction, Dolev *et al.* proposed HBSS [6], a post-quantum, hash-based stateless signature scheme. Although HBSS addresses digital signatures rather than key evolution, its stateless, hash-based nature motivated the design choices of PQ-STAR.

We introduce PQ-STAR, the first rekeying primitive that unifies post-quantum security, stateless fresh rekeying, and an explicit audit mechanism. Auditability is beneficial in various social contexts, as it can provide (sampled) evidence of correct and lawful actions, see, e.g., [5]. We present three progressively stronger variants: plain, commitment-based, and zero-knowledge proof, and argue informally that each satisfies *key-uniqueness*, *index-hiding*, and *forward-secrecy*. Some of the details and proofs are omitted from this short version; see [7] for additional details.

2 Plain PQ-STAR

Construction. PQ-STAR is a stateless rekeying scheme that supports message auditing by a third party. The scheme relies on a large shared secret array from which ephemeral session keys are derived. By design, each encrypted message can be independently verified by an auditor without revealing any long-term secret.

Let $m \geq 1$ and let $A = (a_1, a_2, \ldots, a_m)$ be an array of m secret values shared between Alice and Bob (e.g., via an initial key exchange). The array A functions as a large shared secret key that must remain confidential at all times (like any long-term symmetric key).

In PQ-STAR, each session (each message sent between Alice and Bob) uses a fresh session key derived from A. At a high level, the process is as follows:

1. Alice and Bob establish the shared master array A (or equivalent seeds).
2. For every new message, Alice (or Bob) selects k random indices (with repetition allowed) to assemble an index tuple $\mathcal{I} = (j_1, \ldots, j_k)$ from $[m]$. She then computes the session key $sk = H(a_{j_1} \parallel a_{j_2} \parallel \cdots \parallel a_{j_k})$,
   where H is a cryptographic hash function (e.g., SHA-256). Bob (or Alice), who also knows A, can derive the identical sk from the shared tuple $\mathcal{I}$.
3. Alice uses sk to encrypt the message, and she includes $\mathcal{I}$ in the transmission (in plaintext) so that Bob knows which a_i's were used. An authorized auditor can later verify the ciphertext using $\mathcal{I}$ and sk

PQ-STAR also supports an *audit* mechanism to prove the origin and correctness of an encrypted message. If an authorized auditor needs to verify a ciphertext C and its claimed plaintext M, Alice (or Bob) can reveal the corresponding session key sk without exposing any part of the master key A.

Security Analysis.

The next proofs holds for large enough parameters; for instance $n \geq 256$, $k \geq 16$, and $m \geq 2^{20}$.

Theorem 1 (Key-Uniqueness). *With overwhelming probability, distinct sessions produce distinct session keys.*

Proof (Sketch). Two sessions produce the same sk only if they use the same index tuple $\mathcal{I}$ or if H collides on two different inputs. $\square$

Theorem 2 (Index-Hiding). *No PPT adversary has more than a negligible advantage in determining the session key sk from the public index tuple $\mathcal{I}$.*

Proof (Sketch). The adversary sees $\mathcal{I} = (j_1, \ldots, j_k)$ but not the values $a_{j_1}, \ldots, a_{j_k}$ themselves. $\square$

Theorem 3 (Forward-Secrecy). *No PPT adversary that obtains up to t session keys (and their index tuples) has more than negligible advantage in learning any information about any other session key.*

Proof (Sketch). Suppose the adversary learns $(\mathcal{I}_1, sk_1), \ldots, (\mathcal{I}_t, sk_t)$ for some sessions. These give no direct information about any a_i, since each $sk_j = H(a_{j_1} \| \cdots \| a_{j_k})$ is a one-way hash. $\square$

3   Extensions and Conclusions

Commitment-based PQ-STAR.

• **Construction.** The commitment-based variant of PQ-STAR differs from the plain scheme only in the *Setup* and *Audit* phases. As before, Alice and Bob share a secret array $A = (a_1, \ldots, a_m)$ of random n-bit values. In *Setup*, they derive a *commitment* array $Cmt = (c_1, \ldots, c_m)$ by setting $c_i = SHA(a_i)$,

where SHA is a cryptographic hash function (e.g., SHA-256). The Cmt is published publicly, while the values a_i remain secret. These commitments bind Alice and Bob to their chosen preimages (the a_i): once c_i is public, neither party can later claim a different a_i without detection.

• **Security Analysis.**

In the commitment variant, each audit reveals the specific a_i values used in that session. Thus, the total number of sessions derived from a single master key must be further limited to preserve the unpredictability of new session keys. In practice, one can adopt the technique from HBSS [6] where the parameter α represents the fraction of total array entries that have been exposed so far. This maintains the scheme's index-hiding and forward secrecy.

Zero-Knowledge-Based PQ-STAR.

• **Construction.** We now describe a third variant of PQ-STAR that enables an auditor to *validate* every session key in *zero knowledge*. Instead of exposing the preimages $a_{j_1}, \ldots, a_{j_k}$ during the audit, the sender proves possession of the correct secret material using a post-quantum digital signature equipped with a zero-knowledge proof. We instantiate the signature with *FAEST* [3], but any (post-quantum) scheme providing unforgeability and zero-knowledge suffices.

To verify a transcript $(C, \mathcal{I})$, an auditor proceeds as follows:

1. Alice reveals the session key sk.
2. The auditor decrypts C with sk to obtain $(M', \mathcal{I}', \sigma')$ and aborts if $\mathcal{I}' \neq \mathcal{I}$.
3. The auditor checks that $\mathrm{Verify}_{\mathsf{FAEST}}(\mathsf{pk}_{\mathsf{FAEST}}, sk, \sigma') = 1$.

At no point does the auditor learn any a_i; the zero-knowledge property of FAEST guarantees that the signature discloses no information beyond the validity of sk.

• **Security Analysis**

The construction inherits key-uniqueness, index-hiding, and forward-secrecy. The additional FAEST signature provides:

Authenticity. No PPT adversary can produce an auditor-accepted transcript $(C, \mathcal{I}, \sigma)$ without knowledge of the real session key sk, except with negligible probability. A forged transcript yields a fresh (sk, σ) signature on a message never queried, violating EUF-CMA.

Zero-Knowledge Privacy. For every PPT auditor $\mathcal{A}$ there exists a PPT simulator $\mathcal{S}$ whose output is computationally indistinguishable from $\mathcal{A}$'s real view of any transcript. Because FAEST is zero-knowledge and the encryption is IND-CPA, the resulting distribution is indistinguishable from a real transcript.

Performance and Parameter Selection.

PQ-STAR is parameterised by three integers: n, m and k. For example, for post-quantum security, one may choose $n = 256$, so that the session key has 256 bits and offers roughly 128 bits of quantum security [9]. Selecting $k = 16$ and $m = 2^{20}$, for example, yields a negligible chance of any collision even across billions of sessions.

Deriving a session key requires computing a single hash; hashing a few hundred bytes, therefore, takes on the order of microseconds [15], which is negligible compared with the time to encrypt a message. Consequently, the overhead of session-key derivation is constant per message [8].

PQ-STAR can be integrated into existing protocols with minimal changes. The index tuple $\mathcal{I}$ can be sent in the clear or as associated data in the *authenticated encryption with associated data*. PQ-STAR operates at the application layer and can run atop a secure transport such as TLS 1.3 [14].

Conclusions.

We introduced PQ-STAR, a post-quantum, stateless, auditable rekeying scheme. PQ-STAR allows each message to be encrypted with a fresh session key derived from a shared secret array, while enabling an authorized auditor to verify any ciphertext by revealing only the corresponding one-time session key. The scheme utilizes only hash functions, operates without maintaining state between messages, and does not require any synchronization between sender and receiver. We provide informal arguments that all PQ-STAR variants are secure.

References

1. Abdalla, M., Belaïd, S., Fouque, P.-A.: Leakage-resilient symmetric encryption via re-keying. In: Bertoni, G., Coron, J.-S. (eds.) CHES 2013. LNCS, vol. 8086, pp. 471–488. Springer, Heidelberg (2013). https://doi.org/10.1007/978-3-642-40349-1_27
2. Abdalla, M., Bellare, M.: Increasing the lifetime of a key: a comparative analysis of the security of re-keying techniques. In: International Conference on the Theory and Application of Cryptology and Information Security, pp. 546–559. Springer (2000)
3. Baum, C., et al.: Faest v2: Algorithm specifications (2025)
4. Chen, L., et al.: Report on post-quantum cryptography, vol. 12. National Institute of Standards and Technology ., US Department of Commerce (2016)
5. Dolev, S., Panagopoulou, P.N., Rabie, M., Schiller, E.M., Spirakis, P.G.: Rationality authority for provable rational behavior. Algorithms, Probability, Networks, and Games: scientific Papers and Essays Dedicated to Paul G. Spirakis on the Occasion of His 60th Birthday, pp. 33–48 (2015)
6. Dolev, S., Yagudaev, A., Yung, M.: Hbss:(simple) hash-based stateless signatures–hash all the way to the rescue! Cryptography and Communications, pp. 1–18 (2025)
7. Dolev, S., Yagudaev, A., Yung, M.: PQ-STAR: Post-quantum stateless auditable rekeying. Cryptology ePrint Archive, Paper 2025/1489 (2025). https://eprint.iacr.org/2025/1489
8. Dworkin, M.J.: Sp 800-38d. recommendation for block cipher modes of operation: Galois/counter mode (gcm) and gmac. National Institute of Standards & Technology (2007)
9. Grover, L.K.: A fast quantum mechanical algorithm for database search. In: Proceedings of the twenty-eighth annual ACM symposium on Theory of computing, pp. 212–219 (1996)
10. Kocher, P., Jaffe, J., Jun, B.: Differential power analysis. In: Wiener, M. (ed.) CRYPTO 1999. LNCS, vol. 1666, pp. 388–397. Springer, Heidelberg (1999). https://doi.org/10.1007/3-540-48405-1_25
11. Kocher, P., Jaffe, J., Jun, B., Rohatgi, P.: Introduction to differential power analysis. J. Cryptogr. Eng. 1, 5–27 (2011)
12. Kocher, P.C.: Leak-resistant cryptographic indexed key update (2003). uS Patent 6,539,092
13. Mangard, S., Oswald, E., Popp, T.: Power analysis attacks: Revealing the secrets of smart cards, vol. 31. Springer Science & Business Media (2008)
14. Rescorla, E.: The transport layer security (TLS) protocol version 1.3. Technical Report (2018)
15. Santos Jr, C.E., Silva, L.M.d., Torquato, M.F., Silva, S.N., Fernandes, M.A.: Sha-256 hardware proposal for IoT devices in the blockchain context. Sensors 24(12), 3908 (2024)
16. Smyshlyaev, S.V.: Re-keying Mechanisms for Symmetric Keys. RFC 8645 (2019). https://doi.org/10.17487/RFC8645

On the Computational Power of Mobile Robots Under Sequential Schedulers

Caterina Feletti[1]([⊠]) [iD], Paola Flocchini[2] [iD], and Nicola Santoro[3] [iD]

[1] Università degli Studi di Milano, Milan, Italy
caterina.feletti@unimi.it
[2] University of Ottawa, Ottawa, Canada
paola.flocchini@uottawa.ca
[3] Carleton University, Ottawa, Canada
santoro@scs.carleton.ca

Abstract. We consider distributed systems of autonomous, punctiform, mobile robots that operate in the Euclidean plane by executing an infinite sequence of *Look-Compute-Move* cycles. Robots are anonymous, indistinguishable, homogeneous, and disoriented. In literature, four base models have been proposed to study four different memory-communication settings: $\mathcal{OBLOT}$ (oblivious and silent), $\mathcal{FSTA}$ (finite-state and silent), $\mathcal{FCOM}$ (oblivious and finite-communication), and $\mathcal{LUMI}$ (finite-state and finite-communication). In particular, the research has investigated how the computational power of these models is affected by considering three main classes of robot schedulers: **FSYNCH** (fully synchronous), **SSYNCH** (semi-synchronous), and **ASYNCH** (asynchronous).

This paper focuses on a peculiar type of **SSYNCH** schedulers, the *sequential* ones, which activate only one robot at each round. We consider three subclasses: the general sequential scheduler (**SEQ**), the *permutation* scheduler (**PERM**), and the well-known *round-robin* (**RROBIN**). For each base model, we investigate how the robots' computational power changes as the scheduler class varies, thus providing a first overview of the computational landscape of sequential schedulers.

Keywords: Autonomous mobile robots · Look-Compute-Move · Computational power · Sequential schedulers

1 Introduction

1.1 Framework

Swarm robotics has been the subject of great interest and numerous studies in recent years. This is due to the advantages that such a technology brings (distributed system of tiny and relatively cheap computational units) and, consequently to the variety of applications that include surveillance, exploration, rescue missions, data collection, drug administration in an agricultural field. From a theoretical perspective, the research has been interested in formally modelling

© The Author(s), under exclusive license to Springer Nature Switzerland AG 2026
S. Bonomi et al. (Eds.): SSS 2025, LNCS 16350, pp. 216–232, 2026.
https://doi.org/10.1007/978-3-032-11127-2_18

a swarm of robots as a system of autonomous mobile computational entities, and designing algorithms that such a system can execute. The *Look-Compute-Move* model (LCM) [18] is commonly used to formalize and study the computational capabilities of swarms of robots. According to this model, any robot acts through an infinite sequence of LCM cycles: when active, a robot *looks* at its surroundings obtaining a snapshot of the position of the other robots, it *computes* a destination according to a protocol based on such a snapshot, and it *moves* toward the computed destination. After that, it remains idle until its next activation, and thus its next LCM cycle. Robots are generally assumed to be autonomous, anonymous and indistinguishable (i.e., without internal or external ids), homogeneous (i.e., they execute the same distributed algorithm), and punctiform entities on the Euclidean plane. This research field intends to minimize the robots' capabilities and to design robust algorithms that the swarm can use to solve problems collaboratively. Typically, robots are required to arrange themselves on the space to form some patterns (**Pattern Formation**; e.g., [6,10,12,14,19,26]), to convene at some location not a priori specified (**Gathering**; e.g., [1,7,9]), to cover the space they occupy in some homogeneous fashion (**Scattering**; e.g., [2,15,24]); to travel together in the environment in some formation (**Flocking**; e.g., [5,23]), etc.

Under the umbrella of the LCM model, several (sub)models have been defined in order to analyze the impact of different features on the computational power of robots. The basic model, called $\mathcal{OBLOT}$, assumes robots to be oblivious[1] and silent, i.e., devoid of any persistent storage and communication tools, respectively, except those derived from stigmergy[2] (i.e., movements and positioning of robots). The $\mathcal{LUMI}$ model was introduced to overcome these limits: robots are assumed to be *luminous*, i.e., equipped with a persistent light that can be updated at each cycle using a color from a constant-size palette. Since this light is both internally and externally visible, it plays the role of both an internal state (memory) as well as a communication tool. Halfway between $\mathcal{OBLOT}$ and $\mathcal{LUMI}$, the two intermediate models $\mathcal{FSTA}$ (finite-state) and $\mathcal{FCOM}$ (finite-communication) have been introduced to study how the presence of only memory and only communication affects the computational power of robots, respectively.

The timing of when a robot is active and performs a cycle, as well the duration of each action within a cycle, is determined by an adversary, called *scheduler*, which plays a relevant role in characterizing the computational power of the robots. The literature has studied two main types of schedulers: *synchronous* and *asynchronous*. Under synchronous schedulers, time is divided into logical rounds; in each round, a (non-empty) subset of robots is activated and simultaneously performs an LCM cycle. Under asynchronous schedulers (**ASYNCH**), there is no notion of global time; each robot performs its cycle independently from the others, and the duration of each cycle is arbitrary but finite. Schedulers are assumed to be *fair*, i.e., any robot is activated within finite time and infinitely often.

---

[1] Hence the name $\mathcal{OBLOT}$.
[2] Term coined to refer to termite behaviors.

Intensive research has focused on determining the impact that different scheduler classes have on the computational power of the robots in the various models; this is done by examining the set $\mathcal{P}\left(X^S\right)$ of problems that can be solved by the robots under a given scheduler class S in a given model $X \in \{\mathcal{OBLOT}, \mathcal{FST}\mathcal{A}, \mathcal{FCOM}, \mathcal{LUMI}\}$ and, more importantly, determining the relationship (containment, equivalence, incomparability) between different problem sets, casting some light on the computational landscape (e.g., [3, 4, 11, 16, 20, 21, 25]).

Much of the algorithmic research has been conducted on the class of synchronous schedulers, in particular on the popular *fully synchronous* scheduler (FSYNCH), where all robots are activated in every round, and on the *semi-synchronous* scheduler (SSYNCH), which activates an arbitrary non-empty subset of robots at each round. Observe that, among synchronous schedulers, SSYNCH has the strongest *adversarial power* (in terms of freedom to choose which robots to activate in each round) while FSYNCH has no adversarial power; not surprisingly, $\mathcal{P}\left(X^{\text{SSYNCH}}\right) \subset \mathcal{P}\left(X^{\text{FSYNCH}}\right)$ for all $X \in \{\mathcal{OBLOT}, \mathcal{FST}\mathcal{A}, \mathcal{FCOM}, \mathcal{LUMI}\}$.

An important but less investigated class of synchronous schedulers is that of *sequential* schedulers, which activate only one robot at each round. In this class, we distinguish three subclasses, each a subclass of the previous: the most general (and powerful) sequential class SEQ, where the choice of the activated robot is arbitrary; the *permutation* class (PERM), where the scheduler activates all robots exactly once before the next turn of activations; and the well-known *round-robin* (RROBIN), where the order of activation is the same in each turn.

While extensively studied in other fields such as cellular automata (where they are called *asynchronous*) and self-stabilization (where they are called *centralized*), the investigations on sequential schedulers in swarm robotics have been limited to issues of fault-tolerance [8, 13], and more recently to gathering and pattern formation in $\mathcal{OBLOT}$ [17, 22].

In this paper, we start the investigation of the relationship between the computational power of sequential schedulers, and provide a (near) complete characterization of the computational landscape of the four models ($\mathcal{OBLOT}$, $\mathcal{FST}\mathcal{A}$, $\mathcal{FCOM}$, $\mathcal{LUMI}$) under the sequential classes SEQ, PERM, and RROBIN. This investigation naturally extends the work of [3, 11, 20, 21], which provides the computational landscape of the four models under FSYNCH, SSYNCH, and ASYNCH.

1.2   Contributions

This work studies how the computational power of the robots changes with the change of the sequential scheduler class. In particular, we are interested in determining, for each class of sequential schedulers, the impact of its adversarial capabilities when the robots have different capabilities with respect to (limited or no) persistent memory and/or (limited or no) explicit communication. Given a model X and a class of schedulers S, we denote by $\mathcal{P}\left(X^S\right)$ the set of problems solvable by the robots in model X under the scheduler class S. We trivially have that $\mathcal{P}\left(X^{\text{SEQ}}\right) \subseteq \mathcal{P}\left(X^{\text{PERM}}\right) \subseteq \mathcal{P}\left(X^{\text{RROBIN}}\right)$.

In this paper, we characterize the precise nature of these relationships. More precisely, for any $X \in \{\mathcal{OBLOT}, \mathcal{FSTA}, \mathcal{FCOM}, \mathcal{LUMI}\}$ and $S_1, S_2 \in \{\mathtt{SEQ}, \mathtt{PERM}, \mathtt{RROBIN}\}$, we determine the exact computational relationship (dominance or equivalence) between X^{S_1} and X^{S_2}, thus providing a precise picture of the computational landscape. Among the several results, we show that, in $\mathcal{LUMI}$, the computational power of the robots under the general sequential class $\mathtt{SEQ}$ is the same as under the class $\mathtt{PERM}$. We prove this equivalence constructively, by designing a *simulator* that allows robots to correctly execute under $\mathtt{SEQ}$ any protocol designed to run under $\mathtt{PERM}$. Interestingly, in all other cases, the relationship is one of strict dominance. The proofs are provided by presenting in each case a *separator*, i.e., a problem solvable in one setting but unsolvable in the other. Due to space limitations, some proofs have been omitted.

2   Model and Terminology

2.1   Robots and Schedulers

Core Robot Features. Robot swarms are collaborative systems aimed at solving problems by executing distributed algorithms. In all models considered here, a swarm of robots $\mathcal{R} = \{r_1, \ldots, r_n\}$ is a set of $n > 1$ *punctiform* and *mobile* computational entities operating in synchronous rounds in the Euclidean plane $\mathbb{R}^2$. They are *autonomous, anonymous, indistinguishable, homogeneous*: namely, they act without any central control, they do not have any internal or external identifiers, and they all execute the same algorithm in a distributed way.

Each robot is embedded with a local coordinate system, which may not be the same for the whole swarm; moreover, each robot has a sensor through which it obtains a map, *snapshot*, representing the position (w.r.t. to its coordinate system) of all the robots in the swarm.

Time is marked in atomic rounds ($t = 0, 1, \ldots$). Each robot active in a round executes a *Look-Compute-Move* (LCM) cycle simultaneously with the other active robots: it takes the snapshot of the swarm (*Look*), executes the algorithm using the snapshot as input (*Compute*), and then moves straight towards the computed destination point (*Move*).

Given an absolute coordinate system $\hat{\Xi}$ on $\mathbb{R}^2$ (not known by the swarm) we denote by $\underline{x}_i(t) \in \mathbb{R}^2$ the position of r_i according to $\hat{\Xi}$ at (the beginning of) round t, and by $C(t) = \{\underline{x}_1(t), \ldots, \underline{x}_n(t)\}$ the multiset, called *configuration*, of the positions occupied by the robots in that round. In the following, when no ambiguity arises, we will omit the round number.

More than one robot can occupy the same position at the same time, forming a *multiplicity*. We assume robots have *strong multiplicity detection*: in the *Look* phase, they can distinguish the number of robots in a multiplicity.

Memory and Communication. We consider the well-known models $\mathcal{OBLOT}$, $\mathcal{FSTA}$, $\mathcal{FCOM}$, $\mathcal{LUMI}$, which differ in the possibility for robots to store and/or communicate a $O(1)$ amount of information.

Under $\mathcal{OBLOT}$, robots are assumed to be *oblivious* and *silent*; that is, they are devoid of any direct means to store or communicate some information.

Under $\mathcal{LUMI}$, each robot r is equipped with $O(1)$ bits of persistent memory, called *light* and denoted lig_r, whose value, called *color*, can be updated at the end of the *Compute* phase. We assume the set (or palette) of colors always contains the default color OFF, to which all robots' lights are initialized. The color of each light is visible to any robot in the swarm during its *Look* phase, thus it is adopted both as an internal state and a communication means.

The $\mathcal{FSTA}$ model (silent but finite-state) differs from $\mathcal{LUMI}$ in that the light of each robot r is only internally visible; thus it can be seen only by r in its *Look* phase. Instead, in the $\mathcal{FCOM}$ model (oblivious but finite-communication), the light of r is visible only by the other robots in their *Look* phase.

Observe that, in addition to the location of all the n robots, a snapshot taken by a robot in $\mathcal{LUMI}$ also shows the color of all the n robots, while in $\mathcal{FCOM}$ ($\mathcal{FSTA}$, resp.) it shows the color of only the other $n - 1$ robots (of only the robot itself, resp.).

Additional Features. Our study considers two further relevant features in the robots' models, i.e., rigidity and disorientation, each appearing in two variants.

1. (*rigidity*) With respect to the robot movements, we distinguish between two variants: RIG and NONRIG. Under RIG, the robots always reach their destination in the *Move* phase (*rigid movements*). Under NONRIG, the robots may be stopped during their *Move* phase by an adversary before reaching their destination, but after having traveled at least a fixed non-null distance δ, whose value is unknown to the robots (*non-rigid movements*).
2. (*disorientation*) With respect to the local coordinate system of the robots, we distinguish between two variants: FIXDIS and VARDIS. Under FIXDIS, the local coordinate of each robot does not change cycle by cycle (*fixed disorientation*); under VARDIS, the local coordinate of each robot may change at any activation (*variable disorientation*).

Sequential Schedulers. We study swarms where only one robot is activated at each round, and it executes its LCM cycle in the round. The choice of which robot is activated in a given round is made by an adversary, called *sequential scheduler*. Given a swarm $\mathcal{R} = \{r_1, \ldots, r_n\}$, a sequential scheduler can be defined as the infinite string $\mathfrak{S} = r_{i_0} r_{i_1} \ldots$ describing for any round $t \in \mathbb{N}$ the activated robot $r_{i_t} \in \mathcal{R}$.

The scheduler is fair, i.e., it activates every robot infinitely often. This fairness condition enables us to measure time in terms of successive sequences of activations of robots, called *epochs*: the first epoch starts with the first activation in the system; each epoch ends as soon as all robots have been activated; the next epoch starts with the next activation.

We consider three classes of sequential schedulers, each a subclass of the previous. The most general class, **SEQ**, assumes nothing except that robots are activated sequentially and in a fair fashion. The PERM $\subset$ SEQ class (*permutation*) is composed of the sequential schedulers where, for a swarm of n robots, each epoch consists of n rounds. The RROBIN $\subset$ PERM class (*round-robin*) is composed of the permutation-schedulers where robots are activated following the same order within each epoch.

2.2   Settings and Computational Power

We say that a swarm $\mathcal{R}$ of robots belongs to a setting X^S if $\mathcal{R}$ satisfies all the core features listed above, it has $X \in \{\mathcal{OBLOT}, \mathcal{FSTA}, \mathcal{FCOM}, \mathcal{LUMI}\}$ as memory-communication capability, and it is activated by schedulers of class $S \in \{\text{SEQ}, \text{PERM}, \text{RROBIN}\}$. For a given setting X^S, we will specify along the paper if we consider it under NONRIG or RIG, and FIXDIS or VARDIS.

A problem is a task that a swarm of robots must achieve. Since the robots can only observe the current configuration (including colors, depending on the model) and possibly move forming a new configuration, a *problem* P is specified in terms of the properties that the (possibly infinite sequences of) configurations created by the robots must have; e.g., properties of the initial configurations, intermediate configurations, "terminal" configurations (i.e., those where the robots must no longer move), as well as classes of configurations that must be avoided (e.g., those containing multiplicity, co-linearity, etc.). Formally, these properties are expressed as temporal/geometric predicates that must be satisfied by the sequences. We stress that these predicates cannot impose any restrictions on light colors, absolute times, or the number of rounds needed to achieve a configuration; in fact, P might in principle be solved under $\mathcal{OBLOT}$ and any scheduler class.

An algorithm $\mathbb{A}$ is said to solve a problem P in a given setting $M = X^S$ if, for each swarm of robots belonging to M, in all executions, the sequence of configurations produced by the swarm satisfies the requirements of P. If such an algorithm exists, the problem P is said to be *solvable* in M.

We indicate with $\mathcal{P}(M)$ the *computational power* of setting M, i.e., the set of problems solvable in M. Given two settings M_1, M_2, we say that:

Table 1. Computational relations within the four models.

$\mathcal{OBLOT}$	SEQ	$<$	PERM	$<$	RROBIN
$\mathcal{FSTA}$	SEQ	$<$	PERM	$<$	RROBIN
$\mathcal{FCOM}$	SEQ	$<$	PERM	$<$	RROBIN
$\mathcal{LUMI}$	SEQ	$\equiv$	PERM	$<$	RROBIN

- M_1 is *computationally not less powerful* than M_2, formally $M_1 \geq M_2$, if $\mathcal{P}(M_1) \supseteq \mathcal{P}(M_2)$, i.e., any problem solvable in M_2 is solvable in M_1;
- M_1 is *computationally more powerful* than M_2, formally $M_1 > M_2$, if $\mathcal{P}(M_1) \supset \mathcal{P}(M_2)$, i.e., any problem solvable in M_2 is solvable in M_1 and there exists at least a problem solvable in M_1 that is not solvable in M_2;

- M_1 is *computationally orthogonal* to M_2, formally $M_1 \perp M_2$, if $\mathcal{P}(M_1) \setminus \mathcal{P}(M_2) \neq \varnothing$ and $\mathcal{P}(M_2) \setminus \mathcal{P}(M_1) \neq \varnothing$, i.e., there exists at least a problem solvable in M_1 (M_2, resp.) that is not solvable in M_2 (M_1, resp.);
- M_1 is *computationally equivalent* to M_2, formally $M_1 \equiv M_2$, if $\mathcal{P}(M_1) = \mathcal{P}(M_2)$, i.e., any problem solvable in M_1 (M_2, resp.) is solvable in M_2 (M_1, resp.).

We define the relation $<$ ($\leq$, resp.) as the converse relation of $>$ ($\geq$, resp.).

3   Computational Relations

In this section, we study the computational relations that exist within each base model $X \in \{\mathcal{OBLOT}, \mathcal{FSTA}, \mathcal{FCOM}, \mathcal{LUMI}\}$, by considering the three scheduler classes SEQ, PERM and RROBIN. Specifically, for any pair S_1, S_2 of classes taken from $\{\text{SEQ}, \text{PERM}, \text{RROBIN}\}$, we find the computational relation that holds between the two settings X^{S_1}, X^{S_2}. Moreover, our results (except for one) hold independently from the fact that, under the hypothesis of X, we assume RIG or NONRIG, and FIXDIS or VARDIS.

Indeed, since the three classes (i.e., sets of schedulers) are related in a chain of strict set-inclusion RROBIN $\subset$ PERM $\subset$ SEQ, the following theorem holds:

Theorem 1 (Trivial Relations).   *Independently from* RIG-NONRIG *and* FIXDIS-VARDIS, *for any* $X \in \{\mathcal{OBLOT}, \mathcal{FSTA}, \mathcal{FCOM}, \mathcal{LUMI}\}$, *we have that*

$$X^{SEQ} \leq X^{PERM} \leq X^{RROBIN}.$$

Note that the strict set-inclusion of the scheduler classes does not inevitably lead to a strict dominance for the settings. As we will see, $\mathcal{LUMI}$ robots can use lights to simulate the behavior of a class of schedulers (SEQ) into the subclass (PERM), thus proving that $\mathcal{LUMI}^{\text{SEQ}}$ and $\mathcal{LUMI}^{\text{PERM}}$ are computationally equivalent. For all other models, however, we provide separator problems, demonstrating that the computational power of the robots is inversely related to the adversarial power of the schedulers. Table 1 summarizes the results proved in this paper.

3.1   Relations in $\mathcal{OBLOT}$

In this section, we show that in $\mathcal{OBLOT}$ the relationship between the three scheduler classes is one of strict dominance.

Problem 1 (Pinwheel). The problem is perpetual and it is defined recursively. Refer to Fig. 1. Let $\mathcal{C}$ be a configuration where a swarm of robots forms two different acute scalene triangles, say $\triangle ABC$ and $\triangle abc$, with the same shape, orientation, and circumcenter. Let $\triangle abc$ be completely contained in $\triangle ABC$.

Only one robot lies on each vertex of $\triangle abc$. Let $m \geq 3$ be the total number of robots lying on the vertices of $\triangle ABC$. On the circumcenter of the two triangles, $m + 1$ robots form a multiplicity. Assume that $\angle a > \angle b > \angle c$: let the clockwise orientation be given to all the swarm as $\overset{\frown}{abc}$. The problem requires the swarm to reach a configuration $\mathcal{C}'$ where each multiplicity of $\triangle ABC$ has moved to the adjacent vertex following the common clockwise orientation. From $\mathcal{C}$ to $\mathcal{C}'$, each robot of $\triangle ABC$ must travel only along the edge connecting itself with the related target vertex, following the clockwise direction. Recursively, the problem demands the same request starting from $\mathcal{C}'$.

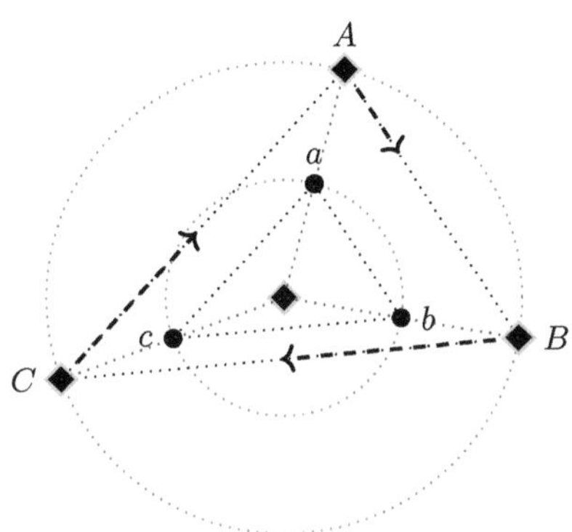

Fig. 1. Pinwheel. Multiplicities are represented through yellow-shaded diamonds.

Lemma 1. *Pinwheel* $\notin \mathcal{P}\left(\mathcal{OBLOT}^{SEQ}\right)$ *even assuming* RIG *and* FIXDIS.

Lemma 2. *Pinwheel* $\in \mathcal{P}\left(\mathcal{OBLOT}^{PERM}\right)$ *even assuming* NONRIG *and* VARDIS.

Theorem 2. *Independently from* RIG-NONRIG *and* FIXDIS-VARDIS:

$$\mathcal{OBLOT}^{SEQ} < \mathcal{OBLOT}^{PERM}.$$

Problem 2 (Kite). Let a, b, c, d be four robots forming a kite with $\angle bac = \frac{\pi}{3}$ and $\angle bdc = \frac{\pi}{6}$ (configuration $\mathcal{C}$ in Fig. 2). Let h be the axis passing through a, d. Robot d is required to move along h and to stop so that a, b, c, d form a diamond (configuration $\mathcal{C}'$). Then, a must move along h and stop in the center of the diamond (final configuration $\mathcal{C}''$).

Lemma 3. *Kite* $\notin \mathcal{P}\left(\mathcal{OBLOT}^{PERM}\right)$ *even assuming* RIG *and* FIXDIS.

Lemma 4. *Kite* $\in \mathcal{P}\left(\mathcal{OBLOT}^{RROBIN}\right)$ *even assuming* NONRIG *and* VARDIS.

Theorem 3. *Independently from* RIG-NONRIG *and* FIXDIS-VARDIS:

$$\mathcal{OBLOT}^{PERM} < \mathcal{OBLOT}^{RROBIN}.$$

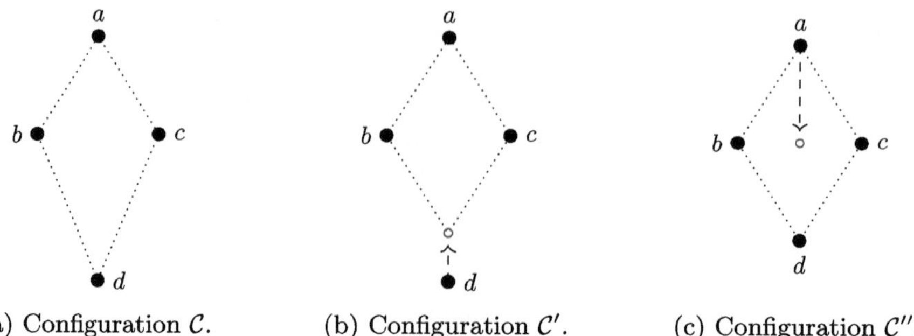

(a) Configuration $\mathcal{C}$.   (b) Configuration $\mathcal{C}'$.   (c) Configuration $\mathcal{C}''$.

Fig. 2. Configurations in `Kite`.

3.2   Relations in $\mathcal{LUMI}$

In this section, we show that the computational power of $\mathcal{LUMI}$ robots under SEQ is the same as under PERM, by proving that any problem solvable in $\mathcal{LUMI}^{\text{PERM}}$ can also be solved under $\mathcal{LUMI}^{\text{SEQ}}$. We then show that there exists a strict dominance between PERM and RROBIN.

$\mathcal{LUMI}$	SEQ	$\equiv$	PERM	$<$	RROBIN

Theorem 4. *Independently from* RIG-NONRIG *and* FIXDIS-VARDIS*:*

$$\mathcal{LUMI}^{SEQ} \equiv \mathcal{LUMI}^{PERM}.$$

Proof. By Theorem 1, we know that $\mathcal{LUMI}^{\text{SEQ}} \leq \mathcal{LUMI}^{\text{PERM}}$. We now prove that $\mathcal{LUMI}^{\text{SEQ}} \geq \mathcal{LUMI}^{\text{PERM}}$, i.e., that any problem P solvable in $\mathcal{LUMI}^{\text{PERM}}$ can be solved under $\mathcal{LUMI}^{\text{SEQ}}$. Let $\mathbb{A}$ be any algorithm solving P under $\mathcal{LUMI}^{\text{PERM}}$ and let $\mathfrak{L}$ be the palette of colors used by $\mathbb{A}$. We design a simulator algorithm, $\mathbb{A}_{\text{SEQ}}$, which solves P under $\mathcal{LUMI}^{\text{SEQ}}$ using the palette $\mathfrak{L}_{\text{SEQ}} = \mathfrak{L} \times \{\text{OFF}, \text{MOVED}, \text{RESET}\}$. Thus, w.l.o.g., we assume that each robot r is equipped with an auxiliary light aux_r, in addition to the default one lig_r, initialized with the color OFF. The idea of our simulator is to make the robots use the auxiliary lights to simulate an epoch in $\mathcal{LUMI}^{\text{PERM}}$ through a *mega-epoch* in $\mathcal{LUMI}^{\text{SEQ}}$. With mega-epoch, we intend a time frame composed of three epochs: in the first one, each robot performs one LCM cycle by executing the original algorithm $\mathbb{A}$, and sets its light from OFF to MOVED. This color indicates that the robot has already executed an LCM cycle within the first epoch; therefore, it will skip its turn if it is activated further times during this epoch. When all the robots are set to MOVED, they set their auxiliary lights to RESET in the second epoch, and then they use the third epoch to reset the lights to OFF and restart the mega-epoch (OFF $\rightarrow$ MOVED $\rightarrow$ RESET)$^\infty$. Let us detail the behavior of a robot r executing $\mathbb{A}_{\text{SEQ}}$, according to the color of its aux_r (see Fig. 3):

- (Case $\mathsf{aux}_r = $ OFF) If r sees no RESET-colored robots, it means that r is activated for the first time during the first epoch of the mega-epoch. So, r takes the snapshot σ and computes the reduced version $\sigma_{\mathbb{A}}$ removing the colors of the aux variables; then r runs the original algorithm $\mathbb{A}(\sigma_{\mathbb{A}})$, and it moves to the computed position, possibly updating the color of its lig_r. To mark the fact that it has already performed its LCM cycle during the epoch, r sets $\mathsf{aux}_r \leftarrow$ MOVED. Otherwise (i.e., r sees some RESET-colored robots), r does nothing.
- (Case $\mathsf{aux}_r = $ MOVED) If r sees that other robots have their aux light set to OFF, it means that r has already been activated during the current epoch and that other robots (those with the aux light to OFF) must still be activated in this epoch. Thus, r skips its turn (i.e., it does nothing). Otherwise, it updates $\mathsf{aux}_r \leftarrow$ RESET, doing nothing else.
- (Case $\mathsf{aux}_r = $ RESET) If r sees other MOVED robots, then it does nothing. Otherwise, it updates $\mathsf{aux}_r \leftarrow$ OFF, doing nothing else.

It is easy to see that a mega-epoch under SEQ simulates the behavior on an epoch (i.e., permutation of the swarm) under PERM. In fact, the only epoch in the mega-epoch where robots execute the main algorithm $\mathbb{A}$ is the first one where only OFF robots execute $\mathbb{A}$ and set their aux color to MOVED. The MOVED color is used both to memorize the already-done action during the first epoch, and to synchronize robots to proceed with the next epoch. Upon all robots are MOVED, they have to reset the mega-epoch: so they firstly all set to RESET and then to OFF. Note that a third color transition (and thus a third epoch) in the mega-epoch is necessary for the correctness of the simulator: having only two colors (OFF and MOVED), it would be impossible for robots to distinguish a transition configuration OFF $\rightarrow$ MOVED between MOVED $\rightarrow$ OFF. $\square$

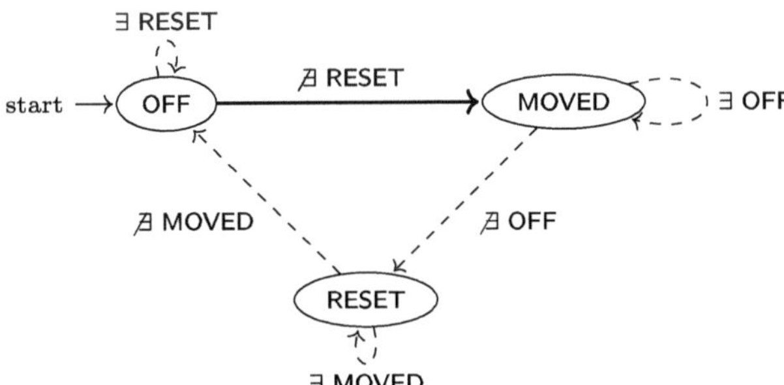

Fig. 3. Transition diagram of the color of the auxiliary light aux_r according to $\mathbb{A}_{\text{SEQ}}$. Transition conditions refer only to auxiliary lights. Dashed edges represent color transition during LCM cycles where r does nothing except change aux_r.

*Problem 3 (**Triline**).* Assume three groups of at least four robots each lie on the vertices of an equilateral triangle $\triangle v_1 v_2 v_3$. The robots of each group v_i are

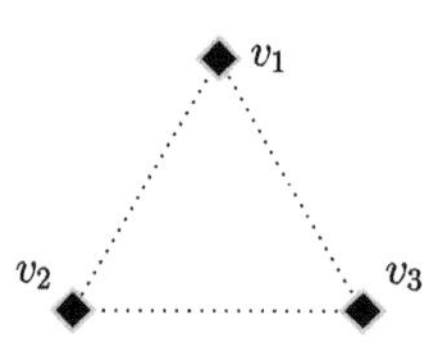

(a) Initial configuration of three multiplicities (diamonds with yellow shadows) forming $\triangle v_1 v_2 v_3$.

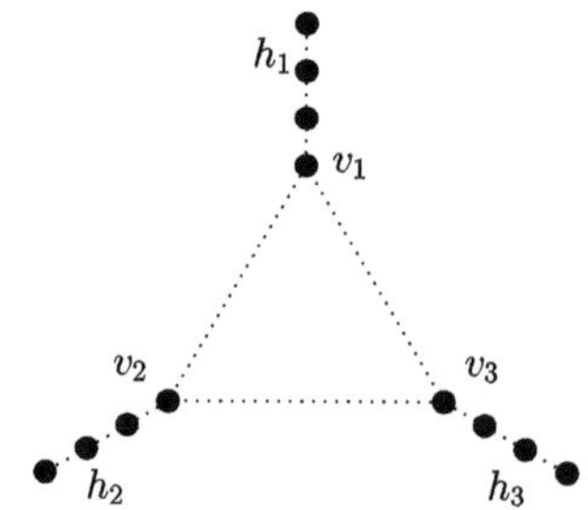

(b) Possible arrangement on lines h_1, h_2, h_3 solving task (1).

Fig. 4. Configurations in `Triline`.

required to (1) arrange along a line h_i, each robot in a distinct point; (2) come back to v_i; (3) form the same line h_i, occupying each the same position as before. The sequence of the three tasks must be performed perpetually. See Fig. 4.

Lemma 5. *`Triline` $\notin P\left(\mathcal{LUMI}^{PERM}\right)$ even assuming* RIG *and* FixDis.

Proof. By contradiction, let $\mathbb{A}$ be a solving algorithm under $\mathcal{LUMI}^{PERM}$ using k colors. Let us consider an instance of the problem with n robots and so that the group on v_1 contains $m \geq 2k + 1$ robots. Let ℓ_1 be the line chosen by $\mathbb{A}$ to arrange the robots of v_1. Let $\mathfrak{S} = \mathcal{E}_1 \mathcal{E}_2 \ldots$ be a **PERM** scheduler where $\mathcal{E}_i$ represents a permutation string of the swarm (and, thus, an epoch). For the sake of simplicity, we assume that each $\mathcal{E}_i$ activates first the m robots in v_1. Let t be the time when all m robots have returned to v_1 and have to rearrange on ℓ_1 in their previous positions. Since $m \geq 2k + 1$, it is possible to partition the multiplicity in v_1 into at most k subsets of robots with the same color. Let t belong to $\mathcal{E}_j$, and let $z = t \bmod n$ (i.e., the round within $\mathcal{E}_j$ corresponding to time t). Since we assume that $\mathcal{E}_j$ activates first the robots of v_1, we have that $z < m$. Suppose that $\mathcal{E}_{j+1} = \mathcal{E}_j$ in $\mathfrak{S}$. We distinguish two scenarios (see Fig. 5):

- if $z \leq k$ (see Fig. 5a), then $m - z > k$. So, at least two robots p, q lie in v_1 with the same color at time t and they will be activated in the next $m - z$ rounds of $\mathcal{E}_j$ from t onward;
- otherwise, (see Fig. 5b), at least two robots r, s lie in v_1 with the same color at time t and they will be activated in the first z rounds of $\mathcal{E}_{j+1}$ (which is equal to $\mathcal{E}_j$).

In both cases, we provide two **PERM** schedulers, namely $\mathfrak{S}'$ (for $z \leq k$) and $\mathfrak{S}''$ (for $z > k$), under which the swarm cannot solve the problem by executing $\mathbb{A}$. In particular, $\mathfrak{S}'$ is identical to $\mathfrak{S}$ except that the activations of p and q are switched from $\mathcal{E}_j$ onward; instead, $\mathfrak{S}''$ is obtained by $\mathfrak{S}$ by switching the activation of r and s from $\mathcal{E}_{j+1}$ onward. According to $\mathfrak{S}'$, from t onward, robot p (q, resp.) will take the same snapshot as q (p, resp.) under $\mathfrak{S}$, and so it will move to the position intended for q (p, resp.). This contradicts the request of the problem. The same contradiction is achieved in the case $z > k$ using $\mathfrak{S}''$. $\qquad\square$

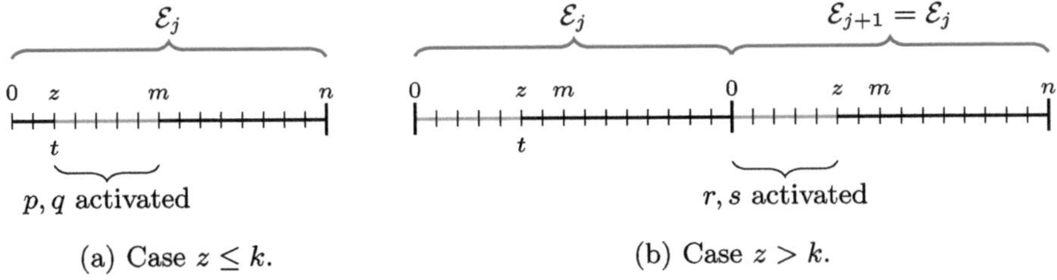

(a) Case $z \leq k$.   (b) Case $z > k$.

Fig. 5. Epochs in $\mathfrak{S}$ for `Triline`.

Lemma 6. *`Triline` $\in \mathcal{P}\left(\mathcal{FCOM}^{RROBIN}\right)$ even assuming* NONRIG *and* VARDIS.

Corollary 1. *`Triline` $\in \mathcal{P}\left(\mathcal{LUMI}^{RROBIN}\right)$ even assuming* NONRIG *and* VARDIS.

Theorem 5. *Independently from* RIG-NONRIG *and* FIXDIS-VARDIS*:*

$$\mathcal{LUMI}^{PERM} < \mathcal{LUMI}^{RROBIN}.$$

3.3   Relations in $\mathcal{FSTA}$

In this section, we show that in $\mathcal{FSTA}$ the relationship between the three scheduler classes is one of strict dominance.

$\mathcal{FSTA}$	SEQ	<	PERM	<	RROBIN

The next corollary follows directly from Lemma 5.

Corollary 2. *`Triline` $\notin \mathcal{P}\left(\mathcal{FSTA}^{PERM}\right)$ even assuming* RIG *and* FIXDIS.

Lemma 7. *`Triline` $\in \mathcal{P}\left(\mathcal{FSTA}^{RROBIN}\right)$ even assuming* NONRIG *and* VARDIS.

Proof. We provide an algorithm to solve `Triline` using {OFF, OUT, M_OUT, IN, LAST} as palette, with OFF being the default light color of any robot r. Let us describe the evolution of our algorithm for a generic vertex v_i, assuming it originally contains m robots. Let ℓ be the length of the edge of the triangle, and let $a > 1$ be a constant of the algorithm. We define h_i as the axis of the triangle passing through v_i. We make the robots in v_i uniformly arrange along external semi-line of h_i starting from v_i: we define the target points for the m robots as the points $U_1, U_2, \ldots, U_m$ on h_i such that $dist(U_j, v_i) = \frac{\ell}{a}(m - j)$ (see Figs. 4b and 6). In the first epoch, all robots stay still and change their state from OFF to OUT, indicating that they have to move out from v_i. Now, starting from the second epoch onward (considering NONRIG, this phase can last multiple epochs according to the number of times and the position where robots are stopped during their movements), the robots in v_i will arrange along the semi-axis h_i: specifically, for any $j = 1, \ldots, m-1$, the j-th robot activated in v_i sets its color to M_OUT (except for m-th activated robot in v_i which never changes its color, as we will soon explain) and heads U_j. Note that if an OUT robot sees that robots are not arranged properly on h_i, it skips its turn since it understands that a M_OUT robot has been stopped during its movement along h_i and thus it has still to reach its target point. The last robot remaining on v_i will set its color to LAST: the LAST robot will never change either its state or its position. Once all the M_OUT robots have uniformly arranged along h_i, they switch their color into IN. Now, starting from the closest IN robot to v_i (i.e., the robot on U_{m-1}), the robots come back in order to v_i setting their color to OFF. Again, if a robot stops during its movement, the other robots will skip their turns waiting for the robot to reach v_i in subsequent epochs. Only the robot on U_1 (which is the last to come back to v_i) sets its color to OUT (instead of OFF) when coming back to v_i. Note that this LIFO scheme allows robots to never create multiplicities with other robots on $h_i \setminus \{v_i\}$ if they are stopped along their trajectories. After the robot on U_1 has reached v_i (note that such a robot has been the first one to be activated), the other robots of v_i will be activated in the same order and thus they will change their color from OFF to OUT, therefore repeating the whole task. Refer to Fig. 7 for the transition diagram of the internal state of a robot r: it is easy to see that the conditions defined on transition edges unambiguously lead r to understand its next action. □

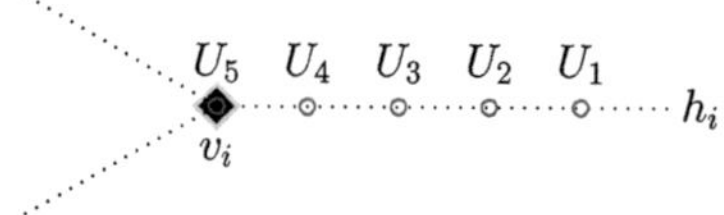

Fig. 6. Triline. Target points along h_i of the m robots belonging to v_i (here, $m = 5$).

Theorem 6. *Independently from* Rig-NonRig *and* FixDis-VarDis*:*

$$\mathcal{FSTA}^{PERM} < \mathcal{FSTA}^{RROBIN}.$$

Problem 4 (Go-If-Odd). Let a, b, c be three aligned points so that $dist(a, b) = 2dist(b, c)$. Only one robot, say r_a, lies on a, while a group of robots lies on b and a group lies on c. The problem requires all the robots in b to move to c. Then, if the original number of robots in b was odd, r_a must move to c as well. Otherwise, r_a must stay still. See Fig. 8.

Lemma 8. *Go-If-Odd* $\notin \mathcal{P}\left(\mathcal{FSTA}^{SEQ}\right)$ *even assuming* Rig *and* FixDis*.*

Lemma 9. *Go-If-Odd* $\in \mathcal{P}\left(\mathcal{FSTA}^{PERM}\right)$ *even assuming* NonRig *and* VarDis*.*

Theorem 7. *Independently from* Rig-NonRig *and* FixDis-VarDis*:*

$$\mathcal{FSTA}^{SEQ} < \mathcal{FSTA}^{PERM}.$$

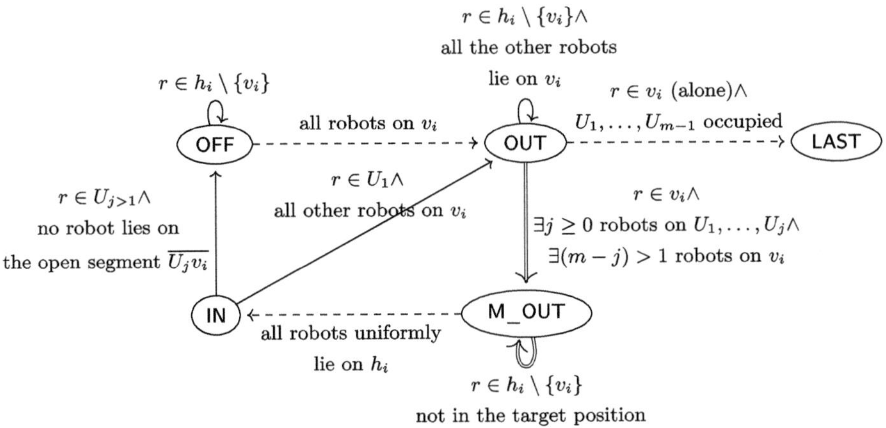

Fig. 7. Transition diagram of the light color of r while solving **Triline** under $\mathcal{FSTA}^{\text{RROBIN}}$. The dashed edges represent color transition occurring in LCM cycles where r executes a null-movement. Single-line (double-line, resp.) edges represent movements inward (outward, resp.) v_i.

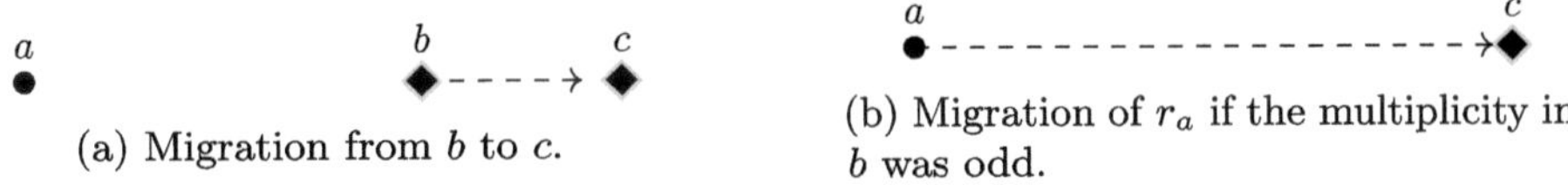

Fig. 8. Go-If-Odd problem.

3.4 Relations in $\mathcal{FCOM}$

In this section, we show that in $\mathcal{FCOM}$ the relationship between the three scheduler classes is one of strict dominance.

By Lemma 5, it straightforwardly follows that:

Corollary 3. $\mathtt{Triline} \notin \mathcal{P}\left(\mathcal{FCOM}^{\text{PERM}}\right)$ *even assuming* RIG *and* FIXDIS.

Theorem 8. *Independently from* RIG-NONRIG *and* FIXDIS-VARDIS:

$$\mathcal{FCOM}^{\text{PERM}} < \mathcal{FCOM}^{\text{RROBIN}}.$$

*Problem 5 (**ApproachTwice**).* Let $\mathcal{C}$ be a configuration with three robots p, q, r, where p, q lie on the same point while r lies on a distinct point. Let $d = dist(p, r) > 0$ be their distance (according to an absolute coordinate system), and let γ be the line where the robots lie. Robot r is required to move along γ and stop to form a new configuration $\mathcal{C}''$ where the new distance of the robots, say d'', is such that $0 < d'' < d$. From $\mathcal{C}$ to $\mathcal{C}''$, r is allowed to possibly stop once and form an intermediate configuration $\mathcal{C}'$ where the distance of the robots, say d', is such that $d'' < d' < d$. See Fig. 9.

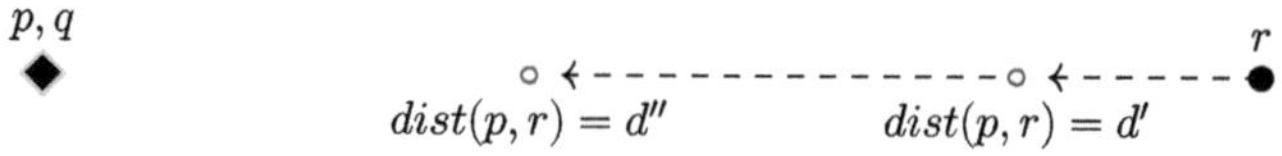

$$p, q \qquad\qquad\qquad\qquad\qquad\qquad\qquad\qquad\qquad\qquad r$$

$$dist(p, r) = d'' \qquad\qquad\qquad dist(p, r) = d'$$

Fig. 9. `ApproachTwice` problem.

Lemma 10. *$ApproachTwice \notin \mathcal{P}\left(\mathcal{FCOM}^{SEQ}\right)$ even assuming* RIG, *but assuming* VARDIS.

Lemma 11. *$ApproachTwice \in \mathcal{P}\left(\mathcal{FCOM}^{PERM}\right)$ even assuming* NONRIG *and* VARDIS.

Theorem 9. *Independently from* RIG-NONRIG, *but assuming* VARDIS*:*

$$\mathcal{FCOM}^{SEQ} < \mathcal{FCOM}^{PERM}.$$

4   Conclusions

In this paper, we have conducted a preliminary study on the computational power of robot swarms under sequential schedulers. In particular, we have considered the four base models of robots $\mathcal{OBLOT}$, $\mathcal{FST\!A}$, $\mathcal{FCOM}$, $\mathcal{LUMI}$, and we have studied how their power is affected when robots are activated by three classes of sequential schedulers—namely RROBIN (*round-robin*), PERM (*permutation*) and SEQ (*sequential*)—each subclass of the other. We have proven that for the settings $\mathcal{LUMI}^{SEQ}$ and $\mathcal{LUMI}^{PERM}$, the use of lights allows robots to simulate the evolution of a $\mathcal{LUMI}^{PERM}$ algorithm under the more adversarial class of schedulers SEQ; as a consequence, the two settings result to be computationally equivalent. For the other settings, we have proven that the computational power of each model decreases if robots are activated by more adversarial schedulers. Notably, we have provided five problems that play the role of separators between the considered settings. It is worth mentioning that, except for the dominance $\mathcal{FCOM}^{PERM} > \mathcal{FCOM}^{SEQ}$, all the computational dominances proved in this paper hold *independently* of the assumptions on the rigidity of robot movements and the disorientation of robots. However, the (in)dependence of the two features (i.e., rigidity and disorientation) seems difficult to prove for the relation $\mathcal{FCOM}^{PERM}$ vs. $\mathcal{FCOM}^{SEQ}$. For that pair of settings, we have demonstrated the related dominance assuming that the disorientation of robots is variable (i.e., not persistent cycle by cycle). A first question arises after this observation: if we drop this hypothesis, is it still valid that $\mathcal{FCOM}^{PERM} > \mathcal{FCOM}^{SEQ}$? We feel that such an investigation may hide an interesting study on the communication power of the model $\mathcal{FCOM}$, under different assumptions.

Future works may complete the computational relation map of the models here considered, finding the cross-model relations (e.g., what is the relation between $\mathcal{FCOM}^{SEQ}$ and $\mathcal{FST\!A}^{PERM}$?), or may expand the taxonomy of the sequential schedulers by defining and studying other interesting subclasses. Eventually, it would also be interesting to characterize the sets of problems (un)solvable under a certain class of sequential schedulers: such a characterization may be preliminary for finding algorithmic solutions for well-known problems under sequential schedulers.

Acknowledgments. This work was partly supported by NSERC through the Discovery Grant program. This research was done while Caterina Feletti was visiting Profs. Flocchini and Santoro.

References

1. Agmon, N., Peleg, D.: Fault-tolerant gathering algorithms for autonomous mobile robots. SIAM J. Comput. **36**(1), 56–82 (2006). https://doi.org/10.1137/050645221
2. Bramas, Q., Tixeuil, S.: The random bit complexity of mobile robots scattering. Int. J. Found. Comput. Sci. **28**(2), 111–117 (2017)
3. Buchin, K., Flocchini, P., Kostitsyna, I., Peters, T., Santoro, N., Wada, K.: Autonomous mobile robots: refining the computational landscape. In: Proceedings of 35th International Parallel and Distributed Processing Symposium Workshops (IPDPSW), pp. 576–585. IEEE (2021). https://doi.org/10.1109/IPDPSW52791.2021.00091
4. Buchin, K., Flocchini, P., Kostitsyna, I., Peters, T., Santoro, N., Wada, K.: On the computational power of energy-constrained mobile robots. Inf. Comput. **303**, 105280 (2025). https://doi.org/10.1016/J.IC.2025.105280
5. Canepa, D., Dèfago, X., Izumi, T., Potop-Butucaru, M.: Flocking with oblivious robots. In: Proceedings of 18th International Symposium on Stabilization, Safety, and Security of Distributed Systems (SSS), pp. 94–108 (2016). https://doi.org/10.1007/978-3-319-49259-9_8
6. Cicerone, S., Stefano, G.D., Navarra, A.: Solving the pattern formation by mobile robots with chirality. IEEE Access **9**, 88177–88204 (2021)
7. Cieliebak, M., Flocchini, P., Prencipe, G., Santoro, N.: Distributed computing by mobile robots: gathering. SIAM J. Comput. **41**(4), 829–879 (2012). https://doi.org/10.1137/100796534
8. Clemente, S., Feletti, C.: Fault detection and identification by autonomous mobile robots. In: Proceedings of 4th Symposium on Algorithmic Foundations of Dynamic Networks (SAND). LIPIcs, vol. 330, pp. 10:1–10:20. Schloss Dagstuhl - Leibniz-Zentrum für Informatik (2025). https://doi.org/10.4230/LIPICS.SAND.2025.10
9. Courtieu, P., Rieg, L., Tixeuil, S., Urbain, X.: Certified universal gathering in R^2 for oblivious mobile robots. In: Proceedings of 30th International Symposium on Distributed Computing (DISC), pp. 187–200 (2016)
10. Das, S., Flocchini, P., Prencipe, G., Santoro, N.: Forming sequences of patterns with luminous robots. IEEE Access **8**, 90577–90597 (2020)
11. Das, S., Flocchini, P., Prencipe, G., Santoro, N., Yamashita, M.: Autonomous mobile robots with lights. Theor. Comput. Sci. **609**, 171–184 (2016). https://doi.org/10.1016/J.TCS.2015.09.018
12. Das, S., Flocchini, P., Santoro, N., Yamashita, M.: Forming sequences of geometric patterns with oblivious mobile robots. Distrib. Comput. **28**(2), 131–145 (2014). https://doi.org/10.1007/s00446-014-0220-9
13. Défago, X., Potop-Butucaru, M., Tixeuil, S.: Fault-tolerant mobile robots. In: Chapter 10 of [18], pp. 234–251. Springer, Heidelberg (2019). https://doi.org/10.1007/978-3-030-11072-7_10

14. Dieudonné, Y., Labbani-Igbida, O., Petit, F.: Circle formation of weak mobile robots. ACM Trans. Auton. Adapt. Syst. **3**(4), 1–20 (2008). https://doi.org/10.1145/1452001.1452006

15. Dieudonné, Y., Petit, F.: Scatter of robots. Parallel Process. Lett. **19**(1), 175–184 (2009). https://doi.org/10.1142/S0129626409000146

16. Feletti, C., Mambretti, L., Mereghetti, C., Palano, B.: Computational power of autonomous robots: transparency vs. opaqueness. Theor. Comput. Sci. **1036**, 115153 (2025). https://doi.org/10.1016/J.TCS.2025.115153

17. Flocchini, P., Navarra, A., Pattanayak, D., Piselli, F., Santoro, N.: Oblivious robots under sequential schedulers: universal pattern formation. In: Proceedings of 32nd International Colloquium On Structural Information and Communication Complexity (SIROCCO). LNCS, vol. 15671, pp. 297–314. Springer, Heidelberg (2025). https://doi.org/10.1007/978-3-031-91736-3_18

18. Flocchini, P., Prencipe, G., Santoro, N. (eds.): Distributed Computing by Mobile Entities, Current Research in Moving and Computing, LNCS, vol. 11340. Springer, Heidelberg (2019). https://doi.org/10.1007/978-3-030-11072-7

19. Flocchini, P., Prencipe, G., Santoro, N., Viglietta, G.: Distributed computing by mobile robots: uniform circle formation. Distrib. Comput. **30**(6), 413–457 (2016). https://doi.org/10.1007/s00446-016-0291-x

20. Flocchini, P., Santoro, N., Sudo, Y., Wada, K.: On asynchrony, memory, and communication: separations and landscapes. In: Proceedings of 27th International Conference on Principles of Distributed Systems (OPODIS). LIPIcs, vol. 286, pp. 28:1–28:23 (2023). https://doi.org/10.4230/LIPICS.OPODIS.2023.28

21. Flocchini, P., Santoro, N., Wada, K.: On memory, communication, and synchronous schedulers when moving and computing. In: Proceedings of 23rd International Conference on Principles of Distributed Systems (OPODIS). LIPIcs, vol. 153, pp. 25:1–25:17 (2019). https://doi.org/10.4230/LIPICS.OPODIS.2019.25

22. Frei, F., Wada, K.: Brief announcement: distinct gathering under round robin. In: Proceedings of 38th International Symposium on Distributed Computing (DISC). LIPIcs, vol. 319, pp. 48:1–48:8. Schloss Dagstuhl – Leibniz-Zentrum für Informatik (2024). https://doi.org/10.4230/LIPIcs.DISC.2024.48

23. Gervasi, V., Prencipe, G.: Coordination without communication: the case of the flocking problem. Disc. Appl. Math. **144**(3), 324–344 (2004). https://doi.org/10.1016/j.dam.2003.11.010

24. Izumi, T., Kaino, D., Gradinariu Potop-Butucaru, M., Sébastien, T.: On time complexity for connectivity-preserving scattering of mobile robots. Theor. Comput. Sci. **738**, 42–52 (2018). https://doi.org/10.1016/j.tcs.2018.04.047

25. Kirkpatrick, D.G., Kostitsyna, I., Navarra, A., Prencipe, G., Santoro, N.: On the power of bounded asynchrony: convergence by autonomous robots with limited visibility. Distrib. Comput. **37**(3), 279–308 (2024). https://doi.org/10.1007/S00446-024-00463-7

26. Suzuki, I., Yamashita, M.: Distributed anonymous mobile robots: formation of geometric patterns. SIAM J. Comput. **28**(4), 1347–1363 (1999). https://doi.org/10.1137/S009753979628292X

Brief Announcement: Cross-Chain Consensus

Sucharita Jayanti$^{(\boxtimes)}$ and Maurice Herlihy

Brown University, Providence, RI, USA
sucharita_jayanti@brown.edu

Abstract. In a cross-chain task, m active, untrustworthy parties attempt to trade assets using n passive but trustworthy smart contracts residing on multiple blockchains. The decentralized finance industry has developed a variety of tasks, including asset swaps, auctions, futures, etc.. In this paper, we show that there is a particular task, called *cross-chain consensus*, that is universal in the sense that any protocol that solves cross-chain consensus can be adapted to solve any well-formed cross-chain task. Like classical consensus, cross-chain consensus provides the basis for universal constructions and impossibility results. Nevertheless, cross-chain consensus differs structurally from its classical counterpart by splitting participants into two distinct roles: active, but untrustworthy (Byzantine) parties provide inputs, while passive but trustworthy smart contracts decide outputs.

1 Introduction

Today, blockchain-based decentralized finance (DeFi) is a multi-billion dollar industry [10]. Multiple mutually-untrusting parties trade electronic assets via networks of trusted automata ("smart contracts") that reside on multiple tamper-proof distributed ledgers ("blockchains"). At first glance, this industry, based as it is on fault-tolerant distributed algorithms, seems like an ideal proving ground for the models and algorithms developed over the years by the distributed computing community. Further inspection, however, shows that despite superficial similarities between the two domains, classical models of distributed computing fall short of capturing the realities of modern DeFi. While some classical results carry over, many others require non-trivial revision and extension.

For example, classical shared memory distributed computing models encompass a bewildering array of inter-process synchronization mechanisms, including queues, test-and-set, compare-and-swap, etc. One classical result at the heart of modern distributed computing is that there is one particular primitive, called *consensus* [3], that is the *strongest* synchronization primitive in the sense that any algorithm that implements consensus can be adapted to implement any other well-formed primitive. This reduction forms the core of universal constructions and a range of impossibility results.

S. Bonomi et al. (Eds.): SSS 2025, LNCS 16350, pp. 233–238, 2026.
https://doi.org/10.1007/978-3-032-11127-2_19

Likewise, DeFi applications encompass a bewildering array of cross-chain tasks, including asset swaps, auctions, options, etc., as well as an equally bewildering array of mechanisms, including hashed timelocks, token bridges, automated market makers, etc. It is the contribution of this paper to define a particular, novel cross-chain task we call *cross-chain consensus*, by analogy to classical consensus, that is universal for trading assets within and across multiple blockchains. Despite the deliberate similarity in names, we will see that classical and cross-chain consensus differ in non-trivial ways.

2   Model

Overview. Let us start with an informal description of the smart contract model how cross-chain coordination works in practice, with the goal of motivating the formal model presented in the next section. We start with an informal summary of the *smart contract* model proposed by Amoussou *et al.* [1].

Multiple agents, called *parties* trade *assets* among themselves. Parties may be people, organizations, or algorithmic agents. Parties may be *honest*, meaning they follow agreed-upon protocols, or *dishonest*, meaning they can depart arbitrarily from protocols, even irrationally. Assets may be cryptocurrencies, financial instruments, NFTs, or any electronic item of value. Asset ownership is controlled by *contracts* (sometimes "smart contracts"). A contract is an automaton residing on a tamper-proof distributed databases called *chains* (sometimes "blockchains"). Contract state is public and tamper proof. Contracts are passive, deterministic, and trustworthy. Communication proceeds in synchronous rounds. In each round, each party sends messages to contracts, then observes the resulting messages and contract state changes.

Honest parties do not communicate directly, but dishonest parties may communicate with one another through hidden channels. Contracts on distinct chains cannot send messages directly to one another, nor can they directly observe one another's states[1].

A (cross-chain) *task* is series of asset exchanges following an agreed-upon set of conditions. For example, parties Alice and Bob may agree to swap some of Alice's (electronic) dollars for Bob's (electronic) euros at an agreed-upon exchange rate. Here, each currency is managed by a distinct contract on a distinct chain. A *protocol* is a distributed algorithm that solves a task. A protocol must guarantee that if both parties are honest, the exchange takes place, and also that a dishonest party cannot cheat an honest party.

Comparison with Classical Models. In the classical *shared-memory* model[2] multiple sequential *processes* share multiple *objects* in a shared memory. In a classical *task* [4], processes start with private input values, communicate by applying operations to the shared objects, and decide output values according to a task specification. Processes run at arbitrary relative speeds. Processes can fail by

---

[1] In practice, contracts residing *on the same chain* can call one another.

[2] Similar observations apply to the classical message-passing models.

crashing, halting at any point in the protocol. A consensus protocols is required to be *wait-free*, tolerating as many as $m - 1$ crash failures out of m processes.

At first glance, there may be a temptation to view the smart contract model as a variant of the classical shared-memory model, where parties assume the role of processes, and contracts assume the role of shared objects. Like processes, parties are active agents subject to failures, and like shared-memory objects, smart contracts are passive elements encapsulating state that can be manipulated by active agents. Nevertheless, such an analogy is misleading at best.

Setting aside the obvious differences in timing (asynchronous versus synchronous) and failure modes (crash versus Byzantine), the principal difference between the two models is structural. For classical tasks, processes provide inputs and decide outputs. For cross-chain tasks, parties provide inputs, *but contracts, not parties, decide outputs*. The role of processes in the classical model is split between parties and contracts, with subtle but far-reaching consequences.

For example, smart contract code must be *deterministic* because each contract's state is replicated and its function calls are re-executed multiple times by mutually-suspicious parties. It follows that a contract on one chain cannot call or observe contracts on other chains, because remote, cross-chain calls cannot be deterministically replayed. As a result, cross-chain protocol designers face a novel challenge: if a contract on one chain cannot directly observe contracts on other chains, then it must rely on untrustworthy parties for that information. As a result, cross-chain protocols tend to center around on cryptographic gadgets such as timed hashlocks [7], token bridges [8,11], state proofs [6] (zero-knowledge or otherwise) etc. The resulting protocols end up looking nothing like classical shared-memory or message-passing protocols.

Formal Definitions. Formally, a task is a tuple $(\mathcal{I}_P, \mathcal{I}_C, \mathcal{O}_C, U)$, where $\mathcal{I}_P$ is a set of m-element *party input vectors*, representing each party's input to the task, $\mathcal{I}_C$ is a set of n-element *contract input vectors*, representing each contract's state before executing the task, $\mathcal{O}_C$ is a set of n-element *contract output vectors*, representing each contract's state after executing the task, and $U : \mathcal{I}_P \times \mathcal{I}_C \times \mathcal{O}_C \to \mathbb{R}^m$ is a *utility function* that characterizes how each party values each possible outcome.

A *transition* is a triple $(I_P, I_C, O_C) \in \mathcal{I}_P \times \mathcal{I}_C \times \mathcal{O}_C$. Any transition where no contract changes state ($I_C = O_C$) has utility 0 to each party: $U(I_P, I_C, I_C) = 0$. Each party considers transitions with positive utility to be more desirable than the *status quo*, and transitions with negative utility to be less desirable. A transition is *acceptable* to party P if $U(I_P, I_C, O_C)[P] \geq 0$, and *preferred* if $U(I_P, I_C, O_C)[P] > 0$. A transition is *acceptable* if it is acceptable to all parties, and *preferred* if it is preferred by all parties.

For a task to be *feasible*, it must satisfy certain common-sense conditions.

– Because dishonest parties can always obstruct a protocol, the null transition must always be acceptable.
– To avoid trivialities, for every pair of party input vectors $I_P \in \mathcal{I}_P, I_C \in \mathcal{I}_C$, there must exist a preferred transition

- Dishonest parties who lie about their inputs should not be able to trick honest parties into a negative utility.
- A hidden input can shift a honest party's utility from positive to zero, but not from positive to negative.

When describing a protocol execution, we indicate which parties are honest by a *compliance set* $\mathcal{B} \subseteq \mathcal{P}$.

A protocol execution is a tuple $(\mathcal{I}_P, \mathcal{I}_C, \mathcal{B}, \mathcal{O}_C, \Xi)$, where $\mathcal{I}_P$ is a set of m-element *party input vectors*, $\mathcal{I}_C$ is a set of n-element *contract input vectors*, $\mathcal{O}_C$ is a set of n-element *contract output vectors*, $\Xi : \mathcal{I}_P \times \mathcal{I}_C \times 2^{\mathcal{P}} \rightarrow 2^{\mathcal{O}_C}$, the *execution function*, is a map that carries a party input vector, a contract input vector, and a compliance set to a set of output contract vectors representing possible outcomes.

A protocol *implements* a task if the following conditions are satisfied.

- Each transition permitted by the execution function must be acceptable
- If all parties are honest, that is $\mathcal{B} = \mathcal{P}$, then the protocol's transitions are preferred

3   Cross-Chain Consensus

In the *m-party, n-contract cross-chain consensus task*, each party starts with an input value from a finite domain $\mathcal{V}$, and each contract eventually decides a value from $\mathcal{V}$. The conditions for cross chain consensus are as follows:

- **liveness:** All contracts must eventually decide a value
- **agreement:** If all parties are honest, all contracts decide the same value, which must be some party's input
- **validity:** If some parties are dishonest, all contracts decide the same value from $\mathcal{V}$, which need not be any party's input.

Formally, the (m, n)-XC consensus task is defined by the tuple $(\mathcal{I}_P, \mathcal{I}_C, \mathcal{O}_C, U)$, where each $I_P \in \mathcal{I}_P$ is a vector of inputs from $\mathcal{V}$, one for each party. $\mathcal{I}_C$ is a single initial contract state vector where each contract has value $\perp$. Each $O_C \in \mathcal{O}_C$ is a vector of values from $\mathcal{V}$. The task's utility function U is:

$$U(I_P, I_C, O_C)[P] = \begin{cases} 1 & \exists Q \ \forall C, O_C[C] = I_P[Q]: \text{All contracts decide one party's input.} \\ 0 & \forall C, O_C[C] = v \in \mathcal{V}: \text{All contracts decide the same value from } \mathcal{V}. \\ -1 & \text{Otherwise.} \end{cases}$$

4   Cross-Chain Versus Classical Consensus

In this section, we compare classical asynchronous shared-memoryconsensus with cross-chain consensus. (As noted, similar observations apply to related models, such as message-passing.) There are m sequential processes, each starting with an arbitrary proposal value, and each halting with a decision value.

- **liveness**: All non-faulty processes must eventually decide on some value.
- **agreement**: All processes decide on the same value.
- **validity**: Every decided value is some process's proposal value.

It is well-known [3] that any protocol for classical consensus can solve any shared-memory task (via a universal construction), and that universality is a useful tool for deriving impossibility results.

Here are some of the more important ways cross-chain consensus differs from classical consensus. While classical processes can crash, they are honest. By contrast, cross-chain parties can be dishonest (even irrational). In classical consensus, processes choose both proposal values and decision values, while in cross-chain consensus, untrustworthy parties choose proposal values, and trustworthy contracts choose decision values. As previously noted, this distinction between untrustworthy parties and honest contracts imposes a challenge absent from the classical model: how to convince one contract that another contract has changed state, given that contracts cannot communicate directly. This structural difference in the models gives cross-chain protocols a flavor distinct from their classical counterparts.

5   Related Work

Most work connecting consensus to blockchains is concerned with the consensus protocol used to secure the blockchain itself (see surveys by Cachin and Vukolic [2]), Singh et al. [9]).

The model considered here is incomparable to the classical *Byzantine agreement problem* [5], as are the failure assumptions (super majority correct vs one party correct).

References

1. Amoussou-Guenou, Y., Herlihy, M., Jayanti, S., Potop-Butucaru, M., Rajsbaum, S.: The smart contract model (2025). arXiv Preliminary version appeared in as "Invited paper: The smart contract model" in the proceedings of the The 26th International Symposium on Stabilization, Safety, and Security of Distributed Systems, Aichi, Japan, October 2024
2. Cachin, C., Vukolić, M.: Blockchain Consensus Protocols in the Wild. arXiv (2017). Publisher: arXiv Version Number: 2. https://arxiv.org/abs/1707.01873
3. Herlihy, M.: Wait-free synchronization. ACM Trans. Program. Lang. Syst. **13**(1), 124–149 (1991). https://doi.org/10.1145/114005.102808
4. Herlihy, M., Shavit, N., Luchangco, V., Spear, M.: The art of multiprocessor programming, 2nd edn. Morgan Kaufmann, Cambridge (2021)
5. Lamport, L., Shostak, R., Pease, M.: The byzantine generals problem. ACM Trans. Program. Lang. Syst. **4**(3), 382–401 (1982). https://doi.org/10.1145/357172.357176
6. Micali, S.: State proofs (2022). https://medium.com/algorand/state-proofs-e8c7c2dcb131. As of 16 May 2024

7. Nolan, T.: Atomic swaps using cut and choose (2016). https://bitcointalk.org/index.php?topic=1364951
8. Axelar Research. Axelar network: Connecting applications with blockchain ecosystems (2022). As of 8 May 2023. https://arxiv.org/pdf/2011.12783.pdf
9. Singh, A., Kumar, G., Saha, R., Conti, M., Alazab, M., Thomas, R.: A survey and taxonomy of consensus protocols for blockchains. J. Syst. Archit. **127**, 102503 (2022). https://linkinghub.elsevier.com/retrieve/pii/S1383762122000777. https://doi.org/10.1016/j.sysarc.2022.102503
10. Uniswap. Zero-knowledge rollups (2025). https://blog.uniswap.org/what-is-defi. As of 10 September 2025
11. Zarick, R., Pellegrino, B., Banister, C.: Layerzero: trustless omnichain interoperability protocol (2022). As of 8 May 2023. https://layerzero.network/pdf/LayerZero_Whitepaper_Release.pdf

Invited Paper: Towards Demand-Aware Peer Selection with XOR-Based Routing

Qingyun Ji, Darya Melnyk, Arash Pourdamghani$^{(\boxtimes)}$, and Stefan Schmid

TU Berlin, Berlin, Germany
`pourdamghani@tu-berlin.de`

Abstract. Peer-to-peer networks, as a key enabler of modern networked and distributed systems, rely on peer-selection algorithms to optimize their scalability and performance. Peer-selection methods have been studied extensively in various aspects, including routing mechanisms and communication overhead. However, many state-of-the-art algorithms are oblivious to application-specific data traffic. This mismatch between design and demand results in underutilized connections, which inevitably leads to longer paths and increased latency.

In this work, we propose a novel demand-aware peer-selection algorithm, called *Binary Search in Buckets* (BSB). Our demand-aware approach adheres to a local and greedy XOR-based routing mechanism, ensuring compatibility with existing protocols and mechanisms. We evaluate our solution against two prior algorithms by conducting simulations on real-world and synthetic communication network traces. The results of our evaluations show that BSB can offer up to a 43% improvement compared to two other algorithms.

Keywords: demand-aware networks · peer selection · local routing

1 Introduction

Many modern networked and distributed systems rely on peer-selection algorithms, including (but not limited to) Amazon Dynamo [11], Apache Cassandra [19], and Ethereum [18]. Since peer-selection affects the latency and throughput of the system, designing efficient and simple peer-selection algorithms has been a major research topic over the past decades [24]. While many different algorithms have been proposed in the literature already, they are typically oblivious to the inherent patterns in demand. Recently, Avin et al. [4] have demonstrated that different types of applications have their own unique data traffic characteristics. These characteristics manifest, for example, in significant spatial (non-temporal) locality. Roughly speaking, high spatial locality means that communication happens frequently between specific pairs of nodes within the network.

This work takes a step toward demand-aware peer-selection, by introducing a novel approach that utilizes information about the communication demand to optimize peer selection. This change from demand-oblivious to demand-aware design promises a more efficient, scalable, and cost-effective network topology.

© The Author(s), under exclusive license to Springer Nature Switzerland AG 2026
S. Bonomi et al. (Eds.): SSS 2025, LNCS 16350, pp. 239–252, 2026.
https://doi.org/10.1007/978-3-032-11127-2_20

1.1 Our Contribution

In this paper, we present two novel demand-aware peer-selection algorithms, under the class of *Binary Search in Buckets* (BSB). These algorithms use the fact that communication between peers reveals patterns that can be exploited by the peer-selection process. The algorithms are based on the structure of Kademlia [23]. From the perspective of a sending peer, the peers of the network are divided into buckets, where the number of peers in each bucket increases exponentially. The algorithms differ in how sending peers select which peers to communicate with inside the buckets.

- The HALF-SPLIT strategy considers the demand in each bucket and connects to a central peer based on the demand.
- The MAX-DEMAND strategy connects to the peer with the maximum demand in each bucket.

By design, our algorithms respect the local and greedy XOR-based routing and rely on a local view of the demand. Hence, they can replace most of the existing peer-selection algorithms without requiring a change in other parts of the system. We show the benefits of our algorithms by comparing them to previous algorithms on a range of synthetic and real-world datasets. We show that for the cases where the spatial complexity of demand is not high, our algorithms can perform better than the previous algorithms, which is observed in some real-world instances.

1.2 Related Work

In this section, we first review the traditional algorithms for peer selection, and then discuss works related to demand-aware designs

Peer-Selection Algorithms. One of the key results in peer-to-peer (P2P) network design is Chord [29], a simple network design for efficient network location. The peer identifiers are arranged on a ring, each peer maintains up to $\log n$ connections, and key location is implemented via greedy routing. The Kademlia protocol [23] improves the key location by allowing the peers to select their connections in a randomized manner. Many more demand-oblivious P2P network designs have been proposed in the literature, including Viceroy [21], CAN [27], Pastry [28], and Tapestry [31]. For a detailed overview of P2P network architectures, please refer to [3] and [20].

Demand-Aware Network Design. Demand-aware network design has been pioneered by the work of [6], which gives a glimpse of the potential and challenges of demand-aware network design. Following this work, demand-aware network design has been studied in the context of datacenter networks [13,16,25], distributed hash tables [26], and other settings [1,8]. One direction closely related to our work is presented in [5,10,12]. There, the authors consider the problem of designing *bounded-degree* demand-aware networks, i.e., demand-aware networks where each peer can only be connected to a constant number of other peers. More recently, authors of [14] provided algorithms for demand-aware augmentation of an existing network with a matching.

Table 1. A summary of selected peer-selection algorithms compared to our work.

Protocol	Demand-awareness	XOR-based routing
Chord [29]	No	Yes
Permutations [30]	Yes	No
BSB [this paper]	Yes	Yes

Demand-Aware Peer Selection. In one of the early works on demand-aware peer selection, the authors of [2] focus on accounting for the underlying Internet topology, and thus leveraging the cooperation with internet service providers. Among the most recent related works, we note [30], which introduces a demand-aware peer-selection algorithm for training deep neural networks. In this work, we refer to the algorithm in [30] as the *Permutations algorithm*. Intuitively, this algorithm consists of $\log n$ overlapping rings of peers on top of each other, and depends on coin-change routing. This routing is not in line with the traditional XOR-based routing, hence can not be easily integrated with existing protocols. Furthermore, as this work only focuses on overlapping rings, it does not fully utilize many other possible peer-selection scenarios. Lastly, we want to point out that in the context of blockchain systems, works such as [7,22] optimize the peer-selection procedure to minimize latency between the broadcast of a transaction and its confirmation. Compared to these works, our objective of adjusting to any network traffic demand is more general.

In the empirical evaluation, we compare our work to Chord and to the Permutations algorithms. A comparison of the theoretical guarantees of these algorithms is presented in Table 1.

2   Model

In this section, we discuss the underlying model used in the design of our algorithm and in the empirical analysis.

The peer-to-peer network considered in this paper is an overlay topology operating on top of an underlying network. We consider an overlay network topology N consisting of n peers/nodes[1] (e.g., computers) with identifiers $0, 1, \ldots, n-1$. Without loss of generality, we assume that n is a power of two. Note that the algorithms presented in this paper assign nodes to buckets of different sizes. If n was not a power of two, the algorithms would simply assign the additional nodes to the largest bucket. Further, we assume that the nodes of the

---

[1] From now on, we use nodes and peers interchangeably.

network communicate with each other by sending messages over the neighboring links and forwarding other nodes' messages.

Overlay Network. We consider an overlay network that consists of two parts. First, to ensure the connectivity of the underlying network, we consider the existence of an overlay network structure where the n nodes are already connected to each other such that they form a ring. This is a common assumption in the literature [29]. Then, to achieve an efficient peer-to-peer network, we allow nodes to locally augment the network by adding up to $\log n$ additional links to their peers. We assume that the links of these two parts have the same properties: all having equal capacity and delay between their two endpoints. We denote the distance between nodes i and j by $\mathrm{dist}_{i,j}$. The exact value of the distance depends on the underlying routing strategy, detailed below.

Routing Mechanisms. We define routing on the overlay network (including the ring and the augmented links) and discuss two possible routing mechanisms. The first considered routing strategy uses the *shortest paths*. This is a routing strategy that always uses the smallest number of edges for communication. When computing the shortest path distance, we assume that all nodes are aware of the whole network. We only consider this type of routing as a benchmark for other routing strategies, since assuming the knowledge of the whole network is unrealistic.

The second routing strategy is *local routing*. Here, we assume that each node has a local view of the network. The nodes forward messages based on the messages they receive and their own connections to neighbors. As many local routing strategies depend on the peer-selection algorithm, we will specify these strategies later in the paper. Observe that local routing, unlike shortest path routing, only require nodes to keep track of their neighbors.

Demand Matrix. To be able to design demand-aware peer-selection algorithms, we assume that the communication demand of the network is fully described by a demand matrix $D \in \mathbb{R}^{n \times n}$. Each entry i and j of the demand matrix shows the amount (or the percentage of) traffic between nodes i and j. Observe that the elements on the diagonal of the demand matrix always have zero values, as the nodes do not communicate with themselves. Moreover, since we consider directed networks, the demand matrix is not required to be symmetric.

Cost Function. The objective of our model is to minimize the overall communication cost. This cost depends on the communication matrix and on the routing strategy in the augmented network. Formally, the cost function C_N for a network N is defined as

$$C_N = \sum_{\forall i,j} \mathrm{dist}_{ij} \times D_{ij}$$

where dist_{ij} is the distance between the nodes i and j under the applied routing strategy.

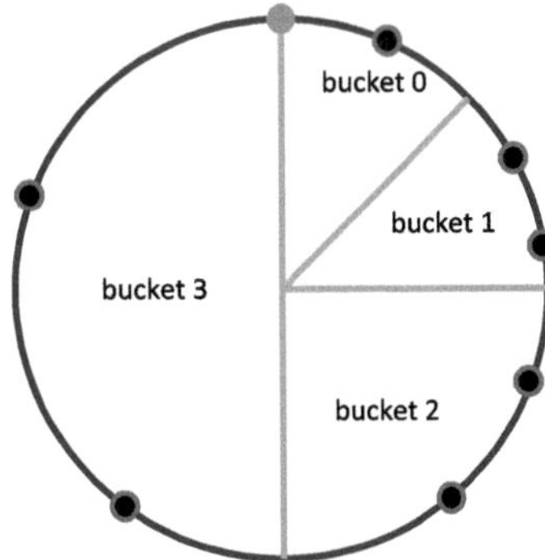

Fig. 1. This figure shows the layout of the buckets on a cycle from the blue node's perspective. We assume that the blue node has the key 0000. The buckets from the farthest to the closest relative to the source node (the blue one) are correspondingly indexed from 0 up to $\log n - 1$. They are made up of all nodes with prefixes 0001, 001, 01, and 1 respectively. (Color figure online)

Algorithm 1. BSB algorithm

1: $degree \leftarrow \log n$
2: **for** $i \leftarrow 1 \, to \, n$ **do**
3:    **for** $j \leftarrow 1 \, to \, degree$ **do**
4:       $startpoint \leftarrow (i \oplus (1 \ll j)) \gg j \ll j$
5:       $endpoint \leftarrow startpoint + (1 \ll j) - 1$
6:       $demand \leftarrow D[i][startpoint : endpoint]$
7:       $peer \leftarrow LocateNode(n, i, demand)$
8:       $peer \leftarrow (indexOfMax(demand) + i)\%n$
9:       $G[i][peer] \leftarrow 1$

---

3   Peer-Selection Algorithms

We now describe our demand-aware peer-selection algorithms. The overarching part of our algorithms is inspired by the Kademlia protocol [23]. In the routing tables of Kademlia, nodes are organized in a binary tree-like structure. For our algorithms, we instead use the cyclic structure.

From the perspective of a single source node, the entire key-space is first divided into $\log n$ buckets, as shown in an example in Fig. 1. Bucket 0 only contains the successor node, while the largest and farthest bucket contains half of the nodes inside the network. As the nodes in a bucket always have keys with a specific length of common prefix, a bucket can be simply represented by a key range with its start point and end point. Algorithm 1 presents this overarching part as pseudocode. The next step is to select a peer in each bucket. We have designed two demand-aware methods for peer selection in a bucket. A summary of these two node selection algorithms is described below and is presented as pseudocode in Algorithm 2.

Algorithm 2. LocateNode($n, i, demand$)

1: **if** HALF-SPLIT **then**
2:     $total\_demand \leftarrow 0$
3:     **for** $k \leftarrow 1\ to\ n$ **do**
4:         $total\_demand \mathrel{+}= demand[k]$
5:     $half\_demand \leftarrow \frac{total\_demand}{2}$
6:     $cumulative\_demand \leftarrow 0$
7:     **for** $k \leftarrow 1\ to\ n$ **do**
8:         $cumulative\_demand \mathrel{+}= demand[k]$
9:         **if** $cumulative\_demand \geq half\_demand$ **then**
10:             $node = (i + k)\ \mod n$
11: **if** MAX-DEMAND **then**
12:     $max\_demand \leftarrow -1$
13:     **for** $k \leftarrow 1\ to\ n$ **do**
14:         **if** $demand[k] > max\_demand$ **then**
15:             $max\_demand = demand[k]$
16:             $node = (i + k)\ \mod n$
17: **return** $node$

3.1   Half-Split

Given the local communication demand from the source node s, from which the corresponding bucket-level demand can be extracted, node s chooses the peer in each bucket such that it splits the bucket-level demand approximately evenly. Note that bucket 0 only contains the successor node of the source node, so node s will directly add its successor in bucket 0 to the routing table. In other buckets, node s determines the node along the cycle that has approximately 50% of the total bucket-level demand on both sides, i.e., splits the bucket-level demand in half. We call this node the *mid-node*, and a link to this node will be added to the source's routing table.

3.2   Max-Demand

Another demand-aware method for peer selection is to directly establish links to the node that has the highest inbound demand from the source node. Intuitively, this MAX-DEMAND method could help shorten the routing path length between node pairs that communicate frequently, since it could at least reduce the communication cost from the source node to peers within its routing table. Note, however, that finding the peer with the highest bucket-level demand is more time-consuming than the previous method.

3.3   XOR-Based Greedy Routing

Since the peers are chosen in a demand-aware manner, the global routing information is not available to the nodes. Therefore, a node can only determine the next hop based on the destination key. In other words, the message will be

---

Algorithm 3. XOR-based geedy routing

Input n : Number of nodes in the network ($n = 2^m, m \in \mathbf{N}$)
Input G : Final topology with directed edges ($G \in \{0,1\}^{n \times n}$)
Output R : XOR-based routing path lengths in a matrix($R \in \mathbf{R}^{n \times n}$)
 1: ▷ *Compute the path length from each node-pair and add the corresponding routing cost to total cost*
 2: **for** $i \leftarrow 1$ *to* n **do**
 3:   **for** $j \leftarrow 1$ *to* n **do**
 4:     $curr \leftarrow i$
 5:     $next\_hop \leftarrow curr$
 6:     $max\_common\_prefix\_len \leftarrow 0$
 7:     $path\_length \leftarrow 0$
 8:     ▷ *Stop when the destination node j is arrived*
 9:     **while** $curr \neq j$ **do**
10:       **for** $k \leftarrow 1$ *to* n **do**
11:         **if** $G[curr][k] = 1$ **then**
12:           $common\_prefix\_len \leftarrow log_2 n - bit\_length(k \oplus j)$
13:           **if** $common\_prefix\_len > max\_common\_prefix\_len$ **then**
14:             $max\_common\_prefix\_len \leftarrow common\_prefix\_len$
15:             $next\_hop \leftarrow k$
16:       $curr \leftarrow next\_hop$
17:       $path\_length += 1$
18:     ▷ *Add the routing cost from node i to node j*
19:     $R[i][j] \leftarrow path\_length$
20:
21: **return** R

---

forwarded to the node whose key has the longest common prefix with the destination key in the current node's routing table. Unlike the shortest path routing in a peer-to-peer network, the greedy routing does not require individual nodes to know all links. At each step of forwarding a message, the node only considers the destination key to determine the next hop.

The pseudo-code for XOR-based greedy routing is presented in Algorithm 3. In this algorithm, each node at each step checks the peers in its routing table and chooses the peer with the longest common prefix with the destination key as the next hop, until the message arrives at the destination. Consequently, the length of the XOR-based routing path is upper bounded by $\log n$, as the key is represented by a binary value of $\log n$ bits.

4   Experimental Evaluation

The main goal of the evaluation section is to answer the following questions:

- **Question 1:** What is the running time of the presented peer-selection algorithms?
- **Question 2:** Under which parameters do the algorithms perform best?

- **Question 3:** How do our algorithms perform in a wide range of real-world datasets?

4.1   Datasets

In our work, we consider two variants of datasets, a synthetic dataset that was created based on Zipf distribution, and also the realistic dataset from [4].

Zipf Distribution. We generated 16 synthetic network communication data samples with different α-values, following the rules of the random Zipf distribution [9]. α is a hyperparameter in generating sequences based on Zipf distribution, which must be greater than 1. The probability density of every random variable becomes smaller as the α-value increases. That is, the skewness of the distribution becomes weaker with larger α-values.

Real-World Dataset. We conducted our simulations using two of the three datasets available: cluster A and cluster C. They are collected from two different applications, namely Database and Hadoop applications. Every split data chunk includes network packets in a 20-min time interval, where the time intervals of different data chunks do not overlap with each other.

Besides the Facebook main datasets, we evaluated our work on datasets of smaller sizes from various applications, like Microsoft, pFabric, and ProjecToR.

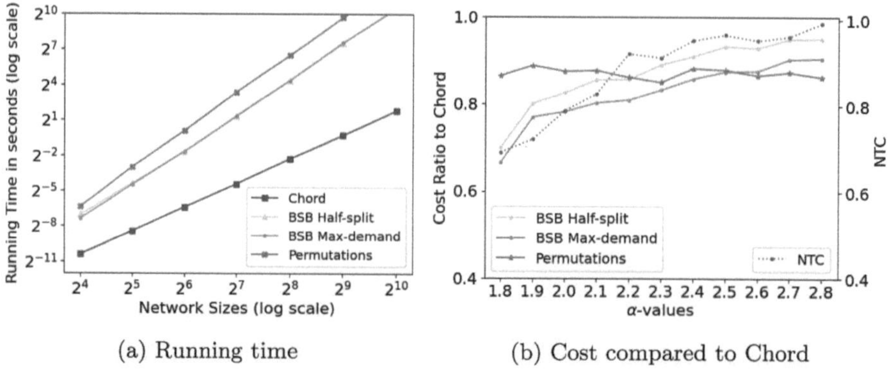

(a) Running time　　　　　(b) Cost compared to Chord

Fig. 2. Figure 2a demonstrates the running time of the peer-selection algorithms that we focus on. Figure 2a shows the correlation between the communication costs calculated with different peer-selection algorithms and the non-temporal complexity (NTC) of the network traffic data.

4.2   State-of-the-Art Algorithms

In the following, we provide a short overview of how we implemented two peer-selection algorithms from the literature, Chord [29] and the Permutations algorithm [30], and also how the coin-change routing algorithm works.

Chord. Chord employs a cyclic topology of nodes. In this case, nodes simply select peers using the bitwise XOR operator. That is, the k-th peer of node with ID i is $(i \oplus 2^k)$. Moreover, the routing between any pair of nodes on a cyclic network of size n is guaranteed within $\log n$ steps.

Permutations Algorithm. This algorithm is similar to the standard Chord protocol in the sense that the peers are selected based on a group of predefined unidirectional distances to the source node. The difference is that while the distances in the Chord protocol are powers of two, those in the permutation algorithm are calculated based on the given demand matrix. The permutations are selected from the integers in $[1, n-1]$. Then the GCD-filter (greatest common divisor filter) is applied to reduce the number of candidate permutations. The final permutations that define the augmented communication links are chosen one by one from the candidate list. The network is then updated with additional edges after each selection. The GCD-filter ensures the connectivity of the network during the process of permutation selection, which is essential for the calculation of communication costs.

Coin Change Algorithm. The goal of the coin change algorithm is to find a way to deliver a target amount of money using the minimum number of coins. The available coins refer to the selected permutations, and the target amount of money refers to the unidirectional distance from the source node to the destination node. The coin change algorithm computes the minimum number of permutations needed to have all possible distances ranging from 1 to $n - 1$. It correspondingly stores the combinations of permutations that make up the routing paths associated with valid distances. Dynamic programming is used to reduce the computational cost of repeatedly calculating the routing paths for each source-destination pair. Once the calculation is finished, routing becomes quite fast by looking up the associated linear combination of permutations based on distances in a list of $n - 1$ elements.

4.3   Results

To simulate our system, we have used python 3.10 with networkx [15] and Matplotlib [17] libraries. Our programs were running on 16 AMD Opteron[TM] processors with a clock frequency of 2.0GHz, and 15 GB of RAM.[2]

Answer 1. Running Time. As shown in Fig. 2a, our BSB algorithms have an advantage over the Permutations algorithm due to faster computational speed, which makes them stand out in some specific domains.

Answer 2. Non-temporal Complexity. In order to compute the non-temporal complexity of the datasets, we prepared three data files per network and compressed the files, respectively, following the methodology of [4]. Firstly, the original file stores the source-destination pairs associated with the packets transmitted across the network in their original sequence. On the basis of the

---

[2] The source code of this paper can be found in https://github.com/inet-tub/BSB.

original file, another file, the shuffled file, is generated by randomly shuffling the rows in the original file. At this point, the temporal structure of the network traffic is eliminated. The third file consists of the same number of rows as the original and shuffled file, while the source-destination pairs are randomly generated from the node set using the uniform distribution. During this step, the non-temporal structure is then removed. We name it the random file. Therefore, the non-temporal complexity is measured as the ratio of the size of the compressed shuffled file to that of the compressed random file. Figure 2b depicts a strong correlation between the cost ratios of BSB algorithms and the non-temporal complexity of communication demand, which indicates that our BSB algorithms are targeting cases of low non-temporal complexity.

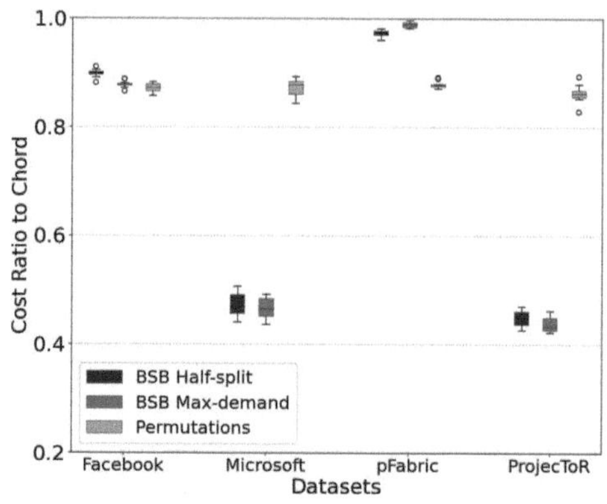

Fig. 3. Communication cost ratio of BSB and Permutations algorithms compared to Chord, considering various small networks (with 64 nodes in each).

Answer 3.1. Small Real-World Dataset. Figure 3 exhibits the simulation results on smaller datasets. We can see the four groups on the x-axis, namely Facebook, Microsoft, pFabric, and ProjecToR. The original datasets are all based on a 100-node peer-to-peer network. To simplify the process of XOR-based routing, we reduced the number of nodes to 64 in each group. We performed the filtering as follows: each of the original 100 nodes is randomly assigned a node identifier that ranges from 0 to 99. Then the nodes with identifiers from 0 to 63 were selected. After that, we walked through the whole dataset to remove the rows (i.e., network packets) where either the source node or the destination node is out of the filtered scope. For each application, we performed 10 random ID assignments to avoid the situation where a single result is not sufficiently representative for our study.

Similarly, the performance of the Permutations algorithm does not vary much on different datasets, with a cost ratio of 0.86 on average. Nevertheless, the simulations of our BSB algorithms show completely different results in different types of applications. On Microsoft and ProjecToR networks, our BSB algorithms clearly outperform the Permutations algorithm with an average cost reduction of 55% (compared to Chord). On the other hand, they have only a slight advantage over Chord on Facebook and pFabric networks, underperforming compared to the Permutations algorithm. This phenomenon is again related to our hypothesis that the non-temporal complexity of the network flow could affect the efficiency of our BSB peer selection. In other words, we infer that the frequency distribution of network communications in Microsoft and ProjecToR datasets seems to be more skewed than that in Facebook and pFabric datasets.

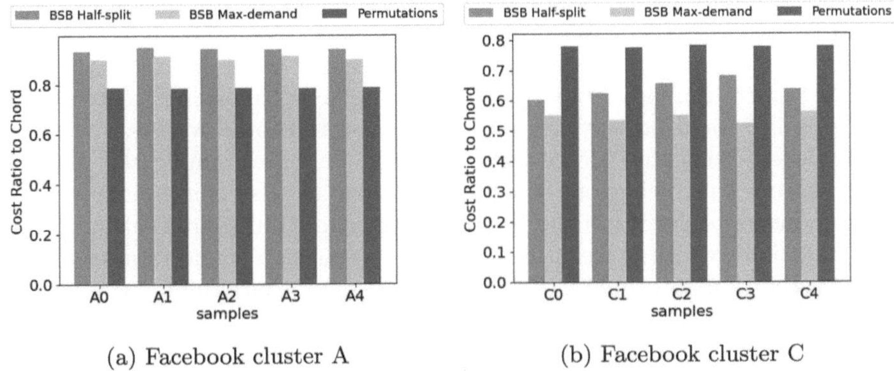

(a) Facebook cluster A   (b) Facebook cluster C

Fig. 4. This figure shows the communication cost ratio of BSB and Permutations algorithms compared to the Chord algorithm, which are simulated on 4096-node networks generated from Facebook data samples. Figure 4a shows results for the database cluster (A). Figure 4b shows results for Hadoop cluster (C).

Answer 3.2. Real-World Dataset. Each full dataset is partitioned into data chunks based on the timestamps of the network packets, where each data chunk includes the network traffic data within a non-overlapping 20-min time interval. Although the network size is reduced to 4096 in each data chunk using a key-filter, the cost calculation is still a time-consuming task, especially for the Permutation peer-selection algorithm. Therefore, we conducted simulations only on five data chunks per Facebook cluster due to the time limit.

Although our BSB algorithms and the Permutations algorithm are both based on the real-world network traffic, their peer-selection logics are quite different. As we discussed in the previous section, the peer-selection method is highly dependent on the routing mechanism. What we can now observe from the two plots is that the Permutations algorithm shows a stable performance on both datasets. Its cost ratio stays around 0.78, which indicates that it can reduce the communication cost by 22% compared to the Chord algorithm.

However, our BSB algorithms do not present results similar to the Permutations algorithm. They perform much worse than the Permutations algorithm on the Database cluster, whereas they perform much better on the Hadoop cluster. The BSB HALF-SPLIT algorithm exhibits a cost ratio of approximately 93%

on the former dataset and 60% on the latter, where the BSB MAX-DEMAND algorithm shows lower cost ratios of around 89% and 55% respectively. A reason for the unstable performance of our BSB algorithms could be the different frequency distributions of communication demand in various applications.

4.4   Discussion

BSB Algorithms Versus the Permutations Algorithm. Intuitively, if there is a high communication demand between a source-destination pair, the easiest way to ensure low cost is to connect the source node to the destination nodes directly. Since the BSB algorithms use the "local view", simply connecting such source-destination pairs would be possible. Especially in the MAX-DEMAND version, the direct channels from the source node to its most frequent contacts in each bucket could be very helpful for reducing the total overhead. However, selecting peers using the "global view" seems to be more complicated, as nodes have to ensure that behaviors benefit themselves are also compatible with the collective interests. Nodes in the Permutations algorithm share the same group of permutations, which makes it restrictive to establish a connection for a single source-destination pair, unless those pairs have significantly higher communication demand among other pairs of nodes.

Comparing Two BSB Algorithms. The reason why MAX-DEMAND performs better than HALF-SPLIT in most of the cases seems to be straightforward: MAX-DEMAND tends to build a direct link from a source node to the node that it has highest demand to. Recall that the total communication cost can be viewed as the sum of weighted path lengths. Hence, by shortening the path lengths between these pairs of nodes that communicate frequently, the communication cost is expected to be heavily reduced. Although MAX-DEMAND is more beneficial for improving the efficiency of P2P network communication, it has longer runtime than HALF-SPLIT does on selecting peers at the bucket level.

5   Conclusion

In this work, we took initial steps toward a demand-aware peer-selection algorithm, that supports XOR-based routing, and aims to minimize the shortest path, weighted by the communication demand between peers. In particular, we introduced a class of algorithms under the general name of BSB, and gave a comparative performance analysis of this class considering a wide range of real-world and synthetic inputs. Our results indicated that when the communication demand is skewed, BSB algorithms achieved significantly reduced communication cost compared to state-of-the-art algorithms, by up to 43%. In the future, we aim to study extended variants of BSB algorithms, in particular, algorithms that incorporate randomization. We also aim to extend our evaluation to other applications where peer selection is relevant, for example, in blockchain networks.

Acknowledgments. This project has received funding from the European Research Council (ERC) under grant agreement No. 864228 (AdjustNet), 2020–2025.

References

1. Addanki, V., Pacut, M., Pourdamghani, A., Rétvári, G., Schmid, S., Vanerio, J.: Self-adjusting partially ordered lists. In: INFOCOM, pp. 1–10. IEEE (2023)
2. Aggarwal, V., Feldmann, A., Scheideler, C.: Can ISPS and P2P users cooperate for improved performance? ACM SIGCOMM Comput. Commun. Rev. (2007)
3. Androutsellis-Theotokis, S., Spinellis, D.: A survey of peer-to-peer content distribution technologies. ACM Comput. Surv. (2004)
4. Avin, C., Ghobadi, M., Griner, C., Schmid, S.: On the complexity of traffic traces and implications. Proc. ACM Meas. Anal. Comput. Syst. (2020)
5. Avin, C., Mondal, K., Schmid, S.: Demand-aware network designs of bounded degree. Distrib. Comput. **33**(3–4), 311–325 (2020)
6. Avin, C., Schmid, S.: Toward demand-aware networking: a theory for self-adjusting networks. Comput. Commun. Rev. **48**(5), 31–40 (2018)
7. Babel, K., Baker, L.: Strategic peer selection using transaction value and latency. In: DeFi@CCS, pp. 9–14. ACM (2022)
8. Bentert, M., Franke, M., Melnyk, D., Pourdamghani, A., Schmid, S.: Demand-aware multi-source IP-multicast: minimal congestion via link weight optimization. In: International Federation for Information Processing Networking Conference (IFIP Networking) (2025)
9. Breslau, L., Cao, P., Fan, L., Phillips, G., Shenker, S.: Web caching and zipf-like distributions: evidence and implications. In: IEEE INFOCOM (1999)
10. Dallot, J., Caldeira, C., Pourdamghani, A., Goussevskaia, O., Schmid, S.: Laslin: a learning-augmented peer-to-peer network (2025). https://arxiv.org/abs/2509.11904
11. DeCandia, G., et al.: Dynamo: amazon's highly available key-value store. In: ACM SOSP (2007)
12. Figiel, A., Korhonen, J.H., Olver, N., Schmid, S.: Efficient algorithms for demand-aware networks and a connection to virtual network embedding. In: OPODIS. LIPIcs, vol. 324, pp. 38:1–38:24. Schloss Dagstuhl - Leibniz-Zentrum für Informatik (2024)
13. Figiel, A., Melnyk, D., Milentijevic, T., Schmid, S.: Distributed construction of demand-aware datacenter networks. In: IPDPS, pp. 162–172. IEEE (2025)
14. Figiel, A., Melnyk, D., Nichterlein, A., Pourdamghani, A., Schmid, S.: Spiderdan: matching augmentation in demand-aware networks. In: SIAM Symposium on Algorithm Engineering and Experiments (ALENEX) (2025)
15. Hagberg, A.A., Schult, D.A., Swart, P.J.: Exploring network structure, dynamics, and function using networkx. In: Proceedings of the 7th Python in Science Conference (2008)
16. Hanauer, K., Henzinger, M., Schmid, S., Trummer, J.: Fast and heavy disjoint weighted matchings for demand-aware datacenter topologies. In: INFOCOM, pp. 1649–1658. IEEE (2022)
17. Hunter, J.D.: Matplotlib: a 2D graphics environment. Comput. Sci. Eng. (2007)
18. Kiffer, L., Salman, A., Levin, D., Mislove, A., Nita-Rotaru, C.: Under the hood of the ethereum gossip protocol. In: Borisov, N., Diaz, C. (eds.) FC (2021)

19. Lakshman, A., Malik, P.: Cassandra: a decentralized structured storage system. ACM SIGOPS Oper. Syst. Rev. (2010)
20. Lua, E.K., Crowcroft, J., Pias, M., Sharma, R., Lim, S.: A survey and comparison of peer-to-peer overlay network schemes. IEEE Commun. Surv. Tutor. (2005)
21. Malkhi, D., Naor, M., Ratajczak, D.: Viceroy: a scalable and dynamic emulation of the butterfly. In: ACM PODC (2002)
22. Mao, Y., Deb, S., Venkatakrishnan, S.B., Kannan, S., Srinivasan, K.: Perigee: efficient peer-to-peer network design for blockchains. In: PODC, pp. 428–437. ACM (2020)
23. Maymounkov, P., Mazières, D.: Kademlia: a peer-to-peer information system based on the XOR metric. In: Druschel, P., Kaashoek, M.F., Rowstron, A.I.T. (eds.) IPTPS (2002)
24. Naor, M., Wieder, U.: Novel architectures for P2P applications: the continuous-discrete approach. In: SPAA. ACM (2003)
25. Pourdamghani, A., Avin, C., Sama, R., Schmid, S.: Seedtree: a dynamically optimal and local self-adjusting tree. In: INFOCOM, pp. 1–10. IEEE (2023)
26. Pourdamghani, A., Avin, C., Sama, R., Shiran, M., Schmid, S.: Hash & adjust: competitive demand-aware consistent hashing. In: OPODIS. LIPIcs, vol. 324, pp. 24:1–24:23. Schloss Dagstuhl - Leibniz-Zentrum für Informatik (2024)
27. Ratnasamy, S., Francis, P., Handley, M., Karp, R.M., Shenker, S.: A scalable content-addressable network. In: Cruz, R.L., Varghese, G. (eds.) ACM SIGCOMM. ACM (2001)
28. Rowstron, A.I.T., Druschel, P.: Pastry: scalable, decentralized object location, and routing for large-scale peer-to-peer systems. In: Guerraoui, R. (ed.) IFIP/ACM Middleware (2001)
29. Stoica, I., et al.: Chord: a scalable peer-to-peer lookup protocol for internet applications. IEEE/ACM Trans. Netw. (2003)
30. Wang, W., et al.: Topoopt: co-optimizing network topology and parallelization strategy for distributed training jobs. In: USENIX NSDI (2023)
31. Zhao, B.Y., Huang, L., Stribling, J., Rhea, S.C., Joseph, A.D., Kubiatowicz, J.: Tapestry: a resilient global-scale overlay for service deployment. IEEE J. Sel. Areas Commun. (2004)

Label Leakage in Regression Federated Learning Using Cryptographic Tools

Pierre Jobic$^{(\boxtimes)}$![ORCID], Aurélien Mayoue ![ORCID], and Sara Tucci-Piergiovanni ![ORCID]

Université Paris-Saclay, CEA List, 91120 Palaiseau, France
`pierre.jobic@cea.fr`

Abstract. Federated Learning (FL) enhances data privacy by enabling users to collaboratively train neural networks without sharing raw data. However, FL does not guarantee model privacy because clients share gradients of the model. This paper introduces a novel attack against FL in the regression setting, uncovering a previously unexplored vulnerability. We show that, in regression tasks where labels are continuous real numbers, gradient equations can be reduced to a hidden subset sum problem (HSSP, a cryptographic problem originally studied for integer values). By adapting cryptographic techniques for solving the HSSP, we demonstrate that labels can be accurately recovered, posing a significant privacy risk. Unlike analytical approaches for classification such as iDLG [22] that do not extend to regression, our method is specifically designed for this setting. Through extensive experiments on eight datasets with three different deep learning models, we demonstrate that label recovery achieves very low reconstruction error (e.g., MSE below 0.1 in some scenarios), significantly outperforming prior gradient inversion attacks such as DLG [23].

Keywords: Machine Learning · Federated Learning · Analytical Attack · Privacy · Cryptography

1 Introduction

Federated Learning (FL) [15] enables multiple clients to collaboratively train a machine learning model without sharing raw data, promising enhanced privacy for sensitive applications such as healthcare and finance. Despite these benefits, FL is vulnerable to gradient-based privacy attacks [8,9], where an adversary leverages shared gradients to infer private information from clients.

Reconstruction attacks in FL can target either the input data or the corresponding labels. While much of the literature has focused on recovering inputs (e.g., images, texts), label leakage has gained attention due to its potential to enhance input reconstruction attacks. However, in regression FL (RFL), this problem is particularly critical: regression labels are *continuous*, often represent directly sensitive information (e.g., predicting an employee's salary in a federated HR system), and lack the discrete structure that analytical attacks exploit in classification. These characteristics make regression leakage both more sensitive in practice and technically more challenging to analyze.

Analytical label attacks [6,22] have proven highly effective in classification FL (CFL), especially with cross-entropy loss and one-hot encoded labels. Such methods are fast, architecture-independent, and require only the gradients of the last layers. However, their success relies on the discreteness and sparsity of classification labels, which does not extend to regression settings that use continuous labels and losses such as mean squared error (MSE).

In contrast, optimization-based reconstruction attacks [9,23] can in principle handle regression tasks. They simulate the gradient computation process with dummy inputs and labels, iteratively optimizing until the dummy gradients match the observed ones. While versatile, these methods are slow, heuristic, and often yield poor label recovery, especially for continuous values [18].

In this paper, we introduce the first analytical attack for regression label reconstruction in FL. We show that, under certain assumptions (mean squared error loss, batch size $K \leq 18$, and models ending with two fully connected layers separated by a ReLU activation), and in the common *Federated Stochastic Gradient Descent (FedSGD)* setting, where clients compute and share per-batch gradients with the server at each round, the regression label recovery problem can be reduced to an instance of the *Hidden Subset Sum Problem* (HSSP), a well-studied problem in cryptography. In the HSSP, one observes linear combinations of hidden subsets of values and aims to recover both the subsets and the values, a problem known to be computationally hard in general. By adapting lattice-based algorithms originally designed for integer instances of HSSP to the real-valued gradient setting, we derive an efficient, mathematically grounded attack that achieves high accuracy in practice.

We evaluate our attack on eight datasets and three neural networks (two Convolutional Neural Networks (CNNs) and a Multilayer Perceptron (MLP)). Across most configurations, our method significantly outperforms the joint optimization baseline [23] in terms of root mean squared error (RMSE) while running orders of magnitude faster. The attack also provides an explicit indicator of success, allowing the adversary to determine when label recovery is exact.

Contributions. Our main contributions[1] are:

- We present the first analytical label reconstruction attack for regression FL, addressing a gap in the adversarial FL literature that has focused almost exclusively on classification.
- We reduce regression label leakage to the Hidden Subset Sum Problem (HSSP) and adapt cryptographic techniques to operate on real-valued gradients, bridging privacy analysis in FL with lattice-based cryptography.
- We conduct extensive experiments across eight datasets and three models, showing that our method achieves RMSE below 1 in most scenarios and consistently outperforms optimization-based approaches.

---

[1] See [12] for the full paper version.

This work reveals a new and significant privacy vulnerability in RFL and motivates the design of defenses specifically tailored to continuous labels.

2   Background

This section introduces the technical foundations for our attack. We first recall the FL training protocol (FedSGD), then briefly summarize existing approaches to label reconstruction, and finally present the cryptographic concepts relevant to our method. Our novel contribution, the reduction of regression label leakage to HSSP, is presented later in Sect. 5.

The Federated Learning Stochastic Gradient Descent (FedSGD). [15] procedure is as follows. The server defines a model f and initializes its parameters θ randomly which gives f_θ. During each round t, a subset S_t of clients, indexed by k, receive the model f with its parameters θ_t from the server. Then, the selected clients are asked to optimize these parameters using their local datasets (X_k, Y_k). Each selected client optimizes the model parameters by doing a single step of gradient descent using the loss function $\mathcal{L}$ and a learning rate η (i.e. $\theta_t^k = \theta_t - \eta \nabla_\theta \mathcal{L}(f_{\theta_t}(X_k), Y_k)$). The clients communicate back their parameters to the server that aggregates them to compute the updated global parameters:

$$\theta_{t+1} \leftarrow \theta_t - \eta \sum_{k \in S_t} \frac{|X_k|}{\sum_{k \in S_t} |X_k|} \nabla_\theta \mathcal{L}(f_{\theta_t}(X_k), Y_k)$$

These back and forth communication steps between the server and the clients last until model convergence.

The parameters shared by the selected clients are often interpreted as the true gradient of the loss function computed on their respective local datasets: $-\frac{\theta_t^k - \theta_t}{\eta} = \nabla_\theta \mathcal{L}(f_{\theta_t}(X_k), Y_k)$. That is why the privacy attacks in the literature often consider that the attacker has access to the gradients (when using FedSGD). We consider the same in this paper.

FedSGD is a simplistic procedure that is more theoretical than practical. It suffers from communication issues [15], robustness to data heterogeneity [13]. However it is a starting point for all privacy attacks in FL [19,22]. We focus on FedSGD because it is the standard setting in which gradient-based privacy attacks are formulated, and it provides the strongest leakage signals.

Label Reconstruction Attacks. Gradient leakage attacks attempt to reconstruct private client data from shared gradients. Two main approaches exist.

Optimization-based attacks (e.g., DLG [23], improved variants [9]) reconstruct private data by creating dummy samples $(\hat{X}, \hat{Y})$ and iteratively optimizing them so that the resulting gradients $\nabla_\theta \mathcal{L}(f_\theta(\hat{X}), \hat{Y})$ match the observed gradients $\nabla_\theta \mathcal{L}(f_\theta(X), Y)$. This optimization typically involves gradient descent over $(\hat{X}, \hat{Y})$, combined with regularization terms to encourage plausible reconstructions (e.g., image priors or total variation). Such methods are flexible and, unlike analytical attacks, are directly applicable to regression FL (RFL) since they make no assumptions about label discreteness. However, they are computationally expensive, sensitive to hyperparameters, and often converge to poor local minima, leading to inaccurate or unstable label recovery in practice.

Analytical attacks exploit structural properties of classification with cross-entropy loss and one-hot encoded labels. In this setting, the gradient of the final layer directly reveals information about the ground-truth class, enabling very high accuracy recovery with negligible computation. These approaches are simple, architecture-agnostic, and highly effective for classification. However, their success relies on the discreteness of labels: the one-hot encoding isolates a single class dimension, effectively embedding the label in the gradient. In regression FL, where labels are continuous real values, this structure vanishes, rendering existing analytical methods inapplicable.

So far, no analytical approach has been designed for regression FL (RFL), where labels are continuous and the output space is unbounded.

Challenges in Regression. In regression FL, models predict real-valued quantities (e.g., salary, age) using losses such as mean squared error (MSE). Unlike classification, gradients do not contain discrete or sparse patterns but reflect residual errors between predictions and targets. This weaker, more diffuse signal makes label inversion substantially more difficult and leaves analytical attacks unexplored until now.

The Hidden Subset Sum Problem (HSSP). Our method draws a connection between regression gradients and the HSSP, a well-studied problem in cryptography [4,16]. In HSSP, the goal is to recover hidden subsets of numbers from observed linear combinations.

Formal Definition. Given a set of m secret vectors $s_i \in \{0,1\}^d$ and a secret set of numbers $a_1, \ldots, a_d \in \mathbb{Z}$, we observe m noisy linear combinations of the form:

$$y_i = \sum_{j=1}^{d} s_{ij}\, a_j + e_i, \tag{1}$$

where e_i is an optional error term. The problem is to recover the hidden binary vectors s_i (i.e., which subset of $\{a_j\}$ participates in the sum) and the secret vector a given the observations y_i. This problem generalizes the classical Subset Sum Problem, where the numbers a are known but where there is only one observation y (thus only one secret vector s), and is believed to be computationally hard in general.

Problem Density. The hardness of solving HSSP instances is often characterized by their *density*, defined as the ratio between the number of unknowns and the logarithm of the magnitude of the numbers involved [16]. Intuitively, high-density instances are harder to solve because many different subsets can produce the same sum, whereas low-density instances tend to admit unique solutions. In our regression setting, we adapt this notion by defining the density for secret set of numbers of size d as:

$$d_{10}(d) = \frac{d}{\log_{10}(\max |a|)},$$

where a is the secret vector from (Eq. 1). As in the cryptographic literature, values of $d_{10} > 1$ indicate that the instance lies in the high-density regime, where lattice-based solvers are known to fail with high probability.

Toy Example. Given secret numbers $a = \{312354, 584645, 754654\}$ and the observation of target sums of $y_1 = 896999(= 312354 + 584645), y_2 = 1339299(= 584645 + 754654), y_3 = 1651653(= 312354 + 584645 + 754654)$, then the solution is unique and is a with $s = \{(1,1,0),(0,1,1),(1,1,1)\}$. In addition, $d_{10} = 3/log_{10}(754654) = 0.51 \leq 1$

In later sections, we show how HSSP can be adapted to regression FL to enable label recovery. The interested reader can refer to [12] for broader treatment of HSSP algorithms; here we provide only the necessary background. Our novel contribution: the reduction of regression label leakage to HSSP, is presented in Sect. 5.

3   Related Work

We group our related work section in four categories: gradient inversion attacks, analytical label leakage in classification FL, defenses against such attacks, and cryptographic approaches related to our method. Below, we summarize the most relevant works and highlight why regression label leakage remains unexplored.

Gradient Inversion Attacks. Optimization-based attacks reconstruct private data by optimizing dummy inputs and labels until their gradients match the observed ones. Deep Leakage from Gradients (DLG) [23] introduced this approach, later improved by incorporating image priors and regularizers [9]. These methods are general and applicable to regression tasks but are computationally expensive (compared to analytical attacks), sensitive to hyperparameters, and often converge to poor local minima, leading to unstable label recovery.

Analytical Label Leakage in Classification FL. A line of work shows that, under cross-entropy loss with one-hot labels, last-layer gradients reveal labels in closed form. iDLG [22] and its extensions demonstrate very high accuracy recoveries in various classification scenarios, often with negligible computation. These methods, however, fundamentally rely on discrete one-hot encodings and therefore do not extend to regression FL, where labels are continuous. For a more detailed overview of analytical attacks in classification FL, including their extensions to larger batches, alternative optimizers, and defenses please refer to [12].

HSSP. The Hidden Subset Sum Problem (HSSP) is a niche cryptographic problem, with only a handful of works addressing it. Nguyen and Stern [16] first formalized HSSP and showed that it can be attacked using lattice-based techniques, establishing worst-case hardness results. Coron [4] proposed polynomial-time heuristics for specific instances, showing that certain structured cases are easier to solve. Coron et al. [5] later provided provably polynomial-time algorithms, refining the boundary between hard and tractable cases. More recently, Notarnicola et al. [17] reformulated HSSP within a new lattice framework, enabling effective handling of noisy instances.

To the best of our knowledge, our work is the first to connect regression label leakage in FL to HSSP, enabling the import of these cryptographic techniques into the analysis of FL privacy.

Summary. Existing analytical attacks focus exclusively on classification FL and cannot be applied to regression because continuous labels lack the discrete structure needed for closed-form recovery. Optimization-based methods remain the only available baseline for regression but are slow and unstable for labels. Our work closes this gap by introducing the first analytical attack for RFL, leveraging its reduction to HSSP.

In our experiments, we compare our method against the baseline [23] of optimization-based attack, which, although fundamentally different in approach and assumptions, serves as the only available baseline for RFL label reconstruction. For interested readers, we refer to [12] for an overview of optimization-based label leakage attacks.

4   Threat Model

We assume a standard federated learning (FL) setup where the adversary is the **central server**, modeled as *honest-but-curious*. This means the server correctly follows the FL protocol (e.g., model distribution, gradient aggregation), but simultaneously tries to infer additional private information from the messages it receives. Such an adversarial role has been widely adopted in prior gradient leakage and label inference works [9,10,22,23].

The attack relies on the following explicit assumptions:

- **Adversary role.** The central server is honest-but-curious: it does not deviate from the protocol, but inspects shared gradients to reconstruct private labels.
- **Model knowledge.** The server has full white-box access to the model architecture and parameters at each training round.
- **Gradient access.** The server observes per-client gradients of the loss function with respect to local data, i.e., $\nabla_\theta \mathcal{L}(f_\theta(X), Y)$.
- **Batch size.** Clients use a batch size K that is typically small (e.g., $K \leq 18$), since larger batches reduce the feasibility of our reconstruction attack.

Table 1. Table of Notations

Notation	Space	Description
K	$\mathbb{N}$	Batch size
d	$\mathbb{N}$	Dimension of the input space
x, y	$\mathbb{R}^{K \times d}, \mathbb{R}^{K}$	Input data and corresponding labels
m, n	$\mathbb{N}$	Dimensions of the last and second-to-last fully connected layers (FCLs)
z_2	$\mathbb{R}^{n}$	Input to the second-to-last FCL
$z_1 = \sigma(w_1^T z_2 + b_1)$	$\mathbb{R}^{m}$	Output of the second-to-last FCL for a single input
$z_0 = w_0^T z_1 + b_0$	$\mathbb{R}$	Final output of the network for a single input
w_0, w_1	$\mathbb{R}^{m,1}, \mathbb{R}^{n,m}$	Weights of the last and second-to-last FCLs
b_0, b_1	$\mathbb{R}, \mathbb{R}^{m}$	Biases of the last and second-to-last FCLs
$e^k = z_0^k - y^k$	$\mathbb{R}$	Error between the model's output and the true label
$\delta_{condition}$	$\{0,1\}$	Equals 1 if the condition is true; 0 otherwise

- **Training setting.** The attack is compatible with **FedSGD** (one local step per round) or fully distributed SGD settings, where gradients are communicated at each iteration.
- **Model assumptions.** The target model ends with two fully connected layers separated by a ReLU activation and is trained with mean squared error (MSE) loss. These assumptions can be relaxed (see Sect. 5.4).

Under these assumptions, the server's objective is to reconstruct the client's private regression labels $Y \in \mathbb{R}^{K}$ from observed gradients and model information:

$$\text{Given: } \theta, \ \nabla_\theta \mathcal{L}(f_\theta(X), Y), \ K \ \implies \ \text{Recover: } Y.$$

4.1 Notations

Table 1 provides an overview of all used notations. For a variable v, we denote by $V = (v^k)_{k \in [1..K]}$ its values in a batch of size K. To simplify notation, we omit the model when writing gradients of the loss, i.e., $\nabla \mathcal{L}(X, Y) \doteq \nabla_\theta \mathcal{L}(f_\theta(X), Y)$.

5 Method

Our method builds on classical gradient computations in regression networks but introduces a novel reduction of regression label recovery to the Hidden Subset Sum Problem (HSSP). We then design an analytical attack (LLR: Label Leakage in Regression) that leverages cryptographic tools to efficiently solve this reduction. Below we first recall the gradient structure in regression models, then show the reduction to HSSP, and finally describe our attack algorithm and its complexity.

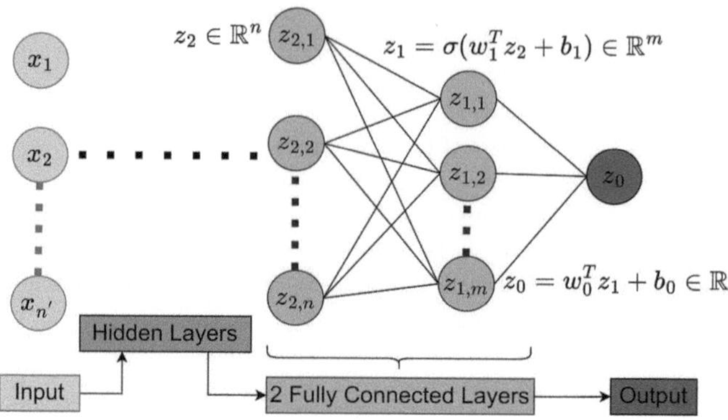

Fig. 1. Architecture considered with two fully connected layers at the end and a ReLU (σ) activation.

5.1   Warm-up: Single-Sample Case ($K = 1$)

We begin with the simplest case, where the batch size equals one. This allows us to highlight the algebraic link between gradients and labels before considering more realistic batch settings.

Theorem 1. *If $K = 1$ and the gradient of the last layer's bias is non-zero ($\nabla b_0 \neq 0$), then the regression label y can be uniquely recovered from the gradients. For interested readers, we refer to [12] for the proof.*

This theorem is reminiscent of iDLG [22] in classification FL, where batch size one also enables exact label recovery. However, as we discuss next, extending to $K > 1$ is substantially more challenging in regression due to the continuous label space and weaker gradient signal.

5.2   Gradients for Batch Size $K > 1$

Based on the notations provided in Tab. 1, for a batch of size K, the mean squared error (MSE) loss is

$$l(X,Y) = \frac{1}{K} \sum_{k=1}^{K} (z_0^k - y^k)^2,$$

where z_0^k denotes the model output for input x^k and label y^k. The gradients of the last fully connected layer (FCL) are:

$$\nabla w_0 = \frac{1}{K} \sum_k 2(z_0^k - y^k) z_1^k, \quad \nabla b_0 = \frac{1}{K} \sum_k 2(z_0^k - y^k). \tag{2}$$

Compared to classification, where one-hot labels and cross-entropy loss yield rich discrete gradient patterns, regression only exposes low-dimensional, continuous signals. In particular: (i) The output is scalar rather than categorical, reducing the amount of label information carried in ∇b_0. (ii) The system in Eq. (2)

is underdetermined, as the number of unknown labels exceeds the number of observed gradient constraints. Thus, without additional structural assumptions, label recovery is infeasible. This motivates considering models with two final FCLs separated by a ReLU activation (Fig. 1), which induces a binary structure exploitable by our method.

5.3   Reduction to HSSP

Let $e^k = z_0^k - y^k$ be the prediction error for sample k. For the second-to-last bias b_1 and last bias b_0, the gradients can be expressed as:

$$\nabla b_1 = \tfrac{2}{K}\operatorname{Diag}(w_0)\,\Delta\,E, \qquad \nabla b_0 = \tfrac{2}{K}\sum_{k=1}^{K} e^k,$$

where $\Delta \in \{0,1\}^{m\times K}$ encodes the ReLU activation mask, and $E = (e^1,\ldots,e^K)$ is the error vector. Proof of these equations are provided in [12].

Theorem 2 (Reduction to HSSP). *Define*

$$P = \tfrac{K}{2}\operatorname{Diag}(w_0)^{-1}\nabla b_1.$$

Then $P = \Delta E$, which is exactly an instance of the Hidden Subset Sum Problem (HSSP), with P observable, Δ binary and hidden, and E the unknown error vector to recover. Proof is provided in [12].

Since gradients are real-valued, we apply a scaling factor C (least common multiple of denominators) to transform the system into integers, which allows us to apply classical HSSP's algorithms to solve our HSSP. For interested readers, that want a detailed transformation from real-valued to integers, we refer to [12].

5.4   Label Leakage in Regression (LLR) Attack

Our proposed attack proceeds in two main steps:

Step 1: Recovering Errors. We construct the HSSP instance $P = \Delta E$, scale it to integers, and solve it using lattice-based algorithms. This yields a candidate error vector E_{atk}. A built-in correctness check is available: if $Error := |\tfrac{K}{2}\nabla b_0 - \sum_k e^k_{\mathrm{atk}}| = 0$, then E_{atk} is very likely (see Nguyen et al. [16]) equal to the true E.

Step 2: Recovering Labels. Given E, labels are reconstructed as $\hat{Y} = \hat{Z}_0 - E$, where $\hat{Z}_0$ is an estimator of model outputs. We consider (i) $\hat{Z}_0 = \widehat{\mathbb{E}[Z_0]}$ using dummy inputs or auxiliary data, or (ii) $\hat{Z}_0 = 0$ as a distribution-free fallback. Accuracy depends primarily on how well $\hat{Z}_0$ approximates Z_0. In particular, when $\hat{Z}_0 = \mathbb{E}[Z_0]$, the reconstruction error reduces to the variance of Z_0, i.e., $\mathrm{MSE}(\hat{Y}) \propto \mathrm{Var}(Z_0)$.

A Verifiable Success Test for Step 1. Solving the HSSP instance yields a candidate error vector $E_{\mathrm{atk}} \in \mathbb{R}^K$. Let us define the known scalar S and the discrepancy:

$$S :- \frac{K}{2} \nabla b_0 = \sum_{k=1}^{K} e^k \qquad \text{and} \qquad Error :- \Big| S - \sum_{k=1}^{K} e_{\mathrm{atk}}^k \Big|. \tag{3}$$

These quantities are computable by the attacker since S is obtained from ∇b_0 and $\sum_k e_{\mathrm{atk}}^k$ from the solver's output.

Proposition 1 (One-sided certificate). *If $Error > 0$, then $E_{atk} \neq E$ (the recovery failed).*

Proof (sketch). By definition $S = \sum_k e^k$. If $E_{\mathrm{atk}} = E$, then $\sum_k e_{\mathrm{atk}}^k = \sum_k e^k = S$, hence $Error = 0$. The contrapositive gives the claim. $\square$

When $Error = 0$, E_{atk} satisfies the necessary sum constraint. While $Error = 0$ is not a formal proof of correctness in all pathological cases (distinct solutions could, in principle, share the same sum), we find it to be an extremely sharp practical indicator in our regime (low density, integer scaling): in our experiments, $Error = 0$ coincides with exact recovery ($E_{\mathrm{atk}} = E$) in 99.98% of successful solver returns (see Sect. 6). We therefore use $Error = 0$ as a *practical certificate* to proceed to Step 2; otherwise, we declare Step 1 unsuccessful.

Step 2: Recovering the labels. Once the error vector E is obtained, the true labels can be reconstructed as

$$\hat{Y} = \hat{Z}_0 - E_{\mathrm{atk}},$$

where $\hat{Z}_0$ is an estimator of the model outputs Z_0. We consider two practical options:

- **Mean estimator.** $\hat{Z}_0 = \widehat{\mathbb{E}[Z_0]}$, computed by feeding the model with dummy inputs (e.g., sampled from a normal distribution) or using an auxiliary dataset.
- **Zero estimator.** $\hat{Z}_0 = 0$, which requires no auxiliary data. While less accurate, this is motivated by the fact that the average network output is often close to zero in practice, i.e., $\mathbb{E}_\theta[\mathbb{E}_X[z_0]] \approx 0$.

The accuracy of label recovery depends on how well $\hat{Z}_0$ approximates Z_0. Writing the reconstruction error explicitly:

$$\mathrm{MSE}(\hat{Y}) = \frac{1}{K} \sum_{k=1}^{K} \mathrm{MSE}(\hat{y}^k) = \frac{1}{K} \sum_{k=1}^{K} \mathbb{E}[(\hat{y}^k - y^k)^2] = \mathrm{MSE}(\hat{y})$$
$$= \mathbb{E}[((\hat{z}_0 - z_0) + (e - e_{\mathrm{atk}}))^2]$$
$$= \mathbb{E}[(\hat{z}_0 - z_0)^2] + \mathcal{O}(\mathbb{E}[Error])$$

When Step 1 succeeds ($Error = 0$), this reduces to

$$\begin{cases} \mathrm{MSE}(\hat{Y}) \approx \mathrm{Var}(z_0), & \text{if } \hat{Z}_0 = \mathbb{E}[Z_0], \\ \mathrm{MSE}(\hat{Y}) \approx \mathrm{Var}(z_0) + \mathbb{E}[z_0]^2, & \text{if } \hat{Z}_0 = 0. \end{cases} \tag{4}$$

Thus, if the mean estimator is used, the error is essentially the variance of the model output Z_0.

Confidence Intervals. From this characterization, the attacker can also construct a statistical confidence interval (CI). Using Markov's inequality and assuming access to $\mathrm{Var}[z_0]$, for a confidence level γ (e.g., 95%) we obtain:

$$\mathbb{P}\left[Y \in [\hat{Y} - \sqrt{\tfrac{\mathrm{Var}[z_0]}{1-\gamma}}, \ \hat{Y} + \sqrt{\tfrac{\mathrm{Var}[z_0]}{1-\gamma}}]\right] \geq \gamma. \tag{5}$$

In practice, the attacker does not know $\mathrm{Var}[z_0]$ exactly, but can estimate it empirically by feeding dummy inputs through the model and computing sample statistics. If the dummy distribution is close to the client's data distribution, the resulting CI is informative in practice.

When is the attack effective? The preceding analysis shows that the residual error of label reconstruction is essentially $\mathrm{Var}(Z_0)$ (under the mean estimator). Therefore, the attack is most effective when the label values are large compared to this variance:

$$\text{effective attack} \iff \mathrm{Var}(Z_0) \ll \|Y\|. \tag{6}$$

This condition often holds in practice when models are untrained (since $\mathrm{Var}(Z_0)$ is typically of order 1 at initialization) or when labels take relatively large numerical values (e.g., tens or hundreds for salaries, ages, or medical measurements). In such scenarios, the variance becomes negligible compared to the label scale, yielding highly accurate reconstructions.

Generalization Beyond ReLU+MSE. While we presented the reduction under the assumptions of ReLU activations and MSE loss, the same reasoning can be extended to other loss and activation functions. For a generalization to other loss functions or activation functions, we refer to [12].

6   Experiments

6.1   Setup

Reproducibility. Each experiment is repeated 10 times with different random seeds. Our code is publicly available on GitHub[2]. Experiments ran on a cluster with SKYLAKE 3GHz CPUs and P5000 GPUs.

Assumptions. Unless otherwise specified, all experiments follow the assumptions of our threat model: (i) the adversary observes FedSGD updates at the first round, (ii) models are untrained, (iii) the last two layers are fully connected with a ReLU in between, and (iv) training uses mean squared error (MSE) loss. These conditions represent the setting in which our attack is provably most effective.

---

[2] https://github.com/PierreJobic/LLR-Attack.

Datasets. We evaluate across eight datasets, covering both synthetic benchmarks and real-world regression tasks: (i) **Salary Prediction (S)**[3]: 203 samples, 5 tabular features (age, gender, experience, job role, education). Labels: salary ($25k–$250k). (ii) **Boston Housing (BH)** [11]: 506 samples, 13 features (crime rate, teacher ratio, etc.). Labels: median house value ($5k–$50k). (iii) **Liver Disorders (L)** [1]: 345 samples, 6 medical features. Labels: weekly alcohol consumption (0–20). (iv) **UTKFace (Face)** [21]: 23k face images, resized to 28×28, grayscale. Labels: age (0–100). (v) **MNIST** [7] (regression form): 60k training, 10k test images. Labels: digit class (0–9) treated as continuous. (vi) **Productivity (PPGE)** [3]: 691 samples, tabular features of workplace context. Labels: productivity (0–1.2). (vii) **Real Estate Valuation (REV)** [20]: 414 samples, 6 tabular features (location, distance to metro, etc.). Labels: price per unit area (0–200). (viii) **Apartment Rent Classified (ARC)** [2]: 39k samples, text+tabular features of listings. Labels: rental price (0–200k). Data preprocessing steps (normalization, resizing, splits) are detailed in [12].

Models and Initialization. We evaluate three architectures: (1) a multilayer perceptron (MLP) with two hidden layers for tabular data, (2) the simple CNN from [15] for MNIST, and (3) a VGG variant for UTKFace and MNIST; all models terminate with two fully connected layers and a ReLU activation. Unless otherwise noted, layers use PyTorch's default initialization; for CNNs, we adopt Kaiming initialization, which we found improves attack performance by increasing the number of distinct values in P (consistent with [14]). Full architectural specifications are provided in [12].

Evaluation Metrics. Step 1 accuracy is defined as:

$$\text{Accuracy}(E_{\text{atk}}, E) = \begin{cases} 1 & \text{if } E_{\text{atk}} = E, \\ 0 & \text{otherwise.} \end{cases}$$

Since solving HSSP either recover E exactly or fail, accuracy is binary. Step 2 label recovery is evaluated by RMSE:

$$\text{RMSE}(\hat{Y}, Y) = \left(\frac{1}{K} \sum_{i=1}^{K} (\hat{y}^i - y^i)^2 \right)^{\frac{1}{2}}.$$

We define **near-perfect recovery** as RMSE $\leq 10^{-1}$, which corresponds to exact label reconstruction up to floating-point precision.

Baselines and Fair Comparison. We compare against Deep Leakage from Gradients (DLG) [23], the most common optimization-based method applicable to regression, while noting that classification-specific baselines such as iDLG, LLG, and IBL [14,18,22] are not comparable because they rely on cross-entropy loss and one-hot labels. For DLG, we implemented Adam with learning rate 0.1, 4,800 steps, and Gaussian initialization of dummy data. To ensure fairness, we evaluate at matched success rates: for batch size $K=16$, we measure our Step 1

---

success rate $x\%$ and compute our RMSE only over successful runs; we then sort DLG RMSE values (same scenario and seeds) in ascending order and report the *average RMSE of the top $x\%$ DLG reconstructions*. We use $K=16$ for this matching because our RMSE (conditional on Step 1 success) is independent of K—it depends on the quality of $\hat{Z}_0$ and $\mathrm{Var}(Z_0)$—whereas Step 1 success varies with K. This protocol aligns percentiles across methods and avoids penalizing DLG for occasional outliers while reflecting its best achievable accuracy under the same conditions.

6.2   Results

Table 2. Step 1 results of the proposed HSSP-based attack, reporting the mean accuracy (%) for different batch sizes K. Accuracy is defined as exact recovery of the error vector E (i.e., Accuracy = 1 if $E_{\mathrm{atk}} = E$, and 0 otherwise). All models are trained with MSE loss and share the same final architecture: two fully connected layers of size $n=m=256$ with a ReLU in between. Values are averaged over 10 independent random seeds, and variability is reported as mean $\pm$ standard deviation across seeds. Thus, higher values indicate that the attack consistently achieves exact recovery. The last row reports the average density d_{10} of the induced HSSP instance, which depends mainly on the batch size K and predicts the hardness of recovery.

Accuracy ↑		Batch size (K)					
Dataset	Model	8	10	12	14	16	18
S	MLP	0.85 ± 0.02	0.75 ± 0.02	0.64 ± 0.03	0.54 ± 0.04	0.42 ± 0.03	0.20 ± 0.04
L	MLP	0.94 ± 0.05	0.86 ± 0.05	0.80 ± 0.10	0.70 ± 0.09	0.68 ± 0.09	0.42 ± 0.10
BH	MLP	0.98 ± 0.02	0.95 ± 0.02	0.91 ± 0.03	0.85 ± 0.06	0.79 ± 0.05	0.46 ± 0.06
REV	MLP	0.98 ± 0.02	0.95 ± 0.03	0.92 ± 0.03	0.85 ± 0.09	0.82 ± 0.08	0.37 ± 0.14
PPGE	MLP	1.00 ± 0.01	1.00 ± 0.01	0.99 ± 0.02	0.98 ± 0.01	0.99 ± 0.02	0.21 ± 0.12
ARC	MLP	0.65 ± 0.11	0.53 ± 0.11	0.33 ± 0.15	0.16 ± 0.09	0.14 ± 0.08	0.08 ± 0.09
Face	MLP	1.00 ± 0.01	1.00 ± 0.01	1.00 ± 0.01	0.99 ± 0.01	0.99 ± 0.01	0.03 ± 0.02
Face	CNN	1.00 ± 0.01	1.00 ± 0.01	0.99 ± 0.01	0.99 ± 0.01	0.99 ± 0.01	0.99 ± 0.01
Face	VGG	0.72 ± 0.14	0.56 ± 0.12	0.33 ± 0.11	0.32 ± 0.19	0.17 ± 0.17	0.12 ± 0.25
MNIST	MLP	1.00 ± 0.01	1.00 ± 0.01	1.00 ± 0.01	1.00 ± 0.01	1.00 ± 0.01	0.03 ± 0.02
MNIST	CNN	1.00 ± 0.01	1.00 ± 0.01	1.00 ± 0.01	0.99 ± 0.01	0.99 ± 0.01	0.98 ± 0.01
MNIST	VGG	0.53 ± 0.17	0.30 ± 0.13	0.28 ± 0.19	0.14 ± 0.14	0.06 ± 0.07	0.04 ± 0.05
Average d_{10}		0.43	0.56	0.67	0.77	0.86	1.1

Step 1: Accuracy of Error Recovery. Table 2 reports the accuracy of recovering the prediction error vector E across datasets and models. Accuracy is binary: a run counts as success if $E_{\mathrm{atk}} = E$, and 0 otherwise. Thus, an accuracy of $x\%$ means that in $x\%$ of batches, the full error vector was recovered exactly. (i) **Accuracy decreases with batch size.** As expected, larger batch sizes K make the HSSP instance harder to solve. Each additional label increases the dimensionality of the unknown vector E, while the amount of gradient information available to the attacker remains fixed. This imbalance raises the *density* of the problem, and we observe a consistent drop in accuracy as K grows. The decline is

particularly sharp around $K = 18$, which aligns with the known phase transition of HSSP solvers: once the density exceeds 1, classical lattice-based algorithms struggle to recover the hidden subset. This empirical threshold reinforces that our reduction captures the computational hardness of HSSP. (ii) **Model architecture matters more than dataset choice.** The vulnerability of a configuration appears to depend primarily on the neural network's structure rather than the dataset. For example, CNNs on MNIST or UTKFace consistently yield near-perfect recovery rates, whereas VGG-based architectures on the same datasets exhibit much lower success. This difference reflects how architectural choices affect gradient sparsity, activation patterns, and ultimately the hardness of the induced HSSP instance. In contrast, changing datasets while keeping the architecture fixed produces only minor variations once inputs are normalized. This highlights that defending against regression label leakage may require rethinking model design, not just curating the training data. (iii) **Verifiable success through the discrepancy test.** A key strength of our attack is that it is *self-verifiable*. In 99.98% of all runs, whenever the discrepancy measure *Error* equals zero, the recovered error vector exactly matches the ground truth ($E = E_{\mathrm{atk}}$). Conversely, nonzero *Error* reliably indicates failure. This property provides the adversary with a certificate of correctness: unlike optimization-based methods, which may converge to approximate solutions without clear stopping criteria, our HSSP-based attack informs the adversary when exact recovery has been achieved.

Step 2: Label Recovery. Table 3 reports RMSE between reconstructed labels $\hat{Y}$ and true labels Y, conditional on Step 1 success ($Error = 0$). For fairness, we compare against DLG [23] following the percentile-matching protocol described in Sect. 6.1. Classification-only baselines (e.g., iDLG [22]) are not included, as they require cross-entropy loss and one-hot labels, which do not apply in regression.

Main observations: (i) **High accuracy.** For most configurations, RMSE ≤ 1 (often ≤ 0.1), allowing an attacker to infer salaries, ages, or housing prices with very low error. Several cases (e.g., ARC, Salary, REV) achieve near-perfect recovery. (ii) **Attack stronger than DLG.** Our method consistently outperforms DLG, except in the {MNIST, CNN} setting, where high model output variance benefits the optimization approach. (iii) **RMSE matches variance bound.** As predicted by Eq. 5.4, reconstruction error is upper bounded by $\mathrm{Var}(z_0)$, and indeed our observed RMSE closely tracks this bound. (iv) **Architecture > dataset.** RMSE is more sensitive to network depth/width than to dataset specifics, consistent with the analysis in Sect. 5. (v) **Real-world vulnerability.** On real datasets such as Salary, Boston Housing, and Apartment Rent, labels are reconstructed with errors far below the scale of the values, exposing severe privacy risks.

Runtime and Scalability. Our attack runs in 1.24 s on average (min 0.007 s, max 6.04 s), largely independent of architecture. In contrast, DLG requires minutes on VGG and hours on larger models, limiting practicality. Thus, our method offers both stronger accuracy and higher efficiency.

Table 3. Step 2 results: RMSE of reconstructed labels when Step 1 succeeds. Values are reported as mean $\pm$ std across 10 independent seeds. The third column reports $\sqrt{\mathrm{Var}(z_0)}$ (predicted upper bound from Eq. 5.4). The last column shows the best-case DLG baseline [23], matched by the percentile protocol in Sect. 6.1. Real-world datasets are highlighted in bold (Salary, Boston Housing, REV, ARC).

$RMSE$ $\downarrow$		$\sqrt{\mathrm{Var}(Z_0)}$	Batch size (K)			[23]
Dataset	Model	Mean	8	12	18	N/A
S	MLP	0.01 ± 0.01	0.01 ± 0.01	0.01 ± 0.01	0.01 ± 0.01	$(1.208\,57 \pm 0.020\,16) \times 10^5$
L	MLP	0.07 ± 0.02	0.07 ± 0.02	0.07 ± 0.03	0.07 ± 0.02	1.51 ± 0.09
BH	MLP	0.07 ± 0.02	0.06 ± 0.01	0.06 ± 0.01	0.07 ± 0.01	3.66 ± 0.09
REV	MLP	0.07 ± 0.02	0.06 ± 0.01	0.06 ± 0.02	0.06 ± 0.01	$(3.070 \pm 0.149) \times 10^1$
PPGE	MLP	0.03 ± 0.01	0.03 ± 0.01	0.02 ± 0.01	0.02 ± 0.01	$(2.760 \pm 0.001) \times 10^{-1}$
ARC	MLP	0.01 ± 0.01	0.00 ± 0.01	0.00 ± 0.01	0.00 ± 0.01	$(8.670 \pm 3.828) \times 10^2$
MNIST	MLP	0.06 ± 0.01	0.06 ± 0.01	0.06 ± 0.01	0.06 ± 0.01	9.26 ± 0.47
MNIST	CNN	7.6 ± 1.50	3.97 ± 0.83	4.36 ± 0.46	4.15 ± 1.20	$(2.660 \pm 0.132) \times 10^1$
MNIST	VGG	1.8 ± 0.89	1.16 ± 0.27	1.07 ± 0.38	0.86 ± 0.15	$(1.010 \pm 0.083) \times 10^1$
Face	MLP	0.07 ± 0.01	0.06 ± 0.01	0.06 ± 0.01	0.06 ± 0.01	1.39 ± 0.05
Face	CNN	6.5 ± 1.7	5.06 ± 1.25	5.32 ± 1.06	3.75 ± 0.40	4.27 ± 0.14
Face	VGG	2.1 ± 1.00	1.38 ± 0.33	2.03 ± 0.81	1.94 ± 0.64	2.33 ± 0.43

More Experiments Available. For additional experiments, including: the weaker estimator $\hat{Z}_0 = 0$, varying the dimensions (n, m) and alternative activations/losses, we refer to [12].

7 Conclusion and Future Works

Most gradient leakage studies in federated learning (FL) have focused on classification (CFL), where one-hot labels and cross-entropy loss make analytical attacks effective. Regression FL (RFL), with continuous labels, has been assumed more resilient. In this work, we introduced the first analytical attack on RFL. Our main insight is a reduction of regression label recovery to the Hidden Subset Sum Problem (HSSP). By adapting lattice-based cryptographic techniques to real-valued gradients, we designed a two-step attack: recovering prediction errors, then inferring labels. A computable discrepancy test further provides the attacker with a practical certificate of success. Experiments on eight datasets and three models show that our method achieves RMSE often below 0.1 on real-world tasks such as salary and housing prediction, significantly outperforming optimization-based baselines. However, our attack has clear limitations: it assumes access to gradients under FedSGD at early rounds, requires models ending with two fully connected layers and a ReLU, and its success decreases sharply for batch sizes $K \geq 18$, in line with HSSP density thresholds. These results expose concrete risks for regression tasks in finance, healthcare, or HR. Future work should explore defenses such as differential privacy and gradient pruning.

References

1. Liver Disorders (2016). https://doi.org/10.24432/C54G67
2. Apartment for Rent Classified (2019). https://doi.org/10.24432/C5X623
3. Productivity Prediction of Garment Employees (2020). https://doi.org/10.24432/C51S6D
4. Coron, J.-S., Gini, A.: A polynomial-time algorithm for solving the hidden subset sum problem. In: Micciancio, D., Ristenpart, T. (eds.) CRYPTO 2020. LNCS, vol. 12171, pp. 3–31. Springer, Cham (2020). https://doi.org/10.1007/978-3-030-56880-1_1
5. Coron, J.S., Gini, A.: Provably Solving the Hidden Subset Sum Problem via Statistical Learning (2021)
6. Dang, T., Thakkar, O., Ramaswamy, S., Mathews, R., Chin, P., Beaufays, F.: Revealing and Protecting Labels in Distributed Training (2021). arXiv:2111.00556
7. Deng, L.: The mnist database of handwritten digit images for machine learning research [best of the web]. IEEE Signal Process. Mag. **29**(6), 141–142 (2012). https://doi.org/10.1109/MSP.2012.2211477
8. Du, J., et al.: Sok: on gradient leakage in federated learning. arXiv preprint arXiv:2404.05403 (2024)
9. Geiping, J., Bauermeister, H., Dröge, H., Moeller, M.: Inverting Gradients – How easy is it to break privacy in federated learning? arXiv:2003.14053 (2020). https://doi.org/10.48550/arXiv.2003.14053
10. Geng, J., Mou, Y., Li, F., Li, Q., Beyan, O., Decker, S., Rong, C.: Towards General Deep Leakage in Federated Learning (2022). arXiv:2110.09074
11. Harrison, D., Rubinfeld, D.L.: Hedonic housing prices and the demand for clean air. J. Environ. Econ. Manag. **5**(1), 81–102 (1978). https://doi.org/10.1016/0095-0696(78)90006-2
12. Jobic, P., Mayoue, A., Tucci Piergiovanni, S.: Label Leakage in Regression Federated Learning using Cryptographic Tools (2025). https://cea.hal.science/hal-05268925
13. Karimireddy, S.P., Kale, S., Mohri, M., Reddi, S.J., Stich, S.U., Suresh, A.T.: SCAFFOLD: stochastic controlled averaging for federated learning (2021). arXiv:1910.06378
14. Ma, K., Sun, Y., Cui, J., Li, D., Guan, Z., Liu, J.: Instance-wise batch label restoration via gradients in federated learning (2023)
15. McMahan, H.B., Moore, E., Ramage, D., Hampson, S., Arcas, B.A.: Communication-Efficient Learning of Deep Networks from Decentralized Data (2017). arXiv:1602.05629. https://doi.org/10.48550/arXiv.1602.05629
16. Nguyen, P., Stern, J.: The hardness of the hidden subset sum problem and its cryptographic implications. In: Wiener, M. (ed.) CRYPTO 1999. LNCS, vol. 1666, pp. 31–46. Springer, Heidelberg (1999). https://doi.org/10.1007/3-540-48405-1_3
17. Notarnicola, L., Wiese, G.: The Hidden Lattice Problem (2021)
18. Wainakh, A., et al.: User-Level Label Leakage from Gradients in Federated Learning (2022). arXiv:2105.09369
19. Wang, Z., Lee, J., Lei, Q.: Reconstructing training data from model gradient, provably. In: International Conference on Artificial Intelligence and Statistics, pp. 6595–6612. PMLR (2023)
20. Yeh, I.C.: Real Estate Valuation (2018). https://doi.org/10.24432/C5J30W
21. Zhang, Z., Song, Y., Qi, H.: Age progression/regression by conditional adversarial autoencoder. In: IEEE Conference on Computer Vision and Pattern Recognition (CVPR). IEEE (2017)

22. Zhao, B., Mopuri, K.R., Bilen, H.: iDLG: Improved Deep Leakage from Gradients. arXiv:2001.02610 (2020)
23. Zhu, L., Liu, Z., Han, S.: Deep leakage from gradients. In: Wallach, H., Larochelle, H., Beygelzimer, A., d'Alché-Buc, F., Fox, E., Garnett, R. (eds.) Advances in Neural Information Processing Systems, vol. 32. Curran Associates, Inc. (2019). https://proceedings.neurips.cc/paper_files/paper/2019/file/60a6c4002cc7b29142def8871531281a-Paper.pdf

Time and Space-Optimal Silent Self-stabilizing Exact Majority in Population Protocols

Haruki Kanaya[1]([✉])[iD], Ryota Eguchi[1][iD], Taisho Sasada[1][iD], Fukuhito Ooshita[2][iD], and Michiko Inoue[1][iD]

[1] Nara Institute of Science and Technology, Nara, Japan
{kanaya.haruki.kk3,taisho.sasada}@naist.ac.jp,
{ry.eguchi,kounoe}@is.naist.jp
[2] Fukui University of Technology, Fukui, Japan
f-oosita@fukui-ut.ac.jp

Abstract. We address the self-stabilizing exact majority problem in the population protocol model, introduced by Angluin, Aspnes, Diamadi, Fischer, and Peralta (2004). In this model, there are n state machines, called agents, which form a network. At each time step, only two agents interact with each other, and update their states. In the self-stabilizing exact majority problem, each agent has a fixed opinion, A or B, and stabilizes to a safe configuration in which all agents output the majority opinion from any initial configuration.

In this paper, we show the impossibility of solving the self-stabilizing exact majority problem without knowledge of n in any protocol. We propose a silent self-stabilizing exact majority protocol, which stabilizes within $O(n)$ parallel time in expectation and within $O(n \log n)$ parallel time with high probability, using $O(n)$ states, with knowledge of n. Here, a silent protocol means that, after stabilization, the state of each agent does not change. We establish lower bounds, proving that any silent protocol requires $\Omega(n)$ states, $\Omega(n)$ parallel time in expectation, and $\Omega(n \log n)$ parallel time with high probability to stabilize. Thus, the proposed protocol is time- and space-optimal.

Keywords: Population Protocols · Self-Stabilization · Exact Majority

1 Introduction

The population protocol model, introduced by Angluin, Aspnes, Diamadi, Fischer, and Peralta [3], is a model of a passively mobile sensor network. In this model, there are n state machines (called *agents*), and these agents form a network (called the *population*). Each agent lacks a unique identifier and updates its state through pairwise communication (called *interaction*). At each time step, only one pair of agents interacts, chosen by a uniform random scheduler. In this paper, we assume that the network is a complete graph, meaning that each agent

S. Bonomi et al. (Eds.): SSS 2025, LNCS 16350, pp. 270–286, 2026.
https://doi.org/10.1007/978-3-032-11127-2_22

can interact with all other agents. The time complexity is measured in *parallel time*, which is defined as the number of interactions divided by n.

The majority problem requires each agent to have a fixed opinion, either A or B, and to determine the majority opinion. The agents output the majority opinion, A or B, if there is no tie (i.e., the numbers of agents with input A and agents with input B are not equal); otherwise, they output T. There are two types of majority problems: approximate and exact. The approximate majority problem [5,15] allows a small error; that is, a protocol may stabilize to an incorrect answer with low probability. On the other hand, the exact majority problem does not allow errors; that is, a protocol stabilizes to a correct answer with probability 1. The exact majority problem with designated initial states has been widely studied [1,2,4,6,7,9–11,16,17,19,21–23], and a time- and space-optimal protocol [16], which converges within $O(\log n)$ parallel time and uses $O(\log n)$ states, has been proposed.

We address the self-stabilizing exact majority problem, which requires each agent to have a fixed opinion, either A or B, and to determine the exact majority opinion starting from any configuration. In prior research, a study [8] uses loose-stabilization [24], which relaxes the closure property of stabilization. However, to the best of our knowledge, no study has solved the self-stabilizing exact majority problem. We solve this problem with initial knowledge of n.

With initial knowledge of n, there have been some studies on another problem: the self-stabilizing leader election [12,13,18], in which agents elect a unique leader. In [12], Burman, Chen, Chen, Doty, Nowak, Severson, and Xu solve the self-stabilizing ranking problem, in which agents are assigned unique ranks from $[1,n]$. They proposed two protocols: a silent protocol, which is both time- and space-optimal ($O(n)$ parallel time, $O(n)$ states), and a non-silent protocol, which is time optimal ($O(\log n)$ parallel time) but requires $\exp(O(n^{\log n} \cdot \log n))$ states. Here, a silent protocol refers to a protocol in which, after stabilization, the state of each agent does not change.

In the non-silent self-stabilizing ranking protocol [12], agents are ranked by assigning unique names in $[1, n^3]$. Upon stabilization, each agent holds a list of all agents' names and determines its rank in $[1, n]$ from the name list. This protocol works correctly even if the name space is reduced to half its size, while still maintaining the same time complexity and number of states, since Lemma 5.1 of [12] still holds, as the name space is $O(n^3)$. Thus, using this property, the exact majority problem can be solved by assigning agents with input A a name space in $[1, n^3/2]$ and agents with input B a name space in $[n^3/2 + 1, n^3]$. Also, each agent can determine its rank in $[1, n]$ from the name list. This protocol is time-optimal since, in self-stabilization, all agents must interact at least once, and by the coupon collector's problem, $\Omega(n \log n)$ interactions (which corresponds to $\Omega(\log n)$ parallel time) are required.

1.1 Our Contribution

In this paper, we address a silent protocol that solves the self-stabilizing exact majority problem. Our contribution is summarized in Table 1. First, we show the impossibility that any protocol without knowledge of n cannot solve the self-stabilizing exact majority. Second, we also show the lower bounds: any silent protocol with knowledge of n requires at least n states, and any silent protocol requires at least $\Omega(n)$ parallel time in expectation and $\Omega(n \log n)$ parallel time with high probability to stabilize. Here, the phrase "with high probability" refers to with a probability of $1 - O(1/n)$. Finally, we propose a silent protocol, $\mathcal{P}_{EM}$, which stabilizes to a silent configuration[1] within $O(n)$ parallel time in expectation and within $O(n \log n)$ parallel time with high probability using $O(n)$ states, with knowledge of n.

Table 1. Overview of our results. The variable n denotes the number of agents. W.H.P. means with high probability.

	knowledge	states	expected parallel time	parallel time W.H.P.
Impossibility	without n	–	–	–
Lower Bound (space)	n	n	–	–
Lower Bound (time)	n	–	$\Omega(n)$	$\Omega(n \log n)$
Protocol	n	$O(n)$	$O(n)$	$O(n \log n)$

1.2   Organization of This Paper

Section 2 defines the model and problem. Section 3 introduces existing protocols used in our protocol. In Sect. 4, we show that any protocol without knowledge of n cannot solve the self-stabilizing exact majority problem. Section 5 presents lower bounds for any silent protocol that solves the self-stabilizing exact majority problem and proposes a protocol that matches these lower bounds. We conclude and discuss future directions in Sect. 6.

2   Preliminaries

We denote the set of positive integers by $\mathbb{N}$, and the set of non-negative integers by $\mathbb{N}_0$.

---

[1] A silent protocol stabilizes to a silent configuration; Thus, its time complexity is measured by silence time, where silence time is defined as parallel time until reaching a silent configuration. However, comparing the lower bound of stabilization time with the upper bound of silence time is not an issue, since stabilization time is no more than silence time.

2.1   Population Protocols

A *population* is represented by a bidirectional complete graph $G = (V, E)$, where V denotes the set of agents, and $E = \{(u, v) \in V \times V \mid u \neq v\}$ denotes the set of ordered pairs of agents that can interact, and let $n = |V|$. Since G is a bidirectional complete graph in this paper, we simply represent the population as V, the set of agents. A protocol is defined as a 5-tuple $\mathcal{P} = (Q, X, Y, \delta, \pi_{out})$, where Q is the set of agents states, X is the set of input symbols, Y is the set of output symbols, $\delta : (Q \times X) \times (Q \times X) \to Q \times Q$ is the state transition function, and $\pi_{out} : Q \times X \to Y$ is the output function. The variable var of agent u is denoted as u.var. Similarly, the input of each agent u is denoted by u.input. Each agent outputs $\pi_{out}(s, x) \in Y$ at each time step when it is in state $s \in Q$ and its input is $x \in X$. In each interaction, the interacting agents u and v are assigned the roles of initiator and responder, respectively, breaking the symmetry of the interaction. Suppose that two agents u, v interact, with u as the initiator and v as the responder, while they are in states p and q, respectively, and their inputs are x and y respectively. They then update their states to p' and q', respectively, where $\delta((p, x), (q, y)) = (p', q')$.

A *configuration* $C : V \to Q \times X$ represents the state and input of all agents. A configuration C changes to C' via an interaction $(u, v) \in E$, denoted by $C \xrightarrow{(u,v)} C'$ if $(C'(u), C'(v)) = \delta(C(u), C(v))$ and $\forall w \in V \setminus \{u, v\} : C'(w) = C(w)$. When a configuration C changes to C' via an interaction, we say that C can change to C' denoted by $C \to C'$. The set of all configurations by a protocol $\mathcal{P}$ is denoted by $\mathcal{C}_{all}(\mathcal{P})$. A uniform random scheduler $\Gamma = \Gamma_0, \Gamma_1, \ldots$ determines which ordered pair of agents interacts at each step, where $\Gamma_t \in E$ (for $t \geq 0$) is a random variable satisfying $\forall (u, v) \in E, \forall t \in \mathbb{N}_0 : \Pr(\Gamma_t = (u, v)) = \frac{1}{n(n-1)}$. Similarly, a deterministic scheduler $\gamma = \gamma_0, \gamma_1, \ldots$ determines which ordered pair of agents interacts at each time step, where $\gamma_i \in E$ (for $i \geq 0$) is a predetermined value. Note that γ can be a possible value of Γ with positive probability.

An *execution* of $\mathcal{P}$ is defined as an infinite sequence of configurations $\Xi_\mathcal{P}(C_0, g) = C_0, C_1, \ldots$ that starts from an initial configuration $C_0 \in \mathcal{C}_{all}(\mathcal{P})$, and satisfies $\forall i \in \mathbb{N} : C_{i-1} \to C_i$, where g is a uniform random or deterministic scheduler. Note that if g is a uniform random scheduler, then C_i is a random variable. A configuration C' is reachable from C_0 if there exists an execution $\Xi_\mathcal{P}(C_0, \Gamma) = C_0, C_1, \ldots$ such that $\exists i \in \mathbb{N} : C_i = C'$. In an execution under a uniform random scheduler, if a configuration C appears infinitely often, then every configuration reachable from C also appears infinitely often[2]. An execution is said to reach a set of configurations $\mathcal{C}$ if it reaches a configuration in $\mathcal{C}$. Similarly, an execution is said to belong to a set of configurations $\mathcal{C}$ if its initial configuration is in $\mathcal{C}$. We assume that the input remains unchanged during execution, meaning that $\forall v \in V, \forall i \in \mathbb{N} : C_i(v).\text{input} = C_0(v).\text{input}$ holds for an execution $\Xi_\mathcal{P}(C_0, g) = C_0, C_1, \ldots$, where g is a uniform random or deterministic scheduler. A configuration C is safe if and only if, for any execution starting

---

[2] This property holds since a uniform random scheduler is globally fair with probability 1, as shown in [14].

from C, the outputs of all agents never change. Similarly, a configuration C is silent if and only if, for any execution starting from C, the states of all agents never change. A protocol is silent if and only if an execution reaches a silent configuration with probability 1.

The time complexity of a protocol is defined as the number of interactions divided by n. This is referred to as either *parallel time* or simply *time*. The stabilization time is defined as the time until an execution of the protocol under a uniform random scheduler reaches a safe configuration. Similarly, the silence time is defined as the time until an execution of the protocol under a uniform random scheduler reaches a silent configuration. The phrase *"with high probability"* refers to with a probability of $1 - O(1/n)$.

2.2 Self-stabilizing Exact Majority

The self-stabilizing exact majority problem requires that each agent outputs the exact majority opinion from the two types of opinions, A and B, held by agents, starting from any initial configuration. We define V_a as the set of agents with the opinion A, and V_b as the set of agents with the opinion B. Note that we assume the input of each agent does not change during execution; thus, V_a and V_b remain unchanged throughout execution. The exact majority opinion is as follows: If $|V_a| < |V_b|$, the exact majority opinion is B; if $|V_a| > |V_b|$, it is A; otherwise, it is T.

Definition 1. *A protocol $\mathcal{P} = (Q, X, Y, \delta, \pi_{out})$ is a self-stabilizing exact major-ity protocol if and only if $X = \{A, B\}$, $Y = \{A, B, T\}$, and the following conditions hold for any initial configuration $C_0 \in \mathcal{C}_{all}(\mathcal{P})$:*

- *For any safe configuration C_{safe} reachable from C_0, it holds that $\forall v \in V :$ $\pi_{out}(C_{safe}(v)) = y$, where $y \in Y$ is the exact majority opinion of agents.*
- *Any execution starting from C_0 reaches safe configurations with probability 1.*

3 Tools

3.1 Epidemic Protocol

The epidemic protocol proposed by Angluin, Aspnes, and Eisenstat [4] is used in PROPAGATE-RESET described later. In this protocol, each agent has a variable $x \in \{0, 1\}$, and when two agents u and v interact, $u.x$ and $v.x$ are both updated to $\max(u.x, v.x)$. Suppose that all agents have $x = 0$. Once one agent's x becomes 1, all agents' x will become 1 within $\Theta(\log n)$ time in expectation and with high probability.

3.2 Ranking Protocol

In this subsection, we explain the silent self-stabilizing ranking protocol proposed by Burman et al. [12] since we will use it in Sect. 5.2. A self-stabilizing ranking protocol assigns agents unique ranks in $[1, n]$ from any initial configuration, and the agents maintain their ranks forever. Their protocol is divided into two parts, PROPAGATE-RESET and OPTIMAL-SILENT-SSR. In their protocol, an agent has a `role` $\in$ {Resetting, Settled, Unsettled} to reduce the number of states, and they define variables for each role. The number of states becomes the sum of the states required for each `role`. We briefly explain their protocol, but refer to [12] for more details and proofs.

PROPAGATE-RESET. In this protocol, when some inconsistency is detected in the population, all agents are reset to a specific state with high probability. Each agent uses two variables, `role` and `resetcount`, which represent the agent's role and a timer used to determine whether a reset has been completed. When agents detect some inconsistency, they set `role` to Resetting and `resetcount` to $R_{max}(= 60 \log n)$. The role of Resetting propagates to all agents through the epidemic protocol, treating Resetting as 1 and all other roles as 0. Eventually, all agents are reset within $O(\log n)$ time with high probability. After resetting, all inconsistencies are eliminated. We have following lemmas from [12].

Lemma 1 (Lemma 3.2 in [12]). *If some inconsistency is detected, all agents'* `role` *become* Resetting *within* $4 \log n$ *time with high probability.*

Lemma 2 (Corollary 3.5 in [12]). *Starting from any configuration, an execution reaches a configuration in which no agent has* `role` = Resetting *within* $O(n)$ *time with high probability.*

OPTIMAL-SILENT-SSR. In this protocol, agents are assigned unique ranks in $[1, n]$ from any initial configuration. The functions of `role` for agents are as follows: `role` = Settled means the agent has been assigned a rank, `role` = Unsettled means the agent has not been assigned a rank, and `role` = Resetting means the agent is executing PROPAGATE-RESET. When some rank conflict or some inconsistency is detected, PROPAGATE-RESET starts. At this point, as mentioned above, `role` is set to Resetting, `resetcount` is set to R_{max}, and additionally, `leader` is set to L. Here, `leader` $\in \{L, F\}$ is a variable of an agent with `role` = Resetting in this protocol, indicating whether the agent is a leader. While resetting, agents elect a unique leader through a process where, when leaders meet, one of them becomes a follower (F). After resetting, all agents are unranked, and a unique leader agent is elected. The leader's `role` becomes Settled, with its `rank` set to 1. Each agent with `rank` = i creates agents with `rank` = $2i$ and `rank` = $2i+1$ if the newly assigned `rank` does not exceed n. Eventually, all agents are assigned unique ranks. Let s_{rank} be the expected silence time of OPTIMAL-SILENT-SSR. We have the following lemma from [12].

Lemma 3 (Theorem 4.3 in [12]). OPTIMAL-SILENT-SSR *is a silent self-stabilizing ranking protocol with* $O(n)$ *states and* $s_{rank} = O(n)$ *expected time.*

4   Impossibility

In this section, we provide the impossibility. We prove that no protocol can solve the self-stabilizing exact majority problem without knowledge of n, using the same approach as the proof of the impossibility of a self-stabilizing bipartition protocol by Yasumi, Ooshita, Yamaguchi, and Inoue [25].

Theorem 1. *There is no protocol that solves the self-stabilizing exact majority without knowledge of n.*

Proof. For contradiction, we assume there is a protocol $\mathcal{P}$ that solves self-stabilizing exact majority without knowledge of n. Consider a population V, where $|V| \geq 4$, $|V_a| \geq 2$, $|V_b| \geq 2$, and $|V_a| > |V_b|$. From the assumption, $\mathcal{P}$ solves the problem for V and V_b. Let C_0 be any configuration of V, and let $\gamma = \gamma_0, \gamma_1, \ldots$ be a deterministic scheduler on V that makes C_0 eventually reach a safe configuration. For an execution $\Xi_{\mathcal{P}}(C_0, \gamma) = C_0, C_1, \ldots$ on V, there exists a safe configuration C_t in which all agents output A.

Similarly, consider another population V' consisting of all agents in V_b, and let C'_0 be a configuration of V' satisfying $\forall v \in V' : C'_0(v) = C_t(v)$. Let $\gamma' = \gamma'_0, \gamma'_1, \ldots$ be a deterministic scheduler on V' that makes C'_0 eventually reach a safe configuration. For an execution $\Xi_{\mathcal{P}}(C'_0, \gamma') = C'_0, C'_1, \ldots$ on V', there exists a safe configuration $C'_{t'}$ in which all agents output B.

From these executions, let $\gamma'' = \gamma''_0, \gamma''_1, \ldots$ be a deterministic scheduler on V satisfying $\forall i \in [0, t-1] : \gamma''_i = \gamma_i \wedge \forall j \in [0, t'-1] : \gamma''_{t+j} = \gamma'_j$. Then, let $C''_0 = C_0$, and consider the execution $\Xi_{\mathcal{P}}(C''_0, \gamma'')$. In this execution, C''_t is a safe configuration in which all agents output A. However, in $C''_{t+t'}$, all agents that belong to V_b output B. This contradicts the assumption that C''_t is a safe configuration. $\qquad\square$

Note that a deterministic scheduler can appear as a prefix of a uniform random scheduler with nonzero probability. Therefore, this theorem is also applicable to the impossibility result under a uniform random scheduler.

5   Lower Bound and Upper Bound

5.1   Lower Bounds

In this subsection, we provide the lower bounds.

Lemma 4. *Consider a population V where $|V| \geq 4$, $|V_a| \leq |V_b|$. In a silent configuration of a silent self-stabilizing exact majority protocol, the states of agents with input A are all distinct from one another.*

Proof. We prove this lemma by contradiction. We assume that there exists a silent self-stabilizing exact majority protocol $\mathcal{P}$ that uses $|V_a| - 1$ states for agents with input A in a silent configuration. Suppose some execution reaches a silent configuration C. From the assumption, there exist two agents with input A that are in the same state s_A in C. Since C is a silent configuration, the state satisfies $\delta((s_A, \mathtt{A}), (s_A, \mathtt{A})) \to (s_A, s_A)$ and $\pi_{out}(s_A, \mathtt{A}) \in \{\mathtt{B}, \mathtt{T}\}$. Next, consider another population with n agents, where all agents have $\mathtt{input} = \mathtt{A}$, and an execution starting from the configuration in which all agents' states are s_A. Such an execution cannot reach a safe configuration since all agents eternally output $\mathtt{B}$ or $\mathtt{T}$, while the exact majority opinion is $\mathtt{A}$. This is a contradiction. $\quad\square$

Theorem 2. *A silent self-stabilizing exact majority protocol requires n states.*

Proof. Let n be an even number no less than 4. Consider a population V with n agents, where $|V_a| = n/2 - 1$. From Lemma 4, there exist $n/2 - 1$ distinct states of agents with $\mathtt{input} = \mathtt{A}$ that output $\mathtt{B}$. Next, consider another population V' with n agents, where $|V_a'| = n/2$. From Lemma 4, there exist $n/2$ distinct states of agents with $\mathtt{input} = \mathtt{A}$ that output $\mathtt{T}$. Finally, consider a population V'' with n agents, where $|V_a''| = n/2 + 1$. It is clear that there exists at least one state of agents with $\mathtt{input} = \mathtt{A}$ such that they output $\mathtt{A}$. Thus, the number of states is at least $(n/2 - 1) + (n/2) + 1 = n$. $\quad\square$

We prove a lower bound for the stabilization time of a silent self-stabilizing exact majority protocol in the same way as the proof of the lower bound for the stabilization time of a silent self-stabilizing leader election protocol by Burman et al. [12].

Theorem 3. *Any silent self-stabilizing exact majority protocol requires $\Omega(n)$ time in expectation and $\Omega(n \log n)$ time with high probability to reach a safe configuration.*

Proof. Consider a population V where $|V| \geq 5 \wedge |V| \bmod 2 = 1$, $|V_a| = \lfloor n/2 \rfloor$, and $|V_b| = \lceil n/2 \rceil$. From Lemma 4, in a silent configuration C, the states of agents whose input are $\mathtt{A}$ are distinct from one another. Let s_A be the state of an agent w whose input is $\mathtt{A}$ in C.

For an agent $u \in V$ whose input is $\mathtt{B}$, we consider another population $V' = V$ and a configuration C' satisfying $\forall v \in V' \setminus \{u\} : C'(v) = C(v)$ and $C'(u) = (s_A, \mathtt{A})$. In C', all agents output $\mathtt{B}$, and the exact majority opinion of V' is $\mathtt{A}$. Thus, C' is not safe. Since C is a silent configuration, in C', unless u and w interact with each other, their states will not be changed. The probability that u and w interact is $\frac{2}{n(n-1)}$. Thus, the expected number of interactions until the interaction occurs is $n(n-1)/2 = \Omega(n^2)$. Therefore, divided by n, the expected time is $\Omega(n)$.

The probability that u and w do not interact in $\alpha \frac{n(n-1)}{2} \log_e n - 1$ interactions is $\left(1 - \frac{2}{n(n-1)}\right)^{\alpha \frac{n(n-1)}{2} \log_e n - 1} > e^{-\alpha \log_e n} = n^{-\alpha}$ for some constant value $\alpha > 0$. Thus, the number of interactions until u and w interact with each other is $\frac{n(n-1)}{2} \log_e n - 1 = \Omega(n^2 \log n)$ with probability $1 - O(1/n)$ $(\alpha = 1)$. Therefore, divided by n, the time with high probability is $\Omega(n \log n)$. $\quad\square$

5.2   Matching Upper Bound

In this subsection, we propose a silent self-stabilizing exact majority protocol, $\mathcal{P}_{EM}$, using $O(n)$ states and stabilizing within $O(n)$ time in expectation and within $O(n \log n)$ time with high probability, with knowledge of n. In $\mathcal{P}_{EM}$, each agent has an input and a state. The input of an agent a is $a.\mathtt{input} \in \{\mathtt{A}, \mathtt{B}\}$ (read-only). The state of each agent is represented by separate variables. Since we use the silent self-stabilizing ranking protocol OPTIMAL-SILENT-SSR [12], each agent has the variables used in OPTIMAL-SILENT-SSR. Additionally, each agent a has two other variables: $a.\mathtt{answer} \in \{\phi, \mathtt{T}, \mathtt{A}, \mathtt{B}\}$ and $a.\mathtt{timer} \in [0, 7(t_{rank} + 4)]$. Here, t_{rank} is a constant satisfying $s_{rank} \leq t_{rank} \cdot n$ and $t_{rank} = O(1)$, where $t_{rank} \cdot n$ denotes a sufficient amount of time for OPTIMAL-SILENT-SSR to stabilize. The number of states of $\mathcal{P}_{EM}$ is $O(n)$ since the number of states of $\mathtt{answer}$ and $\mathtt{timer}$ is $O(1)$, and the number of states of the variables used in OPTIMAL-SILENT-SSR is $O(n)$. Each agent a outputs $\mathtt{T}$ if $a.\mathtt{answer} = \phi$; otherwise, it outputs $a.\mathtt{answer}$.

Our protocol $\mathcal{P}_{EM}$, described in Algorithm 1, can be summarized as follows:

- **Ranking.** Execute OPTIMAL-SILENT-SSR (line 1).
- **Swapping.** Swap the values of all variables between interacting agents, except for **input**, if the **rank** of an agent with input B is less than the **rank** of an agent with input A (lines 10–11). This eventually places agents with input A in the lower ranks and agents with input B in the higher ranks. Note that even if the states of pairs of agents are swapped, this does not affect the correctness or the silence time of the ranking protocol, due to the symmetry among agents.
- **Decision.** If n is even, the agent with $\mathtt{rank} = n/2$ decides the exact majority opinion based on the inputs of both itself and the agent with $\mathtt{rank} = n/2 + 1$ (lines 12–16). Otherwise, the agent with $\mathtt{rank} = \lceil n/2 \rceil$ decides the exact majority opinion based on its own input (lines 17–18).
- **Propagation.** After completing the above steps, the agent with $\mathtt{rank} = \lceil n/2 \rceil$ checks the opinions of the other agents. If there is an agent with a different opinion, PROPAGATE-RESET is triggered, and the correct opinion is propagated during resetting. (lines 2–8 and 19–24).

In what follows, we explain the details of the Decision part and the Propagation part.

In the Decision part, agents decide the exact majority opinion. We consider the situation where the Ranking part and the Swapping part are completed. If n is even, the exact majority opinion is determined by the agents with $\mathtt{rank} = n/2$ and $\mathtt{rank} = n/2 + 1$. If the inputs of both agents are the same, then the agents with that input are in the majority, making that input the exact majority opinion. If not, the number of agents with input A and input B is the same, so the exact majority opinion is T. If n is odd, the exact majority opinion is the input of the agent with $\mathtt{rank} = \lceil n/2 \rceil$ since the number of agents with that input constitutes a majority.

Algorithm 1: Silent Self-Stabilizing Exact Majority Protocol $\mathcal{P}_{EM}$

---

when *an agent a_0 interacts with an agent a_1* **do**

1   Execute OPTIMAL-SILENT-SSR [12].

2   **for** $i \in \{0,1\}$ **do**

3   **if** a_i.role becomes Resetting in this interaction **then**

4   a_i.answer $\leftarrow \phi$

5   **if** a_i.role becomes Settled in this interaction $\wedge\ a_i$.rank $= \lceil n/2 \rceil$ **then**

6   a_i.timer $\leftarrow 7(t_{rank} + 4)$

7   **if** $\exists i \in \{0,1\} : a_0$.role $= a_1$.role $=$ Resetting $\wedge\ a_i$.answer $= \phi \wedge a_{1-i}$.answer $\neq \phi$ **then**

8   a_i.answer $\leftarrow a_{1-i}$.answer

9   **if** a_0.role $= a_1$.role $=$ Settled **then**

10   **if** a_0.rank $< a_1$.rank $\wedge\ a_0$.input $=$ B $\wedge\ a_1$.input $=$ A **then**

11   Swap the states between a_0 and a_1.

12   **if** $n \bmod 2 = 0 \wedge \exists i \in \{0,1\} : a_i$.rank $= a_{1-i}$.rank $- 1 = n/2$ **then**

13   **if** a_i.input $= a_{1-i}$.input **then**

14          $(a_i$.answer, a_{1-i}.answer$) \leftarrow (a_i$.input, a_i.input$)$

15   **else**

16          $(a_i$.answer, a_{1-i}.answer$) \leftarrow ($T, T$)$

17   **else if** $n \bmod 2 = 1 \wedge \exists i \in \{0,1\} : a_i$.rank $= \lceil n/2 \rceil$ **then**

18   a_i.answer $\leftarrow a_i$.input

19   **if** $\exists i \in \{0,1\} : a_i$.rank $= \lceil n/2 \rceil$ **then**

20   **if** a_{1-i}.rank $= n$ **then** a_i.timer $\leftarrow \max(0, a_i$.timer $- 1)$

21   **if** a_i.timer $= 0 \wedge a_i$.answer $\neq a_{1-i}$.answer **then**

22   a_{1-i}.answer $\leftarrow a_i$.answer

23   **for** $j \in \{0,1\}$ **do**

24            $(a_j$.role, a_j.leader, a_j.resetcount$) \leftarrow$ (Resetting, L, R_{max})

In the Propagation part, the agent with rank $= \lceil n/2 \rceil$ propagates its answer to all agents. However, if the agent with rank $= \lceil n/2 \rceil$ shares its answer to every agent by a direct interaction, it would take $\Theta(n \log n)$ time, since it is equivalent to the coupon collector's problem. To achieve propagation within $O(n)$ time in expectation, we use the following method. The agent with rank $= \lceil n/2 \rceil$ checks the opinions of the other agents (line 22). It detects that ranking, swapping, and decision are completed when its timer reaches 0. To detect this, the agent with rank $= \lceil n/2 \rceil$ sets its timer to $7(t_{rank} + 4)$ when it is ranked (lines 5–6) and decreases it by 1 when it interacts with the agent with rank $= n$ (line 20). If there is an agent with a different opinion, it can be detected within $O(n)$ time in expectation and triggers PROPAGATE-RESET by setting its role to Resetting (lines 21–24). Note that the variables leader and resetcount in the pseudocode are used in the ranking protocol (see Sect. 3.2). During PROPAGATE-RESET, that is, while the $\mathtt{role}$ of agents is $\mathtt{Resetting}$, the correct opinion is propagated to all agents. Specifically, any agent with $\mathtt{role} = \mathtt{Resetting}$ other than the agent with $\mathtt{rank} = \lceil n/2 \rceil$ sets its $\mathtt{answer}$ to ϕ when its $\mathtt{role}$ becomes $\mathtt{Resetting}$ (lines 3–4). Then, the exact majority opinion overwrites ϕ via the epidemic protocol[3] (lines 7–8). After the succeeding ranking, swapping, and decision, the agent with $\mathtt{rank} = \lceil n/2 \rceil$ again decides the correct opinion (lines 12–18). At this time, all agents have the correct opinion; thus, PROPAGATE-RESET will not occur. These process takes $O(n)$ time in expectation.

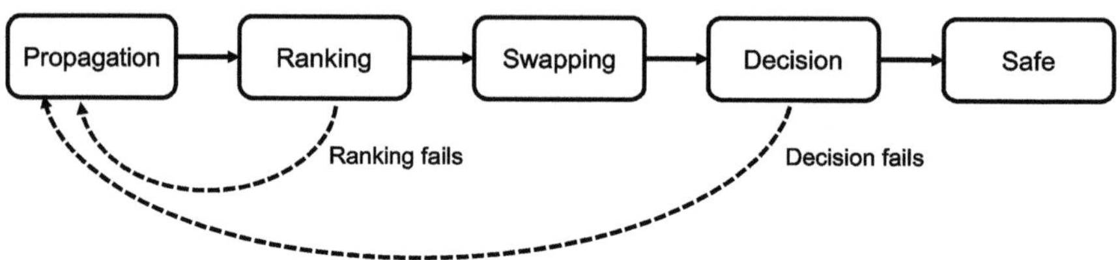

Fig. 1. Flow of $\mathcal{P}_{EM}$.

To summarize the above, $\mathcal{P}_{EM}$ is as shown in Fig. 1. First, agents execute ranking, swapping, and decision. After that, if $\mathtt{answer}$ of some agent is incorrect, the output of the agent with $\mathtt{rank} = \lceil n/2 \rceil$ is propagated to all agents by restarting the ranking. After re-ranking, re-swapping, and re-decision, the outputs of all agents are identical and correct. Note that these transitions in the figure are probabilistic. Thus, if ranking or decision fails, agents also restart the protocol from the beginning by resetting via PROPAGATE-RESET.

Correctness and Analysis. In this part, we prove that an execution of $\mathcal{P}_{EM}$ under a uniform random scheduler reaches a silent configuration within $O(n)$ time in expectation and $O(n \log n)$ time with high probability. To prove it, we define the sets of configurations as follows: $\mathcal{S}_{rank} = \{C \in \mathcal{C}_{all}(\mathcal{P}_{EM}) \mid \forall v \in V : C(v).\mathtt{role} = \mathtt{Settled} \wedge \forall u, \forall v \in V : u \neq v \Rightarrow C(u).\mathtt{rank} \neq C(v).\mathtt{rank}\}$, $\mathcal{S}_{swap} = \{C \in \mathcal{S}_{rank} \mid \forall u, \forall v \in V : C(u).\mathtt{input} = \mathtt{A} \wedge C(v).\mathtt{input} = \mathtt{B} \Rightarrow C(u).\mathtt{rank} < C(v).\mathtt{rank}\}$, $\mathcal{T}_{swap} = \{C \in \mathcal{S}_{rank} \mid \exists v \in V : C(v).\mathtt{rank} = \lceil n/2 \rceil \wedge C(v).\mathtt{timer} \geq 28\}$, $\mathcal{S}_{dec} = \{C \in \mathcal{S}_{swap} \mid \forall v \in V : C(v).\mathtt{rank} = \lceil n/2 \rceil \Rightarrow C(v).\mathtt{answer}$ is the exact majority opinion$\}$, $\mathcal{S}_{out} = \{C \in \mathcal{S}_{rank} \mid \forall v \in V : C(v).\mathtt{answer}$ is the exact majority opinion$\}$, $\mathcal{S}_{tim} = \mathcal{S}_{swap} \cap \mathcal{S}_{out}$, and $\mathcal{S}_{em} = \{C \in \mathcal{S}_{tim} \mid \forall v \in V : C(v).\mathtt{rank} = \lceil n/2 \rceil \Rightarrow C(v).\mathtt{timer} = 0\}$.

Lemma 5. $\mathcal{S}_{em}$ *is a set of silent configurations.*

Proof. In a configuration $C_0 \in \mathcal{S}_{em}$, ranking, swapping, and decision are completed, the $\mathtt{timer}$ of the agent with $\mathtt{rank} = \lceil n/2 \rceil$ is 0, and all agents' $\mathtt{answer}$ are the same. Thus, throughout a execution from C_0, no agent changes its state. $\square$

Table 2. Stabilization Steps of $\mathcal{P}_{EM}$

Step	Time	Success Probability	Lemmas
$\mathcal{C}_{all}(\mathcal{P}_{EM}) \to \mathcal{S}_{rank}$	$O(n)$	1/10	Lemma 6
$\mathcal{S}_{rank} \to \mathcal{T}_{swap} \cup \mathcal{S}_{tim}$	$O(n)$	1/20	Lemma 8
$\mathcal{T}_{swap} \to \mathcal{S}_{dec}$	$O(n)$	1/8	Lemma 9
$\mathcal{S}_{dec} \to \mathcal{S}_{tim}$	$O(n)$	1/1280	Lemma 11
$\mathcal{S}_{tim} \to \mathcal{S}_{em}$	$O(n)$	1/2	Lemma 12
$\mathcal{C}_{all}(\mathcal{P}_{EM}) \to \mathcal{S}_{em}$	$O(n)$	1/4096000	Lemma 13

We analyze $\mathcal{P}_{EM}$ by dividing it into several steps, as shown in Table 2.

Lemma 6. *Let $C_0 \in \overline{\mathcal{S}_{rank}}$ and $\Xi_{\mathcal{P}_{EM}}(C_0, \Gamma)$. The probability that the execution reaches $\mathcal{S}_{rank}$ within $O(n)$ time is at least $1/10$.*

Proof. First, we consider the case that the agent with $\mathtt{rank} = \lceil n/2 \rceil$ does not start PROPAGATE-RESET by interacting with an agent that has a different $\mathtt{answer}$ until the execution reaches $\mathcal{S}_{rank}$. In this case, from Lemma 3, agents complete the ranking within s_{rank} expected time. From Markov's inequality, the probability that ranking has completed within $2 \cdot s_{rank}$ time is at least $1/2$.

Second, we consider the case where the agent with $\mathtt{rank} = \lceil n/2 \rceil$ starts PROPAGATE-RESET by interacting with an agent that has a different $\mathtt{answer}$ until the execution reaches $\mathcal{S}_{rank}$. In this case, after PROPAGATE-RESET, an agent will be assigned the rank $\lceil n/2 \rceil$ within $2 \cdot s_{rank}$ time with at least $1/2$ probability from Markov's inequality. At the time, the agent sets $\mathtt{timer}$ to $7(t_{rank}+4)$. We analyze the probability that the $\mathtt{timer}$ does not reach 0 before the ranking is completed. Since the silence time of the ranking is $s_{rank} \leq t_{rank} \cdot n$, we analyze the probability that $\mathtt{timer}$ does not reach 0 during $t_{rank} \cdot n^2$ interactions. Let $Z \sim B(n^2 \cdot t_{rank}, 2/(n(n-1)))$ be a binomial random variable representing the number of times the agent with $\mathtt{rank} = \lceil n/2 \rceil$ interacts with the agent with $\mathtt{rank} = n$ during $n^2 \cdot t_{rank}$ interactions. From the Chernoff bound (Eq. (4.2) in [20], with $\delta = 2/3$), $\Pr(Z \geq (1 + 2/3)E[Z]) \leq e^{-2/3(2/3)^2} < 4/5$. Since $(1 + 2/3)E[Z] \leq \frac{20}{3}t_{rank} < 7t_{rank}$, that probability is at last $1/5$. Thus, the execution reaches $\mathcal{S}_{rank}$ with probability at least $1/10$. $\square$

Lemma 7. *Let $C_0 \in \mathcal{S}_{rank}$ and $\Xi_{\mathcal{P}_{EM}}(C_0, \Gamma)$. If PROPAGATE-RESET does not occur, the expected time for the execution to reach $\mathcal{S}_{swap}$ is at most n.*

Proof. Let m be the number of agents with input $\mathtt{A}$. Let V_B be the set of agents with input $\mathtt{B}$ and $\mathtt{rank}$ at most m, and let V_A be the set of agents with input $\mathtt{A}$ and $\mathtt{rank}$ greater than m at initial configuration. Note that $|V_A| = |V_B|$. The sizes of V_A and V_B each decrease by 1 if and only if an interaction occurs between $a \in V_A$

---

[3] In the epidemic protocol, the exact majority opinion is treated as 1, and ϕ is treated as 0.

and $b \in V_B$. Additionally, in any other interactions, the sizes of V_A and V_B remain unchanged. The probability that an interaction occurs between $a \in V_A$ and $b \in V_B$ for each interaction is $\frac{2|V_A||V_B|}{n(n-1)}$. Thus, the number of interactions until $|V_A|$ and $|V_B|$ become 0 is $\sum_{i=1}^{|V_A|} \frac{n(n-1)}{2i^2} = \frac{n(n-1)}{2} \sum_{i=1}^{|V_A|} \frac{1}{i^2} < \frac{n(n-1)}{2} \cdot \frac{\pi^2}{6} < n^2$. Therefore, after dividing by n, the expected time for the execution to reach $\mathcal{S}_{swap}$ is at most n. □

Lemma 8. *Let $C_0 \in \mathcal{S}_{rank} \cap \overline{\mathcal{T}_{swap}}$ and $\Xi_{\mathcal{P}_{EM}}(C_0, \Gamma)$. The probability that the execution reaches $\mathcal{T}_{swap} \cup \mathcal{S}_{tim}$ within $O(n)$ time is at least $1/20$.*

Proof. First, we show that the expected time until either PROPAGATE-RESET occurs or the execution reaches $\mathcal{S}_{dec}$ is at most $1/2 + (n-1)/2$. We will derive this by analyzing the time to reach $\mathcal{S}_{dec}$ without PROPAGATE-RESET occurring. From Lemma 7, the execution reaches $\mathcal{S}_{swap}$ within n expected time. If n is even, for the execution starting from a configuration belonging to $\mathcal{S}_{swap}$ to reach $\mathcal{S}_{dec}$, it is sufficient for the agent with $\mathtt{rank} = n/2$ and the agent with $\mathtt{rank} = n/2 + 1$ to interact. Since the probability that those two agents interact in each interaction is $2/(n(n-1))$, the expected time for this to happen is $(n-1)/2$. If n is odd, for the execution starting from a configuration belonging to $\mathcal{S}_{swap}$ to reach $\mathcal{S}_{dec}$, it is sufficient for the agent with $\mathtt{rank} = \lceil n/2 \rceil$ to participate in an interaction. Since the probability of that agent participating in an interaction is $2/n$, the expected time is $1/2$. Thus, the execution starting from C_0 reaches $\mathcal{S}_{dec}$ within $1/2 + (n-1)/2$ in expectation.

After the execution reaches $\mathcal{S}_{dec}$, if there are agents with an $\mathtt{answer}$ that is not the exact majority opinion, an interaction between the agent with $\mathtt{rank} = \lceil n/2 \rceil$ and such an agent will trigger PROPAGATE-RESET. The probability of this happening is at least $2/(n(n-1))$, so the expected time is $(n-1)/2$. If the $\mathtt{answer}$ of all agents are the exact majority opinion, the execution has reached to $\mathcal{S}_{tim}$. Therefore, the expected time the execution starting from C_0 reaches $\mathcal{S}_{tim}$ is bounded by $1/2 + (n-1)/2 + (n-1)/2 = n - 1/2$. From Markov' inequality, the probability that the execution reaches $\mathcal{S}_{tim}$ within $2n - 1$ time is at least $1/2$.

Now, we analyze the time and probability for the execution to reach $\mathcal{T}_{swap}$ after PROPAGATE-RESET occurs. From Lemma 3 and Markov's inequality, if the agent with $\mathtt{rank} = \lceil n/2 \rceil$ does not trigger PROPAGATE-RESET again, the execution reaches $\mathcal{S}_{rank}$ within $2 \cdot s_{rank}$ time with probability at least $1/2$. During the ranking, when the agent with $\mathtt{rank} = \lceil n/2 \rceil$ is created, its $\mathtt{timer}$ is set to $7(t_{rank} + 4)$. Thus, we analyze the probability that its $\mathtt{timer} \geq 28$ when the execution reaches $\mathcal{S}_{rank}$. Since $s_{rank} \leq t_{rank} \cdot n$, we analyze the probability that $\mathtt{timer} \geq 28$ after $n^2 \cdot t_{rank}$ interactions. Let $Z \sim B(n^2 \cdot t_{rank}, 2/(n(n-1)))$ be a binomial random variable representing the number of times the agent with $\mathtt{rank} = \lceil n/2 \rceil$ interacts with the agent with $\mathtt{rank} = n$ during $n^2 \cdot t_{rank}$ interactions. From the Chernoff bound (Eq. (4.2) in [20], with $\delta = 2/3$), $\Pr(Z \geq (1 + 2/3)E[Z]) \leq e^{-2/3(2/3)^2} < 4/5$. Since $(1 + 2/3)E[Z] \leq \frac{20}{3}t_{rank} < 7t_{rank}$, that probability is at last $1/5$. Therefore, after PROPAGATE-RESET occurs, the execution reaches $\mathcal{T}_{swap}$ within $O(n)$ time with probability at least $1/10$.

From the above, the execution reaches $\mathcal{T}_{swap} \cup \mathcal{S}_{tim}$ within $O(n)$ time with probability at least $1/20$. $\qquad\square$

Lemma 9. *Let $C_0 \in \mathcal{T}_{swap}$ and $\Xi_{\mathcal{P}_{EM}}(C_0, \Gamma)$. The probability that the execution reaches $\mathcal{S}_{dec}$ within $3n - 1$ time is at least $1/8$.*

Proof. We analyze the case where the execution reaches $\mathcal{S}_{dec}$ without PROPAGATE-RESET occurring.

First, we analyze the time and probability for the execution to reach $\mathcal{S}_{dec}$ if PROPAGATE-RESET does not occur. From Lemma 7 and Markov's inequality, agents finish swapping within $2n$ time with probability at least $1/2$. After swapping, decision is finished within $(n - 1)/2$ time in expectation as shown in the proof of Lemma 8. From Markov's inequality, agents finish decision within $n - 1$ time with probability at least $1/2$. Therefore, agents finish swapping and decision within $3n - 1$ time with probability at least $1/4$.

Next, we analyze the probability that PROPAGATE-RESET does not occur within $4n$ time. Let $Z \sim B(4n^2, 2/(n(n - 1)))$ be a binomial random variable representing the number of times the agent with $\mathtt{rank} = \lceil n/2 \rceil$ interacts with the agent with $\mathtt{rank} = n$ during $4n^2$ interactions. From the Chernoff bound (Eq. (4.2) in [20], with $\delta = 2/3$) and $E[Z] \geq 8$, $\Pr(Z \geq (1+2/3)E[Z]) \leq e^{-8/3(2/3)^2} \leq 1/2$. Since $(1 + 2/3)E[Z] \leq 80/3 < 27$, the $\mathtt{timer}$ of the agent with $\mathtt{rank} = \lceil n/2 \rceil$ does not reach 0 during $4n$ time with probability at least $1/2$.

From the above, the probability for the execution to reach $\mathcal{S}_{dec}$ within $3n - 1$ time is at least $1/8$. $\qquad\square$

Lemma 10. *Let $C_0 \in \mathcal{S}_{rank}$ and $\Xi_{\mathcal{P}_{EM}}(C_0, \Gamma)$. The variable $\mathtt{timer}$ of the agent with $\mathtt{rank} = \lceil n/2 \rceil$ reaches 0 within $O(n)$ time with probability at least $1/2$.*

Proof. Let x be the value of the $\mathtt{timer}$ of the agent with $\mathtt{rank} = \lceil n/2 \rceil$. We analyze the expected time for the agent with $\mathtt{rank} = \lceil n/2 \rceil$ to interact with the agent with $\mathtt{rank} = n$ exactly x times. Since the probability that those two agents interact is $2/(n(n - 2))$, the expected time to interact once is $(n - 1)/2$. By the linearity of expectation, the expected time for the $\mathtt{timer}$ to reach 0 is $x(n - 1)/2$. From Markov's inequality, the $\mathtt{timer}$ reaches 0 within $x(n - 1) \leq 7(t_{rank} + 4)(n - 1) = O(n)$ time with probability at least $1/2$. $\qquad\square$

Lemma 11. *Let $C_0 \in \mathcal{S}_{dec}$ and $\Xi_{\mathcal{P}_{EM}}(C_0, \Gamma)$. The probability that the execution reaches $\mathcal{S}_{tim}$ within $O(n)$ time is at least $1/1280$.*

Proof. If there is no agent whose $\mathtt{answer}$ is not the exact majority opinion in C_0, then $C_0 \in \mathcal{S}_{tim}$.

Now, we consider the case where there are agents whose $\mathtt{answer}$ is not the exact majority opinion. In this case, after the $\mathtt{timer}$ of the agent with $\mathtt{rank} = \lceil n/2 \rceil$ reaches 0, when the agent interacts with an agent whose $\mathtt{answer}$ is not the exact majority opinion, PROPAGATE-RESET occurs, and the exact majority opinion propagates to all agents. The variable $\mathtt{timer}$ of the agent with $\mathtt{rank} = \lceil n/2 \rceil$ reaches 0 within $O(n)$ time with probability at least $1/2$, from Lemma 10. Since the probability that agent with $\mathtt{rank} = \lceil n/2 \rceil$ interacts with the agent whose **answer** is not the exact majority opinion is at least $2/(n(n-1))$, the expected time is $(n-1)/2$. From Markov's inequality, such an interaction occurs within $n-1$ time with probability at least $1/2$. Thus, PROPAGATE-RESET occurs within $O(n)$ time with probability at least $1/4$.

During PROPAGATE-RESET, the exact majority opinion propagates to all agents within $O(\log n)$ time with high probability from Lemma 1. Since PROPAGATE-RESET occurs, agents also execute OPTIMAL-SILENT-SSR. Note that PROPAGATE-RESET finishes within $O(n)$ time with high probability, and OPTIMAL-SILENT-SSR finishes within $2 \cdot O(n)$ time with probability at least $1/2$ from Lemma 2 and Markov's inequality. After both protocols finished correctly, the **timer** of the agent with **rank** $= \lceil n/2 \rceil$ is no less than 28 with probability at least $1/10$ from the latter part of the proof of Lemma 8. Thus, the execution reaches $\mathcal{T}_{swap} \cap \mathcal{S}_{out}$ within $O(n)$ time with probability at least $1/4 \cdot 1/10 \cdot (1-1/n)^2 \geq 1/160$.

From Lemma 9, the execution belonged $\mathcal{T}_{swap}$ reaches $\mathcal{S}_{dec}$ within $3n-1$ time with probability $1/8$. Thus, since there is no agent whose **answer** is not the exact majority opinion, the execution reaches $\mathcal{S}_{dec} \cap \mathcal{S}_{out} \subseteq \mathcal{S}_{tim}$. Therefore, the probability that the execution reaches $\mathcal{S}_{tim}$ within $O(n)$ time is at least $1/160 \cdot 1/8 = 1/1280$. $\qquad\square$

Lemma 12. *Let $C_0 \in \mathcal{S}_{tim}$ and $\Xi_{\mathcal{P}_{EM}}(C_0, \Gamma)$. The probability that the execution reaches $\mathcal{S}_{em}$ within $O(n)$ time is at least $1/2$.*

Proof. For the execution to reach $\mathcal{S}_{em}$, it is necessary for the **timer** of the agent with **rank** $= \lceil n/2 \rceil$ to reach 0. From Lemma 10, the **timer** of the agent with **rank** $= \lceil n/2 \rceil$ reaches 0 within $O(n)$ time with probability at least $1/2$. Thus, the execution reaches $\mathcal{S}_{em}$ within $O(n)$ time with probability at least $1/2$. $\qquad\square$

Lemma 13. *$\mathcal{P}_{EM}$ reaches $\mathcal{S}_{em}$ with probability 1, and the silence time of $\mathcal{P}_{EM}$ is $O(n)$ in expectation, and $O(n \log n)$ with high probability.*

Proof. Let $C_0 \in \mathcal{C}_{all}(\mathcal{P}_{EM})$ and $\Xi_{\mathcal{P}_{EM}}(C_0, \Gamma)$. From Lemma 6, 8, 9, 11, and 12, the execution reaches $\mathcal{S}_{em}$ with probability at least $1/4096000$. Since an execution reaches $\mathcal{S}_{em}$ with nonzero probability, the execution eventually reaches $\mathcal{S}_{em}$ if it runs long enough. Thus, there exists an execution that reaches a silent configuration from any initial configuration; therefore, $\mathcal{P}_{EM}$ reaches $\mathcal{S}_{em}$ with probability 1.

Let $E_{\mathcal{P}_{EM}}$ be the expected time for the execution to reach $\mathcal{S}_{em}$. From the above, $E_{\mathcal{P}_{EM}} \leq O(n) + (1 - 1/4096000)E_{\mathcal{P}_{EM}}$ holds. Thus, $E_{\mathcal{P}_{EM}} = O(n)$. From Markov's inequality, the execution reaches $\mathcal{S}_{em}$ within $2 \cdot O(n)$ time with probability at least $1/2$. Thus, the execution reaches $\mathcal{S}_{em}$ within $2 \cdot O(n) \log n = O(n \log n)$ time with probability at least $1 - (1/2)^{\log n} = 1 - 1/n$. $\qquad\square$

From Theorem 2, 3, and Lemma 13, we derive the following theorem.

Theorem 4. *$\mathcal{P}_{EM}$ is a time- and space-optimal silent self-stabilizing exact majority protocol that uses $O(n)$ states and reaches a silent configuration within $O(n)$ time in expectation and within $O(n \log n)$ time with high probability.*

6   Conclusion and Discussions

We addressed a silent protocol that solves the self-stabilizing exact majority problem. We showed that no protocol can solve this problem without knowledge of n. We proposed a silent protocol within $O(n)$ time in expectation and $O(n \log n)$ time with high probability, using $O(n)$ states, with knowledge of n. We established lower bounds, proving that any protocol requires $\Omega(n)$ states, $\Omega(n)$ time in expectation, and $\Omega(n \log n)$ time with high probability to reach a safe configuration. Thus, the proposed protocol is time- and space-optimal.

We propose the following open problems: An extension of the self-stabilizing majority problem to the exact plurality consensus problem [6], in which agents have k different opinions, can also be considered. Whether this problem can achieve the lower bound shown in this paper remains an intriguing open question. Additionally, it is worth considering a non-silent self-stabilizing exact majority protocol with fewer states.

Acknowledgments. This work was supported by JSPS KAKENHI Grant Numbers JP22H03569, JP23K28037, and JP25K03101.

References

1. Alistarh, D., Aspnes, J., Gelashvili, R.: Space-optimal majority in population protocols. In: Proceedings of the Twenty-Ninth Annual ACM-SIAM Symposium on Discrete Algorithms, pp. 2221–2239 (2018)
2. Alistarh, D., Gelashvili, R., Vojnović, M.: Fast and exact majority in population protocols. In: Proceedings of the 2015 ACM Symposium on Principles of Distributed Computing, pp. 47–56 (2015)
3. Angluin, D., Aspnes, J., Diamadi, Z., Fischer, M.J., Peralta, R.: Computation in networks of passively mobile finite-state sensors. Distrib. Comput. **18**(4), 235–253 (2006)
4. Angluin, D., Aspnes, J., Eisenstat, D.: Fast computation by population protocols with a leader. In: Distributed Computing, pp. 61–75 (2006)
5. Angluin, D., Aspnes, J., Eisenstat, D.: A simple population protocol for fast robust approximate majority. Distrib. Comput. **21**(2), 87–102 (2008)
6. Bankhamer, G., et al.: Population protocols for exact plurality consensus: How a small chance of failure helps to eliminate insignificant opinions. In: Proceedings of the 2022 ACM Symposium on Principles of Distributed Computing, pp. 224–234 (2022)
7. Ben-Nun, S., Kopelowitz, T., Kraus, M., Porat, E.: An O(log3/2 n) parallel time population protocol for majority with O(log n) states. In: Proceedings of the 39th Symposium on Principles of Distributed Computing, PODC '20, New York, NY, USA, pp. 191–199 (2020)
8. Berenbrink, P., Biermeier, F., Hahn, C., Kaaser, D.: Loosely-stabilizing phase clocks and the adaptive majority problem. In: 1st Symposium on Algorithmic Foundations of Dynamic Networks (SAND 2022), vol. 221, pp. 7:1–7:17 (2022)

9. Berenbrink, P., Elsässer, R., Friedetzky, T., Kaaser, D., Kling, P., Radzik, T.: A population protocol for exact majority with O(log5/3 n) stabilization time and theta(log n) states. In: 32nd International Symposium on Distributed Computing (DISC 2018), vol. 121, pp. 10:1–10:18 (2018)

10. Berenbrink, P., Elsässer, R., Friedetzky, T., Kaaser, D., Kling, P., Radzik, T.: Time-space trade-offs in population protocols for the majority problem. Distrib. Comput. **34**(2), 91–111 (2021)

11. Bilke, A., Cooper, C., Elsässer, R., Radzik, T.: Brief announcement: population protocols for leader election and exact majority with O(log2 n) states and O(log2 n) convergence time. In: Proceedings of the ACM Symposium on Principles of Distributed Computing, pp. 451–453 (2017)

12. Burman, J., et al.: Time-optimal self-stabilizing leader election in population protocols. In: Proceedings of the 2021 ACM Symposium on Principles of Distributed Computing, PODC'21, pp. 33–44 (2021)

13. Cai, S., Izumi, T., Wada, K.: How to prove impossibility under global fairness: On space complexity of self-stabilizing leader election on a population protocol model. Theory Comput. Syst. **50**(3), 433–445 (2012)

14. Chatzigiannakis, I., Dolev, S., Fekete, S., Michail, O., Spirakis, P.: On the fairness of probabilistic schedulers for population protocols. In: Algorithmic Methods for Distributed Cooperative Systems, vol. 9371, pp. 1–23 (2010)

15. Condon, A., Hajiaghayi, M., Kirkpatrick, D., Maňuch, J.: Approximate majority analyses using tri-molecular chemical reaction networks. Nat. Comput. **19**(1), 249–270 (2020)

16. Doty, D., Eftekhari, M., Gᴀsieniec, L., Severson, E., Uznański, P., Stachowiak, G.: A time and space optimal stable population protocol solving exact majority. In: 2021 IEEE 62nd Annual Symposium on Foundations of Computer Science (FOCS), pp. 1044–1055 (2022)

17. Draief, M., Vojnovic, M.: Convergence speed of binary interval consensus. In: 2010 Proceedings IEEE INFOCOM, pp. 1–9 (2010)

18. Gąsieniec, L., Grodzicki, T., Stachowiak, G.: Near-state and state-optimal self-stabilising leader election population protocols (2025). https://arxiv.org/abs/2502.01227

19. Kosowski, A., Uznanski, P.: Brief announcement: population protocols are fast. In: Proceedings of the 2018 ACM Symposium on Principles of Distributed Computing, PODC '18, pp. 475–477 (2018)

20. Mitzenmacher, M., Upfal, E.: Probability and Computing: Randomized Algorithms and Probabilistic Analysis. Cambridge University Press, Cambridge (2005)

21. Mocquard, Y., Anceaume, E., Aspnes, J., Busnel, Y., Sericola, B.: Counting with population protocols. In: 2015 IEEE 14th International Symposium on Network Computing and Applications (2015)

22. Mocquard, Y., Anceaume, E., Sericola, B.: Optimal proportion computation with population protocols. In: 2016 IEEE 15th International Symposium on Network Computing and Applications (NCA), pp. 216–223 (2016)

23. Perron, E., Vasudevan, D., Vojnovic, M.: Using three states for binary consensus on complete graphs. In: IEEE INFOCOM 2009, pp. 2527–2535 (2009)

24. Sudo, Y., Nakamura, J., Yamauchi, Y., Ooshita, F., Kakugawa, H., Masuzawa, T.: Loosely-stabilizing leader election in a population protocol model. Theor. Comput. Sci. **444**, 100–112 (2012)

25. Yasumi, H., Ooshita, F., Yamaguchi, K., Inoue, M.: Space-optimal population protocols for uniform bipartition under global fairness. IEICE Trans. Inf. Syst. **E102.D**(3), 454–463 (2019)

Invited Paper: Setchain Algorithms for Blockchain Scalability

Arivarasan Karmegam[1,3(✉)] , Gabina Luz Bianchi[6],
Margarita Capretto[2,4] , Martín Ceresa[5] , Antonio Fernández Anta[1,2] ,
and César Sánchez[2]

[1] IMDEA Networks Institute, Madrid, Spain
arivarasan.karmegam@networks.imdea.org
[2] IMDEA Software Institute, Madrid, Spain
[3] Universidad Carlos III de Madrid, Madrid, Spain
[4] Universidad Politécnica de Madrid, Madrid, Spain
[5] Input-Output, Madrid, Spain
[6] Universidad Nacional de Rosario, Rosario, Argentina

Abstract. Setchain has been proposed to increase blockchain scalability by relaxing the strict total order requirement among transaction. Setchain organizes elements into a sequence of sets, referred to as *epochs,* so that elements within each epoch are unordered. In this paper, we propose and evaluate three distinct Setchain algorithms that leverage an underlying block-based ledger. *Vanilla* is a basic implementation that serves as a reference point. *Compresschain* aggregates elements into batches and compresses these batches before appending them as epochs in the ledger. *Hashchain* converts batches into fixed-length hashes, which are appended as epochs in the ledger. This requires Hashchain to use a distributed service to obtain the batch contents from its hash. To allow light clients to safely interact with only one server, the proposed algorithms maintain, as part of the Setchain, proofs for the epochs. An *epoch-proof* is the hash of the epoch, cryptographically signed by a server. A client can verify the correctness of an epoch with $f+1$ epoch-proofs (where f is the maximum number of Byzantine servers assumed). All three Setchain algorithms are implemented on top of the CometBFT blockchain application platform. We conducted performance evaluations across various configurations, using clusters of four, seven, and ten servers. Our results show that the Setchain algorithms reach orders of magnitude higher throughput than the underlying blockchain, and achieve finality with latency below 4 s.

Keywords: Blockchain · Setchain · Scalability · CometBFT

1 Introduction

A blockchain is a *reliable distributed object* containing the list of transactions performed on behalf of users, totally ordered and packed into blocks [8,9]. Real-

This work is funded in part by a research grant from Nomadic Labs and the Tezos Foundation, and by MICIU/AEI /10.13039/501100011033/, ERDF, and the ESF+ under predoctoral training grant PREP2022-000373, grant DRONAC (PID2022-140560OB-I00), and grant DECO (PID2022-138072OB-I00).

world blockchains are maintained by multiple servers (without a central authority) that leverage a *Byzantine-tolerant consensus algorithm* to order the transactions and that compute their effects [6,7].

Challenges. One of the main challenges for practical blockchains is to append transactions to the chain at a fast rate. This is called *throughput* and is measured in transactions per second (TPS). Throughput is constrained by multiple factors, including the speed at which the consensus is solved, network latency, and the computational cost of executing transactions to update or validate the blockchain state. To maintain efficiency and decentralization, blockchains often impose limits on block size or execution complexity, which further impact throughput. A second challenge is to reduce the *latency* of the blockchain, i.e., the time spent from when users send their transactions until those transactions are added to the blockchain. The latency in some blockchains can be of hours. Moreover, in some blockchains (e.g., Bitcoin), a block that has been reported to be added to the blockchain may end up being removed due to a fork in the chain. Hence, a third challenge is to achieve *finality* in the blockchain, i.e., reach a point in time when users know their transactions are in the chain, never to be removed. Finally, a fourth challenge is to provide users with *proofs* that their transactions are in the chain. This can be achieved by having each user contact several servers and ask them for individual proofs (e.g., a copy of the whole chain). This is rarely done, and users typically trust that the server with which they interact is honest. It would be desirable that solid proofs that the transactions are appended to the chain are provided to users who interact with one single server. We address these challenges with Setchain [4], which is a *reliable distributed object* that implements Byzantine-tolerant distributed grown-only sets with barriers and has been shown to have a large throughput, and employs *epoch-proofs,* which are cryptographic artifacts that allow users to interact reliably with just one server. Related Works section is deferred to the full version [13] due to space constraints.

Setchain. An approach to improve blockchain scalability is Setchain [4], a Byzantine-tolerant distributed object that implements a sequence of sets (called *epochs*). Setchain relaxes the total order requirement of blockchain and thus can achieve higher throughput and scalability by allowing transactions to be validated in parallel within an epoch. Setchain can be used for applications like digital registries (e.g., MIT Digital Diplomas [3,15], Georgia Government registry) or voting systems (e.g., Follow My Vote, Chirotonia [16]), where different elements in the blockchain need not be ordered except across infrequent epoch barriers. Byzantine-tolerant distributed algorithms that implement Setchain have been proposed, but no efficient real-world implementations exist.

Contributions. In this work, we propose a family of real-world Setchain algorithms built on top of a block-based ledger. We follow an incremental approach, providing several approximations to the final and most complex solution (Sect. 3). We first present **Algorithm Vanilla**, which is a naive implementation of Setchain with similar throughput as the underlying ledger. Then, we present **Algorithm Compresschain** that increases the throughput using compression of epochs. Finally, we present **Algorithm Hashchain**, a solution using hash functions. From these algorithms, our primary contribution is Hashchain, which exploits the succinctness of hashes to reduce the communication necessary during broadcasts and consensus, communicating a fixed-size hash instead of the hundreds or thousands of elements of an epoch. The price to pay is an additional distributed algorithm to obtain the epoch contents of a hash from the corresponding server. Additionally, to allow users to interact with only one server, these algorithms maintain, as part of the Setchain, proofs for its epochs. An *epoch-proof* is the hash of the epoch, cryptographically signed by a server. A user can verify the correctness of an epoch with $f + 1$ epoch-proofs (where f is the maximum number of Byzantine servers). All three Setchain algorithms are implemented on top of the CometBFT blockchain application platform (Sect. 4). We deployed these implementations as Docker nodes in a cluster. We conducted performance evaluations across various configurations, using clusters of four, seven, and ten servers. Our results show that the Setchain algorithms reach orders of magnitude higher throughput than the underlying blockchain (tens of thousands TPS), and achieve finality with latency below 4 seconds.

2   Model and Definitions

System Model. We consider a distributed system consisting of n servers and an unbounded number of clients, which together are the system processes. The system is permissioned but public, also known as "open permissioned" [7], referring to an open model for clients but where servers are known upfront (permissioned). Nevertheless, this model can also be adapted to a permissionless setting with committee sortition [10] without significant modifications. Up to $f < n/2$ of the servers may exhibit Byzantine behavior, while the clients' behavior is not restricted (i.e., they can be Byzantine). We assume that the number f of Byzantine servers is known. We use $f + 1$ as a lower bound on the number of consistent epoch-proofs required to ensure that an epoch is correct and as the number of signatures needed in order to consolidate hashes into epochs (see Sect. 3). We assume that there is a public key infrastructure (PKI) deployed so that each process (server or client) has a pair of private and public keys. Every process also knows the public key of all the other processes. The communication between processes is assumed to be reliable, meaning that messages sent between correct processes are eventually delivered only once, and no spurious messages are generated. Faulty processes can send any message, but thanks to the PKI, they cannot impersonate other processes. The PKI enables authenticated and secure communication between processes, since messages are signed by the sender using their private key, and the receiver verifies the signature using the sender's public key. If a signature is invalid, the message is discarded. Clients create elements and invoke a Setchain operation to add them. The elements created by the clients are signed and can hence be authenticated. They can also be validated by the servers for syntactic and semantic correctness. Only authenticated valid elements are processed by the correct servers. We assume that a server cannot create a valid element by itself, and that clients and servers do not collude.

Setchain. A Setchain [4] is a Byzantine-tolerant distributed object S that implements a sequence of sets called *epochs*. The Setchain defines an order between elements in different epochs, but not between elements in the same epoch. Thus, Setchain relaxes the total order requirement imposed by blockchains, potentially achieving higher throughput and scalability. Let U be the set of elements that client processes can inject into the Setchain. A Setchain S is a reliable distributed object where a collection of servers maintain: (i) a grow-only set *the_set* $\subseteq U$ of elements added; (ii) a natural number *epoch* $\in \mathbb{N}$; (iii) a map *history* : $\{1, \ldots, epoch\} \to 2^{U}$ with the set of elements that have been stamped with a given epoch number[1]; (iv) a set *proofs* of epoch-proofs (see Sect. 2)[2]. Servers support two operations: add and get. Operation $S.add_v(e)$ is used by a client to request a server v to add an element e to the Setchain S, while operation $S.get_v()$ returns the values (*the_set, history, epoch, proofs*) that server v maintains. Servers can use a function valid_element(e) to locally decide if an element e is valid. In a Setchain, when a new epoch is created, the servers collaboratively decide which of the added elements not yet assigned an epoch are included in the new epoch, and increase the epoch number. We call this an *epoch increment*. For the rest of the paper, we assume that at any given time, there will be a future epoch increment. This assumption is reasonable in real-world scenarios, and it can be reliably ensured by utilizing timeouts. A typical workflow from the point of view of a client is as follows: a client invokes $S.add_v(e)$ in a server v to insert a new valid element e in the Setchain. The element e will be propagated among the servers, and when an epoch increment occurs, the servers will attempt to include it in the new epoch. After waiting for some time, the client invokes $S.get_w()$ in a (possibly different) server w to check that the element has been effectively added to the Setchain and included in an epoch. Setchain implementations must satisfy certain properties that guarantee consistency between correct servers and that the valid elements added are eventually included in an epoch [4]. These properties are restricted to correct servers, since Byzantine servers do not provide any guarantee.

1. *Consistent-Sets:* Let $(T, H, h, P) = S.get_v()$ be the result of an invocation to a correct server v. Then, for all $i \in \{1, \ldots, h\}, H[i] \subseteq T$.
2. *Add-Get-Local:* Let $S.add_v(e)$ be an operation invoked in a correct server v and let e be valid. Then, eventually all invocations $(T, H, h, P) = S.get_v()$ satisfy $e \in T$.
3. *Get-Global:* Let v and w be two correct servers, let e be a valid element, and let $(T, H, h, P) = S.get_v()$. If $e \in T$, then eventually all invocations $(T', H', h', P') = S.get_w()$ satisfy that $e \in T'$.
4. *Eventual-Get:* Let v be a correct server, let e be a valid element and let $(T, H, h, P) = S.get_v()$. If $e \in T$, then eventually all invocations $(T', H', h', P') = S.get_v()$ satisfy that $e \in H'$.

---

[1] 2^{U} denotes the power set of U.

[2] This set was not part of the API described in the original Setchain paper [4] and is one of the contributions of this work.

5. *Unique-Epoch:* Let v be a correct server, $(T, H, h, P) = \mathsf{S.get}_v()$, and let $i, i' \in \{1, \ldots, h\}$ with $i \neq i'$. Then, $H[i] \cap H[i'] = \emptyset$.

6. *Consistent-Gets:* Let v, w be correct servers, $(T, H, h, P) = \mathsf{S.get}_v()$, $(T', H', h', P') = \mathsf{S.get}_w()$, and $i \in \{1, \ldots, \min(h, h')\}$. Then $H[i] = H'[i]$.

7. *Add-before-Get:* Let v be a correct server, $(T, H, h, P) = \mathsf{S.get}_v()$, and $e \in T$ be a valid element. Then there was an operation $\mathsf{S.add}_w(e)$ invoked in the past in some server w.

Properties 1, 5, 6, and 7 are safety properties. Properties 2, 3, and 4 are liveness properties.

Setchain Epoch-proofs. The original Setchain [4] proposal assumes that a client interacts with a correct server. However, in practice, clients do not know whether they are communicating with a correct or a Byzantine server. Hence, a client may need to interact with a sufficient number of servers (at least $f + 1$) to guarantee that at least one is correct. In this work, we introduce *epoch-proofs* as a mechanism that allows clients to achieve the same guarantees without needing to contact multiple servers, thereby simplifying the process and improving efficiency. An epoch-proof is the cryptographic signature of an epoch i by a server v. As described above, in this paper, the Setchain maintains a set *proofs* of epoch-proofs. This set is returned by the Setchain when a get operation is invoked. Hence, when *proofs* contain at least $f + 1$ consistent epoch-proofs for a given epoch i, the client can be sure the epoch is correct. The epoch-proof is created by signing the hash of the epoch number and the elements of the epoch: $p_v(i) = \mathsf{Sign}_v(\mathsf{Hash}(i, history[i]))$. To add an element, a client only performs a single $\mathsf{S.add}_v(e)$ request to one server v, hoping it is a correct server. After waiting for some time, the client can invoke a $\mathsf{S.get}_w()$ from a single server w and check whether e is in some epoch, and the set *proofs* returned contains at least $f + 1$ valid epoch-proofs for that epoch to guarantee that at least one correct server has signed it. Clients can verify whether epoch-proofs are valid by generating the hash of a given epoch, and verifying if the signature in the epoch-proof is valid using the hash and the public key of the signing server. Recall that servers' public keys are known to the clients. Note that this process usually requires only one message per add and one message per get. Of course, it is possible that the client is unlucky and servers v or w are Byzantine, and after waiting a reasonable amount of time, has to restart the process. The following is the basic property that any Setchain algorithm must satisfy concerning epoch-proofs.

8. *Valid-Epoch:* Let v be a correct server, $(T, H, h, P) = \mathsf{S.get}_v()$, and $i \in \{1, \ldots, h\}$. Then eventually all invocations $(T', H', h', P') = \mathsf{S.get}_v()$ satisfy that P' contains at least $f + 1$ epoch-proofs of $H[i]$.

Block-based Ledger. In the proposed algorithms to implement a Setchain as presented in this paper, we assume the availability of a Byzantine-tolerant consensus service. For simplicity, we abstract that service as a block-based ledger. While our algorithms only require that the number of Byzantine processes satisfies $f < n/2$, the block-based ledger may have more stringent requirements. For instance, in Sect. 4 we use the ledger CometBFT that requires $f < n/3$. A block-based ledger L is a Byzantine-tolerant distributed object that maintains a sequence of blocks. Each block contains a sequence of transactions that have been appended to the ledger by its clients. We prefer not to call this object a blockchain since its transactions have no semantics. To keep the nomenclature consistent, the term *transaction* is always used to refer to *block-based ledger transactions*, while *element* refers to the elements of the Setchain. Depending on the implementation, a single ledger transaction may contain one or many elements. The block-based ledger provides two endpoints. First, $\mathtt{append}(tx)$ is used to submit a transaction tx to the ledger, which eventually gets included in a block. Then, $\mathtt{new\_block}(B)$ notifies the servers whenever a new block B has been appended. The number of transactions in B is $|B|$, and the ith transaction in B is $B[i]$. Since Setchain algorithms use a block-based ledger, to prove correctness, it is necessary to define properties that the block-based ledger guarantees. In particular, the ledger property that will be used is that any valid transaction appended by a correct server will eventually be included in a block, which is then notified to all the servers. Moreover, we assume that all blocks notified are final.

9. *Ledger-Add-Eventual-Notify:* Let tx be a valid transaction, and let v be a correct server. Then, if v invokes $\mathtt{L.append}_v(tx)$, transaction tx will be eventually and permanently added in a fixed position i within a block B. Moreover, all correct servers w will be eventually notified of the new block B with $\mathtt{L.new\_block}_w(B)$.
10. *Ledger-Consistent-Notification:* All correct servers w are notified with $\mathtt{L}.new\_block_w(B)$ of the same set of blocks, and in the same order.
11. *Notification-Implies-Append:* If a correct server w is notified with $\mathtt{L}.new\_block_w(B)$ containing a valid element e, then some server v had invoked $\mathtt{L.append}_v(e)$.

3   Setchain Algorithms

In this section, we propose practical Setchain algorithms built on top of a block-based ledger. In particular, we present three different implementations, beginning with a naive but trivially correct implementation and ending with a more complex algorithm implementing Setchain using hashing. As described in Sect. 2, a Setchain S provides two methods, $\mathtt{add}$ and $\mathtt{get}$, defined in each algorithm.

Vanilla. In Vanilla, we propose a basic implementation of Setchain S using a block-based ledger L. The algorithm for Vanilla is deferred to the full version [13] due to space constraints. In this algorithm, each server v maintains three local sets: *the_set, history,* and *proofs,* and the counter *epoch,* as defined in Sect. 2. Whenever a client queries Setchain S using $\mathtt{S.get}_v()$, the server returns the current value of *the_set, history, epoch,* and *proofs* (Line 9). On the other hand, clients add an element e by invoking $\mathtt{S.add}_v(e)$. The server only accepts valid elements not already present in *the_set* (see Line 4). If so, the element is added to *the_set,* and the server invokes a $\mathtt{L.append}_v(e)$ operation, which eventually adds the element e to ledger L. Whenever a new block is appended to the ledger L, server v is notified through $\mathtt{L.new\_block}(B)$ (see Line 10). Ledger

Algorithm Compresschain. Code executed by server v. Uses block-based ledger L shared by all servers.

```
 1 Init: the_set ← ∅, epoch ← 0, history ← ∅,        18 upon (L.new_block(B)) do
       proofs ← ∅, batch ← ∅                         19    for i = 1 to |B| do
 2 function add(e)                                    20       batch_original ← Decompress(B[i])
 3    assert valid_element(e) ∧ (e ∉ the_set)         21       if batch_original = ∅ then continue
 4    the_set ← the_set ∪ {e}                         22       np ← {ep ∈ batch_original :
 5    add_to_batch(e)                                             ep = ⟨j, p, w⟩ is an epoch-proof
 6    return                                                      ∧ valid_proof(j, p, w, history[j])}
 7 function add_to_batch(e)                           23       proofs ← proofs ∪ np
 8    batch ← batch ∪ {e}                             24       G ← {e ∈ batch_original :
 9    return                                                     e is an element ∧ valid_element(e)
10 function get()                                                ∧(e ∉ history)}
11    return (the_set, history, epoch, proofs)        25       the_set ← the_set ∪ G
12 upon (isReady(batch)) do                           26       epoch ← epoch + 1
13    assert batch ≠ ∅                                27       history[epoch] ← G
14    b ← Compress(batch)                             28       p ← Sign_v(Hash(epoch, G))
15    L.append(b)                                     29       add_to_batch(⟨epoch, p, v⟩)
16    batch ← ∅                                       30 end upon
17 end upon
```

L is used to disseminate the epoch-proofs in the three algorithms proposed. Hence, the server starts by extracting the valid epoch-proofs from B and adding them to the set *proofs* (Line 12). Then, the server extracts the elements from B that are valid and not yet in an epoch, and places them into batch G (Observe that checking whether an element is valid cannot be avoided because a Byzantine server may have added invalid elements to the ledger). The set G of valid elements is added to *the_set*, and is used to form a new epoch. This new epoch is added to the *history* map. Then, the epoch-proof containing the epoch number, the epoch signature, and the server's identity is added to the ledger by invoking an **L.append** operation (Line 18). Vanilla solves the proof challenge via epoch-proofs, allowing a client to interact with only one correct server. However, the throughput and latency of the Setchain implemented are those of the block-based ledger L.

Compresschain. The second algorithm we propose, Compresschain, increases the throughput compared to Vanilla. In Compresschain, each server has a *collector* in which the elements added by clients and epoch-proofs added by servers are held until the collector size is reached (or possibly a timeout expires). Then, the collected batch, which will become a new epoch, is compressed and appended to the ledger as a single ledger transaction. Compresschain adapts Vanilla to handle the compressed batches by adding an additional set *batch* to collect client elements and epoch-proof. When a client adds an element e to the Setchain using $S.add_v(e)$, in addition to adding the element to *the_set*, it is also added to *batch* (the collector; Line 5). When *batch* reaches the collector size (or a timeout is triggered, and the *batch* is not empty), a notification **isReady**(*batch*) is signaled. Then, the batch is compressed and appended to the ledger, and *batch* is reset to ∅. When the creation of a new block B is notified, each compressed batch in B

Algorithm Hashchain. Code executed by server v. Uses block-based ledger L shared by all servers.

```
 1 Init: the_set ← ∅, epoch ← 0, history ← ∅,
         proofs ← ∅, hash_to_batch ← ∅,
         batch ← ∅, hash_to_signers ← ∅
 2 function add(e)
 3   assert valid_element(e) ∧ (e ∉ the_set)
 4   the_set ← the_set ∪ {e}
 5   add_to_batch(e)
 6   return
 7 function add_to_batch(e)
 8   batch ← batch ∪ {e}
 9   return
10 function get()
11   return (the_set, history, epoch, proofs)
12 upon (isReady(batch)) do
13   assert batch ≠ ∅
14   h ← Hash(batch)
15   hash_to_batch[h] ← batch
16   Register_batch(h, batch)
17   s ← Sign_v(h)
18   hb ← ⟨h, s, v⟩
19   L.append(hb)
20   batch ← ∅
21 end upon
22 upon (L.new_block(B)) do
23   for i = 1 to |B| do
24     if (B[i] = ⟨h, s_w, w⟩)∧
          valid_hash(h, s_w, w) then
25       batch_original ← hash_to_batch[h]
26       if batch_original = ∅ then     ▷ h is new
27         batch_original ← Request_batch(h)
28         if (batch_original = ∅) ∨   ▷ Not found
            (Hash(batch_original) ≠ h) then
29           continue
30         hash_to_batch[h] ← batch_original
31         Register_batch(h, batch_original)
32         s_v ← Sign_v(h)
33         hb ← ⟨h, s_v, v⟩
34         L.append(hb)
35       np ← {ep ∈ batch_original :
               ep = ⟨j, p, w⟩ is epoch-proof ∧
               valid_proof(j, p, w, history[j])}
36       proofs ← proofs ∪ np
37       G ← {e ∈ batch_original :
             e is an element ∧ valid_element(e)
             ∧(e ∉ history)}
38       the_set ← the_set ∪ G
39       hash_to_signers[h] ← hash_to_signers[h]
             ∪{w}
40       if |hash_to_signers[h]| = f + 1 then
41         epoch ← epoch + 1
42         G ← {e ∈ batch_original :
               e is an element ∧ valid_element(e)
               ∧(e ∉ history)}
43         history[epoch] ← G
44         p ← Sign_v(Hash(epoch, G))
45         add_to_batch(⟨epoch, p, v⟩)
46 end upon
```

---

is processed in order. Let $B[i]$ be the ith transaction in block B. $B[i]$ is decompressed, and the valid epoch-proofs in $B[i]$ are added to the set *proofs*. Then, the valid elements in $B[i]$ are extracted to a set G to form a new epoch. The elements in G are first added to *the_set*. Then, they are assigned an epoch number, and the epoch information is added to the *history* map. An epoch-proof for the epoch is also created and sent to the collector using $\mathsf{S.add\_to\_batch}_v(\cdot)$, which in turn adds it to the *batch*. The most important difference between Compresschain and Vanilla is that in Compresschain each transaction in a block B is a compressed batch (which potentially has many Setchain elements) that becomes an epoch, whereas in Vanilla, each transaction in block B is an element, and an epoch is the set of valid elements in the block B. This potentially increases the throughput significantly.

Hashchain. Hashchain increases the throughput by using hashing of batches instead of compressing them. While the space reduction of hashing may be enormous, hashes are irreversible. Hence, a non-trivial method has to be provided to recover the contents of the original batch of elements from a hash. In Hashchain, when a batch is ready (Line 12), server v generates the hash h of the batch, signs the hash h, and creates the hash-batch hb as the combination of the hash of the batch, the signature, and its identity. Then, the hash-batch hb is appended as a transaction to the ledger L. As the original batch cannot be recovered from the hash h, the server saves the batch associated with h in a map *hash_to_batch*.

Also, the *batch* with its hash h is registered using `Register_batch`. This will allow serving the contents of the original batch to other servers. The *batch* is then reset to $\emptyset$. Upon a `L.new_block`(B) notification, each valid hash-batch $hb_w = \langle h, s_w, w \rangle \in B$ is processed as follows. If server v does not have the batch associated with the hash h in the map *hash_to_batch*, v requests the batch from server w using `Request_batch` (which must have it since w signed that hash-batch). Since w may be Byzantine, server v waits for a limited amount of time for the answer to its request. If the batch is received correctly, v creates its hash-batch $hb_v = \langle h, s_v, v \rangle$ and appends it to ledger L. Then, the batch is saved in map *hash_to_batch* and *the_set* is updated. Observe that finding a valid hash-batch in ledger L is not enough to assign it an epoch number, because it could be a hash-batch appended by a Byzantine server that refuses to provide the batch that corresponds to the hash. Hence, we decided that a hash has to be signed by at least $f + 1$ individual servers to be consolidated into an epoch. The reason is that a hash with hash-batches from $f + 1$ different servers is guaranteed to be signed by at least one correct server, which will have a copy of the batch and serve it upon request. Then, the processing of hash-batch hb_w is continued by adding w to the set of signers in the map *hash_to_signers*, used to track the set of servers that appended hash-batches with hash h (Line 39). When this set of signers reaches $f + 1$, the server extracts the valid elements from the batch to a set G which is assigned an epoch number. We call this process *epoch consolidation*. Also, the valid epoch-proofs in *batch_original* are added to the set *proofs*. As usual, the epoch is added to the *history*, and an epoch-proof for this epoch is generated and sent to the collector using `S.add_to_batch`$_v(ep)$.

Correctness Proof Sketch. We provide here an intuitive explanation for why the Setchain algorithms presented in Sect. 3 behave correctly, even in the presence of up to f Byzantine faults[3]. The Setchain algorithms ensure that all valid elements are eventually included in an epoch, and that all correct servers maintain a consistent view of this sequence. When a client submits an element to a Setchain server v, it is added to that server's local *the_set* only if the element is valid. Once stored locally, the element is returned as a part of that Setchain server's *the_set*, whenever `S.get`$_v()$ is invoked by the client (Property 2). When it consolidates an epoch, a server adds all valid elements from the epoch to *the_set*, if not already present, before recording the epoch in the *history*. Therefore, *history*, which is the collection of all the epochs, is a subset of *the_set* (Property 1). Each server periodically creates a batch either after collecting enough elements or after a timeout. Since all valid elements in the batch are added to *the_set* and eventually put into an epoch, these elements eventually end up in *history*, again by Properties 9 and 10 (Property 4). In Hashchain, a hash-batch consolidates into an epoch (i.e., is assigned an epoch number) only after the server receives $f + 1$ signatures on the hash of the batch. When a correct server v either generates a hash-batch or receives a hash-batch from another server, it first verifies the batch elements (requesting them from the signer of that batch, if the original batch is not found locally), then signs the batch hash and appends

---

[3] We defer the full proof of correctness to the full version [13] due to space restrictions.

it to the ledger. Eventually, a block containing this hash-batch signed by v is notified to all the servers, and the other correct servers request the original batch from v (if the original batch is not found locally), verify the batch elements, sign the batch hash and append it to the ledger. So, given that there are at least $f + 1$ correct servers, eventually this hash-batch receives $f + 1$ signatures and the batch consolidates into an epoch. All correct Setchain servers use the same logic to construct epochs from observed blocks, all Setchain servers observe the same blocks in the same order, and all the transactions inside the block are totally ordered (Properties 9 and 10). Hence, correct Setchain servers create identical epochs and agree on the epoch content (Property 6). Also, each correct server ensures that new epochs exclude elements already included in *history*, which leads directly to Property 5. Again, by Property 9, every element added to a server's local *the_set* eventually appears in a block that is delivered to all Setchain servers. After that, every S.get() invocation to a correct Setchain server will include this element as part of their *the_set* (Property 3). Since we assume that a server can neither create a valid element by itself nor collude with clients, a server cannot append a valid element with L.append(e) without a client invocation of S.add(e) (Property 7). To ensure that clients can validate the correctness of an epoch without needing to trust a single server, Setchain maintains epoch-proofs, which are Setchain signatures of the hash of the epoch. These proofs are included in the ledger as transactions, enabling any client to retrieve them through a S.get() operation. When a correct Setchain server encounters an epoch, after validation, it signs the hash of the epoch and appends it to the block-based ledger as its epoch-proof. Due to Property 9, this epoch would have been notified to all Setchain servers and eventually the ledger contains $f + 1$ epoch-proofs for the epoch, as the number of correct servers is more than $f + 1$. Therefore, eventually, when a client invokes S.get(), at least $f + 1$ epoch-proofs are returned for an epoch (Property 8) and the client can trust the epoch.

4   Implementation and Performance Evaluation

The three algorithms presented have been implemented in Golang using CometBFT as the block-based ledger. We describe some implementation aspects.

CometBFT. CometBFT [5] (previously known as Tendermint) is a Byzantine-tolerant state machine replication engine. CometBFT is a blockchain middleware that supports replicating arbitrary applications, written in any programming language. More details about CometBFT are given in the full version [13].

Performance Evaluation Platform. We carried out the performance evaluation of the Setchain using a cluster. Each machine in the cluster has an Intel(R) Xeon(R) E-2186G CPU @ 3.80 GHz with 12 cores, 32 GB RAM, and runs Debian GNU/Linux 11 (bullseye). We used Docker Engine version 20.10.5. The Setchain algorithms are implemented on CometBFT v0.38, with each ledger server running in a Docker container and each container running on a separate machine in the cluster. The containers have no limit on CPU or RAM usage. Each Docker container contains one client, one collector module, and one CometBFT server.

Experiment Scenarios. The parameters considered for the experiments in this work are listed in the full version [13]. Each client adds at a rate *sending_rateper server_count* by sending the elements to their local server (the server running in the same Docker). The *network_delay* parameter is an artificial latency that is added to all communications between servers to simulate the impact of moving from a cluster to a wide area deployment. The mempool is an important element of CometBFT, where the unconfirmed ledger transactions are held after validation and before being included in the blockchain by CometBFT. The default mempool setting of CometBFT allows only a maximum of $5,000$ transactions to be present in the mempool. For the Setchain evaluation, we did not want this to be a bottleneck. Hence, after a few trial and error experiments, the mempool size has been set to $10,000,000$ transactions or 2 GB, whichever is reached first. By default, each experiment has been designed so that clients add elements in the Setchain for 50 s. When all the elements have been included in epochs and all the epoch-proofs have been inserted in ledger blocks, the experiment ends. The logs are then collected and analyzed. The injection and throughput metrics are based on the Setchain elements sent by the clients, measured in *elements per second (el/s)*. To keep the experiments realistic, we use transactions downloaded from Arbitrum [12] as Setchain elements. For hashing, we use SHA512 [14]; signing is done with ed25519, which is part of the EdDSA (Edwards-Curve Digital Signature Algorithm) family [2,11], and we use Brotli for compression [1]. The average size of an Arbitrum transaction is approximately 438 bytes with a standard deviation of 753.5. The length of an epoch-proof is 139 bytes. In Compress-chain, the average length of a compressed batch is approximately $16,000$ bytes with a standard deviation of $2,100$ for a collector limit of 100, and an average of $66,000$ bytes with a standard deviation of $15,000$ for a collector limit of 500. So, the compression ratio varies approximately from 2.5 to 3.5. In Hashchain, the length of a hash-batch is 139 bytes. In CometBFT, one ledger block is generated roughly every 1.25 seconds (i.e., the block rate is approximately 0.8 blocks/s). Unless otherwise stated, the ledger block size used in CometBFT is 0.5 MB.

Analysis. We derived an analytical formula to estimate the theoretical highest throughput of each algorithm as a function of system parameters. These formulas are used in this section to compute the theoretical throughput values, which are then compared with our experimental results. Due to space constraints, the full derivation is deferred to the full version [13].

Throughput Comparison. Figure 1 shows the throughput (in elements *committed* per second, el/s) over time of the three Setchain algorithms for different sending rates with 10 servers and with no increase in the network delay. An element added becomes *committed* when the epoch in which the element has been included gets at least $f + 1$ epoch-proofs in the ledger. The throughput values obtained in the analysis are also shown for reference. In Fig. 1 (left), for a sending rate of $5,000$ el/s and collector size 100, it can be observed that both Vanilla

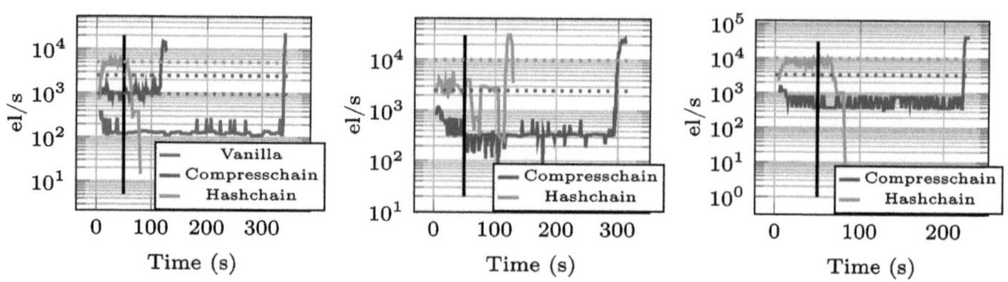

Fig. 1. Throughput over time of the Setchain algorithms for (left) Sending rate $= 5,000$ el/s and Collector size $= 100$, (center) Sending rate $= 10,000$ and Collector size $= 100$, and (right) Sending rate $= 10,000$ el/s and Collector size $= 500$. Solid lines plot the rolling average number of elements committed in 9 s. The vertical bar marks the time clients add the last element (roughly after 50 s). The dotted horizontal lines show the minimum of the sending rate and the analytical throughput computed in the analysis section available in the full version [13].

and Compresschain take a long time to commit all the elements and show the peak at the end, which is a symptom of stress. Moreover, both algorithms show a smaller throughput than the analytical bound. Hashchain, on the other hand, is able to cope and finishes shortly after all elements are added. In Fig. 1 (center), for a sending rate of $10,000$ el/s and collector size 100, we have excluded Vanilla to compare Compresschain and Hashchain more efficiently. In this plot, it can be seen that both Compresschain and Hashchain are stressed, although the former is stressed much more than the latter. As shown in Fig. 1 (right), increasing the collector size to 500 helps to relieve the stress from Hashchain, while it does not help Compresschain much. We present the average throughput achieved up to 50 s for all three algorithms in every experiment in Fig. 1, in a table in the full version [13].

Pushing the Hashchain Limits. As seen in Fig. 1 (right), there is a noticeable gap between the analytical achievable throughput ($147,857$ el/s, see full version [13]) and the actual throughput observed with the implementation of Hashchain, because the experiments did not try large sending rates. To determine the highest achievable throughput, we increased the sending rate. However, we identified a bottleneck around $20,000$ el/s, regardless of further increase in the sending rate beyond that threshold and of the collector size (as seen in Fig. 2 (left)). The most likely cause of this limitation is the hash-reversal process, where batches are exchanged between servers across the network for each batch-hash sent by the collector. In our implementation, Setchain servers handle the distribution of transaction batches. More efficient methods could be employed, such as having only a set of $2f + 1$ servers sign each batch-hash and epoch, utilizing optimistic validation of hash-batches (share a batch only on request), or implementing alternative distributed batch-sharing mechanisms. To assess the impact of hash-reversal, we conducted experiments removing this service and the validation of hash-batches, while assuming that all servers are

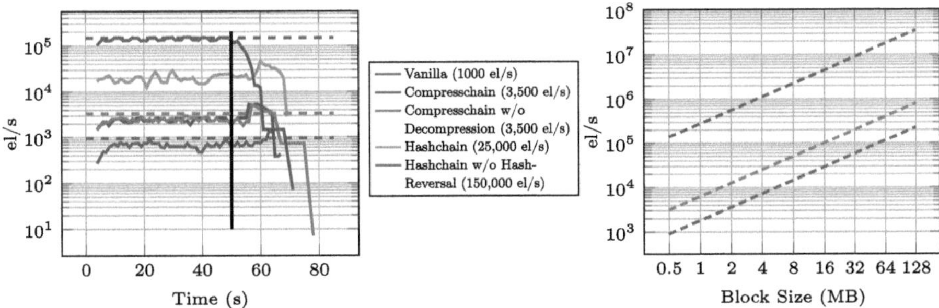

Fig. 2. (left) Highest throughput measured in the evaluation platform with collector size 500. The vertical bar marks the time clients add the last element (roughly after 50 s). Solid lines plot the rolling average number of elements committed in 9 s. The dashed lines show the analytical throughput. (right) Analytical throughput of the Setchain algorithms for various block sizes with collector size 500.

correct—meaning all hash-batches are inherently valid. Figure 2 (left) illustrates the highest achieved throughput by Hashchain with and without hash-reversal. From this, it is clear that hash-reversal is the bottleneck, and with a more efficient implementation of hash-reversal, Hashchain would scale much better in throughput. The results confirm that hash-reversal significantly limits performance. Without it, Hashchain reaches an average throughput of $133,882$ el/s for the first 50 seconds with a sending rate of $150,000$ el/s, compared to an average throughput of $20,061$ el/s for the first 50 seconds with a sending rate of $25,000$ el/s with hash-reversal enabled. It is also important to note that these results were obtained with a collector size of 500, chosen to maintain comparability with other experiments in this work. A larger collector size would likely yield even higher throughput without hash-reversal. For comparison, Fig. 2 (left) also shows the highest achieved throughput by Compresschain and Vanilla. For Compresschain, we ran it with and without decompression and validation to measure the impact of these processes. Observed throughputs are well below Hashchain's throughput, even with hash-reversal. The highest throughputs observed with Vanilla, Compresschain Light, and Hashchain Light are very close to the analytical values. Figure 2 (right) shows the analytical throughput for all three Setchain algorithms for larger block sizes, while keeping the other parameters constant (For Compresschain and Hashchain, Collector size = 500). As can be seen, with the usual 4 MB block size of CometBFT, Hashchain reaches a throughput of 10^6 el/s, and with blocks of 128 MB reaches more than 30 million el/s.

Efficiency. To easily quantify the level of stress of an algorithm, we define and use a metric that we call *efficiency*. The efficiency is obtained by dividing the number of elements committed by the total number of elements added. We calculate the efficiency factor after 50, 75, and 100 seconds. Remember that clients add elements for 50 seconds in every experiment. If the algorithm is not stressed, we expect to observe efficiency close to 1 after 50 seconds, and exactly 1 after

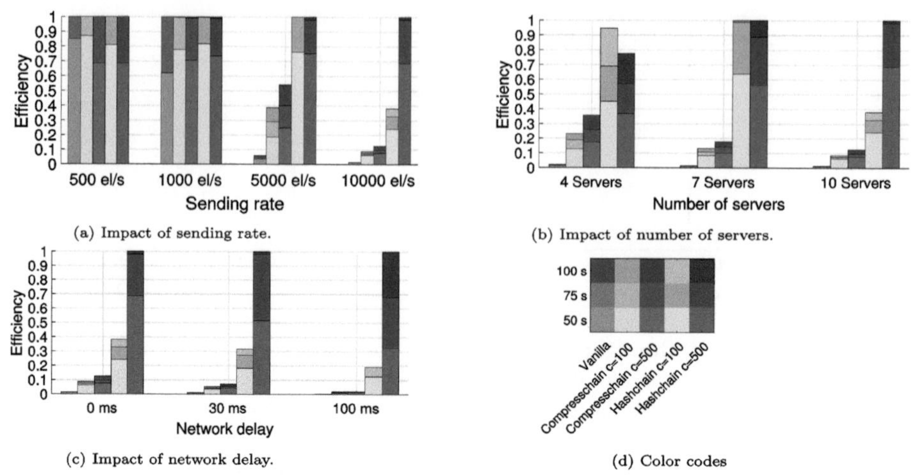

(a) Impact of sending rate.

(b) Impact of number of servers.

(c) Impact of network delay.

(d) Color codes

Fig. 3. Efficiency values observed under different scenarios. The base scenario has 10 servers, a sending rate of 10,000 el/s, and no (0) network delay.

75 seconds. In this section, we use a base scenario with 10 servers, a sending rate of 10,000 el/s, and no (0) network delay, and vary one of these parameters at a time. Figure 3a shows the efficiency for four different sending rates (500, 1000, 5,000, and 10,000), and for collector sizes 100 and 500, when applicable. This figure shows that all algorithms reach full efficiency in 70s for the lower rates 500 and 1,000. Then, for rates 5,000 and 10,000, Vanilla has very low efficiency. Compresschain, on the other hand, also reduces its efficiency significantly, and increasing the collector size from 100 to 500 does not help much. Finally, Hashchain only shows a decrease in efficiency with a rate of 10,000, which is alleviated by increasing the collector size. In Fig. 3b, we compare the change in efficiency observed when the number of servers is varied, maintaining a sending rate of 10,000 el/s. Observe that Vanilla has the lowest efficiency, even for 4 servers. The efficiency of Compresschain is also low, and increasing the collector size does not improve it much. Moreover, it decreases as the number of servers increases. Finally, Hashchain only shows low efficiency with 10 servers and a collector size of 100. Interestingly, it shows a less-than-perfect efficiency for 4 servers, which may be due to having fewer servers available for the reverse hashing process. Figure 3c shows how adding an artificial delay to all communication messages affects efficiency. This models the impact of moving from a local-area network to a wide-area network. As can be observed, the increase in network delay reduces the efficiency. However, even with the largest delay of 100 ms, Hashchain with a collector size of 500 achieves full efficiency in 100 seconds.

Latency. When a valid element is sent by a client to its Setchain server, it is eventually sent in a ledger transaction to a CometBFT server. The transaction, once validated, enters the mempool of the CometBFT server. Then, it is shared with other CometBFT servers via a gossip protocol. These servers validate the transaction and replicate it in their local mempool. Eventually, the transaction

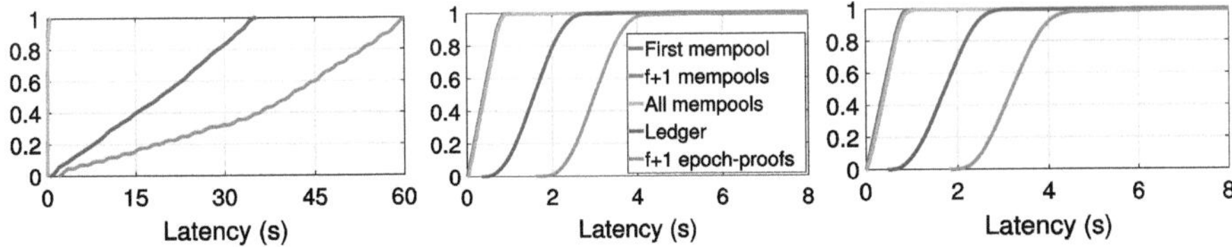

Fig. 4. Cumulative distribution function $F(x)$ of the latency experienced by the elements added to the Setchain to reach several stages in their process for (left) Vanilla, (center) Compresschain and (right) Hashchain. The collector size is 100. The scenario is with 10 servers, a sending rate of $1,250$ el/s, and no network delay.

will be included in a ledger block, and the Setchain servers will send epoch-proofs to the ledger. When $f + 1$ such proofs have been included in ledger blocks, the element has been committed. Figure 4 shows the distribution of the latencies experienced by the elements added to the Setchain to reach five different stages until commit, for each algorithm. It shows the latency until an element reaches (1) a CometBFT server mempool, (2) $f + 1$ CometBFT server mempools, (3) all the CometBFT server mempools, and (4) the ledger. Finally, it shows the (5) commit latency (when $f + 1$ epoch-proofs of the epoch in which the element was included are in the ledger). In Fig. 4 (left), the mempools are reached almost immediately because in Vanilla, the elements are directly sent to the CometBFT server. Compresschain and Hashchain, on the other hand, wait for the collector to be full or a timeout before sending the batch of elements to the CometBFT server as a transaction. This explains the delay in mempool latency in Fig. 4 (center) and (right). Consider now the latency an element takes to reach the ledger and to be committed (with $f + 1$ epoch-proofs). With Vanilla, these two time gaps are tens of seconds in most cases, while for Compresschain and Hashchain, they are usually between one and two seconds. These two algorithms show in this scenario commit latency and finality below 4 s with probability almost 1.

5   Conclusion and Discussions

In this work, we presented three real-world Setchain implementation algorithms built on top of a block-based ledger. We have formally verified that the properties of Setchain hold for these algorithms (proofs are deferred to the full version due to space limitations [13]). We have implemented the three algorithms and performed an empirical evaluation to evaluate the effectiveness of each approach in various scenarios. Among the three, the Hashchain algorithm consistently performed better across all evaluated metrics. Its efficient use of hashing techniques allowed for improved scalability, making it particularly suitable for large-scale applications. Compresschain, while offering some improvements over Vanilla, did not match the performance gains observed with Hashchain.

References

1. Alakuijala, J., Szabadka, Z.: Brotli Compressed Data Format. RFC 7932 (2016). https://doi.org/10.17487/RFC7932. https://www.rfc-editor.org/info/rfc7932
2. Bernstein, D.J., Duif, N., Lange, T., Schwabe, P., Yang, B.Y.: High-speed high-security signatures. J. Cryptogr. Eng. $\mathbf{2}$(2), 77–89 (2012). https://doi.org/10.1007/s13389-012-0027-1
3. Blockcerts: Blockcerts: The open standard for blockchain credentials. https://github.com/blockchain-certificates
4. Capretto, M., Ceresa, M., Fernández Anta, A., Russo, A., Sánchez, C.: Improving blockchain scalability with the setchain data-type. Distrib. Ledger Technol. Res. Pract. $\mathbf{3}$(2) (2024). https://doi.org/10.1145/3626963
5. Cason, D., Fynn, E., Milosevic, N., Milosevic, Z., Buchman, E., Pedone, F.: The design, architecture and performance of the tendermint blockchain network. In: 2021 40th International Symposium on Reliable Distributed Systems (SRDS), pp. 23–33 (2021). https://doi.org/10.1109/SRDS53918.2021.00012
6. Castro, M., Liskov, B.: Practical byzantine fault tolerance and proactive recovery. ACM Trans. Comput. Syst. $\mathbf{20}$(4), 398–461 (2002). https://doi.org/10.1145/571637.571640
7. Crain, T., Natoli, C., Gramoli, V.: Red belly: a secure, fair and scalable open blockchain. In: 2021 IEEE Symposium on Security and Privacy (SP), pp. 466–483 (2021). https://doi.org/10.1109/SP40001.2021.00087
8. Fernández Anta, A., Georgiou, C., Herlihy, M., Potop-Butucaru, M.: Principles of Blockchain Systems. Morgan & Claypool Publishers (2021)
9. Fernández Anta, A., Konwar, K., Georgiou, C., Nicolaou, N.: Formalizing and implementing distributed ledger objects. ACM SIGACT News $\mathbf{49}$(2), 58–76 (2018)
10. Gilad, Y., Hemo, R., Micali, S., Vlachos, G., Zeldovich, N.: Algorand: scaling byzantine agreements for cryptocurrencies. In: Proceedings of the 26th Symposium on Operating Systems Principles, SOSP '17, pp. 51–68. Association for Computing Machinery, New York (2017). https://doi.org/10.1145/3132747.3132757
11. Josefsson, S., Liusvaara, I.: Rfc 8032: Edwards-curve digital signature algorithm (eddsa) (2017)
12. Kalodner, H., Goldfeder, S., Chen, X., Weinberg, S.M., Felten, E.W.: Arbitrum: scalable, private smart contracts. In: 27th USENIX Security Symposium (USENIX Security 18), pp. 1353–1370. USENIX Association, Baltimore (2018). https://www.usenix.org/conference/usenixsecurity18/presentation/kalodner
13. Karmegam, A., et al.: Setchain algorithms for blockchain scalability (2025). https://arxiv.org/abs/2509.09795
14. National Institute of Standards and Technology (NIST): FIPS PUB 180-4: Secure Hash Standard (SHS) (2012). https://nvlpubs.nist.gov/nistpubs/FIPS/NIST.FIPS.180-4.pdf. federal Information Processing Standards Publication
15. News, M.: Digital diploma debuts at mit (2017). https://news.mit.edu/2017/mit-debuts-secure-digital-diploma-using-bitcoin-blockchain-technology-1017
16. Russo, A., Fernández Anta, A., González Vasco, M.I., Romano, S.P.: Chirotonia: a scalable and secure e-voting framework based on blockchains and linkable ring signatures. In: Xiang, Y., Wang, Z., Wang, H., Niemi, V. (eds.) 2021 IEEE International Conference on Blockchain, Blockchain 2021, Melbourne, Australia, 6–8 December 2021, pp. 417–424. IEEE (2021)

Brief Announcement: Dependency-Aware Execution Mechanism in Hyperledger Fabric Architecture

Sanyam Kaul, Manaswini Piduguralla[(✉)], Gayathri Shreeya Patnala, and Sathya Peri

Indian Institute of Technology Hyderabad, Hyderabad, India
`cs20resch11007@iith.ac.in`

Abstract. Hyperledger Fabric is a leading permissioned blockchain framework for enterprise use, known for its modular design and privacy features. While Hyperledger Fabric provides robust mechanisms for configurable consensus and access control, it encounters significant challenges in sustaining high transaction throughput and minimizing rejection rates under heavy workloads. We propose a dependency-aware execution model (Code for our proposed framework is available here: *[Link]*) that: (a) Flags transactions as dependent/independent during endorsement, (b) Prioritizes independent ones in block construction, (c) Embeds a DAG per block to capture dependencies, and (d) Executes transactions in parallel at commit.

1   Introduction

Hyperledger Fabric is a permissioned blockchain platform widely used in enterprise applications [1]. Its modular design, with pluggable consensus and privacy features, enables industrial-grade flexibility. Fabric operates via a three-phase transaction pipeline: endorsement, ordering, and validation.

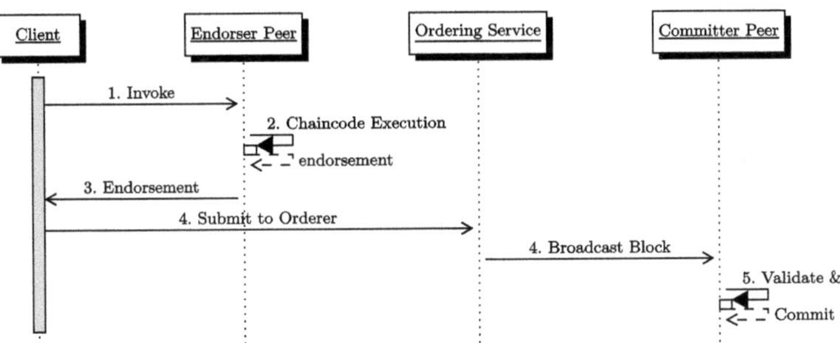

Fig. 1. Fabric transaction flow by Androulaki et al. [1]

Hyperledger Fabric Architecture: The transaction flow in Hyperledger Fabric, as illustrated in Fig. 1, begins when the client (**Step 1**) sends a transaction proposal to one or more *endorser* peers. Each endorser peer then simulates the transaction by executing the specified chaincode (smart contract code, **Step 2**), generating a read/write set and an endorsement signature. The endorser returns this endorsement to the client (**Step 3**), and the client collects enough endorsements as required by the policy. The client then submits the endorsed transaction to the *ordering service* (**Step 4**), which orders the transactions and packages them into blocks. The ordering service broadcasts the new block to all relevant peers, including both endorsers and *committers* (**Step 4**). Finally, each committer peer validates the transactions in the block, checking endorsement policies and for conflicts, and commits valid transactions to the ledger (**Step 5**).

While this architecture performs well in low-contention environments, it exhibits significant limitations under high transaction volume. A key bottleneck is the lack of early dependency detection and parallelism during the commit phase. Fabric employs an optimistic concurrency model with versioning, where all transactions can proceed until final validation. Based on world-state version checks, conflicts are detected only during the commit stage. As a result, conflicting transactions are often rejected later in the commit stage, leading to increased retries and wasted compute resources.

Another constraint is Fabric's strictly sequential commit logic, which executes all transactions in a block one after another, even if many are independent. This approach underutilized multicore systems and limits concurrency.

2   Proposed Solution

We propose a dependency-aware execution mechanism for Hyperledger Fabric that improves parallelism and reduces transaction rejections. Dependencies are detected early during endorsement and leveraged in the commit phase to enable parallel execution while preserving consistency. Please refer to technical report [5] for detailed proposed solution.

As shown in Fig. 2, the client proposal is forwarded to a leader endorser, which simulates execution, flags dependencies, and returns this information with the endorsement. After ordering, the committer builds a DAG of transactions, executing independent ones in parallel and scheduling dependent ones accordingly. This design maximizes throughput via dependency-aware parallelism while ensuring ledger integrity. Key Enhancements of our proposed solution are:

Dependency Flagging at Endorsement: Leader endorser inspects read/write sets and flags transactions as dependent (`flag=1`) or independent (`flag=0`), enabling early conflict detection.

Propagation Through Ordering: Ordering services remain unchanged but preserve dependency metadata in blocks for committers.

DAG-based Parallel Commit: Committers construct a DAG per block; independent transactions run in parallel, dependents follow predecessors. Thread pools process DAG levels, with failed transactions triggering re-evaluation.

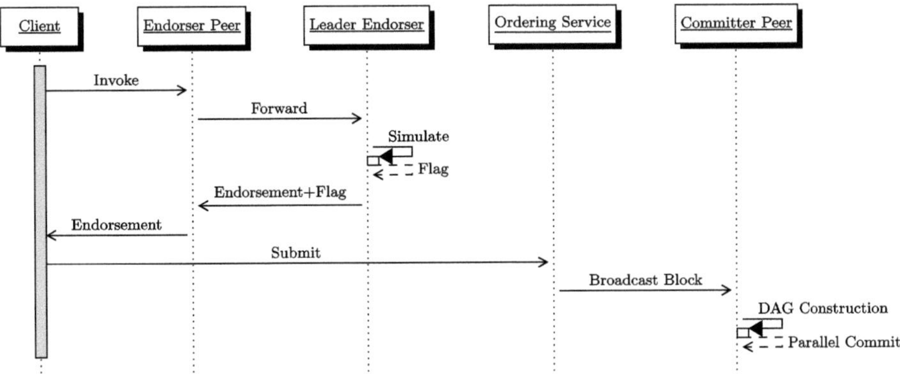

Fig. 2. Proposed Fabric transaction flow

Expiry and Cleanup: Flags and dependency metadata expire after a threshold to prevent performance issues; expired transactions must be resubmitted.

Architecture Overview: Figure 3 provides a high-level view of the modified architecture. The architecture retains Fabric's existing modular design. Clients continue to interact with the peer nodes through standard SDK interfaces. The only change is in how dependencies are internally detected, propagated, and processed. Endorsement flagging, DAG construction, and parallel validation are all encapsulated within the peer logic.

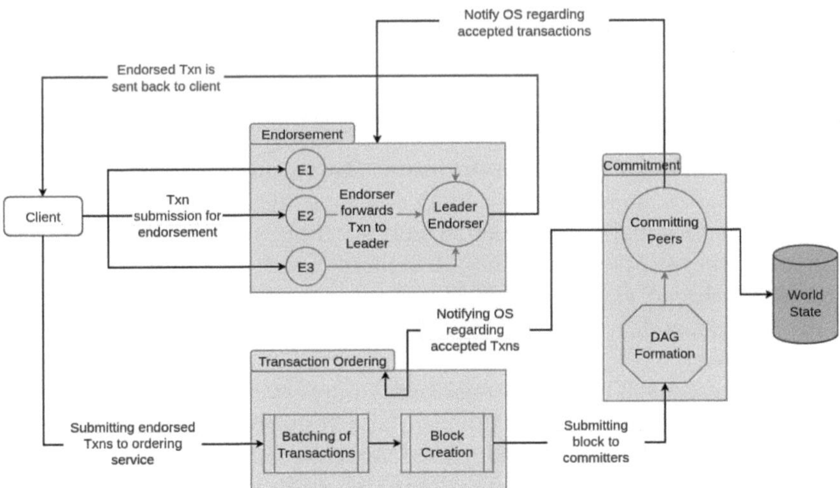

Fig. 3. Proposed DAG-aware Fabric Architecture

3   Performance Analysis

The experiments were conducted on a system with an AMD Ryzen 5 5500U processor (2.1 GHz, 6 cores/12 threads), 14 GB DDR4 RAM, and a 512 GB NVMe SSD, running Ubuntu 24.04 LTS. All results presented here are based on the Voting Contract. Please refer to technical report [5] for extended results.

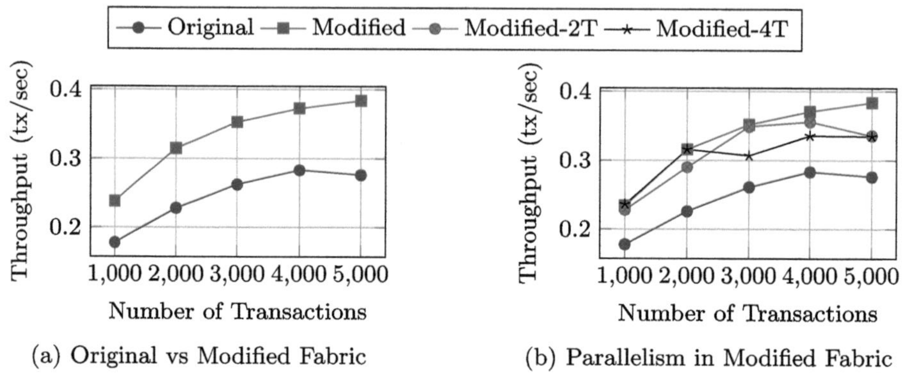

(a) Original vs Modified Fabric   (b) Parallelism in Modified Fabric

Fig. 4. Experiment 1: Impact of Number of Transactions on Throughput

Experiment 1: Impact of Number of Transaction on Throughput
This experiment illustrated in Fig. 4 evaluated the throughput across increasing transaction loads. At 5000 transactions, throughput improved from 0.276 tx/s to 0.384 tx/s (approx. 39% gain). Introduction of fixed-thread variants showed limited or mixed improvements. While 2-threaded execution showed some benefit, the 4-threaded variant often plateaued or underperformed due to thread contention or underutilization depending on DAG width.

Experiment 2: Impact of Dependency Ratio on Latency
Figure 5 evaluates performance under varying dependency ratios (0–0.9). Latency in the Original Fabric rises sharply with higher dependencies, while the Modified Fabric with Dynamic Threads remains lower and more stable. Fixed 2-thread and 4-thread variants achieve mid-range performance, with 4-threading best at high dependencies (e.g., 92 ms at 0.9) but not always surpassing the dynamic strategy.

Overall, Dynamic Threading consistently outperforms the baseline by adapting to DAG width, balancing parallelism and resource use, and demonstrating the benefits of adaptive concurrency in smart contract execution.

4   Related Work

Recent research has proposed various improvements for permissioned blockchains. Androulaki et al. [1] introduced Fabric's execute-order-validate

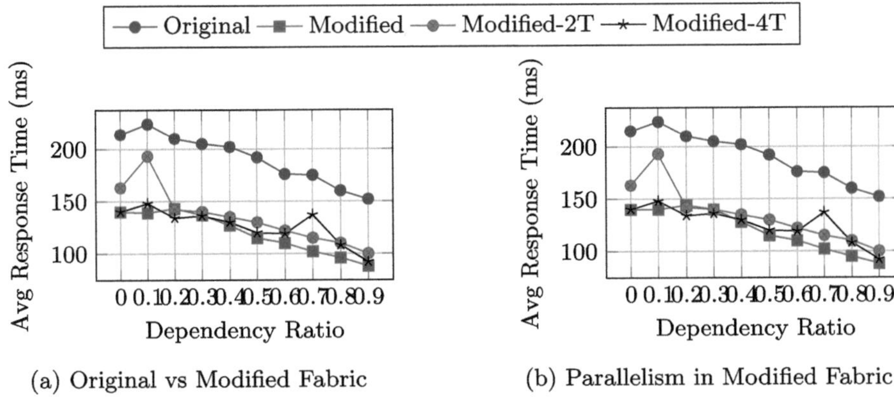

(a) Original vs Modified Fabric   (b) Parallelism in Modified Fabric

Fig. 5. Experiment 2: Impact of Dependency Ratio on Latency

architecture. DAG-based approaches have emerged as effective alternatives to improve concurrency [9]. For instance, BlockPilot [12] and RT-DAG [11] enable concurrent execution using dependency-aware scheduling. Constructing order restricted DAG for execution and sharing the DAG for validation stage are explored for generic blockchains framework [2,8]. Coloring the contructed DAG to rearrange the DAG levels to get optimized performance is introduced throug Batch-schedule-Execute [4] recently.

Particularly for Hyperledger Fabric, Fabric++ [7] and FastFabric [3] enhance throughput via endorsement and validation optimizations, though they retain sequential commitment. Public blockchain systems like Omniledger [6], Rapid-Chain [10], and NutBolt [13] introduce parallel models with distinct trust and consensus assumptions.

Unlike prior approaches that require significant architectural changes or new consensus layers, our method preserves full compatibility with Fabric's chain-code, client APIs, and consensus modules. We implement our modifications on Fabric v2.5 and validate them through experiments.

5   Conclusion and Future Work

Hyperledger Fabric's transaction model lacks concurrency-awareness, limiting performance under at high loads. We proposed a dependency-aware execution scheme that flags transaction dependencies, constructs dependency-preserving blocks, and applies DAG-guided execution, extending Fabric with minimal code changes. The implementation extends the existing Fabric codebase with minimal intrusion, ensuring compatibility and maintainability. Experiments show up to 25% higher throughput, consistently lower latencies, and improved commit rates under diverse loads. Future work will explore dynamic dependency detection through static analysis or machine learning to eliminate reliance on client-declared read/write sets.

References

1. Androulaki, E., et al.: Hyperledger fabric: a distributed operating system for permissioned blockchains. In: Proceedings of the Thirteenth EuroSys Conference, EuroSys 2018, pp. 30:1–30:15. ACM (2018). https://doi.org/10.1145/3190508.3190538
2. Anjana, P.S., Kumari, S., Peri, S., Rathor, S., Somani, A.: An efficient framework for optimistic concurrent execution of smart contracts. In: PDP, pp. 83–92 (2019)
3. Gorenflo, C., Lee, S., Golab, L., Keshav, S.: FastFabric: scaling hyperledger fabric to 20,000 transactions per second. In: Proceedings of the 17th USENIX Conference on Networked Systems Design and Implementation, pp. 125–140. USENIX (2020)
4. Hay, Y., Friedman, R.: Batch-schedule-execute: on optimizing concurrent deterministic scheduling for blockchains. In: 2024 43rd International Symposium on Reliable Distributed Systems (SRDS), pp. 163–174 (2024). https://doi.org/10.1109/SRDS64841.2024.00025
5. Kaul, S., Piduguralla, M., Patnala, G.S., Peri, S.: Dependency-aware execution mechanism in hyperledger fabric architecture (2025). https://arxiv.org/abs/2509.07425
6. Kokoris-Kogias, E., Jovanovic, P., Gailly, L., Gasser, L., Ford, B.: OmniLedger: a secure, scale-out, decentralized ledger. In: Proceedings of the 27th ACM Symposium on Operating Systems Principles, pp. 406–421. ACM (2020). https://doi.org/10.1145/3341301.3359631
7. Nasir, Z., Zhang, X., Yu, L., Ren, K.: Fabric++: optimizing smart contract execution in hyperledger fabric. In: 19th USENIX Symposium on Networked Systems Design and Implementation, pp. 251–266. USENIX (2022)
8. Piduguralla, M., Chakraborty, S., Anjana, P.S., Peri, S.: Dag-based efficient parallel scheduler for blockchains: hyperledger sawtooth as a case study. In: Cano, J., Dikaiakos, M.D., Papadopoulos, G.A., Pericàs, M., Sakellariou, R. (eds.) Euro-Par 2023. LNCS, vol. 14100, pp. 184–198. Springer, Cham (2023). https://doi.org/10.1007/978-3-031-39698-4_13
9. Wang, Q., Yu, J., Peng, Z., Bai, V.C., Guo, Z., Liang, W., Yang, G.: SoK: DAG-based blockchain systems. ACM Comput. Surv. **56**(5), 1–35 (2024). https://doi.org/10.1145/3576899
10. Zamani, M., Movahedi, M., Raykova, M.: RapidChain: scaling blockchain via full sharding. In: Proceedings of the 2018 ACM SIGSAC Conference on Computer and Communications Security, pp. 931–948. ACM (2018). https://doi.org/10.1145/3243734.3243853
11. Zhang, Z., Hong, C., Zhou, J., Ahmad, I., Zheng, Z., Chen, S.: RT-DAG: DAG-based blockchain supporting real-time transactions. IEEE Trans. Parallel Distrib. Syst. **35**(8), 1349–1363 (2024). https://doi.org/10.1109/TPDS.2024.3416751
12. Zhao, H., et al.: BlockPilot: a proposer-validator parallel execution framework for blockchain. In: Proceedings of the 52nd International Conference on Parallel Processing, ICPP 2023, pp. 641–651. ACM (2023). https://doi.org/10.1145/3605573.3605621
13. Zheng, S., Wang, J., Ren, K., Yuan, Y., Wang, B.: Nutbolt: efficient and scalable execution of smart contracts. In: Proceedings of the 13th USENIX Symposium on Operating Systems Design and Implementation, pp. 755–768. USENIX (2018)

Black Hole Search by Scattered Agents on Time-Varying Dynamic Graphs

Tanvir Kaur[1], Ashish Saxena[1], Partha Sarathi Mandal[2], and Kaushik Mondal[1(✉)]

[1] Indian Institute of Technology Ropar, Rupnagar 140001, Punjab, India
`kaushik.mondal@iitrpr.ac.in`
[2] Indian Institute of Technology Guwahati, Guwahati 781039, Assam, India

Abstract. A black hole is a malicious node in a graph that destroys resources entering into it without leaving any trace. The problem of Black Hole Search (BHS) using mobile agents requires that at least one agent survive and terminate after locating the black hole. Recently, this problem is studied on 1-bounded 1-interval connected dynamic graphs, where there is a footprint graph, and at most one edge can disappear from the footprint in a round, provided that the graph remains connected. In this setting, the authors proposed an algorithm that solves the BHS problem when all agents start from a single node (rooted initial configuration). They also proved that at least $2\delta_{BH} + 1$ agents are necessary to solve the problem when agents are initially placed arbitrarily across the nodes of the graph (scattered initial configuration), where δ_{BH} denotes the degree of the black hole. In this work, we present an algorithm that solves the BHS problem using $2\delta_{BH} + 17$ many initially scattered agents. Our result matches asymptotically with the existing rooted algorithm under the same model assumptions.

Keywords: Dynamic Graphs · Time-Varying Graphs · Black Hole Search · Mobile Agents · Distributed Algorithms · Deterministic Algorithms

1 Introduction

Mobile agents are software entities that work in networking environments. In real-life scenarios, these networking environments are vulnerable to various risks. A lot of research is concentrated on safeguarding network sites (hosts) against malicious agents and, conversely, defending agents from attacks by the hosts. In this work, we address a particularly dangerous type of host, referred to as a *black hole*, which is a network site that eliminates visiting agents immediately, leaving no trace. The presence of a black hole in systems that support code mobility is not uncommon. For example, a site can unintentionally become a black hole due to an unnoticed failure or if a virus, installed unknowingly, discards any incoming communication (such as marking it as spam). In these instances, identifying this harmful host is crucial. This problem of identifying such a harmful host in the graph through a team of mobile agents is known as the Black Hole Search (BHS) problem. Precisely, an agent is said to have located the black hole (thereby

S. Bonomi et al. (Eds.): SSS 2025, LNCS 16350, pp. 309–324, 2026.
https://doi.org/10.1007/978-3-032-11127-2_25

solving the BHS problem) if it terminates its algorithm at a safe node[1] v that is adjacent to the black hole and outputs the port number that leads from v to the black hole [4,5,15,17]. The BHS problem is extensively studied under various settings, influenced by agents' communication methods, synchronization levels, and graph knowledge. Most work assumes static graphs [2,6,8,11], but recent research explores dynamic graphs, including dynamic rings [17], cactuses [5], and tori [4]. Recently, this problem is investigated for dynamic graphs with an arbitrary underlying topology [15]. However, in [15], the authors provide an algorithm to solve BHS when the agents are in a rooted initial configuration. In this work, we consider the scattered initial configuration of the agents.

1.1   The Model and the Problem

Dynamic Graph Model: A dynamic network is modeled as a *time-varying graph (TVG)*, denoted by $\mathcal{G} = (V, E, T, \rho)$, where V is a set of nodes, E is a set of edges, T is the temporal domain, and $\rho : E \times T \to \{0, 1\}$ is the *presence function*, which indicates whether a given edge is present at a given time. Here, $\rho(e, t) = 0$ indicates that the edge $e \in E$ is not present at round $t \in T$, and $\rho(e, t) = 1$ indicates that the edge $e \in E$ is present at round $t \in T$. The static graph $G = (V, E)$ is referred to as the *underlying graph* (or *footprint*) of the TVG $\mathcal{G}$, where $|V| = n$ and $|E| = m$. For a node $v \in V$, let $E(v) \subseteq E$ denote the set of edges incident on v in the footprint. The *degree* of node v is defined as $\delta_v = |E(v)|$, and the *maximum degree* of G is given by $\Delta = \max_{v \in V} \delta_v$. The nodes in V are assumed to be *anonymous*, i.e., they don't have identifiers. Each edge incident to a node v is locally labeled with a *port number*. This is defined by a bijective function $\lambda_v : E(v) \to \{0, \ldots, \delta_v - 1\}$, which assigns a distinct label to each incident edge of v. No further assumptions are made about the labeling.

Under the assumption that time is discrete, the TVG $\mathcal{G}$ can be viewed as a sequence of static graphs $\mathcal{S}_{\mathcal{G}} = G_0, G_1, \ldots, G_t, \ldots$, where each $G_t = (V, E_t)$ denotes the *snapshot* of $\mathcal{G}$ at time t, with $E_t = \{e \in E \mid \rho(e, t) = 1\}$. The set of edges not present at time t is denoted by $\overline{E_t} = E \setminus E_t \subseteq E$. Based on this representation, we recall the following definition.

Definition 1 ([4,5,14–16]). *(1-Bounded 1-Interval Connectivity) A dynamic graph $\mathcal{G}$ is 1-interval connected (or always connected) if every $\mathcal{G}_t \in \mathcal{S}_{\mathcal{G}}$ is connected. Furthermore, $\mathcal{G}$ is said to be 1-bounded 1-interval connected if it is always connected and $|\overline{E_t}|$ is at most 1.*

In particular, suppose an edge e disappears from G at a round, resulting in the graph $G \setminus \{e\}$, which remains connected as per the definition of 1-Interval Connectivity. By the time another edge e' disappears, e must reappear as it is 1-Bounded. In this work, we consider 1-bounded 1-interval connected TVGs.

Agent Model: Initially, l many agents are scattered across the safe nodes of the graph. Note that, in a scattered configuration, one node may contain multiple

---

[1] A safe node is a node that is not the black hole.

agents as well. Each agent has a unique identifier assigned from the range $[1, n^c]$, where $c > 1$ is a constant. Each agent knows its ID and has no prior knowledge of l, or any other graph parameters like n, δ_{BH}, Δ. They are equipped with $O(\log n)$ bits of memory and have access to the whiteboard. When an agent visits a node, it knows the degree of that node in the footprint G. However, when an agent is at a node v, it cannot detect if any edge incident to v is missing. To be precise, say an agent a reaches a node v at round t and e_v be an edge associated with v, then the agent a can not determine $\rho(e_v, t)$. It can only understand this by traversing through the edge. If it tries to move through an edge and is unable to do so, it understands that the edge is missing. In other words, it understands that $\rho(e_v, t) = 0$ at the beginning of the next round $t+1$. Otherwise ($\rho(e_v, t) = 1$), it can successfully move through that edge. An agent learns the port number through which it enters into a node. If two agents move via an edge, they can not see each other during the movement.

Our algorithm runs in synchronous rounds. In each round, an agent performs one *Look-Compute-Move* (LCM) cycle at a safe node:

- *Look*: The agent reads the contents of the whiteboard of its occupied node and sees if there are other agents there. It can read all the variable values of all other agents present at the same node. The agent also understands whether it had a successful or an unsuccessful move in the last round.
- *Compute*: On the basis of the information obtained in the Look phase, the agent decides whether to move through the port or not in this round. The agent may write some information on the whiteboard as per the computation.
- *Move*: In compute, if an agent computes a port p to move, it tries to move through p; if the corresponding edge is present in that round, the agent reaches at the adjacent node; otherwise, it remains at the current node.

The time complexity of the algorithm is defined as the number of rounds until an agent finds the black hole in the graph G. Now we provide the problem definition.

Definition 2 *(1-BHS)*. Let $\mathcal{G}$ be a 1-bounded 1-interval connected TVG with a black hole located at some node in its footprint G. Initially, l agents are located at an arbitrary subset of the safe nodes in G. The agents have no prior knowledge of the graph or any of the global parameters. The goal is for at least one agent to survive, locate the black hole, and ultimately terminate.

1.2   Related Work

The BHS problem was originally introduced for static networks [10] and was later extensively investigated under various assumptions on agents' capabilities, memory limitations, and communication models [2,6–12,18]. More recently, research has shifted to dynamic graphs, with works addressing 1-BHS and f-BHS[2] in specific topologies such as rings [17], cactus graphs [5], and tori [4]. Another recent

---

[2] Similar to the 1-Bounded 1-Interval Connectivity, f-bounded 1-Interval connected dynamic graph $\mathcal{G}$ is the graph that is always connected and, for all $t \in T$, the number of missing edges satisfies $|\overline{E}_t| \leq f$. Note that, for f-bounded 1-interval connected

Table 1. A summary of the existing results and our results on general graphs. Here, IC denotes initial configuration and l denotes the number of agents.

	IC	Problem	l	Node storage	Agent memory	Time Complexity
[15]	Rooted	1-BHS	9	$O(\log n)$	$O(\log n)$	$O(m^2)$
		f-BHS	$6f$	$O(\log n)$	$O(\log n)$	$3\Delta^n(\Delta + 1)^{2f+n}$ $\cdot(n-1)^{2f}$
This work	Scattered	1-BHS	$2\delta_{BH} + 17$	$O(\log n)$	$O(\log n)$	$O(m^2)$

research development includes perpetual exploration of rings [13] and arbitrary graphs [3] in the presence of a Byzantine black hole.

Recently, the problems of 1-BHS and f-BHS are studied for general dynamic graphs in [15]. In this work, the authors show that 1-BHS cannot be solved with less than $2\delta_{BH} + 1$ agents arbitrarily placed at safe nodes, even when both agents and nodes have $O(\log n)$ bits of memory, and f-BHS cannot be solved with $2f + 1$ co-located agents despite unbounded memory and storage. While the paper presents algorithms for both models, these solutions assume a *rooted initial configuration*, where all agents begin at a common safe node. In this work, we consider a more general scenario in which agents are arbitrarily placed across safe nodes in the graph. We adopt the same dynamic graph model as in [15].

1.3   Our Contribution

We provide Algorithm 1-BHS_{scatt} which solves the 1-BHS problem using $2\delta_{BH} + 17$ scattered agents within $O(m^2)$ rounds. It requires $O(\log n)$ bits of storage per node, and each agent is equipped with $O(\log n)$ bits of memory. We match the asymptotic time complexity of the solution of 1-BHS from a rooted initial configuration in [15]. Table 1 illustrates a comparison of our results with [15].

1.4   Preliminaries

In this section, we recall the strategy which is used in the existing literature [15], known as the cautious movement of agents. For the sake of completeness, we provide the strategy of cautious movement by three agents.

Cautious Walk: Cautious walk is a movement strategy that ensures that one agent stays alive if a group of agents performs this movement while exploring an edge (u, v) from a node u without knowing that v is the black hole. Let a_1 (leader), a_2 (first helper), and a_3 (second helper) be the three agents at a safe node u and these agents want to traverse through the edge (u, v) to reach v.

---

graphs, $\rho(e, t)$ can take value 0 for at most f many edges $e \in E$ at a round $t \in T$. Analogously, if $\mathcal{G}$ is an f-bounded 1-interval connected dynamic graph with a black hole located at some node in its footprint G, then the black hole search problem is referred to as f-BHS.

Since the agents are not aware whether v is a safe node or a black hole, it is necessary to first verify this. Thus, the first helper a_2 attempts to move through (u, v). If a_2 is successful in moving through (u, v) at some round t, then in the round $t + 1$, a_3 attempts to move through (u, v) while a_2 (if alive) attempts to move through (v, u). If a_3 is successful in its movement and a_2 is not present at the current node, then a_1 concludes that v is a black hole. If a_3 is successful in its movement and a_2 is also present at u, then it is concluded that v is a safe node, and so, a_1 and a_2 move through the edge (u, v). In this way, cautious movement can be executed by three agents.

Idea of the Algorithm for Rooted Agents [15]: The authors first provide an exploration strategy using 3 agents that ensures the exploration of a TVG when there is no black hole. Further, they replace each edge movement of the exploration strategy with the cautious walk. The role of one agent is taken over by three agents that are sufficient to do the cautious movement. Thus, a total of 9 agents can solve 1-BHS on an arbitrary graph from a rooted configuration. Their algorithm solves 1-BHS using 9 agents starting from a rooted initial configuration in $O(m^2)$ rounds, requiring $O(\log \delta_v)$ storage per node.

Challenges with Fewer Agents: An intuitive approach to solve the 1-BHS problem is to first design an exploration strategy that guarantees every node of the graph is visited by at least one agent, even under adversarial edge deletions (when the graph is 1-bounded 1-interval connected). However, since the graph contains a black hole, it becomes essential that every movement performed by an agent is cautious. If agents move without caution, they may all fall into the black hole and die. Therefore, solving the 1-BHS problem requires a combination of two strategies: cautious movement and exploration of dynamic graphs.

We assume the graph to be 1-bounded 1-interval connected. As shown in [15], it is impossible to solve the 1-BHS problem on such graphs using only $2\delta_{BH}$ agents. Let us assume that we have $2\delta_{BH} + 1$ agents. Suppose $2\delta_{BH}$ agents are initially close by the black hole and all die soon after, keeping information on the adjacent nodes. However, since only one agent remains alive in the graph, the problem is solved only if this agent reaches one of those δ_{BH} nodes and reads the information. Unfortunately, the adversary can always prevent this agent's movement by making an edge disappear, and hence it might be impossible to guarantee exploration. Now, let us assume that we have $2\delta_{BH} + 3$ agents, and so there are 3 alive agents even after $2\delta_{BH}$ agents die. Moreover, let us assume that those 3 alive agents are together. We know 3 rooted agents can explore 1-bounded 1-interval graph if there is no black hole as provided in [15]. Note that the death of $2\delta_{BH}$ agents ultimately isolates the black hole, if the information of all the ports corresponding to the edges leading to the black hole is kept at those nodes. The problem is that those 3 agents do not know when to start the algorithm of exploration by 3 agents, as they do not know if and when all the remaining $2\delta_{BH}$ agents die. Further, as we start from a scattered initial configuration, those 3 agents may not be together, which adds to the difficulty. We are unable to provide a tighter bound than that of $2\delta_{BH}$ with scattered agents as provided in [15] as of now.

We start with a number of extra agents c such that the adversary is either bound to form a group of 9 agents, where any such group can deterministically transition to the existing algorithm of rooted 1-BHS without bothering about the situation of the other agents. If such a group is not formed, then at least $2\delta_{BH} + c'$ $(c' \leq c)$ agents explore the graph, where at most $2\delta_{BH}$ may die, and at least one from the remaining agents finds the black hole.

2   Algorithm to Solve 1-BHS

In this section, we provide an algorithm using $2\delta_{BH} + 17$ agents to solve 1-BHS when the agents are initially present arbitrarily at the safe nodes of G.

2.1   High-Level Idea of Our Algorithm

Let us first consider that there is sufficient storage at each node to store the information corresponding to all the $2\delta_{BH} + 17$ agents. Each agent, in this case, simply runs its own DFS unless it is stuck by a missing edge or dies in the black hole. If agents do not move cautiously, all the agents may die in the black hole. The issue is that the agents cannot implement the cautious movement strategy outlined in Sect. 1.4. This is due to the fact that in order to use this strategy, three agents must be co-located, but we consider scattered initial configuration here. Thus, each agent does the cautious movement individually. The movement strategy used by each agent is *'Mark and Move'* - *'Return'* - *'Delete and Move'*. Basically, when an agent a_i decides to move through a port, it writes this information at its current node u and moves through that port (if the corresponding edge is available). After reaching the new node v, if it survives (v is not a black hole), it returns to u (if the corresponding edge is available), removes the marked port information written at u, and finally safely moves to the new node v, provided the corresponding edge is available. Now, if the visited node v is a black hole, then the information written on the previous node u stays put. This is a warning that can be seen by some other agent that visits u. We call this type of movement the Individual Cautious Movement (ICM). The interesting part is that if two agents moved through the same port, say p, in different rounds, and have marked at the node u about p, then the third agent that visits u understands that port p leads to a black hole. However, a single mark of information is not enough for the second agent to understand the black hole, as it might be the case that the first agent could not return to u due to the unavailability of the corresponding edge. However, information about two marked ports is enough for the third agent to understand the presence of the back hole, since if it were due to the unavailability of the port, then the second agent could not have left through the same port p. More specifically, even if the corresponding edge became available only for one round and the second agent left in that round to v, then the first agent must be able to come back from v in that round, and therefore only one marked information should be there at u which is due to the second agent.

The above ICM strategy ensures that at most $2\delta_{BH}$ many agents can die in the black hole. Each individual agent runs its own DFS traversal unless it is stuck by a missing edge. Proceeding in this manner, let us suppose that $2\delta_{BH}$ many agents die in the black hole. Since each agent performs ICM, there are markings at each of the δ_{BH} nodes adjacent to the black hole. Thus, the problem is solved if at least one agent visits one of these δ_{BH} nodes. However, since there are only 17 agents left in the graph and none of them can die, the movement of these agents can only be blocked by the missing edge. Whenever a group of at least 9 agents is formed, they initiate the algorithm for the rooted case as described in [15] and outlined in Sect. 1.4. A missing edge (u, v) can block at most 16 agents, 8 agents at each of the nodes u and v, while ensuring that no group of 9 agents is formed. Hence, in this case, at least one agent proceeds according to its DFS traversal in each round. Thus, it eventually visits one of the δ_{BH} nodes that have some sort of marked information by the two dead agents, leading to the solution of the 1-BHS problem.

Now, our objective is to limit the storage at each node to $O(\log n)$ bits. Thus, information corresponding to only a constant number of agents can be stored at each node. We utilize the same idea as mentioned above to search for the black hole. The difference is to restrict the number of DFS traversals being run by the agents on the graph. This is done in the following manner. When an agent visits a node, it checks the ID corresponding to the information written on the whiteboard. If no such information is present, then it continues its traversal. However, if it finds that some smaller ID agent has already visited the current node, then it dismisses its own DFS traversal and follows the path followed by this smaller ID agent. But if the information corresponds to some larger ID agent than its own ID, then it continues its traversal by overwriting the information on the whiteboard. By proceeding according to this, either a group of 9 agents is formed eventually or at least one agent reaches one of the δ_{BH} nodes. In both scenarios, we have our solution to the problem of 1-BHS. This explains the overall idea of our algorithm.

2.2   The Algorithm: 1-BHS_{scatt}

We first provide a description of the parameters maintained by each agent a_i.

- $a_i.state$: This is a binary variable. It can take values either *explore* or *backtrack*. It is initialized to *explore*.
- $a_i.success$: This is a binary variable that can take values either *True* or *False*. This is required to identify whether the movement of a_i in the previous round was successful or not. It is initially set to *True*.
- $a_i.pout$: It stores the port that will be used by the agent to exit from the node in the current round. It can take values from $\{-1, 0, 1, \ldots, \delta_v - 1\}$, where δ_v is the degree of the current node v. Initially, it is set to -1.

- $a_i.pin$: It stores the port used by the agent to enter into the node at the beginning of the current round. It can take values from $\{-1, 0, \ldots, \delta_v - 1\}$ where δ_v is the degree of the current node v. It is initially set to -1.
- $a_i.ID$: This variable stores the ID of the agent.
- $a_i.grp$: If an agent is part of a group of at least 9 agents, it updates $a_i.grp$ to $True$. It is initialized to $False$.
- $a_i.grp\_ID$: If an agent is part of a group of at least 9 agents, it updates $a_i.grp\_ID$ to the smallest ID agent present in the group. Initially set to $\perp$.

Agent a_i stores and updates information on the whiteboard. The parameters maintained by the agent a_i on the whiteboard are as follows:

- $wb_v(a_i.parent)$: This parameter stores the port at v from which from which a_i entered node v for the first time while running its DFS. This can take value from $\{-1, 0, 1, \ldots, \delta_v - 1\}$.
- $wb_v(a_i.recent)$: This parameter stores the last port at v used by the agent a_i to continue its traversal.
  The information $wb_v(a_i.parent)$ and $wb_v(a_i.recent)$ together constitutes the DFS information corresponding to the agent a_i. For the sake of simplicity, we call this information the <u>travel information</u>.
- $wb_v(marked_1) = (a_i.pout, a_i.ID)$: Agent a_i writes its ID and its computed value of $pout$ that it uses to exit from node v. Initially, it is set to $\perp$. We denote the agent a_i whose information is written at $wb_v(marked_1)$ with A_1.
- $wb_v(marked_2) = (a_i.pout, a_i.ID)$: The agent a_i writes its ID and computed value of $pout$ that it uses to exit from v. When there is already some information written at $wb_v(marked_1)$, and some other agent wants to either move through the same port or any other port, it writes its information at $marked_2$. Note that the values of $recent$ port and $marked$ port may be the same for an agent a_i. However, we use these two parameters for two separate purposes in the algorithm. We denote the agent a_i whose information is written at $wb_v(marked_2)$ with A_2.
  For the sake of simplicity, we call this information the <u>marked port information</u>.
- $wb_v(grp)$: This parameter can take values either $True$ or $False$. It helps to identify whether a group of at least 9 agents is formed or not. Initially, it is set to $False$. Once the value is set to $True$, it never changes to $False$.
- $wb_v(grp\_ID)$: If a group consisting of at least 9 agents is formed, then whenever any agent a_i from this group visits node v, it updates this parameter with the value of its $a_i.grp\_ID$.

Let $2\delta_{BH} + 17$ many agents be initially positioned arbitrarily at the safe nodes of G. If there is at least one group of at least 9 agents in the initial configuration, then they can proceed with the rooted algorithm from [15]. The problem of 1-BHS is solved in this scenario. Hence, let us assume that there is no group of at least 9 agents in the initial configuration.

Our algorithm operates in alternating odd and even rounds. In odd rounds, agents perform a specific set of actions, while in even rounds, they execute a different set of operations. Let an agent a_i be present at a node v in round t. If

$t \bmod 2 = 1$, then a_i updates its $a_i.success = True$ if its attempt to move in the round $t - 1$ was successful. Otherwise, if its attempt to move in the round $t - 1$ was unsuccessful, then it sets $a_i.success = False$. On the other hand, if $t \bmod 2 = 0$, then a_i performs its movement based on the various cases as described below. Before proceeding with the specific cases, we describe ICM.

Individual Cautious Movement (ICM): Let a_i be present at v. Let us suppose that a_i decides to exit from v by performing ICM via port p_i at round t to reach the adjacent node, say v'. It writes its marked port information at v and moves through the port p_i at the end of round t. In case the move fails due to a missing edge, it removes the marked port information in the subsequent odd round and starts ICM freshly from the next even round. After successfully arriving at the adjacent node v' at some round $t' \geq t$, a_i (if alive) returns to v and deletes its marked port information at v. After deleting the marked port information, a_i finally moves through port p_i to reach the adjacent node safely. Once a_i reaches v' safely, it is said to complete ICM. Now we are ready to provide the details of our algorithm.

(I) We first describe the following four cases when an agent a_i is not yet a part of a group of at least 9 agents, i.e., $a_i.grp = False$.

(A) If a_i is at node v in *explore* state after completing its ICM and $a_i.success = True$. In this scenario, we have the following two cases:

A.1 *There are no marked port information at the current node v-* This implies that the values of both $marked_1$ and $marked_2$ are $\perp$. Agents check if there is any travel information corresponding to a_j such that $a_j.ID$ is smaller than all the agents present at the current node. If information corresponding to such a_j is present, then a_i moves through the port $a_j.recent$, where this value is extracted from the whiteboard. Otherwise, if there is no such travel information present at v, then a_i proceeds as follows. If $a_i.ID$ is the smallest among all the agents at v, then a_i proceeds with its own DFS traversal via ICM.[3] On the other hand, if $a_i.ID$ is not the smallest among all the agents at v^4, then a_i waits at the current node.

A.2 *There is one marked port information at the current node v-* Without loss of generality, let $marked_1 \neq \perp$ and the information is written there corresponding to an agent A_1. If A_1 is not present at v, then a_i proceeds based on the following cases. If $a_i.ID$ is smaller than all the agents at the current node, as well as smaller than $A_1.ID$, then a_i continues its DFS via ICM. If $a_i.ID$ is the smallest among all the agents at the current node, but $a_i.ID > A_1.ID$, then a_i sets $a_i.pout = A_1.pout$ (where $A_1.pout$ is extracted from $wb_v(marked_1)$), and attempts to move through this port by writing

---

[3] Note that by "proceeding with DFS via ICM," we mean that the agent writes its traversal information at the current node according to the DFS traversal of a_i. Moreover, if the agent needs to explore from the current node, the use of ICM is required; however, if it needs to backtrack, ICM is not necessary as it is definite that the node is a safe node.

[4] The comparison among agent IDs is performed only among those agents that are present at the current node after completing their ICM.

its information in $wb_v(marked_2)$. If a_i is not the smallest ID agent at the current node, then it waits at the current node.

On the other hand, if A_1 is present at v after verifying that the adjacent node it had to visit is safe (i.e., $A_1.success = True$ and A_1 has entered into v via $A_1.recent$ port), then a_i proceeds as follows. If A_1 is the smallest ID agent from all the agents at v and the ID of the agent whose travel information is written at v (if any), then a_i sets $a_i.pout = A_1.pin$ and moves through $a_i.pout$. Otherwise, if there is travel information at v corresponding to an agent a_j such that $a_j.ID <$ all the agents at v, then a_i sets $a_i.pout = a_j.recent$ where the value of $a_j.recent$ is extracted from wb_v. Agent a_i moves through its computed value of $a_i.pout$. If no such travel information corresponding to a_j is found at v, then if a_i is the smallest ID agent at v, it continues its DFS via ICM. Else, a_i waits at the current node.

A.3 *Both the marked port information are present at v-* This implies both $marked_1 \neq \bot$ and $marked_2 \neq \bot$. Let the information written in $marked_1$ and $marked_2$ correspond to A_1 and A_2 respectively. We have three cases.

→ If neither of A_1 and A_2 is present at v (to delete their respective *marked* port information), then a_i checks if $marked_1 = marked_2$, then black hole is found. Otherwise, if $marked_1 \neq marked_2$, then a_i waits at the current node.

→ If one of A_1 and A_2 is present at v to delete its marked port information, then a_i proceeds according to the following. Without loss of generality, let A_1 be present at v. If $A_1.ID$ is the minimum among all the agents at v and the travel information present at v, then a_i moves along with A_1 by setting $a_i.pout = A_1.pin$. If $A_1.ID > A_2.ID$ and $A_2.ID$ is smaller than all the agents present at the current node, then a_i proceeds based on the following two cases, (a) If a_i is the smallest ID agent among all the agents at the current node, then a_i sets $a_i.pout$ same as the $marked_2$ (corresponding to A_2). Agent a_i then writes its information in $marked_1$ and attempts to move through its port $a_i.pout$ by performing ICM; (b) If a_i is not the smallest ID agent among all the agents at the current node, then a_i waits at node v. Now, if $A_1.ID > A_2.ID$ and $A_2.ID$ is not the smallest among all the agents at the current node, then if a_i is the smallest ID agent among all the agents at v, then it proceeds with its own DFS via ICM by writing its information in $marked_1$. Otherwise, a_i waits at the current node.

→ If both A_1 and A_2 are present at v to delete their marked port information, this means both the *marked* port information will be deleted by the respective agents. This case thus reduces to case 1 when both the *marked* port information are $\bot$.

This completes the algorithm of an agent a_i when it is present at a node in the *explore* state after completing its ICM.

(B) If a_i is at node v in *explore* state after completing its ICM and $a_i.success = False$. For this scenario, we have the following two cases:

B.1 *There is no agent a_j with $a_j.success = True$ and a_j has completed its ICM-* This means no new agent has entered into the current node. Thus, a_i continues its attempt to move through its computed *pout* value.

B.2 *There is at least one agent a_j present at v with $a_j.success = True$ and a_j has completed its ICM-* If a_i has the minimum ID among all agents at v, then it re-attempts to move through $a_i.pout$. Otherwise, it waits at v.

In this case, *marked* port information checking is not required because an agent first arrives at v with $success = True$ and then writes its *marked* port information. If there was already any *marked* port information at v, then the decision was taken by a_i based on that information according to the case (A) above.
(C) If a_i is at node v with $a_i.state = backtrack$ and $a_i.success = True$. For this scenario, we have the following two cases.

C.1 There is no information related to *marked* ports at the current node v- In this case, a_i checks if there is travel information corresponding to an agent a_j such that $a_j.ID <$ IDs of all agents at v. If such travel information is present at v, then a_i sets $a_i.pout = a_j.recent$ where the value of $a_j.recent$ is extracted from wb_v. If such travel information is not present, then a_i compares its ID with the IDs of other agents present at the current node. If $a_i.ID$ is the smallest among all the agents present at v, then a_i continues its DFS via ICM. Otherwise, if a_i is not the smallest ID agent present at v, then a_i waits at the current node.

C.2 There is one information related to *marked* port at v- Without loss of generality, let the information be present in $marked_1$ and it corresponds to the agent A_1. In this scenario, we have two cases.
→ If A_1 is present at v to delete its marked port information at v, then a_i checks if it is the smallest ID agent among all the agents at v. If yes, then a_i continues its own DFS. Otherwise, if a_i is not the smallest ID agent among all the agents at v, then it waits at the current node.
→ If A_1 is not present at v, then a_i checks if it is the smallest ID agent among all agents at v and smaller than $A_1.ID$ as well, then a_i continues its DFS via ICM. Otherwise, if $a_i.ID > A_1.ID$ and $a_i.ID$ is smaller than all agents at v, then a_i stops its DFS, sets $a_i.pout = A_1.pout$ (value of $A_1.pout$ is extracted from $wb_v(marked_1)$), and attempts to move through the port $a_i.pout$ by performing ICM and writing its information in $marked_2$. Otherwise, if a_i is not the smallest ID agent, then a_i waits at v.

C.3 Both the *marked* port information are present at v- Let the information of $marked_1$ correspond to A_1 and the information of $marked_2$ correspond to A_2. We have the following three subcases.
→ If neither of A_1 or A_2 is present at v to delete their marked port information, then a_i waits at v if $marked_1 \neq marked_2$. If $marked_1 = marked_2$, then black hole is found.
→ If one of A_1 and A_2 is present at v to delete its marked port information, then a_i decides based on the following scenarios. Without loss of generality, let A_1 be present at v. If $A_1.ID$ is the minimum from all agents at v, including the IDs of the agents written at travel information at v (if any), then a_i sets $a_i.pout = A_1.pout$ (where $A_1.pout$ is extracted from $wb_v(marked_1)$) and moves along with A_1. Otherwise, if $A_2.ID$ is the smallest among all agents at v, then a_i moves through $A_2.pout$ (where the value of $A_2.pout$ is

extracted from $wb_v(marked_2)$) by writing its information in $marked_1$ if a_i is the smallest ID agent among all the agents at v. If a_i is not the smallest ID agent at v, then it waits at the current node. The final case is if there is another agent a_j at v after completing its ICM that is smaller than all the agents at v, including the travel information, then a_i stops the execution of its DFS and waits at the current node v. If a_i is the smallest ID agent at v, it performs its DFS via ICM by writing marked port information at $marked_1$.

$\rightarrow$ If both A_1 and A_2 are present at v to delete their marked port information, and either of them is the minimum from all the agents present at the current node, as well as the travel information, then a_i moves along with it. Otherwise, if a_i is the smallest ID agent present at v, then it continues its DFS via ICM. Otherwise, a_i waits at the current node.

(D) If a_i is at node v with $a_i.success = False$ and $a_i.state = backtrack$: In this case, a_i continues to execute its own DFS if there is no new agent with $success = True$ after completing its ICM present at v with ID smaller than $a_i.ID$. Otherwise, a_i waits at the current node.

(II) Algorithm of agent a_i when it is part of a group with at least 9 agents ($a_i.grp = True$): At a round, say t, during the execution of our algorithm, at least 9 agents (say x many agents) are present at a node v. From these x agents, let p many agents be such that they have to return to their previous nodes to delete their *marked* port information on their respective nodes (i.e., they have not yet completed their ICM). If $x - p \geq 9$, then there is a group of at least 9 agents that have completed their ICM. These agents initiate the Rooted Algorithm as described in [15]. Each agent within this group sets $a_i.grp = True$ and, while writing its travel information at any visited node v, it includes this information on the whiteboard by writing $wb_v(grp) = True$. When another agent, say a_j with $a_j.grp = False$ (indicating that it has not started the rooted algorithm and thus is not a part of any group that is executing the rooted algorithm) visits such a node where information is written corresponding to an agent that has $grp = True$ value, a_j understands that a group of 9 agents has initiated the rooted algorithm. The inclusion of this information serves an important purpose: if an agent that is either alone or part of a group with less than 9 agents encounters a node containing this information (travel information corresponding to an agent a_i with $a_i.grp = True$), it terminates its execution of the algorithm, as it is definite that the group of at least 9 agents will solve the 1-BHS problem. It may be the case that more than one group of 9 agents are formed that have initiated the rooted algorithm to solve the problem of 1-BHS. Since each node has $O(\log n)$ bits of memory, the information corresponding to each of these groups cannot be stored at the whiteboard. Thus, to resolve this issue, when a group of 9 agents is formed, each agent a_i that is a part of the group stores the ID of the agent that has the smallest ID among the group in the parameter $a_i.grp\_ID$. When these agents update their information on the whiteboard (as per the rooted algorithm of 1-BHS), they also update the value of grp_ID on the whiteboard of the visited node. When a group (say G_1) visits a node where information corresponding to another group (say G_2) is written, such that grp_ID of the agents in G_1 is less than the grp_ID of the agents in G_2, then G_2 stops its execution of its algorithm.

Now suppose that $x - p < 9$. In this case, all agents at the current node understand that a group of 9 agents can be formed. Thus, these $x - p$ agents remain at current node. The remaining p agents move to their previous nodes (say at round t_1) in order to delete their respective *marked* port information. Since at most one edge can disappear, at most one agent from these p many agents may not be able to move from the current node to its previous node to delete its *marked* port information. Even if all the p agents successfully move to their previous nodes, at most one agent may not be able to return to the current node. Thus, the $x - p$ agents wait at the current node unless a group of 9 agents is formed at the current node, consisting of all the agents that have completed their ICM. This completes the description of Algorithm 1-BHS_{scatt}. Let us proceed with the correctness and analysis of Algorithm 1-BHS_{scatt}.

Remark 1. Unlike the standard 1-BHS problem definition, one may argue that the problem of 1-BHS is solved only when it is guaranteed that no more alive agents will die in the black hole. This stronger requirement can be met by ensuring that all ports leading to the black hole are marked and that at least one agent remains alive. Assuming 3-connectivity of the graph and that the agents know the degree $\delta_{v_{BH}}$ of the black hole node, this stronger version can be solved using $14\delta_{v_{BH}} + 5$ agents. The details are deferred to the full version.

2.3   Correctness and Analysis

In this section, we show the correctness of Algorithm 1-BHS_{scatt}. Let the total number of agents in the graph be l, i.e., $l = 2\delta_{BH} + 17$. In the analysis, we denote the ith smallest ID agent with α_i. Hence, $\alpha_1.ID < \alpha_2.ID < \ldots < \alpha_l.ID$. As per our algorithm, if agent a_i meets a_j or it sees the information written by some agent a_j, and $a_i.ID > a_j.ID$, then agent a_i stops its own DFS and replicates the movement of agent a_j. Based on this, we have the following results. The proofs of Lemmas 1, 2, 5 and 6 are deferred to the full version.

Lemma 1. *Algorithm 1-BHS_{scatt} never instructs more than one agent to write travel information on the whiteboard of any node at any round.*

Lemma 2. *As per Algorithm 1-BHS_{scatt}, an agent writes its marked port information on the whiteboard only if at least one of its marked ports is $\perp$.*

Observation 1. *Suppose a node v has non-empty information on both of its marked ports at round t. If, in the next even round following t, none of the agents associated with these information return, then it must be the case that one of the two marked ports at v leads to the black hole.*

Lemma 3. *The agent α_1 never deviates from its original DFS path unless it becomes part of a group comprising at least 9 agents. The agent α_1 may encounter one of the following three scenarios: (i) it dies upon entering the black hole, (ii) it gets stuck at a node due to a missing edge, or (iii) it waits at a node if both marked ports at that node are different from $\perp$.*

Proof. Since α_1 is the agent with the smallest ID, it never follows the DFS traversal of any other agent. Therefore, it continues its own DFS traversal unless it becomes part of a group of 9 agents. During its traversal, α_1 may enter the black hole, leading to case (i). Alternatively, it may get stuck at a node due to a missing edge, which corresponds to case (ii). The final possibility arises when α_1 is at a node where the whiteboard contains information about both marked ports, each corresponding to another agent. By Lemma 2, Lemma 1, and Observation 1, at most two marked port entries can be present on any node's whiteboard. When both entries are occupied, α_1 cannot record its own marked port information and thus waits at the current node, satisfying case (iii). Note that even if α_1 is later able to record its own information at such a node in a subsequent round, it continues its DFS traversal without deviation from its original DFS path. This completes the proof. $\qquad\square$

Observation 2. *If the movement of more than 16 agents is blocked due to a missing edge, then a group of 9 agents is definitely formed.*

Remark 2. By <u>completing an agent's movement</u>, we mean that either the agent died in the black hole or the agent determined the location of the black hole by reaching a neighbor v of the black hole and outputting the correct port that leads from v to the black hole.

Lemma 4. *An agent either dies in the black hole or finds the black hole by performing at most $12lm$ successful movements.*

Proof. While performing DFS, an agent traverses each edge at most four times [1,15]. Therefore, an agent requires at most $4m$ successful movements to complete its DFS traversal. Each movement in the DFS is executed using a cautious movement, which consists of 3 individual movements. Therefore, an agent completes its DFS traversal, including the cautious movement strategy, in at most $12m$ successful movements. According to our algorithm, an agent α_i may need to perform its traversal up to i times because there are $i-1$ agents with smaller IDs than α_i, each executing their own DFS traversals. In the worst case, agent α_i may sequentially follow the traversals of all these preceding agents before eventually following α_1. By Observation 3, α_1 never follows any other agent. Consequently, in the worst case, agent α_l may require at most $12m \cdot l$ successful movements to complete its traversal. This concludes the proof of the lemma. $\square$

Remark 3. Since agents perform their movements only during even rounds in our algorithm, an agent either dies in the black hole or identifies it within at most $24lm$ rounds, where each round corresponds to a successful movement attempt.

Theorem 1. *Algorithm 1-BHS$_{scatt}$ solves 1-BHS in $O(m^2)$ rounds.*

Proof. During the execution of our algorithm, there are two possible scenarios for the agents: (i) no such group is formed throughout the execution, or (ii) at least one group of 9 agents is formed.

Case (i): Let us suppose that a group of 9 agents is not formed during the run of the algorithm. In this scenario, we claim that 1-BHS is solved in $152m\delta_{BH}$ rounds. In other words, if a group of 9 agents is not formed in $152m\delta_{BH}$ many rounds, the black hole is found. To prove the claim, we guarantee that at least $2\delta_{BH} + 1$ many agents complete their movement by $152m\delta_{BH}$ rounds. As per Lemma 4, an agent can have at most $12lm$ successful movements. Hence, in the worst case, cumulatively $(2\delta_{BH} + 1) \cdot 12lm$ successful movements are needed for $2\delta_{BH} + 1$ many agents to either die in the black hole or to locate it. Since the remaining 16 agents may not complete their movement, this implies at most $16(12lm - 1)$ many successful movements cumulatively can be achieved by these 16 agents. Hence, the cumulative successful movement of all agents must be at most $(2\delta_{BH} + 1) \cdot 12lm + 16 \cdot (12lm - 1)$. However, the adversary can make an edge disappear to delay our algorithm. As per Observation 2, a missing edge can stop at most 16 agents. Hence, at every even round, $2\delta_{BH} + 17 - 16 = 2\delta_{BH} + 1$ successful movements occur. Let T be the required number of even rounds in the worst case to complete the traversal of $2\delta_{BH} + 1$ many agents. Hence, $T = \frac{(2\delta_{BH}+1)\cdot 12lm + 16\cdot(12lm-1)}{2\delta_{BH}+1} \leq 12lm + \frac{16(12lm)}{3} = 76lm$. Thus, if no group of 9 agents is formed, then 1-BHS is solved in $152m\delta_{BH}$ rounds. A black hole is located in this scenario when $2\delta_{BH}$ agents die in the black hole by marking its neighbors, and one agent visits one of these marked neighbors.

Case (ii): When a group of 9 agents is formed, the agents initiate the algorithm proposed in [15], which completes in $O(m^2)$ rounds. If this group encounters information about any other group of 9 agents that has also initiated the algorithm of [15], a comparison of the corresponding grp_ID values is performed. The group with the smaller grp_ID continues the execution while disregarding any information written by the group with the larger grp_ID. Conversely, the group with the larger grp_ID terminates its execution. This mechanism ensures that the group with the smallest grp_ID value ultimately locates the black hole. Hence, in the worst case, agents require $O(m\delta_{BH} + m^2) = O(m^2)$ rounds to solve the problem of 1-BHS. $\qquad\qquad\qquad\qquad\qquad\qquad\qquad\qquad\qquad\qquad\qquad\square$

Lemma 5. *Storage of $O(\log n)$ bits is required to run the Algorithm 1-BHS_{scatt}.*

Lemma 6. *Each agent requires $O(\log n)$ bits of memory to execute the Algorithm 1-BHS_{scatt}.*

Hence, our main theorem follows from Theorem 1, Lemma 5, and Lemma 6.

Theorem 2. *Algorithm 1-BHS_{scatt} solves the 1-BHS problem using $2\delta_{BH} + 17$ agents within $O(m^2)$ rounds. It requires $O(\log n)$ bits of storage per node, and each agent is equipped with $O(\log n)$ bits of memory.*

3   Conclusion

Our Algorithm 1-BHS_{scatt} solves the 1-BHS problem starting with $2\delta_{BH} + 17$ many scattered agents. Solving the f-BHS problem in polynomial time is another direction of study. One can also look for solutions on weakly-connected dynamic graph models like temporal graphs.

References

1. Augustine, J., Moses Jr., W.K.: Dispersion of mobile robots: a study of memory-time trade-offs. In: ICDCN, pp. 1:1–1:10 (2018)
2. Balamohan, B., Flocchini, P., Miri, A., Santoro, N.: Time optimal algorithms for black hole search in rings. Discret. Math. Algorithms Appl. **3**(4), 457–472 (2011)
3. Bhattacharya, A., Goswami, P., Bampas, E., Mandal, P.S.: Perpetual exploration in anonymous synchronous networks with a Byzantine black hole. In: DISC 2025, Berlin, Germany, 27–31 October. LIPIcs (2025)
4. Bhattacharya, A., Italiano, G.F., Mandal, P.S.: Black hole search in dynamic tori. In: SAND, vol. 292, pp. 6:1–6:16 (2024)
5. Bhattacharya, A., Italiano, G.F., Mandal, P.S.: Searching for a black hole in a dynamic cactus. J. Graph Alg. Appl. **29**(2), 127–166 (2025)
6. Chalopin, J., Das, S., Labourel, A., Markou, E.: Black hole search with finite automata scattered in a synchronous torus. In: DISC, pp. 432–446 (2011)
7. Czyzowicz, J., Kowalski, D.R., Markou, E., Pelc, A.: Complexity of searching for a black hole. Fundam. Informaticae **71**(2–3), 229–242 (2006)
8. Czyzowicz, J., Kowalski, D.R., Markou, E., Pelc, A.: Searching for a black hole in synchronous tree networks. Comb. Probab. Comput. **16**(4), 595–619 (2007)
9. Dobrev, S., Flocchini, P., Prencipe, G., Santoro, N.: Searching for a black hole in arbitrary networks: optimal mobile agent protocols. In: PODC, pp. 153–162 (2002)
10. Dobrev, S., Flocchini, P., Prencipe, G., Santoro, N.: Searching for a black hole in arbitrary networks: optimal mobile agents protocols. Distributed Comput. **19**(1), 1–35 (2006)
11. Dobrev, S., Flocchini, P., Santoro, N.: Improved bounds for optimal black hole search with a network map. In: SIROCCO, pp. 111–122 (2004)
12. Dobrev, S., Santoro, N., Shi, W.: Using scattered mobile agents to locate a black hole in an un-oriented ring with tokens. Int. J. Found. Comput. Sci. **19**(6), 1355–1372 (2008)
13. Goswami, P., Bhattacharya, A., Das, R., Mandal, P.S.: Perpetual exploration of a ring in presence of Byzantine black hole. In: OPODIS 2024, Lucca, Italy, 11–13 December, volume 324 of LIPIcs, pp. 17:1–17:17 (2024)
14. Gotoh, T., Flocchini, P., Masuzawa, T., Santoro, N.: Exploration of dynamic networks: tight bounds on the number of agents. J. Comput. Syst. Sci. **122**, 1–18 (2021)
15. Kaur, T., Saxena, A., Mandal, P.S., Mondal, K.: Black hole search in dynamic graphs. In: ICDCN (2025)
16. Kuhn, F., Lynch, N., Oshman, R.: Distributed computation in dynamic networks. In: Proceedings of the Forty-Second ACM Symposium on Theory of Computing, pp. 513–522 (2010)
17. Luna, G.A.D., Flocchini, P., Prencipe, G., Santoro, N.: Locating a black hole in a dynamic ring. J. Parallel Distributed Comput. **196**, 104998 (2025)
18. Markou, E., Paquette, M.: Black hole search and exploration in unoriented tori with synchronous scattered finite automata. In: OPODIS, pp. 239–253 (2012)

Invited Paper: Distributed Rhombus Formation of Sliding Squares

Irina Kostitsyna[1] [iD], David Liedtke[2(✉)] [iD], and Christian Scheideler[2] [iD]

[1] KBR at NASA Ames Research Center, Moffett Field, USA
`irina.kostitsyna@nasa.gov`
[2] Paderborn University, Paderborn, Germany
`liedtke@mail.upb.de`, `scheideler@upb.de`

Abstract. The sliding square model is a widely used abstraction for studying self-reconfigurable robotic systems, where modules are square-shaped robots that move by sliding or rotating over one another. In this paper, we propose a novel distributed algorithm that enables a group of modules to reconfigure into a rhombus shape, starting from an arbitrary side-connected configuration. It is connectivity-preserving and operates under minimal assumptions: one leader module, common chirality, constant memory per module, and visibility and communication restricted to immediate neighbors. Unlike prior work, which relaxes the original sliding square move-set, our approach uses the unmodified move-set, addressing the additional challenge of handling locked configurations. Our algorithm is sequential in nature and operates with a worst-case time complexity of $\mathcal{O}(n^2)$ rounds, which is optimal for sequential algorithms. To improve runtime, we introduce two parallel variants of the algorithm. Both rely on a spanning tree data structure, allowing modules to make decisions based on local connectivity. Our experimental results show a significant speedup for the first variant, and a linear average runtime for the second variant, which is worst-case optimal for parallel algorithms.

Keywords: modular robots · distributed algorithms · sliding squares

1 Introduction

Edge-connected configurations of square modules, which can reconfigure by sliding over or rotating around one another (see Fig. 1a), are a well-established theoretical model for self-reconfigurable robotic systems. There are numerous applications such as search-and-rescue missions or autonomous construction, where modules collectively navigate complex terrains or form shapes while preserving connectivity. Reconfiguration planning in the sliding square model has primarily focused on centralized approaches. In contrast, we propose a fully distributed algorithm that enables modules to autonomously self-organize into a rhombus shape under minimal assumptions about their initial positions, computational

This work was supported by the DFG Project SCHE 1592/10-1. Full version: [26].

capabilities, and sensory range. The rhombus serves as an efficient intermediate shape due to its compactness, minimal bounding box, and lack of holes.

1.1 Our Contribution

We introduce a distributed reconfiguration algorithm for the sliding square model that transforms any initially side-connected configuration with a designated leader and common chirality into a rhombus shape. These two assumptions are minimal but necessary to break initial symmetries. The algorithm has a worst-case runtime of $\mathcal{O}(n^2)$ rounds, which is optimal among solutions that build up any shape sequentially. It operates under strict local communication and visibility constraints, requires only constant memory per module, and maintains connectivity throughout execution. In contrast to previous distributed solutions, which relax the original sliding square move-set, our algorithm utilizes the unmodified move-set, introducing the additional challenge of handling locked configurations. While our main focus is the rhombus, the algorithm generalizes to any shape that can be sequentially constructed, provided that at every step, any two cells in the shape remain connected by a shortest path within the shape.

To improve efficiency, we introduce two parallel variants of the algorithm. Both variants rely on an auxiliary spanning tree structure maintained throughout execution, allowing modules to detect certain moves that preserve connectivity. The first variant permits modules to move only if such moves improve their depth relative to the spanning tree (without explicitly computing or storing depth), resulting in a significant speedup over the sequential version. However, some configurations remain locked, limiting parallelization and requiring sequential progression. The second variant allows modules to move whenever possible, showing an average-case linear runtime, as demonstrated by our experiments.

1.2 Related Work

Reconfiguration in the sliding square model, introduced by Fitch et al. [15], typically proceeds via canonical intermediate shapes. Dumitrescu and Pach [12] gave the first universal 2D reconfiguration algorithm with optimal $\mathcal{O}(n^2)$ moves and a straight line as canonical shape, later adapted to an in-place version by Moreno and Sacristán [28]. Akitaya et al. [2] presented the first input sensitive algorithm, achieving $\mathcal{O}(nP)$ complexity, where P is the maximum perimeter of the bounding boxes of the input and target configuration, and proved NP-hardness of move minimization. Abel et al. [3] gave the first algorithm for sliding cubes in three and higher dimensions with optimal $\mathcal{O}(n^2)$ moves, while Kostitsyna et al. [27] achieved asymptotically optimal in-place reconfiguration in 3D, with the number of moves proportional to the total coordinate sum over all cubes.

All above algorithms are centralized and sequential. The first distributed approach was by Dumitrescu et al. [11], who reconfigured horizontally convex shapes into a line under global communication. Hurtado et al. [21] gave the first universal distributed algorithm with $\mathcal{O}(n)$ rounds, though under a relaxed move-set, a leader with $\mathcal{O}(n)$ memory, $\mathcal{O}(\log n)$ memory in other modules, and vision

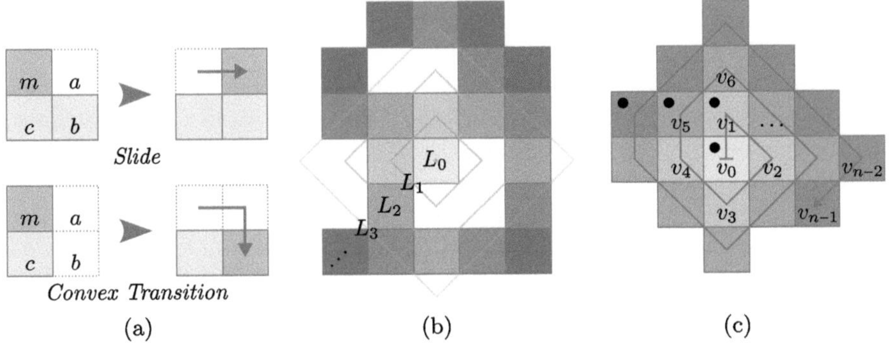

Fig. 1. (a) A slide is viable if cell a is empty; a convex transition is viable if both a and b are empty. (b) Example with the leader module in the center; modules belonging to the same layer L_i are depicted in the same color. (c) The rhombus $\mathcal{R} = (v_0, \ldots, v_{n-1})$ with $n = 20$; connector cells are marked with a dot.

and communication range 2. Wolters [32] proposed parallel methods without correctness proofs, and Akitaya et al. [5] provided a worst-case optimal centralized parallel result of $\mathcal{O}(P)$ and NP-completeness for constant-makespan decisions.

In the limited sliding cube model, modules are restricted to sliding moves only, and reconfiguration is achieved through meta-modules, small groups of modules that act as single units. Kawano proposed a series of increasingly efficient algorithms in this model, culminating in a linear-time solution [23–25].

Pivoting models, which allow only corner rotations, are significantly harder. Akitaya et al. [1,6] showed PSPACE-hardness for several reconfiguration problems, but universal reconfiguration is possible with five auxiliary modules [4]. Non-universal strategies focus on certain forbidden substructures [14,30].

Besides the sliding square model, shape reconfiguration is studied in a variety of other models such as the Amoebot model [8–10], and hybrid programmable matter [16,17,19,20]. Other related areas include coordinated motion planning [13,31], nest-building [7], and bounding-box construction [29]. These theoretical results are complemented by a growing body of real-world modular robotics implementations, such as [18,22].

1.3 Model Definition and Problem Statement

Consider the infinite square grid with set of cells C. Two distinct cells $c_1, c_2 \in C$ are called *adjacent* if they share at least one corner, *side-adjacent* if they share an edge, and *corner-adjacent* if they share a corner but no edge. The *neighborhood* $N(c)$ of a cell c consists of four side-adjacent cells in directions north (N), east (E), south (S), and west (W), and four corner-adjacent cells in directions northeast (NE), southeast (SE), southwest (SW), and northwest (NW).

We consider a set of n square modules, each with the computational power of a deterministic finite automaton. Each module occupies precisely one cell on the grid, and each cell contains at most one module. For simplicity, we refer interchangeably to modules and the cells they occupy. A *configuration* (M, s, l) consists of the set of all modules $M \subseteq C$, a state function $s : M \to \mathcal{S}$ assigning each module $m \in M$ a state $s(m) \in \mathcal{S}$ from the finite set $\mathcal{S}$ of possible states, and a unique cell $l \in C$ initially occupied by the leader module.

Modules move by performing *slides* and *convex transitions* (see Fig. 1a). Let (m, a, b, c) form a 4-cycle of side-adjacent cells, where $m \in M$. If $a \notin M$ and $b, c \in M$, then module m can *slide* into the empty cell a. If $a, b \notin M$ and $c \in M$, then module m can perform a *convex transition* into the empty cell b.

We assume a fully synchronous activation schedule consisting of discrete rounds. In each round, every module simultaneously executes a *Look-Compute-Move* cycle: In the *look* phase, a module m observes its neighborhood $N(m)$. For each cell $c \in N(m)$, it determines whether c is empty or occupied, and if occupied, it observes the state of the module in c. We assume that all modules have the same sense of chirality. In the *compute* phase, m may transition to a new state. It may also change the states of its neighbors, based solely on the information gathered during the look phase. In the *move* phase, the module performs an action associated with its current state, which is either a slide, a convex transition, or staying in place.

Conflicts may arise during the compute phase, when multiple modules attempt to update the state of the same neighbor, or during the move phase, if movement trajectories intersect. We resolve both types of conflicts arbitrarily: a module may take on any of the proposed new states, and in the case of movement conflicts, an arbitrary module proceeds while the others remain in place. While such conflicts could, in principle, lead to a livelock (e.g., if the conflicts are cyclic), our algorithm is designed to ensure that progress is always made.

We assume that the initial configuration (M_0, s_0, l) satisfies the following: The initial set of all modules M_0 is side-connected. Exactly one module occupies the leader cell $l \in M_0$. All other modules share the same initial state, i.e., $s_0(m) = s_0(m')$ for all $m, m' \in M_0 \setminus \{l\}$. The leader's state is distinct from all other modules, i.e., $s_0(l) \neq s_0(m)$ for all $m \in M_0$.

By executing a distributed algorithm, the system goes through a sequence of configurations $(M_0, s_0, l), (M_1, s_1, l) \ldots, (M_T, s_T, l)$ where each pair of consecutive configurations corresponds to the execution of one synchronous round. We require that in each round, the subset of modules that do not move remains side-connected. The goal is to reach a final configuration (M_T, s_T, l) where each module is in a terminated state, and the sum $\sum_{m \in M_T} d(m, l)$ is minimized, where $d(m, l)$ denotes the Manhattan (L_1) distance between m and l. Minimizing this sum in the square grid naturally results in modules being arranged in the shape of a rhombus, centered at l, whose outermost layer may be partially filled.

While modules are technically finite automata, we describe our distributed algorithm from a higher-level of abstraction, assuming each module stores only a constant number of variables, each of constant size domain.

Additional Terminology. Denote by $d(c_1, c_2)$ the Manhattan distance between two cells $c_1, c_2 \in C$. A *layer* $L_i \subset C$ consists of all cells at Manhattan distance i

from the leader cell l, i.e., $L_i = \{c \in C \mid d(l,c) = i\}$ (see Fig 1b). For each layer, we define a unique *connector* cell. The connector of L_0 is l, and for any $i > 0$, the connector of L_i is the cell north of the westernmost cell in L_{i-1}.

Our goal is to form a rhombus shape consisting of consecutive layers arranged around the leader module. Formally, we define a *rhombus* $\mathcal{R} = (v_0, \ldots, v_{n-1})$ of size n, which is constructed by appending the cells of each layer clockwise, starting from their respective connector cells (see Fig. 1c). Our algorithm aims to fill each cell in $\mathcal{R}$ with a module in this sequential order.

A *path* (or *corner-path*) of length k from cell c_0 to cell c_k is a tuple $(c_0, c_1, \ldots, c_k)$ such that c_i is side-adjacent (or adjacent) to c_{i+1} for all $0 \le i < k$. A set of cells is *side-connected* (or *corner-connected*) if there is a path (or corner-path) between any two cells from that set, and that path is fully contained within the set. We call a module $m \in M$ *removable* if $M \setminus \{m\}$ is side-connected. For simplicity, we slightly abuse notation and treat tuples as sets when referring to their contents, e.g., we write $m \in P$ if module m appears in path P.

A *hole* H is a maximal corner-connected subset of empty cells, and a *pseudo-hole* $\tilde{H}$ is a maximal side-connected subset of empty cells. Since the set of modules is finite, there is exactly one infinite hole and one infinite pseudo-hole. For example, the configuration in Fig. 1b has one finite hole of size three, and two finite pseudo-holes of sizes two and three.

Any hole can be partitioned into one or more pseudo-holes. Two distinct pseudo-holes $\tilde{H}$ and $\tilde{H}'$ are *adjacent* if there exist adjacent cells $e \in \tilde{H}, e' \in \tilde{H}'$. In that case, their common neighborhood $N(e) \cap N(e')$ contains precisely two occupied cells c, c'. We refer to (c, c') as a *critical pair*, to cell e as the *bridge* from $\tilde{H}$ to $\tilde{H}'$, and to cell e' as the *bridge* from $\tilde{H}'$ to $\tilde{H}$. Notably, for any two pseudo-holes $\tilde{H}, \tilde{H}'$ there can be at most one critical pair (c, c'), and consequently, at most one bridge in each direction. Otherwise, $\tilde{H} \cup \tilde{H}'$ would contain a corner-path cycle fully enclosing either c or c', contradicting that M is side-connected.

Finally, the *boundary* of a set of cells V is defined as $B(V) = (\bigcup_{v \in V} N(v)) \setminus V$. Note that by definition, a hole's boundary contains only occupied cells, whereas a pseudo-hole's boundary may include empty cells from adjacent pseudo-holes.

2   A Distributed Rhombus Formation Algorithm

We now describe our distributed rhombus formation algorithm. Figure 2 shows the algorithm's execution on an example configuration. Some technical details are deferred to the full version [26].

In our algorithm, modules transition through four distinct phases: HEAD, TAIL, ACTIVE, and PASSIVE. From a high level, the HEAD module identifies the next cell to fill in the rhombus, the ACTIVE module moves to occupy this cell, TAIL modules have already permanently joined the rhombus, and PASSIVE modules include all modules that have yet to join. Initially, the leader module is the HEAD while all other modules are PASSIVE. In any given round, there is precisely one HEAD and at most one ACTIVE module. Any module that is not ACTIVE remains stationary and solely participates in communication.

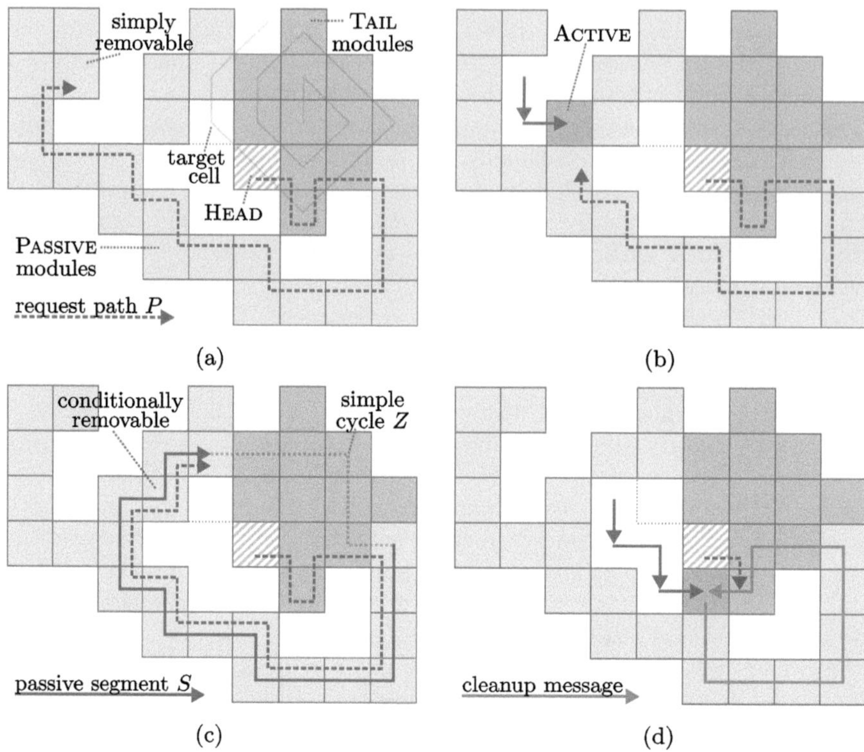

Fig. 2. (a) The HEAD initiates a request, which is forwarded anti-clockwise along the target boundary to a simply removable module. (b) The module becomes ACTIVE, moves along the request path P in reverse until blocked by a critical pair, then becomes PASSIVE as the blocking module is not within P. (c) The request continues until the next forwarding direction is not a PASSIVE module. Going backwards in the request path, the first conditionally removable module becomes ACTIVE. (d) The ACTIVE module is blocked again, this time the blocking module is within P, and its successor equals its predecessor. The ACTIVE module sends a cleanup message to erase the requests stored at modules that it cannot reach. Afterwards, it moves directly to the target cell (not visualized).

The HEAD first identifies its immediate successor within the rhombus $\mathcal{R}$, which we refer to as the *target cell*, and stores the direction to this cell in its constant memory. If the target cell is already occupied by some module, then that module becomes the new HEAD while the previous HEAD becomes a TAIL. If it is empty, the HEAD initiates a request message to find a PASSIVE module that can become ACTIVE and move towards the target cell. Messages are only exchanged between side-adjacent modules, and they travel along the boundary of the hole that contains the target cell, which we refer to as *target boundary* and *target hole*, respectively. Moreover, messages carry no information besides their type. A request message always travels anti-clockwise along the target boundary.

If a request fully traverses the target boundary without seeing PASSIVE modules, the HEAD terminates. All other modules terminate in subsequent rounds upon observing a terminated neighbor. Otherwise, if a request reaches a PASSIVE module m, that module locally checks whether moving may break connectivity, which we formalize as follows:

Definition 1. *A module m is* simply removable *if (1) it is side-adjacent to the target hole, and (2) for every pair m_1, m_2 of modules side-adjacent to m, there exists a path from m_1 to m_2 within $N(m) \cap M$. (Note that $m \notin N(m)$.)*

If m is simply removable, it becomes ACTIVE; otherwise, it forwards the request anti-clockwise along the target boundary. Note that simple removability is a sufficient but not necessary condition for a module to be removable. It may occur that no PASSIVE module that receives a request is simply removable, in which case local information alone does not suffice to identify a removable module. Modules maintain an ordered list of the requests they receive, together with the direction from which each request was received. A module can receive at most four requests for the same target cell, one for each side-adjacent module. Additionally, we ensure that all memory is reset before a new request is initiated. Hence, constant memory suffices to store all requests.

Definition 2. *A request path $P = (m_1, m_2, ..., m_k)$ is a maximal directed path along the target boundary such that for each $1 < i \leq k$, module m_i stores a request (or receives a request at the start of the current round) from module m_{i-1}, and each directed edge (m_{i-1}, m_i) appears at most once in P.*

Definition 3. *Given a request path $P = (m_1, m_2, ..., m_k)$, a* passive segment *is a maximal contiguous sub-path $S = (m_i, ..., m_j) \subseteq P$, where $1 < i \leq j \leq k$, consisting entirely of PASSIVE modules.*

Note that the request path is not necessarily simple in terms of nodes, but it contains any directed cycle at most once. In contrast, in the analysis we show that following our algorithm, the request path P contains at most one passive segment S, that this segment is simple (in terms of nodes), and that all PASSIVE modules storing a request belong to S. As a result, any PASSIVE module m that receives a request in some round must be the last module on both P and S in that round, i.e., $m = m_k = m_j$. Since m can observe the state of its neighbors, it knows which of them belong to the request path. We use this additional information to identify simple cycles within the set of all modules M. This gives us an additional criterion to check for removability:

Definition 4. *Let $Z = (m_0, m_1, ..., m_{k-1})$ be a simple cycle within the set of modules M. A module m_i with $0 \leq i < k$ is* conditionally removable, *if (1) m_i is side-adjacent to the target hole, and (2) for every module m side-adjacent to m_i, there exists a path from m to m_{i-1} or a path from m to m_{i+1} within $N(m) \cap M$ (where indices are taken modulo k).*

2.1 Activating Conditionally Removable Modules

The procedure is initiated whenever the last module m_k on the request path P is PASSIVE and the module in the next forwarding direction is either already on P or not PASSIVE. There are two cases to consider:

Case 1: Request Cycles Back to Itself. Suppose the module in the next forwarding direction from m_k is a PASSIVE neighbor $m_l \in P$, i.e., a module that already belongs to the request path P with $l < k$. Since the passive segment S is unique and simple, $(m_l, m_{l+1}, ..., m_k)$ is a simple cycle. In this case, the algorithm proceeds as follows: Starting with module m_k, traverse the request path in reverse order. At each step $i = k, k-1, \ldots, l$, the current module m_i checks whether it is conditionally removable. If so, it becomes ACTIVE and the process stops. Otherwise, it deletes the request it has received from m_{i-1} from its memory and sends an activation message to its predecessor m_{i-1}, which then continues the process. We will show that this process always finds a conditionally removable module before the activation message reaches m_l.

Case 2: Request Reaches a Non-PASSIVE Module. Now suppose the module in the next forwarding direction from m_k is not PASSIVE, i.e., a HEAD or TAIL module. Since m_k is PASSIVE, the passive segment $S = (m_i, ..., m_k)$ is non-empty. Moreover, by maximality of S, the module m_{i-1} must also be non-PASSIVE. Let P' be a shortest path from m to m_{i-1} within the set of all non-PASSIVE modules. Since all shortest paths are simple and S contains only PASSIVE modules, the paths S and P' are disjoint. Therefore, their concatenation forms a simple cycle. In this case, the algorithm proceeds identical to Case 1. Starting at m_k, each module along S is checked in reverse order for conditional removability while deleting its stored request. The first such module becomes ACTIVE and stops the process. We will show that this process always finds a conditionally removable module before the activation message reaches m_{i-1} unless the target hole has size one, an edge case resolved in the full version [26].

Eventually, some PASSIVE module becomes ACTIVE and deletes its stored request. The ACTIVE module moves via slides and convex transitions through the target hole, following the request path P in reverse direction. During this process, it deletes requests stored at adjacent modules such that it always remains side-adjacent to the last module of P.

For simplicity, we treat the cell occupied by the ACTIVE module as empty w.r.t. all definitions and terminology. For example, if after moving it is side-adjacent to cells c_1 and c_2 that formed a critical pair in the previous round, we still consider (c_1, c_2) a critical pair, even though they are now technically connected through the ACTIVE module.

2.2 Moving Past Critical Pairs

It may happen that the movement of the ACTIVE module is blocked by a critical pair (m_k, c), where m_k is the last module on the request path P. At this point, the current position m of the ACTIVE module is one of the bridge cells of the

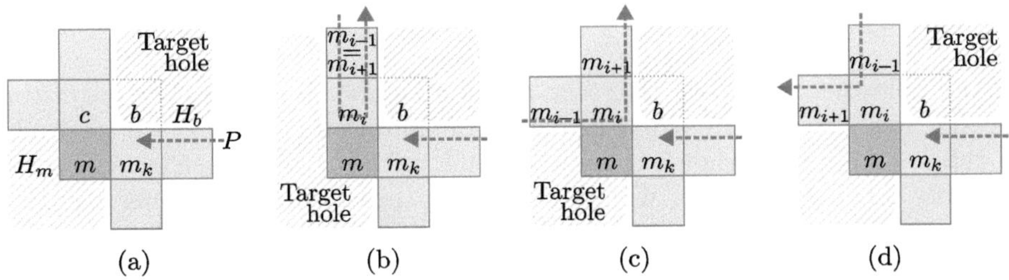

Fig. 3. Case distinction for determining which pseudo-hole contains the target cell when the ACTIVE module m is blocked by a critical pair (m_k, c). The decision is based on whether c occurs on the request path and, if so, on the relative position of its successor and predecessor. A variant of (b) where $m_{i+1} = m_{i-1}$ is corner-adjacent to m, as well as the case where $m_i = m_1$, are omitted.

critical pair. Let $b \neq m$ be the other bridge cell. By placing a module at b, we obtain a hole H_m containing m, and by placing a module at m, we obtain a hole H_b containing b (see Fig. 3). Since any two adjacent pseudo-holes are separated by a unique critical pair, the holes H_m and H_b are disjoint. As a result, exactly one of them contains the target cell.

To proceed, the ACTIVE module determines whether H_m or H_b contains the target cell, a decision that can be made locally based on the following case distinction: If $c \notin P$ (see Fig. 3a), then the target cell lies in H_b. Otherwise, let $m_i \in P$ be the last occurrence of c on the request path, i.e., the module $m_i = c$ with maximum index i. Since requests are stored in ordered lists, this information can be read from the module's memory.

If $m_i = m_1$, i.e., it has no predecessor and initiated the request, then m can read the position of the target cell from m_i's memory and directly determine whether it lies in H_m or H_b. Since modules share a common chirality, the request can carry the HEAD's orientation, so that m and m_i agree on the north direction.

If m_i has a predecessor $m_{i-1} \in P$ and a successor $m_{i+1} \in P$, then the target cell lies in H_m if $m_{i+1} = m_{i-1}$ or $m_{i+1} \notin N(m)$ (see Figs. 3b and 3c); otherwise, it lies in H_b (see Fig. 3d). We establish these claims in the proof of Lemma 3.

In our algorithm, the ACTIVE module m proceeds as follows: It first determines, using the above case distinction, whether H_m or H_b contains the target cell. If H_m contains the target cell, then m sends a cleanup message to m_k (e.g., see Fig. 2d). Each module receiving this message forwards it to its predecessor on the request path and then deletes its last stored request. Eventually, the message reaches c, at which point it is dropped and m can resume its movement along the boundary of H_m. Otherwise, if H_b contains the target cell, then m becomes PASSIVE and acts as if it had received a request from m_k (e.g., see Figs. 2b and 2c). Note that by becoming PASSIVE and continuing the request from m_k, all pseudo-holes contained in H_m are pruned from the target hole. This behavior is essential for showing that all requests terminate in linear time.

3   Analysis

In the following, we prove that our distributed rhombus formation algorithm reconfigures any initially side-connected configuration (M_0, s_0, l) with a leader in cell $l \in M_0$ into a rhombus centered at l, with its outermost layer possibly partially filled. For clarity, we sketch the key proofs here and defer technical lemmas and complete proofs to the full version [26].

Lemma 1. *Let $\mathcal{R} = (v_0, v_1, \ldots, v_{n-1})$ be the rhombus centered at $v_0 = l$. If the* HEAD *occupies cell v_i where $0 \le i < n$, then the set of all* TAIL *modules is exactly $\{v_0, \ldots, v_{i-1}\}$.*

Proof. The lemma holds initially, as there are no TAIL modules and the initial position of the HEAD is v_0. Assume the lemma holds at the start of an arbitrary round. There is exactly one HEAD module at all times, and all TAIL modules are contained within $\{v_0, \ldots, v_{i-1}\}$. This implies that v_i is empty or occupied by a PASSIVE or ACTIVE module. As the HEAD module does not move, and TAIL modules neither move nor change phase, the lemma holds until v_i is occupied. Once v_i is occupied, that module becomes the HEAD while the previous HEAD transitions to a TAIL. In the next round, the HEAD occupies cell v_i and the set of all TAIL module expands to $\{v_0, \ldots, v_i\}$. The lemma follows inductively.   □

The following lemma was proven by Dumitrescu and Pach in [12]. For ease of reference, we state it here using our terminology.

Lemma 2. *Let $\tilde{H}$ be an arbitrary pseudo-hole, let $a, b \in \tilde{H}$ be any distinct cells side-adjacent to its boundary $B(\tilde{H})$. A module placed at cell a can move into cell b while remaining side-adjacent to cells in $B(\tilde{H})$ throughout its movement.*

We now establish essential invariants that hold throughout the algorithm.

Lemma 3. *Let $P = (m_1, m_2, \ldots, m_k)$ be the request path. The following properties hold in every round during the execution of the rhombus formation algorithm:*

1. *For every module m that stores a request from some module m', the directed edge (m', m) is contained in P.*
2. *The request path P contains at most one passive segment $S = (m_i, \ldots, m_j)$, and if S is non-empty, then $j = k$, i.e., S ends at the last module of P.*
3. *The passive segment S is simple, i.e., it contains each module at most once.*

Proof Sketch. The lemma is shown by induction over the rounds of the algorithm. Before any request is initiated, the properties hold trivially. In each round, we analyze how the request path P changes. Forwarding a request appends a single edge to P; a simple case distinction shows that all invariants are preserved. A PASSIVE module becoming ACTIVE removes only the last edge of P. The ACTIVE module then moves by following P in reverse. As long as it stays within the same pseudo-hole, by Lemma 2 it can remain side-adjacent to the last module in P while iteratively deleting the request corresponding to the last edge in P.

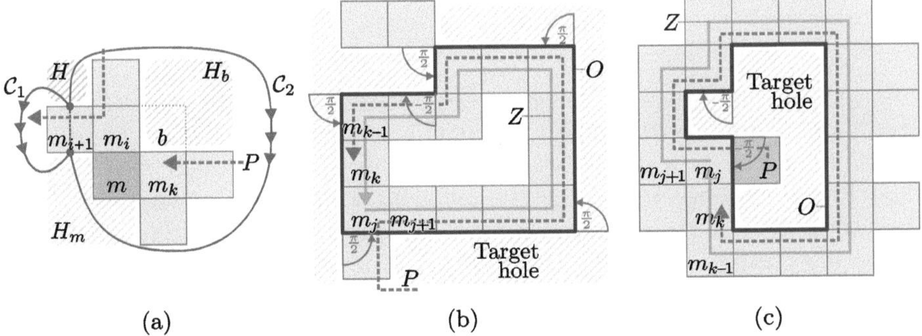

Fig. 4. (a) The ACTIVE module m side-adjacent to a critical pair (m_k, m_i) with $m_i \in P$ and $m_{i+1} \in N(m)$. Assuming $H = H_m$, $\mathcal{C}_1$ and $\mathcal{C}_2$ are closed simple curves in $H \cup \{m_{i+1}\}$. (b) The simple polygon O does not enclose the target hole: at least four vertices have exterior angle $\frac{\pi}{2}$. (c) Polygon O encloses the target hole: at least two vertices have exterior angle $-\frac{\pi}{2}$, or a side has length ≥ 2.

Whenever the last edge of P is removed, the invariants carry over from the previous round.

Special care is needed when the ACTIVE module m is blocked by a critical pair. The difficult case is where $m_{i+1} \neq m_{i-1}$ and $m_{i+1} \in N(m)$ (see Fig. 4a). After m becomes PASSIVE, hole H contains the target cell. We can show that $H \neq H_m$ by considering an anti-clockwise and a clockwise closed simple curves $\mathcal{C}_1$ and $\mathcal{C}_2$ within $H \cup \{m_{i+1}\}$, one of which must exist if $H = H_m$. $\mathcal{C}_1$ encloses all successors of m_{i+1} but not m_k, and $\mathcal{C}_2$ encloses m_k but no successor of m_{i+1}. Both contradict that m_k is a successor of m_{i+1}. Thus it holds that $H = H_b$. $\square$

We call the request path *well-formed* if it satisfies all properties from Lemma 3. Assuming a well-formed request path, we can show that a conditionally removable module exists whenever the passive segment cycles back on itself.

Lemma 4. *Let* $P = (m_1, m_2, \ldots, m_k)$ *be a well-formed request path with non-empty passive segment* $S = (m_i, \ldots, m_k)$. *Suppose the next anti-clockwise module along the target boundary after* m_k *is some module* m_j *with* $i \leq j \leq k-2$. *Then at least one of the modules* $m_j, m_{j+1}, \ldots, m_k$ *is conditionally removable.*

Proof Sketch. Let $Z := (m_j, m_{j+1}, \ldots, m_k)$ be the part of the request path closing the loop. Since Z lies entirely within S, it forms a simple cycle by Lemma 3.

If Z contains a critical pair module, then that module is conditionally removable. Otherwise, we analyze the geometry of Z on the grid. Consider the region outlined by Z, including the cells in Z if Z does not enclose the target hole (see Fig. 4b), and excluding them otherwise (see Fig. 4c). Since Z contains no critical pair, the boundary of this region forms a simple polygon O.

If Z does not enclose the target hole, then by standard angle-sum arguments for simple polygons, O has at least four vertices with exterior angle $\frac{\pi}{2}$, at least one of which corresponds to a conditionally removable module in Z. If Z does enclose the target hole, then it must also enclose the HEAD module, implying the existence of two vertices with exterior angle $-\frac{\pi}{2}$ or a side of length at least two, both of which imply a conditionally removable module in Z. $\square$

In the previous proof, we relied on the fact that the simple cycle Z consisted entirely of PASSIVE modules. This no longer holds if the search for a conditionally removable module is triggered when the next forwarding direction of a request is not PASSIVE. Here, we rely on the following property: A set of cells $V \subset C$ is *weakly convex* if for any $c_1, c_2 \in V$, there exists a shortest path from c_1 to c_2 that lies entirely within V. If the set of all non-PASSIVE modules is weakly convex (as is the case for the rhombus), then we can show that a conditionally removable module exists whenever the target hole has size at least two. The edge case where the target hole has size one is handled in the full version [26].

Lemma 5. *Let $P = (m_1, \ldots, m_k)$ be a well-formed request path with non-empty passive segment $S = (m_i, \ldots, m_k)$. Suppose the next anti-clockwise module along the target boundary after m_k is not* PASSIVE, *and m_k is not simply removable. If no module in S is conditionally removable, then the target hole has size one.*

The following lemma shows that each request fills or nearly fills the target cell within a linear number of rounds, which is crucial for termination and runtime.

Lemma 6. *Consider a round where a module initiates a request for some target cell, and assume at least one* PASSIVE *module exists. Then, within $\mathcal{O}(n)$ rounds, either the target cell is filled or the target hole consists of just the target cell.*

Proof Sketch. One can show that whenever PASSIVE modules exist, eventually some module becomes ACTIVE. If the ACTIVE module m lies in the same pseudo-hole as the target cell, it can reach the target cell in $O(n)$ rounds. Otherwise, it becomes PASSIVE at a bridge cell, effectively disconnecting the target hole. This reduces the number $\mathcal{S}$ of sides shared between the target hole and its boundary by an amount proportional to the length of the request path segment erased before m becomes PASSIVE. Since $\mathcal{S} = O(n)$ initially and $\mathcal{S}$ decreases proportional to the number of rounds it takes for m to become ACTIVE, move, and then become PASSIVE again, after $O(n)$ rounds an ACTIVE module is in the target pseudo-hole, or $\mathcal{S} \leq 4$, in which case the target hole has size one. $\square$

Corollary 1. *Let m be a module that becomes the* HEAD *at some round during the execution of the algorithm. After $\mathcal{O}(n)$ rounds, either m becomes a* TAIL *and another module takes over as the new* HEAD, *or m terminates.*

The corollary follows from Lemma 6, together with the handling of the edge case (see full version [26]) where the target hole has size one and no conditionally removable exists in its boundary. In this edge case, a request for a different cell outside the target hole is sent. One can show that regardless of whether this second cell is filled or becomes a new hole of size one, the target boundary subsequently contains a conditionally removable module. Hence, two consecutive requests suffice to fill the target cell. If no PASSIVE modules are left, the request traverses the entire target boundary in $O(n)$ rounds and the HEAD terminates.

Theorem 1. *Let (M_0, s, l) be an arbitrary, initially side-connected configuration of n modules with a leader $l \in M_0$, and all modules except for the leader being in the same state. Let $\mathcal{R} = (v_0, \ldots, v_{n-1})$ be a rhombus centered at cell $v_0 = l$. After $\mathcal{O}(n^2)$ rounds of the rhombus formation algorithm, all cells in $\mathcal{R}$ are occupied by a terminated module, and side-connectivity is preserved in each of those rounds.*

Proof. By Lemma 1, the rhombus is filled incrementally with each transition of some module into the HEAD phase. Since TAIL modules never change phase again, there are at most n HEAD transitions in total. By Corollary 1, each transition happens within $\mathcal{O}(n)$ rounds of each other. It follows that it takes $\mathcal{O}(n^2)$ rounds until a HEAD module terminates. Afterwards, termination propagates through all TAIL modules within $\mathcal{O}(n)$ rounds. Since TAIL modules never move, all rhombus cells eventually contain terminated modules.

A module becomes ACTIVE only if it is simply removable (see Definition 1) or conditionally removable as part of a simple cycle Z within M (see Definition 4), both of which preserve connectivity. Moreover, whenever Z consists only of PASSIVE modules, we have shown in Lemma 4 that such a conditionally removable module is always found before leaving Z. It follows that all module movements maintain side-connectivity in every round.   □

4   Utilizing Parallelization

No sequential algorithm can solve reconfiguration in fewer than $\mathcal{O}(n^2)$ rounds, not even centralized algorithms. To improve runtime, modules must therefore move in parallel, but preserving connectivity during parallel movement is non-trivial. While simple removability can be detected locally, identifying conditionally removable modules requires extensive communication, and not all such modules can move simultaneously without disconnecting the configuration. Coordinating mutual exclusion would already require $\Omega(n^2)$ rounds under constant memory, so we follow a different approach: We first construct a spanning tree $\mathcal{T}$, rooted at the leader, via a distributed breadth-first search in $\mathcal{O}(n)$ rounds. Leaves then move clockwise along hole boundaries whenever doing so reduces their depth in $\mathcal{T}$. Depth values are never computed; each module maintains only a parent pointer, and all movement decisions are based on its local neighborhood.

Despite this strategy, configurations may still become locked in two ways. First, no leaf is side-adjacent to a hole, which cannot be resolved locally even with logarithmic memory to compute and store depth values (see Fig. 5a). Second, no leaf can locally determine whether a move decreases its depth in $\mathcal{T}$ (see Fig. 5b).

To guarantee progress, we run our sequential algorithm on top of the spanning-tree strategy. The sequential layer also benefits from the additional information from $\mathcal{T}$: whenever a PASSIVE module checks whether it is simply removable, it also checks whether it is a leaf. Three issues must be resolved to preserve the theoretical guarantees (see Theorem 1) in the parallel execution. A detailed description of these issues and their resolution is provided in the full version [26].

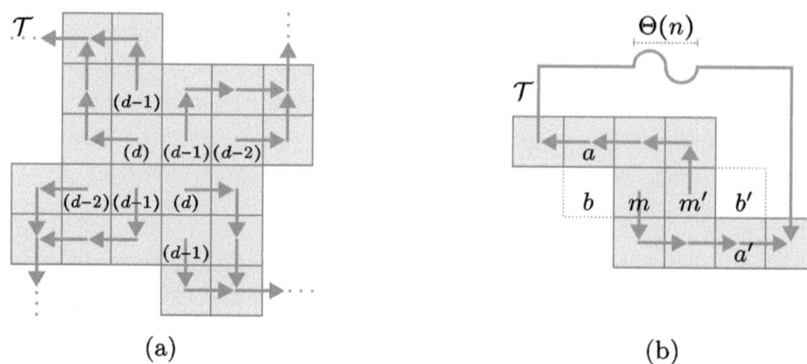

Fig. 5. (a) A configuration that is locked even under logarithmic memory: no leaf is side-adjacent to a hole, and all leaves have minimum depth. Leaves are labeled by their depth in $\mathcal{T}$. (b) A mutual blocking situation: m cannot decide locally whether sliding to b improves its depth, and the same holds for m' with b'.

1. Requests or activation messages may activate modules that are not leaves in $\mathcal{T}$, after which the tree structure must be restored. This is the most challenging issue: locally, a module m cannot determine whether a neighbor is a descendant in $\mathcal{T}$, and redirecting the parent pointers of m's children toward a descendant would create a cycle. Restoring the tree therefore requires additional structure. We maintain the invariant that the subtree induced by all HEAD and TAIL modules forms a shortest-path tree toward the rhombus center and reconfigure parent pointers along the request path during forwarding. This allows a module becoming active to identify its ancestors (since they lie on the request path) and redirect its children's parent pointers to an ancestor; a similar strategy applies to conditionally removable modules.
2. The movement of an ACTIVE module m may be blocked by a moving leaf. We handle these cases using local state swaps: instead of moving, m delegates its role to the blocking leaf.
3. Finally, a technical issue: a moving leaf may temporarily violate Lemma 3 when occupying a cell side-adjacent to the request path. Treating such leaves as empty cells, analogous to the ACTIVE module, resolves the issue.

5 Parallel Variants and Experimental Evaluation

To further improve performance, we developed two variants, V1 and V2, that increase the degree of parallelism. V1 allows leaves to perform state swaps similar to those of the ACTIVE module. This enables progress in configurations such as Fig. 5b without waiting for requests. However, since leaves are restricted to moves that reduce their depth in $\mathcal{T}$, configurations can still become locked, in which case the runtime is asymptotically not better than the sequential algorithm. V2 relaxes this restriction: leaves move as in V1 but proceed with any available move, even if it cannot be locally verified to reduce their depth.

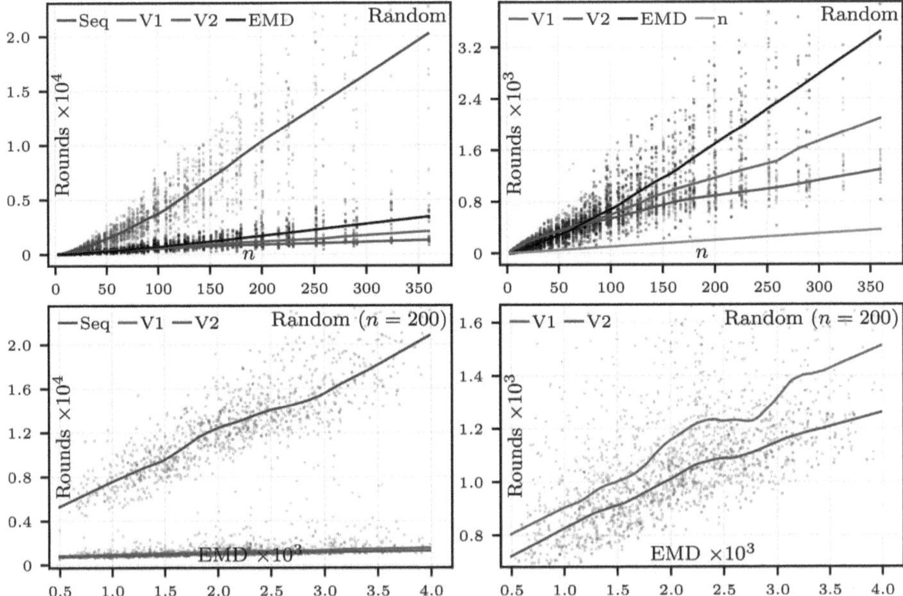

Fig. 6. Runtime comparison of the sequential algorithm (Seq) and its parallel variants (V1, V2). Top: runtimes on random configurations of varying sizes; bottom: runtimes vs. Earth Mover's Distance (EMD) for fixed size $n = 200$. Plots on the right replicate plots on the left but omit Seq for better resolution.

We evaluated all three algorithms on a variety of structured and random configurations. Figure 6 summarizes the main results: the runtime of the sequential algorithm grows quadratically, while both parallel variants scale much better. V2 achieves near-linear runtime on average and has lower variance than V1.

6 Conclusion

We presented a distributed reconfiguration algorithm for the sliding square model that transforms side-connected initial configurations into a rhombus. It runs sequentially in worst-case optimal time $\mathcal{O}(n^2)$ under strict local communication, constant memory, and connectivity constraints. Although we focused on the rhombus, the algorithm extends to any sequentially constructible, weakly convex shape such as lines or rectangles. To improve runtime, we use a spanning tree strategy in which leaves move in parallel, and evaluated two parallel variants in simulation; both outperform the sequential algorithm, with the second achieving near-linear average runtime. Future work includes non-weakly convex shapes, reconfiguration from the rhombus to arbitrary configurations, and lower bounds for distributed reconfiguration in this model.

References

1. Akitaya, A., et al.: Characterizing universal reconfigurability of modular pivoting robots. In: Buchin, K., Colin de Verdière, E. (eds.) 37th International Symposium on Computational Geometry (SoCG 2021). Leibniz International Proceedings in Informatics (LIPIcs), vol. 189, pp. 10:1–10:20. Schloss Dagstuhl – Leibniz-Zentrum für Informatik, Dagstuhl, Germany (2021). https://doi.org/10.4230/LIPIcs.SoCG.2021.10

2. Akitaya, H.A., et al.: Compacting squares: input-sensitive in-place reconfiguration of sliding squares. In: Czumaj, A., Xin, Q. (eds.) 18th Scandinavian Symposium and Workshops on Algorithm Theory (SWAT 2022). Leibniz International Proceedings in Informatics (LIPIcs), vol. 227, pp. 4:1–4:19. Schloss Dagstuhl – Leibniz-Zentrum für Informatik, Dagstuhl, Germany (2022). https://doi.org/10.4230/LIPIcs.SWAT.2022.4

3. Abel, Z., Akitaya, H.A., Kominers, S.D., Korman, M., Stock, F.: A universal in-place reconfiguration algorithm for sliding cube-shaped robots in a quadratic number of moves. In: 40th International Symposium on Computational Geometry (SoCG 2024). Leibniz International Proceedings in Informatics (LIPIcs), vol. 293, pp. 1:1–1:14. Schloss Dagstuhl – Leibniz-Zentrum für Informatik, Dagstuhl, Germany (2024). https://doi.org/10.4230/LIPIcs.SoCG.2024.1

4. Akitaya, H.A., et al.: Universal reconfiguration of facet-connected modular robots by pivots: the O(1) musketeers. Algorithmica **83**(5), 1316–1351 (2021). https://doi.org/10.1007/s00453-020-00784-6

5. Akitaya, H.A., et al.: Sliding squares in parallel (2024). https://doi.org/10.48550/arXiv.2412.05523

6. Akitaya, H.A., Stock, F.: Reconfiguration of 3d pivoting modular robots (2023). https://doi.org/10.48550/arXiv.2304.09990

7. Czyzowicz, J., Dereniowski, D., Pelc, A.: Building a nest by an automaton. Algorithmica **83**(1), 144–176 (2021). https://doi.org/10.1007/s00453-020-00752-0

8. Derakhshandeh, Z., Gmyr, R., Richa, A.W., Scheideler, C., Strothmann, T.: An algorithmic framework for shape formation problems in self-organizing particle systems. In: Proceedings of the Second Annual International Conference on Nanoscale Computing and Communication, pp. 21:1–21:2 (2015). https://doi.org/10.1145/2800795.2800829

9. Derakhshandeh, Z., Gmyr, R., Richa, A.W., Scheideler, C., Strothmann, T.: Universal shape formation for programmable matter. In: Proceedings of the 28th ACM Symposium on Parallelism in Algorithms and Architectures, pp. 289–299 (2016). https://doi.org/10.1145/2935764.2935784

10. Di Luna, G.A., Flocchini, P., Santoro, N., Viglietta, G., Yamauchi, Y.: Shape formation by programmable particles. Distrib. Comput. **33**(1), 69–101 (2020). https://doi.org/10.1007/s00446-019-00350-6

11. Dumitrescu, A., Suzuki, I., Yamashita, M.: Motion planning for metamorphic systems: feasibility, decidability, and distributed reconfiguration. IEEE Trans. Robot. Autom. **20**(3), 409–418 (2004). https://doi.org/10.1109/TRA.2004.824936

12. Dumitrescu, A., Pach, J.: Pushing squares around. In: Proceedings of the Twentieth Annual Symposium on Computational Geometry, pp. 116–123 (2004). https://doi.org/10.1145/997817.997838

13. Fekete, S.P., Keldenich, P., Kosfeld, R., Rieck, C., Scheffer, C.: Connected Coordinated Motion Planning with Bounded Stretch. In: Ahn, H.K., Sadakane, K. (eds.) 32nd International Symposium on Algorithms and Computation (ISAAC 2021).

Leibniz International Proceedings in Informatics (LIPIcs), vol. 212, pp. 9:1–9:16. Schloss Dagstuhl – Leibniz-Zentrum für Informatik, Dagstuhl, Germany (2021). https://doi.org/10.4230/LIPIcs.ISAAC.2021.9

14. Feshbach, D., Sung, C.: Reconfiguring non-convex holes in pivoting modular cube robots. IEEE Robot. Autom. Lett. **6**(4), 6701–6708 (2021). https://doi.org/10.1109/LRA.2021.3095030

15. Fitch, R., Butler, Z., Rus, D.: Reconfiguration planning for heterogeneous self-reconfiguring robots. In: Proceedings 2003 IEEE/RSJ International Conference on Intelligent Robots and Systems (IROS 2003) (Cat. No.03CH37453), vol. 3, pp. 2460–2467 (2003). https://doi.org/10.1109/IROS.2003.1249239

16. Friemel, J., Liedtke, D., Scheffer, C.: Efficient shape reconfiguration by hybrid programmable matter (2025). https://doi.org/10.48550/arXiv.2501.08663

17. Gmyr, R., et al.: Forming tile shapes with simple robots. Nat. Comput. **19**(2), 375–390 (2020). https://doi.org/10.1007/s11047-019-09774-2

18. Gregg, C.E., et al.: Ultralight, strong, and self-reprogrammable mechanical metamaterials. Sci. Robot. **9**(86), eadi2746 (2024). https://doi.org/10.1126/scirobotics.adi2746

19. Hinnenthal, K., Rudolph, D., Scheideler, C.: Shape formation in a three-dimensional model for hybrid programmable matter. In: Proceedings of the 36th European Workshop on Computational Geometry (EuroCG 2020) (2020)

20. Hinnenthal, K., Liedtke, D., Scheideler, C.: Efficient shape formation by 3D hybrid programmable matter: an algorithm for low diameter intermediate structures. In: Casteigts, A., Kuhn, F. (eds.) 3rd Symposium on Algorithmic Foundations of Dynamic Networks (SAND 2024). Leibniz International Proceedings in Informatics (LIPIcs), vol. 292, pp. 15:1–15:20. Schloss Dagstuhl – Leibniz-Zentrum für Informatik, Dagstuhl, Germany (2024). https://doi.org/10.4230/LIPIcs.SAND.2024.15

21. Hurtado, F., Molina, E., Ramaswami, S., Sacristán, V.: Distributed reconfiguration of 2d lattice-based modular robotic systems. Auton. Robot. **38**(4), 383–413 (2015). https://doi.org/10.1007/s10514-015-9421-8

22. Jenett, B., Abdel-Rahman, A., Cheung, K., Gershenfeld, N.: Material–robot system for assembly of discrete cellular structures. IEEE Robot. Autom. Lett. **4**(4), 4019–4026 (2019). https://doi.org/10.1109/LRA.2019.2930486

23. Kawano, H.: Complete reconfiguration algorithm for sliding cube-shaped modular robots with only sliding motion primitive. In: 2015 IEEE/RSJ International Conference on Intelligent Robots and Systems (IROS), pp. 3276–3283. IEEE (2015). https://doi.org/10.1109/IROS.2015.7353832

24. Kawano, H.: Distributed tunneling reconfiguration of cubic modular robots without meta-module's disassembling in severe space requirement. Robot. Auton. Syst. **124**, 103369 (2020). https://doi.org/10.1016/j.robot.2019.103369

25. Kawano, H.: Linear-time reconfiguration of sliding-only cubic modular robots under severe space requirements. IEEE Robot. Autom. Lett. **8**(12), 8176–8183 (2023). https://doi.org/10.1109/LRA.2023.3315208

26. Kostitsyna, I., Liedtke, D., Scheideler, C.: Distributed rhombus formation of sliding squares (2025). https://doi.org/10.48550/arXiv.2508.09638

27. Kostitsyna, I., Ophelders, T., Parada, I., Peters, T., Sonke, W., Speckmann, B.: Optimal in-place compaction of sliding cubes. In: Bodlaender, H.L. (ed.) 19th Scandinavian Symposium and Workshops on Algorithm Theory (SWAT 2024). Leibniz International Proceedings in Informatics (LIPIcs), vol. 294, pp. 31:1–31:14. Schloss Dagstuhl – Leibniz-Zentrum für Informatik, Dagstuhl, Germany (2024). https://doi.org/10.4230/LIPIcs.SWAT.2024.31

28. Moreno, J., Sacristán, V.: Reconfiguring sliding squares in-place by flooding. In: Proceedings of the 36th European Workshop on Computational Geometry (EuroCG), p. 32 (2020)
29. Niehs, E., et al.: Recognition and reconfiguration of lattice-based cellular structures by simple robots. In: 2020 IEEE International Conference on Robotics and Automation (ICRA), pp. 8252–8259 (2020). https://doi.org/10.1109/ICRA40945.2020.9196700
30. Sung, C., Bern, J., Romanishin, J., Rus, D.: Reconfiguration planning for pivoting cube modular robots. In: 2015 IEEE International Conference on Robotics and Automation (ICRA), pp. 1933–1940 (2015). https://doi.org/10.1109/ICRA.2015.7139451
31. Walter, J., Tsai, E., Amato, N.: Algorithms for fast concurrent reconfiguration of hexagonal metamorphic robots. IEEE Trans. Rob. **21**(4), 621–631 (2005). https://doi.org/10.1109/TRO.2004.842325
32. Wolters, M.: Parallel algorithms for sliding squares. Master's thesis, Utrecht University (2024). https://studenttheses.uu.nl/handle/20.500.12932/45768

Invited Paper: On the Equivalence of Snapshot/Append Objects and Broadcast Abstractions under Byzantine Failures

Vincent Kowalski[1], Achour Mostefaoui[2(✉)], Matthieu Perrin[1],
and Jolan Riallo[1]

[1] LS2N, Nantes Université, Nantes, France
[2] IRISA, Université de Rennes, Rennes, France
achour.mostefaoui@univ-rennes.fr

Abstract. Abstractions play a central role in distributed computing, as they capture essential synchronization properties. Equivalence results among abstractions clarify their relative computational power. While many such equivalences are well established in crash-prone systems, far less is known about Byzantine-prone environments, where faulty processes may behave arbitrarily.

This paper revisits the equivalence landscape in Byzantine systems, with a particular focus on shared registers and broadcast abstractions. We establish three new reductions.

Specifically, we prove that the broadcast abstraction called Byzantine Set-Constrained Delivery Broadcast (BSCD-Broadcast) can implement a Snapshot/Append object and vice versa (I.e., a two-way reduction), and that the FIFO variant of the renowned Byzantine Reliable Broadcast (BRB-Broadcast) can implement BSCD-Broadcast under a majority of correct processes. Interestingly, this assumption mirrors the one required in crash-prone systems for a similar transformation.

Keywords: Broadcast primitive · Byzantine process · Communication abstraction · FIFO broadcast · Message ordering · Message-passing system · Read/Append register · Reliable broadcast · Shared memory · Snapshot/Append object

1 Introduction

A distributed system is typically modeled using one of two fundamental paradigms: shared memory or message passing. In both models, a variety of high-level abstractions have been defined to offer convenient programming interfaces and structure for algorithm design. In shared memory, basic abstractions include read/write registers, stacks, and queues. In message passing, typical abstractions are communication services such as FIFO broadcast, reliable broadcast [13], causal broadcast [5], and atomic broadcast, first formally defined in [13]. These

S. Bonomi et al. (Eds.): SSS 2025, LNCS 16350, pp. 343–358, 2026.
https://doi.org/10.1007/978-3-032-11127-2_27

abstractions capture essential synchronization properties and serve as modular building blocks for higher-level algorithms.

In asynchronous distributed systems prone to process failures (crashes or Byzantine faults), not all abstractions are implementable without additional assumptions on the system (bounds on the number of failures, restrictions on the asynchrony, failure detectors, etc.) [12,19]. A central question in distributed computability is therefore to compare the power of abstractions: Can an abstraction A be implemented using an abstraction B as a building block, and conversely? When both directions hold, A and B are said to be equivalent. Such equivalences clarify the landscape, letting us swap abstractions without changing computability — although they may still differ in programming convenience or implementation cost. Establishing equivalences is particularly challenging under failures, where up to t among n processes may be faulty.

Crash-Prone Setting. Many equivalences have been established in the crash-prone setting. For example, single-writer/multi-reader (SWMR) registers and snapshot objects are equivalent. A snapshot object [2] lets each process write to its SWMR register and take a snapshot that returns the whole array, ensuring that snapshots observe all writes in a single consistent total order, which extends the actual execution order of writes and snapshots. In contrast, a shared stack cannot be implemented from registers alone. In the message-passing world, FIFO broadcast, reliable broadcast, and causal broadcast can be implemented from one another, while atomic broadcast is strictly stronger. Bridges between the two models are also well understood: The ABD simulation [3] emulates shared memory over a crash-prone asynchronous message-passing system when a majority of processes remain correct. Moreover, consensus objects are equivalent to atomic broadcast [8]. More recently, mutual broadcast [10,11] and SCD-broadcast [14] were shown equivalent to each other and to SWMR registers/snapshot objects under crash failures (see Figure 1).

Byzantine Setting. The picture is more complex when considering Byzantine failures[21], a fault model proposed more than forty years ago and revived by the advent of blockchains [1] and Byzantine Fault-Tolerant (BFT) state-machine replication [7,22]. In the Byzantine model, faulty processes may deviate arbitrarily from their specification, either due to malicious attacks or accidental software corruption. In this setting, several crash-model equivalences break down. For instance, reliable broadcast becomes strictly stronger than the basic unreliable broadcast abstraction: implementing it requires the resilience condition $t < n/3$, which is both necessary and sufficient [6]. Furthermore, FIFO reliable broadcast can no longer be implemented from reliable broadcast in the presence of Byzantine processes ($t < n/3$ is necessary) [17].

Adapting shared-memory abstractions to Byzantine settings also demands refined register definitions, because Byzantine writers may misbehave. Three variants of SWMR registers appear in the literature: the classical Read/Write register [9], the Read/Write-Increment register [20], and the Read/Append register [15]. While these registers are computationally equivalent in crash-prone systems, they differ in the Byzantine case: Read/Append is strictly stronger, as it can implement the other two, but not conversely [16]. Moreover, these shared registers over a Byzantine-prone message-passing system all need $t < n/3$. This motivates a fresh look at equivalence results under Byzantine faults.

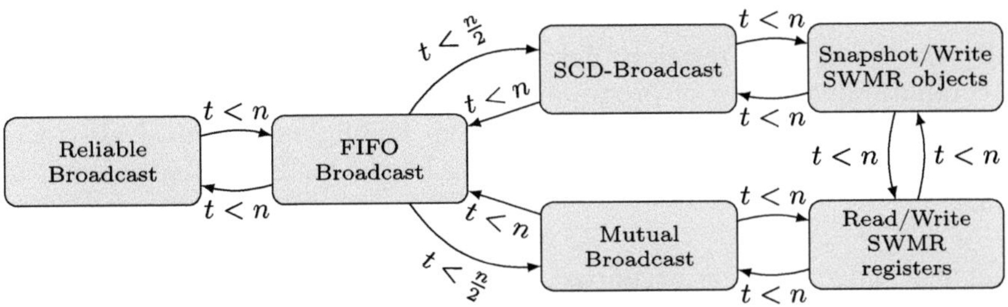

Fig. 1. Equivalence of fundamental abstractions in crash-prone asynchronous systems.

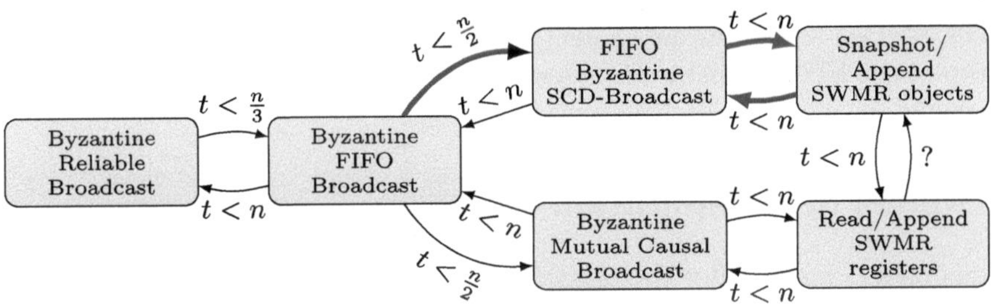

Fig. 2. Equivalence and separation of fundamental abstractions in Byzantine-prone asynchronous systems.

Contributions of the Paper. This paper investigates equivalence relations between fundamental abstractions in Byzantine-prone distributed systems. Figure 1 recalls the crash-prone situation: all four considered abstractions that capture the notion of shared memory (the right part of the figure) are equivalent, while from a FIFO broadcast abstraction, the additional condition $t < n/2$ is required. Figure 2 presents the Byzantine counterpart, comparing the Read/Append register (the strongest register variant), the Snapshot/Append object, Causal Byzantine mutual broadcast (a Byzantine variant of mutual broadcast [10,11] that incorporates the causal ordering of message deliveries [5]), Byzantine SCD-broadcast (BSCD-broadcast) [4,18], and FIFO reliable broadcast (a variant of the well-known Byzantine Reliable Broadcast - BRB - [6] that uses sequence numbers). Some reductions are already known; we establish three new reductions, depicted in Figure 2:

- From FIFO reliable broadcast to BSCD-broadcast under the assumption $t < n/2$ — mirroring the resilience of its crash-prone analogue.
- Equivalence between BSCD-broadcast and Snapshot/Append, established through two reductions: an implementation of Snapshot/Append objects atop BSCD-broadcast and conversely, an implementation of BSCD-broadcast atop Snapshot/Append objects.

By clarifying these relationships, the paper contributes to a unified view of the computability power of programming abstractions in the Byzantine model. Beyond the theoretical interest, such a map highlights canonical abstractions that can serve as reliable, modular building blocks for Byzantine-tolerant algorithms.

Roadmap. This paper is organized into 6 sections. Section 2 introduces the underlying computing model, while Section 3 presents the high-level communication abstractions together with Byzantine shared registers and snapshot objects. The next two sections describe three implementations along with their correctness proofs. Specifically, Section 4 details an implementation of an atomic Snapshot/Append object using the proposed BSCD-broadcast abstraction and, conversely, an implementation of BSCD-broadcast from the Snapshot/Append object. Section 5 presents an implementation of the BSCD-broadcast abstraction on top of the FIFO Reliable Broadcast abstraction. Finally, Section 6 concludes the paper.

2   Model of Computation

In this paper, we consider the classical Byzantine-prone asynchronous message-passing model.

Computing Entities. The system consists of a set of n sequential processes, denoted $p_1, p_2, \ldots, p_n$. Processes are asynchronous: each progresses at its own speed, which may vary arbitrarily during an execution and is unknown to the others. Each process p_i knows its own identifier i, which it can use in its algorithm.

Failure Model. Among the n processes, at most t may exhibit a Byzantine behavior. A Byzantine process deviates arbitrarily from its algorithm [21]: it may start in an arbitrary state, stop at any time (crash), perform arbitrary state transitions, send arbitrary or even distinct values to different processes, etc. A Byzantine process is also said to be *faulty*, while a process that never fails (i.e., is not Byzantine) is called *correct*.

Communication Model. Processes communicate by exchanging messages over bi-directional point-to-point channels that exist between every pair of processes. Each channel allows the receiver to identify the sender, and no process can impersonate another correct process. Message transmission is asynchronous and reliable. "Asynchronous" means that there is no bound on message delay. "Reliable" means that channels do not create, duplicate, lose, or alter messages, and that every message sent by a correct process to another correct process is eventually delivered.

Notation. We use the acronym $\mathcal{BAMP}_{n,t}[\emptyset]$ to denote the Byzantine-prone Asynchronous Message-Passing model without any additional computational power. We write $\mathcal{BAMP}_{n,t}[H]$ for $\mathcal{BAMP}_{n,t}[\emptyset]$ enriched with the additional power H. For instance, $\mathcal{BAMP}_{n,t}[t < \frac{n}{3}]$ denotes the case where at least two-thirds of the processes are correct, namely $t < \frac{n}{3}$.

3   Definitions

This section recalls the high-level abstractions studied in the paper. We first present atomic Snapshot/Append objects, which generalize both snapshot objects and Read/Append registers in shared memory. We then define Byzantine FIFO SCD-Broadcast (BSCD-broadcast), a communication abstraction that extends SCD-broadcast with per-sender FIFO ordering and Byzantine fault tolerance. These definitions provide the foundation for the equivalence results established in the following sections.

3.1   Atomic Snapshot/Append Objects

The snapshot object was introduced to provide a clean abstraction that enables processes in a crash-prone distributed system to obtain a consistent global view of shared memory despite concurrent updates [2]. It consists of an array of n atomic SWMR Read/Write registers, one per process. The object supports two basic operations: a write operation, where process p_i updates its own register, and a snapshot operation, which returns the values of all n registers when invoked.

As discussed in the Introduction, previous work [16] has shown that in Byzantine systems, different register variants (namely, Read/Write, Read/Write-Increment, Read/Append) differ in their computability power. In particular, Read/Append and Snapshot/Append objects are the strongest among them and can be implemented provided that $t < n/2$.

In this paper, we focus on atomic Snapshot/Append objects, as they capture the strongest shared-memory abstraction, which is equivalent to BSCD-broadcast in the Byzantine model. The Snapshot/Append object generalizes both atomic snapshot objects (where a process appends a new value instead of overwriting it) and Read/Append registers (where only a single slot is accessed).

Byzantine Linearizability. The correctness of shared objects in Byzantine systems is defined by *Byzantine linearizability* [9]. Informally, an execution is Byzantine-linearizable if there exists a sequential history that (i) preserves the real-time precedence between operations invoked by correct processes, (ii) is consistent with the sequential specification of the object, and (iii) is indistinguishable to correct processes from the actual execution, even in the presence of arbitrary Byzantine behaviors. This definition ensures that the correct processes observe a behavior compatible with a fault-free sequential execution of the object.

Snapshot/Append Object. An *atomic Snapshot/Append object* consists of an array of n slots (a slot is a SWMR Read/Append register), one per process, with the following operations:

- $append_i(v)$: invoked by process p_i to append value v to its own slot;
- $snapshot()$: invoked by any process to return the full array, where each slot contains the sequence of values appended so far by the corresponding process.

Its sequential specification requires that all operations appear in a single global linearization consistent with real-time precedence: each $append_i(v)$ extends the local sequence of p_i by one value, and each $snapshot()$ returns a vector of sequences reflecting a prefix of all appends.

Informally, the linearizability of the Snapshot/Append object ensures that if a snapshot operation by process p_i returns an array $logs_i$ of Read/Append registers and a snapshot operation by process p_j returns an array $logs_j$, then there do not exist two indices x and y such that $logs_i[x]$ is a prefix of $logs_j[x]$ while simultaneously $logs_j[y]$ is a prefix of $logs_i[y]$. In other words, one of the two operations occurred before the other.

3.2   Byzantine FIFO SCD-Broadcast (BSCD-Broadcast)

On the broadcast side, the definition of SCD-broadcast proposed in [4] must be refined to capture the sequence of messages broadcast by each process. We therefore consider *Byzantine FIFO SCD-Broadcast*, abbreviated BSCD-broadcast.

This communication abstraction provides one broadcast operation and a delivery event generated by the system:

- $bf\_scd\_broadcast(m)$: an operation invoked by a process to scd-broadcast a message m.
- $bf\_scd\_deliver(M)$: an event generated by the system at process p_i, indicating that p_i scd-delivers a non-empty set M of triplets $\langle m, sd, sn \rangle$. Here, m is a message scd-broadcast by process p_{sd}, and sn is its sequence number, i.e., the rank of m in the sequence of messages scd-broadcast by p_{sd}.

Each message is uniquely identified by its triplet: even if two distinct messages carry the same content, they are treated as different if they originate from different senders or with different sequence numbers. For clarity, we write $bf\_scd\_broadcast(m)$ as if it directly takes the content m, although in practice it creates a fresh message tagged with (sd, sn).

We say that process p_i *scd-broadcasts* a message m when it invokes $bf\_scd\_broadcast(m)$, and that p_i *scd-delivers* a set M when the event $bf\_scd\_deliver(M)$ occurs at p_i. By slight abuse of language, we also say that p_i scd-delivers a triplet $\langle m, sd, sn \rangle$ if this triplet belongs to M.

```
1  operation append_i(v) is:
2  |  synch_bf_scd_broadcast APPEND(v);

3  operation snapshot() is:
4  |  synch_bf_scd_broadcast SYNCH();
5  |  return replica_i.snapshot();

6  when the message set M is scd-delivered do:
7  |  foreach ⟨APPEND(v), sd, sn⟩ ∈ M ordered by sn do
8  |  |  replica_i.append_sd(v);
```

Algorithm 1: Implementation of a linearizable Snapshot/Append object using BSCD-broadcast.

Properties. BSCD-broadcast satisfies the following properties:

Validity. If a correct process p_i scd-delivers a triplet $\langle m, sd, sn \rangle$ from a correct sender p_{sd}, then p_{sd} must have invoked bf_scd_broadcast(m) as its sn-th broadcast.

Integrity. No correct process scd-delivers the same triplet more than once.

Termination. If a correct process p_i scd-broadcasts a message m with sequence number sn, then p_i eventually scd-delivers its own triplet $\langle m, i, sn \rangle$.

Consistency. If a correct process scd-delivers a triplet $\langle m, sd, sn \rangle$, then all correct processes eventually scd-deliver $\langle m, sd, sn \rangle$, regardless of whether p_{sd} is correct or Byzantine.

SCD-Ordering. If a correct process p_i scd-delivers a set M containing $\langle m, sd, sn \rangle$ before a set M' containing $\langle m', sd', sn' \rangle$, then no correct process scd-delivers $\langle m', sd', sn' \rangle$ before $\langle m, sd, sn \rangle$.

FIFO-Ordering. For each sender p_{sd}, if $sn < sn'$, then no correct process can scd-deliver $\langle m', sd, sn' \rangle$ before $\langle m, sd, sn \rangle$.

4   Equivalence between BSCD-Broadcast and Snapshot/Append Objects

In this section, we establish the equivalence between atomic Snapshot/Append objects and BSCD-broadcast. We proceed by presenting two complementary reductions. First, we show how BSCD-broadcast can be used to implement a wait-free Snapshot/Append object (Algorithm 1). Then, we demonstrate the reverse direction, namely how an atomic Snapshot/Append object can implement BSCD-broadcast (Algorithm 2). Together, these two reductions prove that the two abstractions are computationally equivalent in the Byzantine-prone model.

4.1   From BSCD-Broadcast to Snapshot/Append Objects

BSCD-broadcast was specifically designed to support the implementation of linearizable Snapshot/Append objects. Algorithm 1 builds such an implementation in a wait-free manner.

Local State. Each process p_i maintains a local replica $\mathsf{replica}_i$ of the object. $\mathsf{replica}_i$ stores, for every process p_j, the sequence of values appended to p_j's slot. Updates are applied locally whenever messages are delivered.

Protocol Messages. Two types of messages are exchanged through the *synchronous* BSCD-broadcast abstraction. "Synchronous" here means that a process blocks until its own message is scd-delivered locally, ensuring it is consistently incorporated into the global order.

- APPEND(v): used when a process appends value v to its slot;
- SYNCH(): used when a process requests a snapshot. Delivering this message ensures that all preceding appends are visible.

Append. To append v, p_i broadcasts APPEND(v) and waits until it scd-delivers the corresponding triplet $\langle \text{APPEND}(v), i, sn \rangle$ within a set. It then applies all triplets in that set to $\mathsf{replica}_i$. The replica ensures separate slots per sender and preserves FIFO order within each slot.

Snapshot. To read the object, p_i broadcasts SYNCH() and waits until it scd-delivers its own message. Any accompanying appends are first applied to $\mathsf{replica}_i$, after which p_i returns the content of $\mathsf{replica}_i$ as the snapshot result. The synchronous delivery of SYNCH() guarantees linearizability with respect to concurrent appends.

We prove that Algorithm 1 implements a wait-free, Byzantine-linearizable Snapshot/Append object (Theorem 1).

Lemma 1 (Per-writer order). *For any writer p_w (correct or Byzantine), all correct processes p_i append the values from p_w into $\mathsf{replica}_i[w]$ in the same order.*

Lemma 2 (Validity). *If a* snapshot *by a correct process returns logs, then for every correct p_i and every $r \in \{1, \dots, |logs[i]|\}$, the entry $logs[i][r]$ is the r-th value appended by p_i.*

Lemma 3 (Snapshot-after-Append). *If a* snapshot *by a correct process returns logs and starts after the r-th* append_w *by a correct p_w completes, then $|logs[w]| \geq r$.*

Definition 1 (Prefix). *For lists A, B, we write $A \lhd B$ if A is a proper prefix of B, and $A \unlhd B$ if A is a (possibly equal) prefix of B.*

Lemma 4 (Inclusion). *Let two* snapshot *operations by correct processes return $logs_i$ and $logs_j$. Then either $\forall w : logs_i[w] \unlhd logs_j[w]$, or $\forall w : logs_j[w] \unlhd logs_i[w]$.*

Lemma 5 (Snapshot-after-Snapshot). *If a* snapshot *s_i completes before a* snapshot *s_j starts (both by correct processes), then $\forall w : logs_i[w] \unlhd logs_j[w]$.*

Theorem 1 (Correctness of Algorithm 1). *Algorithm 1 implements a wait-free and Byzantine-linearizable Snapshot/Append object.*

```
 1  operation bf_scd_broadcast(m) is:
 2  |   shared.append_i(m);
 3  |_  try_deliver();

 4  do regularly: try_deliver() ;
 5  function try_deliver() is:
 6  |   let sent ← shared.snapshot();
 7  |   let M ← ∅;
 8  |   for sd from 1 to n do
 9  |   |   for sn from delivered_i[sd] + 1 to |sent[sd]| do
10  |   |   |_  M ← M ∪ {⟨sent[sd][sn], sd, sn⟩};
11  |   |_  delivered_i[sd] ← |sent[sd]|;
12  |_  if M ≠ ∅ then bf_scd_deliver M ;
```

Algorithm 2: Implementation of BSCD-broadcast using an atomic Snapshot/Append object.

4.2   From Snapshot/Append Objects to BSCD-Broadcast

Algorithm 2 implements BSCD-broadcast using an atomic Snapshot/Append object. The intuition is as follows: each process appends its messages in its own slot, and all processes periodically take snapshots of the whole array to discover new messages. Since snapshots are consistent and prefix-closed, the sets of newly discovered messages can be safely delivered while preserving both consistency across processes and FIFO ordering within each slot.

Shared Variable. All communication relies on the underlying atomic Snapshot/Append object shared, which serves as the only communication medium. Each slot shared[sd] belongs to process p_{sd} and stores the sequence of values appended by p_{sd}.

Local State. Each process p_i maintains an array delivered$_i[1..n]$, where delivered$_i[sd]$ records how many messages from p_{sd} have already been scd-delivered. All entries are initialized to 0.

BSCD-Broadcast Operation. To bscd-broadcast a message m, a process p_i appends m to its own slot in shared (line 2), then calls try_deliver to possibly deliver pending messages (line 3).

Delivery Procedure. The background task periodically invokes try_deliver (line 4). This procedure takes a snapshot of shared (line 6), collects all messages not yet delivered into a set M (lines 8–10), and scd-delivers the corresponding set of triplets (line 12) if M is non-empty. Process p_i finally updates delivered$_i$ accordingly (line 11).

We prove that Algorithm 2 implements BSCD-broadcast using an atomic Snapshot/Append object. Intuitively, each process appends its own messages in a single-writer slot and periodically takes an atomic snapshot to discover new messages across all slots. Because snapshots are atomic and prefix-closed, sets of "new since last snapshot" are safe to deliver, preserving both global SCD-Ordering and per-sender FIFO-Ordering.

Lemma 6 (Validity). *If a correct process p_i scd-delivers a triplet $\langle m, sd, sn \rangle$ from a correct sender p_{sd}, then p_{sd} has invoked* `bf_scd_broadcast`(m) *as its sn-th broadcast.*

Lemma 7 (Integrity). *No correct process scd-delivers the same triplet more than once.*

Lemma 8 (Termination). *If a correct process p_i scd-broadcasts a message m with sequence number sn, then p_i eventually scd-delivers its own triplet $\langle m, i, sn \rangle$.*

Lemma 9 (Consistency). *If a correct process scd-delivers a triplet $\langle m, sd, sn \rangle$, then all correct processes eventually scd-deliver $\langle m, sd, sn \rangle$, regardless of whether p_{sd} is correct or Byzantine.*

Lemma 10 (SCD-Ordering). *If a correct process p_i scd-delivers a set M containing $\langle m, sd, sn \rangle$ before a set M' containing $\langle m', sd', sn' \rangle$, then no correct process scd-delivers $\langle m', sd', sn' \rangle$ before $\langle m, sd, sn \rangle$.*

Lemma 11 (FIFO-Ordering). *For each sender p_{sd}, if $sn < sn'$, then no correct process scd-delivers $\langle m', sd, sn' \rangle$ before $\langle m, sd, sn \rangle$.*

Theorem 2 (Correctness of Algorithm 2). *Algorithm 2 implements BSCD-broadcast.*

5   Implementation of BSCD-Broadcast

5.1   The algorithm

We now present Algorithm 3, which implements BSCD-broadcast using FIFO reliable broadcast (FIFO-broadcast) in an asynchronous message-passing system, resilient to $t < \frac{n}{2}$ Byzantine processes. FIFO-broadcast provides the same guarantees as BSCD-broadcast (Validity, Integrity, Termination, and Consistency), except that it does not enforce SCD-Ordering: for each sender p_{sd}, all correct processes fifo-deliver its messages in the same order (FIFO), regardless of whether p_{sd} is correct or Byzantine, but no constraint is imposed on the relative order of messages from different senders.

Intuition. The algorithm builds on two central ideas. First, every new message is echoed by all processes through a FORWARD message, so that all correct processes are aware not only of the message itself but also of the relative order in which other processes received it. Second, scd-delivery is based on the majority concept: a message can only be scd-delivered once more than half of the processes have forwarded it, ensuring that at least one correct process contributed to this decision. Finally, a filtering step guarantees that two correct processes cannot scd-deliver messages in conflicting orders, thus preserving the SCD-Ordering property of SCD-broadcast.

```
1  operation bf_scd_broadcast(m) is:
2  |    let sn ← delivered_i[i];
3  |    fifo_broadcast INIT(m);
4  |    wait until delivered_i[i] = sn + 1;

5  when INIT(m) is k-th INIT message fifo_delivered from p_sd do:
6  |    wait until delivered_i[sd] = k − 1;
7  |    messages_i[sd] ← m;
8  |    fifo_broadcast FORWARD(sd, k);

9  when FORWARD(sd, sn) is k-th FORWARD message fifo_delivered from p_f do:
10 |    wait until delivered_i[sd] = sn − 1 ∧ messages_i[sd] ≠ ⊥;
11 |    stamps_i[sd][f] ← k;
12 |    try_deliver();

13 function try_deliver() is:
14 |    let ids ← {j : stamped_by_majority(j)};
15 |    while ∃j ∈ ids, ∃k ∉ ids, depends_on(j, k) do   ids ← ids \ {j} ;
16 |    if ids ≠ ∅ then
17 |    |    bf_scd_deliver {⟨messages_i[j], j, delivered_i[j] + 1⟩ : j ∈ ids};
18 |    for j ∈ ids do
19 |    |    delivered_i[j] ← delivered_i[j] + 1;
20 |    |    stamps_i[j] ← [∞, ..., ∞];
21 |    |    messages_i[j] ← ⊥;

22 function stamped_by_majority(j) is:
23 |    return |{f : stamps_i[j][f] < ∞}| > n/2;

24 function depends_on(j, k) is:
25 |    return |{f : stamps_i[j][f] < stamps_i[k][f]}| ≤ n/2;
```

Algorithm 3: Implementation of BSCD-broadcast using FIFO-broadcast provided $t < \frac{n}{2}$.

Local State. Each process p_i maintains the following variables:

- $delivered_i[1..n]$: for each process p_j, the sequence number of the last message from p_j that has been scd-delivered (initially 0).
- $messages_i[1..n]$: a buffer holding the current message expected from each p_j (initially $\perp$).
- $stamps_i[1..n][1..n]$: for each sender sd and each process f, records the local rank k of the FORWARD($sd, \cdot$) messages as fifo-delivered from p_f; initially ∞.

Protocol Messages. Two types of protocol messages are exchanged using FIFO-broadcast:

- INIT(m): announces the intent of a process to scd-broadcast message m.
- FORWARD(sd, sn): indicates that the sender has accepted the sn-th message from process p_{sd} and propagates the local order in which this message was seen.

Together, these messages ensure both the dissemination of the message contents and the propagation of relative ordering information.

Identifiers. Every application message m broadcast by process p_{sd} is associated with an identifier $\langle sd,\ sn \rangle$, where sn is its local sequence number of m, meaning m is the sn-th message broadcast by p_{sd}. This identifier is consistent across all correct processes thanks to the properties of FIFO-broadcast.

Overview of the Algorithm. To scd-broadcast a message m, a process p_i sends an INIT(m) message and waits until m has been scd-delivered locally (lines 2–4).

When a process fifo-delivers INIT(m) from p_{sd}, it records m as the next pending message from p_{sd} and immediately broadcasts a FORWARD(sd, sn) message (lines 6–8). This FORWARD message informs other processes about the relative order in which m was observed at the sender.

When a process fifo-delivers a FORWARD(sd, sn) message from p_f, it records the local rank k of this FORWARD message in $\mathsf{stamps}_i[sd][f]$ (lines 10–12) and invokes `try_deliver()`.

Delivery Rule. The function `try_deliver` computes the set of messages that can be safely scd-delivered:

– `stamped_by_majority(`j`)` holds if process p_i has received FORWARD$(j, \cdot)$ from more than $\frac{n}{2}$ processes, meaning a majority has observed the message from p_j.

– `depends_on(`j, k`)` checks whether p_i has enough evidence to conclude that the message from p_j precedes the message from p_k. Specifically, if at most $n/2$ processes sent a message FORWARD$(j, \cdot)$ before a message FORWARD$(k, \cdot)$, then there is insufficient evidence to scd-deliver p_j's message before p_k's, since another process might derive the opposite order.

The candidate set *ids* consists of all processes p_j such that `stamped_by_majority(`j`)` holds. Messages that are dominated by some non-candidate according to `depends_on` are filtered out (line 15), transitively removing any further dominated candidates. If the resulting set is non-empty, its elements are scd-delivered together as a set of triplets $\langle m, sd, sn \rangle$ (line 17). After scd-delivery, p_i updates its local state. This prepares it to process the next message from each p_j from which it scd-delivered a message (lines 19–21).

5.2   Proof of Algorithm 3

We now prove that Algorithm 3 correctly implements BSCD-broadcast using FIFO-broadcast under the resilience condition $t < \frac{n}{2}$. The proof proceeds property by property, following the specification of BSCD-broadcast.

Lemma 12 (Eventual Delivery). *If a correct process p_i fifo-delivers* INIT(m) *as the sn-th message from p_{sd}, then every correct process eventually scd-delivers* $\langle m, sd, sn \rangle$.

Proof. Suppose, for contradiction, that some correct process never delivers $\langle m, sd, sn \rangle$. Let sn be the smallest sequence number for which this occurs.

By minimality, all earlier messages from p_{sd} are eventually delivered everywhere, so each correct process reaches $\mathsf{delivered}_i[sd] = sn-1$. By FIFO-broadcast termination and consistency, every correct process then fifo-delivers the same $\textsc{init}(m)$ as the sn-th message from p_{sd}, assigns $\mathsf{messages}_i[sd] \leftarrow m$, and broadcasts $\textsc{forward}(sd, sn)$. Since a strict majority of processes are correct, each correct process eventually gathers a majority of such $\textsc{forward}$ messages, so $\mathsf{stamped\_by\_majority}(sd)$ holds permanently until $\langle m, sd, sn \rangle$ is delivered.

As long as delivery does not occur, $\mathsf{try\_deliver}()$ is invoked infinitely often. Indeed, whenever (sd, sn) is filtered out (line 15), this is because some (sd', sn') is considered to precede it. But this requires reception of a $\textsc{forward}(sd', sn')$, which in turn triggers further calls to $\mathsf{try\_deliver}()$. Hence (sd, sn) is reconsidered infinitely many times. Two scenarios remain:

1. Stable $\mathsf{delivered}_i$: If all delivery counters eventually stop increasing, only finitely many identifiers remain pending. Each has a strict majority of $\textsc{forward}$ messages, so none can be blocked forever. In particular, (sd, sn) must eventually be delivered. A contradiction.
2. Unbounded $\mathsf{delivered}_i$: If some $\mathsf{delivered}_i[sd']$ grows without bound, processes keep producing new $\textsc{forward}(sd', \cdot)$. Eventually, for a strict majority of f, $\mathsf{stamps}_i[sd][f] < \mathsf{stamps}_i[sd'][f]$, making $\prec (sd, sd')$ permanently false. Thus $(sd', \cdot)$ can no longer block (sd, sn). Since only finitely many senders exist, all possible blockers are eventually eliminated, so (sd, sn) is delivered. A contradiction.

In both cases we reach a contradiction. Therefore, every correct process eventually delivers $\langle m, sd, sn \rangle$.

Lemma 13 (Validity). *If a correct process p_i scd-delivers a triplet $\langle m, sd, sn \rangle$ from a correct sender p_{sd}, then p_{sd} must have invoked $\mathtt{bf\_scd\_broadcast}(m)$ as its sn-th broadcast.*

Proof. Triplets are created only at line 17 using $\mathsf{messages}_i[sd]$. To have $sd \in ids$, $\mathsf{stamped\_by\_majority}(sd)$ must hold (line 14), i.e., $\mathsf{stamps}_i[sd][f] < \infty$ for a strict majority (line 23). Each such stamp is written at line 11, but only after the guard $\mathsf{messages}_i[sd] \neq \bot$ is satisfied (line 10). Hence, when sd enters ids, necessarily $\mathsf{messages}_i[sd] \neq \bot$. Moreover, $\mathsf{messages}_i[sd]$ is reset to $\bot$ only *after* delivery, at line 21, so the m used in line 17 is well-defined and non-$\bot$.

Now, $\mathsf{messages}_i[sd]$ is assigned only at line 7, upon fifo-delivery of $\textsc{init}(m)$ from p_{sd}, after waiting for $\mathsf{delivered}_i[sd] = k - 1$ (line 6), where k is the local rank of that $\textsc{init}$ from sd. By FIFO-broadcast Validity, this $\textsc{init}(m)$ must have been broadcast by p_{sd} at line 3, hence p_{sd} invoked $\mathtt{bf\_scd\_broadcast}(m)$.

Finally, delivery uses sequence number $\mathsf{delivered}_i[sd] + 1$ (line 17); together with the wait $\mathsf{delivered}_i[sd] = k-1$ in the $\textsc{init}$ handler, this implies $sn = k$. Since the k-th $\textsc{init}$ from sd is the same system-wide (FIFO-Ordering and Integrity), m is exactly the sn-th message broadcast by p_{sd}.

Lemma 14 (Integrity). *No correct process scd-delivers the same triplet more than once.*

Proof. Each delivery at line 17 produces the triplet $\langle \mathsf{messages}_i[j], j, \mathsf{delivered}_i[j] + 1\rangle$. Immediately afterwards, $\mathsf{delivered}_i[j]$ is incremented (line 19) and both $\mathsf{stamps}_i[j]$ and $\mathsf{messages}_i[j]$ are reset (lines 20–21). Hence, the same pair (j, sn) cannot be delivered twice.

Lemma 15 (Termination). *If a correct process p_i scd-broadcasts a message m with sequence number sn, then p_i eventually scd-delivers its own triplet $\langle m, i, sn\rangle$.*

Proof. Process p_i broadcasts $\textsc{init}(m)$ (line 3) and, by FIFO-broadcast termination, eventually fifo-delivers this $\textsc{init}(m)$ as its sn-th $\textsc{init}$ from itself. By Lemma 12, every correct process, including p_i, eventually scd-delivers its own triplet $\langle m, i, sn\rangle$.

Lemma 16 (Consistency). *If a correct process scd-delivers a triplet $\langle m, sd, sn\rangle$, then all correct processes eventually scd-deliver $\langle m, sd, sn\rangle$, regardless of whether p_{sd} is correct or Byzantine.*

Proof. If a correct process scd-delivers $\langle m, sd, sn\rangle$, then it has previously fifo-delivered $\textsc{init}(m)$ as the sn-th $\textsc{init}$ from p_{sd} (as discussed in the proof of Lemma 13). By Lemma 12, all correct processes eventually scd-deliver the triplet.

Lemma 17 (SCD-Ordering). *If a correct process p_i scd-delivers a set M containing $\langle m, sd, sn\rangle$ before a set M' containing $\langle m', sd', sn'\rangle$, then no correct process scd-delivers $\langle m', sd', sn'\rangle$ before $\langle m, sd, sn\rangle$.*

Proof. Let $x = (sd, sn)$ and $y = (sd', sn')$. At p_i, when x is delivered before y, the filtering loop (line 15) guarantees that more than $n/2$ processes p_f have forwarded x before y (i.e., $\mathsf{stamps}_i[x][f] < \mathsf{stamps}_i[y][f]$). Thus, a strict majority supports $x \prec y$.

Assume for contradiction that some correct process p_j delivers y before x. Then p_j must also have obtained a strict majority of forwards in the opposite order ($y \prec x$). But the two majorities would be disjoint, which is impossible with n processes.

Hence, no correct process can deliver y before x. The only alternative is that x and y are delivered together in the same set, which remains allowed by the specification.

Lemma 18 (FIFO-Ordering). *For each sender p_{sd}, if $sn < sn'$, then no correct process can scd-deliver $\langle m', sd, sn'\rangle$ before $\langle m, sd, sn\rangle$.*

Proof. By construction, $\textsc{trydel}$ only delivers for sd the next sequence number $\mathsf{delivered}_i[sd] + 1$ (line 17), then increments $\mathsf{delivered}_i[sd]$ (line 19). Thus, the sequence numbers are strictly increasing.

Theorem 3 (Correctness of Algorithm 3). *Algorithm 3 implements BSCD-broadcast under the condition $t < \frac{n}{2}$.*

Proof. Follows immediately from Lemmas 13–18.

6   Conclusion

This paper clarified the relationships among several fundamental abstractions in Byzantine-prone asynchronous message-passing systems. We established three new reductions, namely: (i) Snapshot/Append objects can be implemented from Byzantine SCD-broadcast, (ii) Byzantine SCD-broadcast can be implemented from Snapshot/Append objects, and (iii) Byzantine SCD-broadcast can be implemented from FIFO reliable broadcast under the resilience bound $t < n/2$. These results extend to Byzantine settings some of the equivalence relations already known in crash-prone systems, while also revealing important separations between register variants.

Overall, our findings contribute to a unified understanding of the computational power of abstractions in the Byzantine model and identify canonical abstractions that can serve as reliable, modular building blocks for distributed algorithms. Future work should refine these equivalence classes to encompass additional classical abstractions in Byzantine systems. In particular, it remains open whether Read/Append and Snapshot/Append objects — equivalent in crash-prone systems — remain equivalent in the Byzantine setting. Addressing this question would further sharpen our understanding of the computability landscape under Byzantine failures.

References

1. Abraham, I., Malkhi, D., Nayak, K., Ren, L., Spiegelman, A.: Solida: a blockchain protocol based on reconfigurable byzantine consensus. In: Proceedings of 21st International Conference on Principles of Distributed Systems (OPODIS). LIPIcs, vol. 95, pp. 25:1–25:19 (2017)
2. Afek, Y., Attiya, H., Dolev, D., Gafni, E., Merritt, M., Shavit, N.: Atomic snapshots of shared memory. J. ACM **40**(4), 873–890 (1993)
3. Attiya, H., Bar-Noy, A., Dolev, D.: Sharing memory robustly in message-passing systems. J. ACM **42**(1), 124–142 (1995)
4. Auvolat, A., Raynal, M., Taïani, F.: Byzantine-tolerant set-constrained delivery broadcast. In: 23rd International Conference on Principles of Distributed Systems, OPODIS 2019, Neuchâtel, Switzerland. LIPIcs, vol. 153, pp. 6:1–6:23 (2019)
5. Birman, K.P., Joseph, T.A.: Reliable communication in the presence of failures. ACM Trans. Comput. Syst. **5**(1), 47–76 (1987)
6. Bracha, G.: An asynchronous $(n-1)/3$-resilient consensus protocol. In: Proceedings of 3rd ACM Symposium on Principles of Distributed Computing (PODC), pp. 154–162. ACM (1984)
7. Castro, M., Liskov, B.: Practical byzantine fault tolerance. In: Seltzer, M.I., Leach, P.J. (eds.) Proceedings of the 3rd USENIX Symposium on Operating Systems Design and Implementation (OSDI), New Orleans, Louisiana, pp. 173–186 (1999)
8. Chandra, T.D., Toueg, S.: Unreliable failure detectors for reliable distributed systems. J. ACM (JACM) **43**(2), 225–267 (1996)
9. Cohen, S., Keidar, I.: Tame the wild with byzantine linearizability: reliable broadcast, snapshots, and asset transfer. In: 35th International Symposium on Distributed Computing (DISC). LIPIcs, vol. 209, pp. 18:1–18:18 (2021)

10. Déprés, M., Mostéfaoui, A., Perrin, M., Raynal, M.: Brief announcement: the MBroadcast abstraction. In: Proceedings of the 2023 ACM Symposium on Principles of Distributed Computing, PODC, Orlando, FL, USA, pp. 282–285 (2023)
11. Déprés, M., Mostéfaoui, A., Perrin, M., Raynal, M.: Send/receive patterns versus read/write patterns in crash-prone asynchronous distributed systems. In: 37th International Symposium on Distributed Computing, DISC 2023, L'Aquila, Italy. LIPIcs, vol. 281, pp. 16:1–16:24 (2023)
12. Fich, F.E., Ruppert, E.: Hundreds of impossibility results for distributed computing. Distrib. Comput. **16**(2–3), 121–163 (2003)
13. Hadzilacos, V., Toueg, S.: Fault-tolerant broadcasts and related problems. In: Mullender, S. (ed.) Distributed Systems, pp. 97–145. ACM Press/Addison-Wesley, 2 edn. (1993)
14. Imbs, D., Mostéfaoui, A., Perrin, M., Raynal, M.: Set-constrained delivery broadcast: a communication abstraction for read/write implementable distributed objects. Theor. Comput. Sci. **886**, 49–68 (2021)
15. Imbs, D., Rajsbaum, S., Raynal, M., Stainer, J.: Read/write shared memory abstraction on top of asynchronous byzantine message-passing systems. J. Parallel Distributed Comput. **93–94**, 1–9 (2016)
16. Kowalski, V., Mostéfaoui, A., Perrin, M.: Atomic register abstractions for byzantine-prone distributed systems. In: 27th International Conference on Principles of Distributed Systems, OPODIS, Tokyo, Japan. LIPIcs, vol. 286 (2023)
17. Kowalski, V., Mostéfaoui, A., Perrin, M.: Invited paper: causal mutual byzantine broadcast. In: Proceedings of the Workshop ApPLIED@PODC 2024 on Advanced Tools, Programming Languages, and PLatforms for Implementing and Evaluating algorithms for Distributed systems, Nantes, France, pp. 1–8. ACM (2024)
18. Kowalski, V., Mostéfaoui, A., Perrin, M.: An optimal byzantine SCD-broadcast protocol. In: 19th European Dependable Computing Conference, EDCC 2024, Leuven, Belgium, pp. 139–146. IEEE (2024)
19. Lynch, N.A.: A hundred impossibility proofs for distributed computing. In: Proceedings of the Eighth Annual ACM Symposium on Principles of Distributed Computing, Edmonton, Alberta, Canada, August 14-16, 1989, pp. 1–28 (1989)
20. Mostéfaoui, A., Petrolia, M., Raynal, M., Jard, C.: Atomic read/write memory in signature-free byzantine asynchronous message-passing systems. Theory Comput. Syst. **60**(4), 677–694 (2017)
21. Pease, M.C., Shostak, R.E., Lamport, L.: Reaching agreement in the presence of faults. J. ACM **27**(2), 228–234 (1980)
22. Schneider, F.B.: Implementing fault-tolerant services using the state machine approach: a tutorial. ACM Comput. Surv. **22**(4), 299–319 (1990)

Improving the Hu-Toueg Construction of a Byzantine Linearizable SWMR Register

Ajay D. Kshemkalyani[1]([⊠])(iD), Manaswini Piduguralla[2], Sathya Peri[2], and Anshuman Misra[3]

[1] University of Illinois Chicago, Chicago, USA
ajay@uic.edu
[2] Indian Institute of Technology Hyderabad, Hyderabad, India
cs20resch11007@iith.ac.in, sathya_p@cse.iith.ac.in
[3] Purdue University Fort Wayne, Fort Wayne, USA
misra47@pfw.edu

Abstract. Recently, Hu and Toueg gave an implementation of the SWMR register from SWSR registers in shared memory distributed systems. While their definition of register linearizability is consistent with the definition of Byzantine linearizability of a concurrent history of Cohen and Keidar, it has several drawbacks. A stronger definition of a Byzantine linearizable register that overcomes these drawbacks has been proposed. In this paper, we give a construction of a Byzantine linearizable SWMR atomic register from SWSR registers that meets the stronger definition. The construction is correct when $n > 3f$, where n is the number of readers, f is the maximum number of Byzantine readers, and the writer can also be Byzantine. The construction relies on a public-key infrastructure.

Keywords: Byzantine fault-tolerance · Register construction · Shared memory · SWMR Register

1 Introduction

We consider the problem of implementing a single-writer multi-reader (SWMR) register from single-writer single-reader (SWSR) registers in a system with Byzantine processes. Recently, Hu and Toueg gave an implementation of the SWMR register from SWSR registers [4,5]. While their definition of register linearizability is consistent with the definition of Byzantine linearizability of a concurrent history of Cohen and Keidar [2], the problem formulation and results in [2,4,5] have the following drawbacks, identified and described in [8].

1. If the writer is Byzantine, the register is vacuously linearizable no matter what values the correct readers return. Reads by correct processes can return any value whatsoever including the initial value while the register meets their

© The Author(s), under exclusive license to Springer Nature Switzerland AG 2026
S. Bonomi et al. (Eds.): SSS 2025, LNCS 16350, pp. 359–375, 2026.
https://doi.org/10.1007/978-3-032-11127-2_28

definition of linearizability. In particular, there is no view consistency. For example, in the Hu-Toueg algorithm, consider a scenario where a Byzantine writer writes a different data value associated with the same counter value to the various readers' SWSR registers. The correct readers will return different data values associated with the same counter value, thus having inconsistent views.

2. Their definition of register linearizability does not factor in, or ignores, those values written by a Byzantine writer, by honestly following the writer protocol for those values. We need a stronger notion of a *correct write operation* that factors in such values as being written correctly. Also, note that the Byzantine writer is in control of the execution both above and below the SWMR register interface and hence the value that it writes in a correct write operation can be assumed to be the value intended to be written (correctly) and not altered by Byzantine behavior.

3. Their definition of register linearizability allows a value written by a Byzantine writer to just a single reader's SWSR register to be returned by a correct process. In order to validate that the writer intended to write that value honestly, we would like a minimum threshold number of readers' SWSR registers to be written that same value to enable that value to become eligible for being returned to a correct reader. This validates the intention of the Byzantine writer to write that particular value.

4. In their definition of register linearizability, their notion of a "current" value returned by a correct reader is not related to the most recent value written by a correct write operation of a Byzantine writer.
We need a more up to date version of the value that can be returned by a correct reader. This helps give a stronger guarantee of progress from the readers' perspective.

The definition of a Byzantine linearizable register given in [8] is stronger than that of [2, 4, 5] and overcomes the above drawbacks. Further, we are interested in implementing the SWMR register over SWSR registers directly in the shared memory model.

Contributions. We first review the stronger definition of a *Byzantine linearizable register* of [8] that overcomes all the above drawbacks of [2, 4, 5]. We examine their concept of a *correct write operation* by a Byzantine writer as one that conforms to the write protocol. We also examine their notion of a *pseudo-correct write operation* by a Byzantine writer, which has the effect of a correct write operation. Only correct and pseudo-correct writes may be returned by correct readers. The correct and pseudo-correct writes are totally ordered by the logical timestamps issued by the writer.

The main contribution is giving a construction of a Byzantine linearizable SWMR atomic register from SWSR atomic registers that meets the stronger definition of [8]. The construction is correct when $n > 3f$, where n is the number of readers, f is the maximum number of Byzantine readers, and the writer can also be Byzantine. The construction relies on a public-key infrastructure (PKI). The construction develops the idea of the readers validating the logical timestamp of the writing of the values set aside for them by the writer. A sufficient number of correct readers will validate this consistently, and that forms the basis of the total order used to ensure Byzantine register linearizability. As compared to the algorithm in [4,5] which can tolerate any number of Byzantine readers, our algorithm requires $f < n/3$. Also, in the algorithm in [4,5], a reader that stops reading also stops taking implementation steps whereas our algorithm requires a reader helper thread to take infinitely many steps even if it has no read operation to apply. The algorithm in [4,5] as well as our algorithm use a PKI. The algorithm in [4,5] uses n^2 shared SWSR registers, whereas our algorithm uses $2n^2 + 2n$ shared SWSR registers.

Related Work. The problem of implementing shared registers from weaker types of registers has been extensively studied, see [8] for a survey. The SWMR register in a Byzantine setting is of great importance in recent research. For example, Mostefaoui et al. [10] prove that in message-passing systems with Byzantine failures, there is a f-resilient implementation of a SWMR register if and only if $f < n/3$ processes are faulty, where f is the number of Byzantine processes and n is the total number of processes. It was the first to give the definition of a linearizable SWMR register in the presence of Byzantine processes and [2] generalized it to objects of any type. Aguilera et al. [1] use atomic SWMR registers to solve some agreement problems in hybrid systems subject to Byzantine process failures. Cohen and Keidar [2] give f-resilient implementations of three objects – asset transfer, reliable broadcast, atomic snapshots – using atomic SWMR registers in systems with Byzantine failures where at most $f < n/2$ processes are faulty. Their implementations were based on their definition of Byzantine linearizability of a concurrent history.

In other related work, a SWMR register was built above a message-passing system where processes communicate using send/receive primitives with the constraint that $f < n/3$ [6,10]. These works do not use signatures. Unbounded history registers were required in [6] whereas [10] used $O(n^2)$ messages per write operation. Although building SWMR registers over SWSR registers or over message-passing systems is equivalent as SWSR registers can be emulated over send/receive and vice versa, this is a round-about and expensive solution and there are fundamental differences between the shared memory paradigm and the message-passing paradigm.

Roadmap: Section 2 gives the system model and preliminaries. Section 3 reviews the characterization and definition of a Byzantine linearizable register, from [8]. Section 4 gives our construction of the SWMR Byzantine linearizable register using SWSR registers. Section 5 gives the correctness proof. Section 6 gives a discussion.

2 System Model

Model Basics. We consider the shared memory model of a distributed system. The system contains a set of asynchronous processes. These processes access some shared memory objects. All inter-process communication is done through an API exposed by the objects. Processes invoke operations that return some response to the invoking process. We assume reliable shared memory but allow for an adversary to corrupt up to f processes in the course of a run. A corrupted process is defined as being *Byzantine* and such a process may deviate arbitrarily from the protocol. A non-Byzantine process is *correct* and such a process follows the protocol and takes infinitely many steps.

We also assume a PKI. Each process has a public-private key pair used to sign data and verify signatures of other processes. A values v signed by process p is denoted $\langle v \rangle_p$.

We give an algorithm that emulates an object O, viz., a SWMR register from SWSR registers. We assume that there is adequate access control such that a SWSR register can be accessed only by the single writer and the single reader between whom the register is set up, and that another (Byzantine) process cannot access it. The algorithm is organized as methods of O. A method execution is a sequence of steps. It begins with the *invoke* step, goes through steps that access lower-level objects, viz., SWSR registers, and ends with a *return* step. The invocation and response delineate the method's execution interval. In an *execution* σ, each correct process invokes methods sequentially, and steps of different processes are interleaved. Byzantine processes take arbitrary steps irrespective of the protocol. The *history* H of an execution σ is the sequence of high-level invocation and response events of the emulated SWMR register in σ. A history H defines a partial order $\prec_H$ on operations. $op_1 \prec_H op_2$ if the response event of op_1 precedes the invocation event of op_2 in H. op_1 is concurrent with op_2 if neither precedes the other.

In our algorithm, we assume that each reader process has a helper thread that takes infinitely many steps even if the reader stops reading the implemented register. These steps are outside the invocation-response intervals of the readers' own operations.

Linearizability of a Concurrent History. *Linearizability*, a correctness condition for concurrent objects, is defined using an object's sequential specification.

Definition 1. *(Linearization of a concurrent history:) A* linearization *of a concurrent history H of object o is a sequential history H' such that:*

1. *After removing some pending operations from H and completing others by adding matching responses, it contains the same invocations and responses as H',*
2. *H' preserves the partial order $\prec_H$, and*
3. *H' satisfies o's sequential specification.*

A SWMR register as well as a SWSR register expose the *read* and *write* operations. The sequential specification of a SWMR and a SWSR register states that a read operation from register Reg returns the value last written to Reg. Following Cohen and Keidar [2], we manage Byzantine behavior in a way that provides consistency to correct processes. This is achieved by linearizing correct processes' operations and offering a degree of freedom to embed additional operations by Byzantine processes.

Let $H|_{correct}$ denote the projection of history H to all correct processes. History H is Byzantine linearizable if $H|_{correct}$ can be augmented by (some) operations of Byzantine processes such that the completed history is linearizable. Thus, there is another history with the same operations by correct processes as in H, and additional operations by at most f Byzantine processes.

Definition 2. *(Byzantine linearization of a concurrent history [2]:) A history H is Byzantine linearizable if there exists a history H' such that $H'|_{correct} = H|_{correct}$ and H' is linearizable.*

An object supports Byzantine linearizable executions if all of its executions are Byzantine linearizable. SWMR registers support Byzantine linearizable executions because before every read from such a register, invoked by a correct process, one can add a corresponding Byzantine write.

Linearizability of Register Implementations. Hu and Toueg defined register linearizability in a system with Byzantine processes as follows [4,5]. They let v_0 be the initial value of the implemented register and v_k be the value written by the kth write operation by the writer w of the implemented register.

Definition 3. *(Register Linearizability [4, 5]:) In a system with Byzantine process failures, an implementation of a SWMR register is linearizable if and only if the following holds. If the writer is not malicious, then:*

- *(Reading a "current" value) If a read operation R by a process that is not malicious returns the value v then:*
  - *there is a write v operation that immediately precedes R or is concurrent with R, or*
  - *$v = v_0$ (the initial value) and no write operation precedes R.*
- *(No "new-old" inversion) If two read operations R and R' by processes that are not malicious return values v_k and $v_{k'}$, respectively, and R precedes R', then $k \leq k'$.*

3   Byzantine Register Linearizability

When a *high-level object (HLO)* is simulated or constructed using a *low-level object (LLO)*, there are two interfaces. A process interacts with the HLO through a *high-level interface (HLI)* through alternating invocations and matching responses. Between such a pair of matching invocation and response, the process interacts with the LLO through a *low-level interface (LLI)* using alternating invocations and responses. Such interactions are in software. For our problem, the HLO is the Byzantine-tolerant SWMR atomic register and the HLI is the read and write operation. The LLO is the SWSR atomic register and the LLI is also the read and write operation.

In the face of Byzantine readers as well as a Byzantine writer, a correct write operation has to be defined. We use u or v to refer to the data value written. If the $write(v)$ at the HLI is converted into multiple serial invocations of $write(v')$ (for different values of v') and the protocol for each of these v' is correctly followed, these various $write(v')$ are considered correct write operations because that sequence of write operations can be taken to be the values the writer writes or intended to write. This is because the invocation/response at the HLI is at a Byzantine process which controls the execution of code above the LLI and above the HLI. In a correct write operation, the code between the HLI and the LLI is followed correctly by the Byzantine process.

Definition 4 ([8]). *A correct (write) operation is a (write) operation that follows the (write) protocol, but possibly with a different value than that passed down at the HLI.*

In the literature [2,4,5], any behavior of a Byzantine writer is allowable in the linearizability definition. A Byzantine writer is accommodated differently in [8] by introducing the concept of a *pseudo-correct write operation* (Definitions 5 and 10), which is a Byzantine write operation that has the effect of a correct write operation, i.e., whose actions that are visible to correct readers cannot be distinguished from the actions of a correct write operation by correct readers.

Definition 5 ([8]). *A pseudo-correct (write) operaton is a (write) operation such that whatever steps the (writer) process performs in it and that results in a value being returned to correct readers, are indistinguishable to correct readers' executions below the HLI from an actual correct (write) operation's steps.*

A stronger definition of a Byzantine linearization of a concurrent history than Definition 2 of Cohen-Keidar is given in [8]. It tames the Byzantine behavior in a stronger way to provide consistency to correct processes. It linearizes the correct processes' operations and offers a *(more) limited* degree of freedom by way of embedding *only* correct and pseudo-correct write operations by Byzantine processes. History H is Byzantine linearizable if $H|_{correct}$ can be augmented by (some) pseudo-correct and correct operations (and not any arbitrary operations) of Byzantine processes such that the completed history is linearizable.

Definition 6 ([8]). *(Byzantine linearization of a concurrent history:) A history H is Byzantine linearizable if there exists a history H' containing correct and pseudo-correct write operations by Byzantine processes and writes and reads by correct processes such that $H'|_{correct} = H|_{correct}$ and H' is linearizable.*

A counter-example to show that the Hu-Toueg algorithm does not have a Byzantine linearization for concurrent histories as per our Definition 6 is given in [8].

In general, there is a many-many mapping from HLO writes by a Byzantine process to pseudo-correct write operations. We define a *virtual invocation (response)* of a pseudo-correct write operation as occurring at the time of invocation (response) of the earliest (lastest) HLO write operation in which some step of the pseudo-correct write operation was taken. To construct H' in Definition 6, we add a virtual invocation-response of a pseudo-correct write operation before an appropriately identified invocation-response of a read by a correct process.

Let v^i be the value written by the ith correct or pseudo-correct write W^i in a Byzantine linearization of a concurrent history (Definition 6), which is used in Definition 7, following the notation in [3]. Note that to determine i, v^i and W^i requires knowing what happened below the HLI and above the LLI because of the nature of pseudo-correct writes; but there is actually no need to determine i, v^i, and W^i.

Definition 7 ([8]). *(Byzantine Linearizable Register:) In a system with Byzantine process failures, an implementation of a SWMR register is linearizable if and only if the following two properties are satisfied in a Byzantine linearization of a concurrent history.*

1. **Reading a current value:** *When a read operation R by a non-Byzantine process returns the value v:*
   (a) if $v = v_0$ then no correct or pseudo-correct write operation precedes R
   (b) else if $v \neq v_0$ then v was written by the correct or pseudo-correct write operation that immediately precedes R.
2. **No "new-old" inversions:** *If read operations R and R' by non-Byzantine processes return values v^i and v^j, respectively, and R precedes R', then $i \leq j$.*

A pseudo-correct (write) operation looks like a correct write operation to correct readers; however the writer may still not follow the write protocol exactly. Taming a Byzantine write and making visible what a Byzantine write does as part of a pseudo-correct write is done by correct readers in their steps below the HLI and above the LLI. As the Byzantine writer may exit its write protocol prematurely, the linearization point of a pseudo-correct write may be after the invocation-response interval(s) of the HLI operations that triggered the pseudo-correct write.

4   The Algorithm

WLOG assume that there are n SWSR registers R_init_{wi} writable by the single writer w and readable by reader $i \in [1, n]$. The Byzantine writer can behave anyhow and can write different values to the SR registers, or write different values to different subsets of SR registers while not writing to some of them at all, or write multiple different values over time to some or all SR registers, as part of the same write operation. We assume that f of the n readers are Byzantine.

The writer writes (k, u), where k is a monotonically increasing sequence number and u is a data value, to the various R_init_{wi}. Although there is a many-many mapping from write operations issued to the HLI object interface to values written to the object, this does not pose any ambiguity because the different values that are returned to the correct readers have different logical timestamps, given by k.

Correct write operations are totally ordered in time. This total order is also the logical time ordering of their timestamps. *To be indistinguishable to correct readers below the HLI from correct write operations (to satisfy Definition 5), write operations whose values can be returned should (i) be written to a sufficient number of R_init_{wi}, and (ii) be totally ordered along with the set of correct write operations, by their logical timestamps.* Based on this above *principle*, we proceed to redefine pseudo-correct write operations. (Note, different algorithms can instantiate Definition 5 in different ways based on their design of read and write operations below the HLI.)

Definition 8 ([8]). *A potential pseudo-correct write operation of value (k, v) is a write operation, timestamped k, that may not follow the write protocol but*

1. *there is a quorum of size $\geq n - 2f$ indices i of correct readers such that (k, v) was written to R_init_{wi}, and*
2. *$k > k'$ for all (k', v') already read from these R_init_{wi}.*

Definition 9 ([8]). *A write operation* stabilizes *if its value can be returned by a correct reader.*

A correct write operation always stabilizes whereas a potential pseudo-correct write may stabilize, depending on run-time dynamic data races due to the asynchronous readers, steps of Byzantine readers and the Byzantine writer, and the algorithm. Only all write operations that stabilize have a linearization point.

Definition 10 ([8]). *A pseudo-correct write operation* is a potential pseudo-correct write operation that stabilizes.

Definition 11 ([8]). *(Monotonicity/Total Order of stabilized write operation timestamps Property:) The set of write operation timestamps that stabilize is totally ordered.*

Satisfying the property is key to prove that the algorithm implements a Byzantine linearizable register.

In the algorithm implementation, write of value v is actually a write of tuple $\langle k, v \rangle$, where k is a sequence number issued by the writer. Only correct and pseudo-correct writes may be returned by correct readers. A correct reader cannot distinguish between a correct and a pseudo-correct write operation. A pseudo-correct write operation $o_1(v_1)$ timestamped k_1 may lose a race due to asynchrony of process executions to a pseudo-correct or correct write operation timestamped k_2 where $k_1 < k_2$, in which case v_1 is not actually returned to any read operation and o_1 is deemed to have an *invisible* linearization point. The correct and pseudo-correct writes are totally ordered by their linearization points, and this order is the total order on the timestamp values k.

4.1   Basic Idea and Operation

Because the writer w is Byzantine, a reader i cannot unilaterally use the value in its register R_init_{wi} but needs to coordinate with other readers. But a reader may not invoke a read operation indefinitely. So we assume that each reader process has a reader helper thread that is always running and participates in this coordination. A correct write operation needs to wait for acknowledgements from the readers so that the write value is guaranteed to get written in the algorithm data structures and stabilize, and is not overwritten before it stabilizes. This guarantees progress by ensuring that the most recent correct write operation advances with time. The algorithm data structures are described in Algorithm 1. The reader helper thread is given in Algorithm 2. w is the writer process and P denotes the set of n readers.

The writer writes $\langle k, u \rangle$, where u is the value to be written and k is a sequence number assigned by the writer, to all the reader registers R_init_{wi} and then waits for $n - f$ of the acknowledgement registers R_ack_{iw} to be written with this value.

The reader helper thread ($\forall p \in P$) loops forever. In each iteration, it reads R_init_{wp} and if the value overwrites the earlier value $\langle k', u' \rangle$ where $k > k'$, it writes $\langle \langle k, u \rangle, p \rangle_p$ to all the registers $R\_witness_{pi}$. It then reads each $R\_witness_{ip}$ and if the logical time of an entry is larger than the previous logical time read, it stores the value in a local variable $T\_witness_i$. (If less than, or equal but the entry is different, i is marked as Byzantine). If there are at least $n - f$ $T\_witness_q$ entries with identical $\langle k, u \rangle$ values then these $T\_witness_q$ entries are placed in the local set $Witness\_Set$. This $Witness\_Set$ is written to all R_final_{pi}. R_ack_{pw} is updated with $\langle k, u \rangle$. All the R_final_{ip} are read and placed in set Z. All the elements in Z, i.e., ($\forall i$) R_final_{ip} are totally ordered by a relation $\mapsto$, (to be defined in Definition 16), as will be proved in Theorem 5. The latest element in Z is identified as Y via a call to $Find\_Latest(Z)$ and if Y is different from the local $Witness\_Set$, i.e., for a different $\langle k, u \rangle$ pair, (a) Y is written to all R_final_{pi} and (b) R_ack_{pw} is updated with the common $\langle k, u \rangle$ occurring in the $T\_witness$ entries $\langle \langle k, u \rangle, l \rangle_l$ for $\geq n - f$ values of l in the $Witness\_Set$ that is Y.

4.2   Instantiating Framework Definitions in the Algorithm

A reader sees the value written by the most recent correct or pseudo-correct write operation that precedes the read operation in a Byzantine linearization of the concurrent history. A write operation *stabilizes* (Definition 14) if the value can be returned by a correct read operation, i.e., when it is written to R_final_{p*} for all values of $*$ (in the form of an $Witness\_Set$ containing at least $n - f$ correctly signed $T\_witness_i$ entries).

A correct write operation always stabilizes because it waits for $n-f$ acknowledgements in R_ack_{*w} before completion, thereby allowing the value written to R_final_{p*} to be eligible for being returned by a read operation. A correct write operation corresponds to a sufficient condition for stabilization. Writing the same value to $n - 2f$ correct readers' R_init_{wi} in a potential pseudo-correct operation

---

Algorithm 1: Constructing a linearizable SWMR atomic register. Data structures and READ, WRITE procedures. Code at process p.

---

1 **Shared variables:**
2 For all processes $i \in P$: R_init_{wi}: atomic SWSR register $\leftarrow \langle 0, u_0 \rangle$
3 For all processes $i \in P$: R_ack_{iw}: atomic SWSR register $\leftarrow \langle 0, u_0 \rangle$
4 For all processes i and j in P: $R\_witness_{ij}$: atomic SWSR register $\leftarrow \langle \langle 0, u_0 \rangle, i \rangle_i$
5 For all processes i and j in P: R_final_{ij}: atomic SWSR register
  $\bigcup_{k \in L} \{ \langle \langle 0, u_0 \rangle, k \rangle_k \}$, where $|L| \geq n - f \wedge L \subseteq P$

6 **Local variables:**
7 variable of w: $c \leftarrow 0$
8 variable of $i \in P$: $k\_init\_latest \leftarrow 0$
9 variables of $j \in P$: For all processes $i \in P$: $T\_witness_i \leftarrow \langle \langle 0, u_0 \rangle, i \rangle_i$
10 variable of $j \in P$: $Witness\_Set \leftarrow \bigcup_{k \in L} \{ \langle 0, u_0 \rangle, k \rangle_k \}$, where
  $|L| \geq n - f \wedge L \subseteq P$

11 $\underline{\text{WRITE}(u):}$
12 $c = c + 1$
13 $W(\langle c, u \rangle)$
14 **return**

15 $\underline{\text{READ}():}$
16 $R()$
17 **return** the value returned by the R call

18 $\underline{W(\langle k, u \rangle):}$
19 **for every** *process* $i \in P$ **do**
20 $\quad \lfloor\ R\_init_{wi} = \langle k, u \rangle$
21 $d = 0$
22 **while** $d < n - f$ **do**
23 $\quad$ **for** *each new* $\langle k, u \rangle$ *in* R_ack_{iw} *among* R_ack *registers* **do**
24 $\quad\quad \lfloor\ d = d + 1$
25 **return done**

26 $\underline{R():}$
27 execute one iteration of the reader helper thread, Algorithm 2
28 **return** the value last written in R_ack_{pw}

---

is a necessary condition for stabilization. For the potential pseudo-correct write operation to become a pseudo-correct write operation, favorable outcome of run-time dynamic data races due to the asynchronous writer and readers and the collaboration by Byzantine readers are needed to allow the value being written to the other correct readers' R_init_{wi} to stabilize. The set of correct and pseudo-correct writes is exactly the set of writes whose values stabilize (follows from Theorem 3).

We show (Corollary 2) that the set of all values that stabilize are totally ordered by the logical times of their writing to the readers' registers

---

Algorithm 2: Reader helper thread for constructing a linearizable SWMR atomic register. Code at process p.

---

1   reader helper thread:

2   **while** *true* **do**

3   **if** $R\_init_{wp}(= \langle k, u \rangle)$ *is newly written* $(k > k\_init\_latest)$ **then**

4 $k\_init\_latest \leftarrow k$

5   **for** *each* $i \in P$ **do**

6 $R\_witness_{pi} \leftarrow \langle\langle k, u \rangle, p \rangle_p$

7   **for** *each* $i \in P$ **do**

8   **if** $R\_witness_{ip}(= \langle\langle k, u \rangle, i \rangle_i)$ *is newly written, i.e.,* $k > T\_witness_i.k$ **then**

9 $T\_witness_i \leftarrow \langle\langle k, u \rangle, i \rangle_i$

10   **if** $\exists \geq n - f$ *correctly signed latest elements* $\langle\langle k, u \rangle, q \rangle_q$ $T\_witness_q$ **then**

11   add all these $\geq n - f$ elements $T\_witness_q$ to the emptyset $Witness\_Set$

12   **for** *each* $i \in P$ **do**

13 $R\_final_{pi} \leftarrow Witness\_Set$

14 $R\_ack_{pw} \leftarrow \langle k, u \rangle$

15 $Z \leftarrow \emptyset$

16   **for** *each* $i \in P$ **do**

17 $Z \leftarrow Z \cup \{R\_final_{ip}\}$

18 $Y \leftarrow Find\_Latest(Z)$

19   **if** $\langle k, u \rangle$ *common to entries in* $Y \neq \langle k, u \rangle$ *common to entries in* $Witness\_Set$ **then**

20   **for** *each* $i \in P$ **do**

21 $R\_final_{pi} \leftarrow Y$

22 $R\_ack_{pw} \leftarrow$ the common $\langle k, u \rangle$ value occurring in identical $T\_witness$ entries $\langle\langle k, u \rangle, l \rangle_l$ for $\geq n - f$ values of l in $Witness\_Set$ that is Y

23 $Find\_Latest(Z):$

24   **for** *each* $Witness\_Set$ $Z_i \in Z$ **do**

25   **for** *each* $T\_witness$ *element* x *in* Z_i *that fails signature test* **do**

26 $Z_i \leftarrow Z_i \setminus \{x\}$; **if** $|Z_i| < n - f$ **then** $Z \leftarrow Z \setminus \{Z_i\}$ and break()

27   **while** $|Z| > 1$ **do**

28   let Z_i, Z_j be any two elements of Z; Z_i (Z_j) has $\geq n - f$ entries and is the $Witness\_Set$ of some reader

29   For the $\geq n - 2f$ readers $q \in Z_i \cap Z_j$, let $e_q^i = \langle\langle k, u \rangle, q \rangle_q$ and $e_q^j = \langle\langle k', u' \rangle, q \rangle_q$ be these $T\_witness$ entries in the $Witness_Set$s that are Z_i and Z_j, resp.

30   **if** $k \geq k'$ *for any/all processes* q **then**

31 $Z \leftarrow Z \setminus Z_j$

32   **else**

33 $Z \leftarrow Z \setminus Z_i$

34   return(element in Z)

R_init_{w*}. The timestamp of the writing of a value $\langle k, u\rangle$ that stabilizes is denoted $\langle k, u\rangle.\overline{T}$ and is defined as k. We say that for values $\langle k, v\rangle$ and $\langle k', v'\rangle$ that stabilize, $\langle k, v\rangle \mapsto \langle k', v'\rangle$ if and only if $\langle k, v\rangle.\overline{T} \mapsto \langle k', v'\rangle.\overline{T}$, i.e., $k < k'$. The total order $\mapsto$ on timestamps of values that stabilize is the total order on the linearization points of correct and pseudo-correct write operations. This follows from the *no "new-old" inversions* clause in Theorem 8.

More than one value can stabilize as part of the same HLO write operation. This can happen when, for example, some of the readers' registers could have been written to in earlier Write operations or a HLO invocation-response of a Byzantine write contains multiple pseudo-correct write operations.

With the introduction of the $\mapsto$ relation, our definition of Byzantine linearizable register (Definition 7) is adapted to the algorithm by rephrasing the *No "new-old" inversions* property as follows.

Definition 12. *(Byzantine Linearizable Register in our algorithm). In a system with Byzantine process failures, an implementation of a SWMR register is linearizable if and only if the following two properties are satisfied in a Byzantine linearization of a concurrent history.*

1. **Reading a current value:** *When a read operation* R *by a non-Byzantine process returns the value* v:
   (a) if $v = v_0$ *then no correct or pseudo-correct write operation precedes* R
   (b) else if $v \neq v_0$ *then* v *was written by the correct or pseudo-correct write operation that immediately precedes* R.
2. **No "new-old" inversions:** *If read operations* R *and* R' *by non-Byzantine processes return values* $\langle k, v\rangle$ *and* $\langle k', v'\rangle$, *respectively, and* R *precedes* R', *then* $\langle k, v\rangle.\overline{T} \stackrel{=}{\mapsto} \langle k', v'\rangle.\overline{T}$, *i.e.,* $k \leq k'$.

5   Correctness Proof

Definition 13. *The witness set that is formed, denoted $WS(k, u)$, is the maximal set of at least $n - f$ $T\_witness$ entries $\langle\langle k, u\rangle, l\rangle_l$ in the $Witness\_Set$ WS.*

Definition 14. *The field/value $\langle k, u\rangle$ common to all the entries of a $Witness\_Set$ is defined to stabilize when the $Witness\_Set$ is written to all the R_final_{pi} for some process p.*

Definition 15. *For a value $\langle k, u\rangle$ of $Witness\_Set$ WS that stabilizes, $\langle k, u\rangle.\overline{T}$ is the value k.*

Only a value that stabilizes may be returned by a correct reader.

Theorem 1. *A correct write operation is guaranteed to stabilize provided $n > 2f$.*

Proof. A correct write operation writes the same $\langle k, u \rangle$ to R_init_{wp} for all correct p and will not complete unless it gets $n - f$ acks in R_ack_{pw}. It may get f acks from Byzantine processes but as $n > 2f$, it will need an ack from at least one correct process. A correct process p gives the ack only after it has written the value to R_final_{p*}, i.e., when the value has stabilized. We now show that at least one correct process will have the value stabilize, before which the value written to the R_init_{w*} will not be overwritten.

For all correct p, the witness timestamps $\langle\langle k, u \rangle, p \rangle$, signed by p, are correctly written to $R\_witness_{p*}$. At least $n - f$ correct reader helper threads of correct processes p will eventually read from $R\_witness_{*p}$, form their $Witness\_Sets$ for that $\langle k, u \rangle$, and write their own $Witness\_Set$ to R_final_{p*}. Such reader threads p will then write $\langle k, u \rangle$ to R_ack_{pw}. After $n - f$ acks have been written by the correct and/or Byzantine reader processes/threads, the correct write operation will complete. Thus, $\langle k, u \rangle$ is guaranteed to have stabilized as the $R\_init_{wi}(\forall i)$ will not be overwritten until then. $\qquad\square$

Theorem 2. *A potential pseudo-correct write operation may stabilize provided* $n > 2f$.

Proof. A value written by a potential pseudo-correct operation writes the same value to at least $n - 2f$ correct readers' R_init_{wp}, across possibly multiple prior write operations and the current write operation. These reader processes write that value, if not overwritten, to $R\_witness_{p*}$. Let the f Byzantine processes b read the value from their $R\_witness_{*b}$ and thereafter behave as though the value had been written to their R_init_{wb} and thenceforth behave correctly. There is now a way that the value may stabilize if the Byzantine writer does not overwrite the values in R_init_{wp} ($\forall p$) until at least one process q forms its $Witness\_Set$ of correctly signed $Witness\_Set$ entries for that value $\langle k, u \rangle$ and writes the $Witness\_Set$ to R_final_{q*}. This condition will be satisfied as per the logic in the second paragraph of the proof of Theorem 1. $\qquad\square$

Theorem 3. *If a value stabilizes, it must have been written by a correct write operation or by a potential pseudo-correct write operation.*

Proof. A correct write operation stabilizes (Theorem 1). So we need to only prove the following contrapositive, namely that if a write is not a potential pseudo-correct write operation, it will not stabilize.

If a write is not a potential pseudo-correct operation, it is not written to at least $n - 2f$ correct processes' R_init_{wi} across possibly multiple write operations. Then there is no way a $Witness\_Set$ of $n - f$ correctly signed $Witness\_Set$ entries can form at any process, correct or Byzantine, and can be written to any R_final_{p*}. If a Byzantine process attempts to write a fake $Witness\_Set$ in R_final_{p*}, that will be detected by correct processes as the entries in that $Witness\_Set$ written in R_final_{p*} will not pass the signature test. Hence that value is deemed not to have stabilized. $\qquad\square$

We next formalize the liveness/progress conditions for the advancement of the latest correct/pseudo-correct write operation. Define an *epoch* to be the interval in which each correct reader helper thread executes an iteration of its main *while* loop at least once.

Theorem 4. *From the time of writing the same $\langle k, u \rangle$ to at least $n - f$ R_init_{wi} for correct processes i and such that $k > k'$ for all previously read $\langle k', u' \rangle$ from these R_init_{wi}, if the Byzantine writer does not overwrite these $n - f$ R_init_{wi} for two epochs then the $\langle k, u \rangle$ tuple is guaranteed to stabilize.*

Proof. In the first epoch, the correct readers i read R_init_{wi} and write the read tuples (after signing the corresponding tuple) to $R\_witness_{i*}$. In the second epoch, the correct readers i read $R\_witness_{*i}$, form the respective $Witness\_Sets$, and write it to R_final_{i*}. At this time the $\langle k, u \rangle$ has stabilized (Definition 14). $\square$

Observation. A correct process forms its successive $Witness\_Sets$ only in increasing order of source witness timestamps for the at least $n - 2f$ common witnesses as it writes these $Witness\_Sets$ to R_final.

Definition 16. *Given $WS1(= WS(k, u))$ and $WS2(= WS(k', u'))$, $WS1 \mapsto WS2$ iff $(k, u).\overline{T} < (k', u').\overline{T}$, i.e., $k < k'$.*

If $WS(k, u) \mapsto WS(k', u')$, we also interchangeably say that for the corresponding values, $\langle k, u \rangle \mapsto \langle k', u' \rangle$.

Definition 17. *Given distinct $WS1(= WS(k, u))$ and $WS2(= WS(k', u'))$, $WS1 \| WS2$ iff $WS1 \not\mapsto WS2 \wedge WS2 \not\mapsto WS1$.*

Theorem 5. *For $Witness\_Sets$ $WS1 = WS(k, u)$ and $WS2 = WS(k', u')$ at any two (possibly different) reader processes, $WS1 \mapsto WS2 \vee WS2 \mapsto WS1$, i.e., $\neg(WS1 \| WS2)$, provided $n > 3f$.*

Proof. There are $n - f$ signed $T\_witness$ entries in each of $WS1$ and $WS2$. For these to be valid, at least one must be from a correct reader, requiring $n - f > f$, implying $n > 2f$. However this is not sufficient.

When $n = 2f + 1$, to prevent f Byzantine processes from giving same witness timestamp inputs but for different values of v (v_α, for $\alpha = [1, n - f]$), forming $n - f = f + 1$ $Witness\_Sets$ with identical common timestamps but different values v_α corresponding to each of $f + 1$ correct processes, that were written to R_init_{wp}, implying $f + 1$ mutually concurrent $Witness\_Sets$, we require that at least one correct process provide witness timestamps for both of any pair of $Witness\_Sets$. This will totally order the $Witness\_Sets$. To achieve this, the minimum number of common reader processes that provided inputs for any pair of $Witness\_Sets$, $n - 2f$ (as f do not provide input to $WS1$ and some other f do not provide input to $WS2$), should be greater than f, implying $n > 3f$. Thus $\neg(WS1 \| WS2)$.

For any two $Witness\_Sets$ that form (and stabilize), there is at least one correct process that provides witness timestamp inputs to both sets. From the pseudo-code, it will necessarily do so in increasing order of timestamps. For either $WS(k, u)$ or $WS(k', u')$, inputs from all processes must necessarily be for the same input ((k, u) or (k', u'), respectively). Thus if $WS1$ and $WS2$ form, all processes that provide input witness timestamps to both $WS1$ and $WS2$ provide first to $WS1$ and then to $WS2$ or all provide first to $WS2$ and then to $WS1$. Furthermore, all such processes provide input first for the smaller of $\{k, k'\}$ and then for the larger. $\qquad\qquad\qquad\qquad\qquad\qquad\qquad\qquad\qquad\qquad\quad\square$

Corollary 1. $WS1 \mapsto WS2$ *implies that for all the at least $n - 2f$ readers z that witnessed values in both $WS1$ and $WS2$, z's $WS1$ timestamp is less than z's $WS2$ timestamp.*

Only value $\langle k, v \rangle$ corresponding to an $Witness\_Set$ that is written to R_final_{p*} could be returned by a correct process if the signed $Witness\_Set$ entries pass the signature test. Each new value returned by a correct reader is a new value genuinely written to at least $n - 2f$ correct reader i's R_init_{wi} and not falsely reported by Byzantine readers, in addition to that same value being the most recent value in a total of $n - f$ readers j's R_init_{wj} registers as per $\langle k, u \rangle.\overline{T}$.

From Theorem 5, all the $\langle k, v \rangle.\overline{T}$ form a total order based on $\mapsto$. This leads to the following corollary.

Corollary 2. *The timestamps $\langle k, u \rangle.\overline{T}$ of $\langle k, u \rangle$ values that stabilize are totally ordered, provided $n > 3f$.*

We give a linearization of an execution to show that the SWMR register we constructed supports Byzantine linearizability of executions (Definition 6). First, read operations of correct readers are linearized in the order of the return values, as per the $\mapsto$ relation. Then an update of a value $\langle k, v \rangle$ by the Byzantine writer is added just before the first read operation that reads that value. In general, the value returned by a read operation is based on the values written by one or more than one HLI write operation as the writer is Byzantine. Further, one HLI write operation by the Byzantine writer may result in different read values being returned by multiple correct readers. To differentiate among the multiple updates of different values over time to the SWMR register, we treat each update of $\langle k, v \rangle$, which has stabilized, as an independent *Byzantine (correct or pseudo-correct) write operation*, associated with its timestamp $\langle k, v \rangle.\overline{T} = k$.

Theorem 6. *Let reads R and R' return $\langle k, v \rangle$ and $\langle k', v' \rangle$, respectively, and let R precede R'. Then $\langle k, v \rangle.\overline{T} \leq \langle k', v' \rangle.\overline{T}$, i.e., $k \leq k'$.*

Proof. When R returns $\langle k, v \rangle$, the $WS(k, v)$ has been written to all R_final_{p*} for some p. From the Algorithm 2 pseudo-code, when R' is invoked, it will be guaranteed to read $WS(k, v)$ into Z and hence the (k', v') returned to R' will be such that $k' \geq k$. $\qquad\qquad\qquad\qquad\qquad\qquad\qquad\qquad\qquad\quad\square$

Theorem 7. *The SWMR register implemented by Algorithm 1 satisfies Byzantine linearizability of executions.*

Proof. Let L_R be a linearization of the correct readers' read operations such that (i) the order of all non-overlapping reads is preserved, and (ii) if read R returns $\langle k, v \rangle$, read R' returns $\langle k', v' \rangle$, and $\langle k, v \rangle.\overline{T} \mapsto \langle k', v' \rangle.\overline{T}$ then R precedes R' in L_R. Such an L_R is guaranteed to exist because Theorem 6 ensures that there is no ordering conflict between (i) and (ii).

For the pseudo-correct writes to form, i.e., be totally ordered along with correct writes, $n > 3f$ by Theorem 5. For any $\langle k, v \rangle$ returned by a read, the value must have been written by a correct or a pseudo-correct write operation. In the linearization L_R, place a (Byzantine) pseudo-correct or correct write operation writing a value $\langle k, v \rangle$ and assigned timestamp $\langle k, v \rangle.\overline{T}$ immediately before the first read operation R in L_R that reads $\langle k, v \rangle$.

The resulting linearization is seen to be a Byzantine linearization that considers all the read operations of correct readers, and includes some correct and pseudo-correct write operations. $\qquad\square$

Theorem 8. *Algorithm 1 implements a Byzantine linearizable SWMR register using SWSR registers, provided $n > 3f$.*

Proof. A Byzantine linearization of the concurrent history was identified in the proof of Theorem 7, and it required $n > 3f$. We now show that the value returned by a read operation satisfies "reading a current value" and "no new-old inversions" with respect to this Byzantine linearization.

- **Reading a current value:** As per the construction of the Byzantine linearization of the concurrent history, a read R of $\langle k, v \rangle$ is preceded by the correct or pseudo-correct write operation that immediately precedes R.
- **No "new-old" inversions:** Let read R by x return $\langle k, u \rangle$ and let read R' by y return $\langle k', u' \rangle$, where R precedes R'. R will write $\langle k, u \rangle$ in R_final_{x*} before returning. R' will read from R_final_{*y}, add these elements to Z and invoke $Find\_Latest(Z)$ which will return the most recent value $\langle k, u \rangle^{max}$ as per $\mapsto$. It is guaranteed that if $\langle k, u \rangle^{max} \neq \langle k, u \rangle$ then $\langle k, u \rangle.\overline{T} \mapsto \langle k, u \rangle^{max}.\overline{T}$ because all the values in Z are totally ordered by $\mapsto$ (Theorem 5, Corollary 2). From Definition 16, $\langle k, u \rangle.\overline{T} \mapsto \langle k, u \rangle^{max}.\overline{T}$ implies that the write of $\langle k, u \rangle^{max}$ had a greater witness timestamp than the write of $\langle k, u \rangle$ for the at least $n - 2f$ common processes that witnessed both writes. The value $\langle k', u' \rangle$ returned by R' is $\langle k, u \rangle^{max}$. As $\langle k, u \rangle.\overline{T} \mapsto \langle k, u \rangle^{max}.\overline{T}$, $i \leq j$ (as defined in Definition 7) as per the Byzantine linearization of the concurrent history. Thus there are no inversions. $\qquad\square$

Space Complexity. The algorithm uses $2n^2$ shared SWSR registers: n^2 $R\_witness$ registers of size $O(1)$, n^2 R_final registers of size $O(n)$. It also uses $2n$ shared SWSR registers: n R_init registers of size $O(1)$, and n R_ack registers of size $O(1)$. The local space at each reader process can be seen to be $O(n)$.

6   Discussion

We gave an algorithm to construct a Byzantine tolerant SWMR atomic register from SWSR atomic registers, that meets the definition of Byzantine register linearizability from [8], and overcomes the drawbacks of the Hu-Toueg algorithm.

One limitation of the proposed algorithm is that if the Byzantine writer uses a high (highest) value of k, no further writes will thenceforth be returnable to the readers, affecting liveness. A solution that overcomes this drawback using vector time [9] is given in [7].

References

1. Aguilera, M.K., Ben-David, N., Guerraoui, R., Marathe, V.J., Zablotchi, I.: The impact of RDMA on agreement. In: Robinson, P., Ellen, F. (eds.) Proceedings of the 2019 ACM Symposium on Principles of Distributed Computing, pp. 409–418. ACM (2019). https://doi.org/10.1145/3293611.3331601
2. Cohen, S., Keidar, I.: Tame the wild with byzantine linearizability: reliable broadcast, snapshots, and asset transfer. In: Gilbert, S. (ed.) 35th International Symposium on Distributed Computing, DISC 2021, 4–8 October 2021, Freiburg, Germany (Virtual Conference). LIPIcs, vol. 209, pp. 18:1–18:18. Schloss Dagstuhl - Leibniz-Zentrum für Informatik (2021). https://doi.org/10.4230/LIPIcs.DISC.2021.18
3. Herlihy, M., Shavit, N.: The Art of Multiprocessor Programming. Morgan-Kaufmann (2008)
4. Hu, X., Toueg, S.: On implementing SWMR registers from SWSR registers in systems with byzantine failures. In: Scheideler, C. (ed.) 36th International Symposium on Distributed Computing, DISC 2022, 25–27 October 2022, Augusta, Georgia, USA. LIPIcs, vol. 246, pp. 36:1–36:19. Schloss Dagstuhl - Leibniz-Zentrum für Informatik (2022). https://doi.org/10.4230/LIPIcs.DISC.2022.36
5. Hu, X., Toueg, S.: On implementing SWMR registers from SWSR registers in systems with byzantine failures. Distrib. Comput. **37**(2), 145–175 (2024). https://doi.org/10.1007/s00446-024-00465-5
6. Imbs, D., Rajsbaum, S., Raynal, M., Stainer, J.: Read/write shared memory abstraction on top of asynchronous byzantine message-passing systems. J. Parallel Distrib. Comput. **93-94**, 1–9 (2016). https://doi.org/10.1016/j.jpdc.2016.03.012
7. Kshemkalyani, A.D., Piduguralla, M., Peri, S., Misra, A.: Construction of a byzantine linearizable SWMR atomic register from SWSR atomic registers. CoRR abs/2405.19457 (2024). https://doi.org/10.48550/arXiv.2405.19457
8. Kshemkalyani, A.D., Piduguralla, M., Peri, S., Misra, A.: On constructing a byzantine linearizable SWMR atomic register from SWSR atomic registers. In: Korman, A., Chakraborty, S., Peri, S., Boldrini, C., Robinson, P. (eds.) Proceedings of the 26th International Conference on Distributed Computing and Networking, ICDCN 2025, Hyderabad, India, 4–7 January 2025, pp. 239–243. ACM (2025). https://doi.org/10.1145/3700838.3700839
9. Kshemkalyani, A.D., Singhal, M.: Distributed Computing: Principles, Algorithms, and Systems. Cambridge University Press (2011). https://doi.org/10.1017/CBO9780511805318
10. Mostéfaoui, A., Petrolia, M., Raynal, M., Jard, C.: Atomic read/write memory in signature-free byzantine asynchronous message-passing systems. Theory Comput. Syst. **60**(4), 677–694 (2017). https://doi.org/10.1007/s00224-016-9699-8

Deterministic Causal Order Under Byzantine Sybil Tolerance: Techniques and Limitations

Ajay D. Kshemkalyani[1]([✉]) [iD] and Anshuman Misra[2] [iD]

[1] University of Illinois Chicago, Chicago, USA
ajay@uic.edu
[2] Purdue University Fort Wayne, Fort Wayne, USA
misra47@pfw.edu

Abstract. A spectrum of solvability and unsolvability results for causal ordering of messages in the presence of Byzantine processes in asynchronous systems for unicast, multicast, and broadcast modes of communication have been shown. The possibility results implicitly assumed that the number of Byzantine processes f was less than $n/3$, where n is the total number of processes in the system. In this paper, we extend these results for the same system assumptions and parameters – mode of communication (unicast/broadcast/multicast), strong safety, weak safety, and liveness, and use of cryptography, to systems with $f < n$. Thus, we show corresponding possibility and impossibility results for the highest degree of Byzantine fault-tolerance. We also give the best-known bounds on f for solvability of causal ordering using deterministic algorithms in synchronous systems under the same combinations of system assumptions and parameters as for asynchronous systems.

Keywords: Byzantine fault-tolerance · Causal Order · Broadcast · Causality · Asynchronous · Message Passing

1 Introduction

Causality is defined by the *happened before* [17] relation on the set of events, and by extension, on the set of messages. Causal ordering of messages is specified by the safety and liveness properties. The *strong safety* property requires that if message m_1 causally precedes m_2 and both are sent to p_i, then m_2 cannot be delivered before m_1 at p_i [2]. The *liveness* property requires that a message from a correct process to another correct process is eventually delivered. Causally related updates to data occur in a valid manner enforcing semantic correctness if causal ordering is enforced [16]. Applications of causal ordering include distributed data stores, fair resource allocation, and collaborative applications such as multiplayer online gaming, social networks, event notification systems, group editing of documents, and distributed virtual environments.

S. Bonomi et al. (Eds.): SSS 2025, LNCS 16350, pp. 376–391, 2026.
https://doi.org/10.1007/978-3-032-11127-2_29

A spectrum of solvability and unsolvability results for causal ordering in the presence of Byzantine processes for unicast, multicast, and broadcast modes of communication were shown in [22,23,25,26]. The possibility results implicitly assumed that the number of Byzantine processes f was less than $n/3$, where n is the total number of processes in the system. This dependency arose because of the reliance on Bracha's Byzantine Reliable Broadcast (BRB) primitive [3,4] in the possibility results. A weakening of the safety condition, termed *weak safety*, was also given in [22,23,25,26]. Weak safety requires that if m_1 causally precedes m_2 and there is a causal path from the send event of m_1 to the send event of m_2 passing through only correct processes in the execution, then m_2 should not be delivered before m_1 at all common destinations of m_1 and m_2. The possibility and impossibility results considered strong safety, weak safety, and liveness, as well as the optional use of cryptography. The results from [23,25,26] for systems satisfying $f < n/3$ are given in Table 1.

In this paper, we consider the highest degree of Byzantine behavior possible in a distributed system by allowing Byzantine processes to control up to all but one process in the system and assess the solvability of the spectrum of causal ordering problems. Typically, distributed algorithms solving problems such as consensus and agreement are subject to a bound, commonly $f < n/3$. Here, we extend the model by allowing Byzantine processes to not only control all but one process but also spawn and create new Byzantine processes, leading to a permissionless setting. Assuming a static bound on the number of Byzantine processes implicitly restricts the system to a permissioned model. Our system model, however, allows for maximal Byzantine behavior, also referred to as Byzantine Sybil Tolerance [9], and provides a critical analysis of causal ordering in permissionless systems, often overlooked in prior work. This relaxation of the traditional $f < n/3$ bound to $f < n$ encapsulates the notion of permissionless Byzantine Sybil tolerance, where the system must remain secure and functional despite an unbounded number of adversarial identities. In such systems, both f and as a result n can change over time; our results are valid as long as the corresponding relationship(s) between f and n hold, e.g., $f < n$. By permitting spawning of new Byzantine processes, this model captures a more generalized adversarial setting. Byzantine Sybil Tolerance has been mooted in [13] and is important in real-world use cases such as the Matrix system [12].

Contributions: Corresponding to the solvability and unsolvability results for various combinations of system assumptions shown in Table 1 for $f < n/3$ in asynchronous systems, we show results assuming only that $f < n$, thus considering systems with the highest degree of Byzantine fault-tolerance. Our results are summarized in Table 2. This study is significant because it is important to understand the limitations on what is solvable under the highest degree of Byzantine fault-resilience and Byzantine Sybil tolerance.

Of particular note on the negative side is the following result.

1. For broadcasts without cryptography, liveness cannot be guaranteed when $f < n$, whereas it could be guaranteed when $f < n/3$.

Of particular note on the positive side are the following results.

Table 1. Solvability of causal ordering using deterministic algorithms in asynchronous systems under different communication modes [23,25,26]. SS = strong safety, WS = weak safety, L = liveness, $\overline{SS}, \overline{WS}, \overline{L}$ represent that strong safety, weak safety, liveness, respectively, cannot be guaranteed. Results are assuming $3f < n$ as all "Yes" results require $3f < n$.

Mode of communication	SS + L	SS + L with cryptography	WS + L	WS + L with cryptography
Unicasts	No $\overline{SS}, \overline{L}$	No $\overline{SS}, L$	No $WS, \overline{L}$	Yes WS, L
Broadcasts	No $\overline{SS}, L$	No $\overline{SS}, L$	Yes WS, L	Yes WS, L
Multicasts	No $\overline{SS}, \overline{L}$	No $\overline{SS}, L$	No $WS, \overline{L}$	Yes WS, L

Table 2. Solvability of causal ordering using deterministic algorithms in asynchronous systems under different communication modes. SS = strong safety, WS = weak safety, L = liveness, $\overline{SS}, \overline{WS}, \overline{L}$ represent that strong safety, weak safety, liveness, respectively, cannot be guaranteed. Results are assuming $f < n$. For each entry in the table, the relation between f and n is explicitly indicated next to the respective property that is satisfiable.

Mode of communic.	SS + L	SS + L with cryptography	WS + L	WS + L with cryptography
Unicasts	No, Theorem 1 $\overline{SS}, \overline{L}$	No, Theorem 8 $\overline{SS}, L(f < n)$	No, Theorem 4 $WS(f < n), \overline{L}$	Yes, Corollary 1 $WS(f < n), L(f < n)$
Broadcasts	No, Theorem 2 $\overline{SS}, \overline{L}$	No, Theorem 9 $\overline{SS}, L(f < n)$	No, Theorem 5 $WS(f < n), \overline{L}$	Yes, Corollary 2 $WS(f < n), L(f < n)$
Multicasts	No, Theorem 3 $\overline{SS}, \overline{L}$	No, Theorem 7 $\overline{SS}, L(f < n)$	No, Theorem 6 $WS(f < n), \overline{L}$	Yes, Theorem 10 $WS(f < n), L(f < n)$

1. Liveness is possible to be guaranteed with the use of cryptography when $f < n$, not just when $f < n/3$.
2. Weak safety can be guaranteed without the use of cryptography when $f < n$, not just when $f < n/3$.
3. Both weak safety and liveness can be guaranteed with the use of cryptography when $f < n$, not just when $f < n/3$.

The above results were for asynchronous systems. We also give the best-known bounds on f for solvability of causal ordering using deterministic algorithms in synchronous systems under the same combinations of system assumptions and parameters as for asynchronous systems. The results are summarized in Table 3.

In particular, on the negative side:

1. Strong safety cannot be guaranteed without the use of cryptography.

On the positive side, we have the following.

1. Strong safety and liveness using cryptography can be guaranteed for $f < n/2$.
2. Weak safety and liveness even without cryptography can be guaranteed for $f < n$.

Outline. Section 2 reviews related work. Section 3 gives the system model. Section 4 gives the main results about the solvability of Byzantine causal unicast, multicast, and broadcast in a deterministic manner in an asynchronous system. Section 5 gives the corresponding results for synchronous systems. Section 6 concludes.

2 Related Work

Algorithms for causal ordering of unicast and multicast messages in an asynchronous setting under a fault-free model have been proposed, e.g., in [6,14, 15,21,30]. An algorithm for causal multicasts in a fault-free setting for mobile systems is given in [7]. There has been some work on causal broadcasts under various failure models. Causal ordering of broadcast messages under crash failures in asynchronous systems was introduced in [2]. This algorithm required each message to carry the entire set of messages in its causal past as control information. An algorithm for causally ordering broadcast messages – providing only a variant of weak safety and liveness – in an asynchronous system with Byzantine failures is proposed in [1]. The feasibility of solving Byzantine causal order for unicasts, multicasts, and broadcasts was analyzed in [23]. Previously, a *probabilistic* algorithm based on atomic (total order) broadcast and cryptography for secure causal atomic broadcast (liveness and strong safety) in an asynchronous system was proposed [5]. This logic used acknowledgements and effectively processed the atomic broadcasts serially. For the client-server configuration, two protocols for crash failures and a third for Byzantine failure of clients based on cryptography were proposed for secure causal atomic broadcast [10]. The third made assumptions on latency of messages, and hence works only in a synchronous system.

A spectrum of solvability and unsolvability results for causal ordering in the presence of Byzantine processes for unicast, multicast, and broadcast modes of communication were shown in [22,23,25,26]. The results were shown for the regular or *strong safety* property, as well as for a weakening of the safety property, termed *weak safety*, defined alongside the above results. The possibility results implicitly assumed that the number of Byzantine processes f was less than $n/3$. This dependency arose because of the reliance on Bracha's Byzantine Reliable Broadcast (BRB) primitive [3] in the possibility results. The above results were for asynchronous systems; an analogous analysis and results for synchronous systems were presented in [28].

3 System Model

This paper deals with a distributed system having Byzantine processes which are processes that can misbehave [18,29]. A correct process behaves exactly as specified by the algorithm whereas a Byzantine process may exhibit arbitrary behaviour by deviating arbitrarily from its protocol during the execution. This subsumes crash failures. A Byzantine process cannot impersonate another process. However, a Byzantine process may spawn other Byzantine processes and an unbounded number of Byzantine processes may join the system. This models a permissionless setting.

The distributed system is modeled as an undirected graph $\mathcal{G} = (P, C)$. Here P is the set of processes communicating asynchronously in the distributed system. The set P can change dynamically. Let n denote $|P|$; thus n can change over time. C is the set of FIFO (logical) communication links over which processes communicate by message passing. The communication links are reliable implying messages cannot get lost or be duplicated, and communication is authenticated.

When new processes are spawned and due to churn, there is a dynamically varying number of processes. Some of the causal ordering protocols do broadcast or multicast and we assume there is a mechanism in place by which processes learn of such view changes.

While stating and proving our solvability results, the system is first assumed to be asynchronous, i.e., there is no fixed upper bound δ on the message latency, nor any fixed upper bound ψ on the relative speeds of processors [11]. We then deal with solvability results for a synchronous system, which is defined as one in which both δ and ψ exist and are known [11].

Likewise our results also deal with the two cases: disallowing or allowing the use of cryptography. For results in asynchronous systems allowing the use of cryptography, we assume a public key infrastructure (PKI) with public and private keys, and group keys. We also assume an on-the-fly key creation and distribution mechanism to deal with a varying number of processes. For results in synchronous systems allowing the use of cryptography, we additionally assume a threshold encryption system.

Let e_i^x, where $x \geq 1$, denote the x-th event executed by process p_i. An event may be an internal event, a message send event, or a message receive event. Let the state of p_i after e_i^x be denoted s_i^x, where $x \geq 1$, and let s_i^0 be the initial state. The *execution* at p_i is the sequence of alternating events and resulting states, as $\langle s_i^0, e_i^1, s_i^1, e_i^2, s_i^2 \ldots \rangle$. The *execution history* at p_i is the finite execution at p_i up to the current or most recent or specified local state. The *happens before* [17] relation, denoted $\rightarrow$, is an irreflexive, asymmetric, and transitive partial order defined over events in a distributed execution that is used to define causality.

Definition 1. *The happens before relation on events consists of the following rules:*

1. ***Program Order:*** *For the sequence of events $\langle e_i^1, e_i^2, \ldots \rangle$ executed by process p_i, $\forall\, j,k$ such that $j < k$ we have $e_i^j \rightarrow e_i^k$.*

2. **Message Order:** *If event e_i^x is a message send event executed at process p_i and e_j^y is the corresponding message receive event at process p_j, then $e_i^x \rightarrow e_j^y$.*
3. **Transitive Order:** *If $e \rightarrow e' \wedge e' \rightarrow e''$ then $e \rightarrow e''$.*

Next, we define the partial order happens before relation $\rightarrow$ on the set of all application-level messages R [25,26].

Definition 2. *The happens before relation $\rightarrow$ on messages in R consists of the following rules [25, 26]:*

1. *If p_i sent or delivered message m before sending message m', then $m \rightarrow m'$.*
2. *If $m \rightarrow m'$ and $m' \rightarrow m''$, then $m \rightarrow m''$.*

Definition 3. *The causal past of message m is denoted as $CP(m)$ and defined as the set of messages in R that causally precede message m under $\rightarrow$.*

To accommodate the possibility of Byzantine behaviour, we use a partial order on messages called *Byzantine happens before*, denoted as $\xrightarrow{B}$, defined on S, the set of all application-level messages that are both sent by and delivered at correct processes in P [25,26].

Definition 4. *The Byzantine happens before relation $\xrightarrow{B}$ on messages in S consists of the following rules [25, 26]:*

1. *If p_i is a correct process and p_i sent or delivered message m (to/from another correct process) before sending message m' to a correct process, then $m \xrightarrow{B} m'$.*
2. *If $m \xrightarrow{B} m'$ and $m' \xrightarrow{B} m''$, then $m \xrightarrow{B} m''$.*

The Byzantine causal past of a message is defined as follows:

Definition 5. *The Byzantine causal past of message m, denoted as $BCP(m)$, is defined as the set of messages in S that causally precede message m under $\xrightarrow{B}$.*

We consider three possible modes of communication: multicast, unicast, and broadcast. In a multicast/unicast/broadcast, a message m is sent at a send event using $\text{send}(m, G)$, $\text{send}(m, \{p_i\})$, $\text{send}(m, P)$, respectively, and is delivered at a receive event via $\text{deliver}(m)$.

Definition 6. *Byzantine Reliable Multicast (BRM) of message m to group G satisfies the following properties:*

1. *(Validity:) If a correct process p_i delivers message m from a correct process $sender(m)$ sent to group G, then $sender(m)$ must have executed $send(m, G)$ and $p_i \in G$.*
2. *(Self-delivery:) If a correct process executes $send(m, G)$, then it eventually delivers m. (Note, a group always contains the sender.)*

3. *(Agreement:) If a correct process delivers a message m from a possibly faulty process, then all correct processes in G will eventually deliver m.*
4. *(Integrity:) For any message m, a correct process p_i delivers m at most once.*
5. *(No Information Leakage:) No process outside the group G sees the content of m.*

It can be seen that Byzantine Reliable Unicast (BRU) and Byzantine Reliable Broadcast (BRB) are special cases of BRM. As a unicast has a single destination, the Agreement property of BRM goes away in BRU. As the destination set of a broadcast is the set of all processes, the No Information Leakage property of BRM goes away in BRB.

The correctness of Byzantine causal order multicast/unicast/broadcast is specified on $(R, \rightarrow)$ and $(S, \xrightarrow{B})$. The definitions of BCM/ BCU/ BCB need to be satisfied in addition to safety and liveness.

Definition 7. *(Strong safety and liveness:) A causal ordering algorithm for unicast/multicast/broadcast messages must ensure the following:*

1. **Strong Safety:** *$\forall m' \in CP(m)$ such that m' and m are sent to the same (correct) process, no correct process delivers m before m'.*
2. **Liveness:** *Each message sent by a correct process to another correct process will be eventually delivered.*

Definition 8. *(Weak safety and liveness:) A causal ordering algorithm for unicast/multicast/broadcast messages must ensure the following [23, 25, 26]:*

1. **Weak Safety:** *$\forall m' \in BCP(m)$ such that m' and m are sent to the same (correct) process, no correct process delivers m before m'.*
2. **Liveness:** *Each message sent by a correct process to another correct process will be eventually delivered.*

The goal is to satisfy both properties in the above definitions. A trivial but unacceptable solution to satisfy strong safety, but no liveness, is to never deliver a message. Likewise, a trivial but unacceptable solution to satisfy liveness, but no strong safety, is to simply deliver any message that is received. Ruling out such trivial but unacceptable solutions, in the sequel when we say that neither safety nor liveness can be guaranteed, or say that one of the two properties but not the other can be guaranteed, we mean using a non-trivial algorithm that attempts to satisfy both properties.

In our solvability results, we focus on only the strong safety or weak safety, and liveness properties. The Validity, Self-delivery, Agreement, Integrity, No Information Leakage properties follow in a straightforward manner.

4   Solvability Results

4.1   Strong Safety and Liveness Without Cryptography

Theorem 1. *It is impossible to solve causal ordering (Definition 7) of unicast messages in an asynchronous message passing system with f Byzantine processes where $f < n$, as neither strong safety nor liveness is guaranteed.*

Proof. It was proved in [23,25,26] that strong safety cannot be satisfied in a system with even one Byzantine process. Hence it cannot be satisfied in a system with f Byzantine processes where $f < n$. It was also proved in [23,25,26] that liveness cannot be satisfied in a system with even one Byzantine process. Hence it cannot be satisfied in a system with f Byzantine processes where $f < n$.   □

Theorem 2. *It is impossible to solve causal ordering (Definition 7) of broadcast messages in an asynchronous message passing system with f Byzantine processes where $f < n$, as neither strong safety nor liveness is guaranteed.*

Proof. It was proved in [23,25,26] that strong safety cannot be satisfied in a system with even one Byzantine process. Hence it cannot be satisfied in a system with f Byzantine processes where $f < n$.

Let a process p_a send a (supposed) broadcast message m that is received by p_b. p_b has two options. (1) p_b does a relay broadcast of m to all other processes as p_a may not have done a true broadcast but sent m selectively to some processes. This tries to ensure that each correct process p_c receives m directly and/or indirectly. But if p_b is Byzantine, it can relay broadcast fake messages by some supposed sender p_z, and this will cause p_c to receive and deliver fake messages. Hence this option cannot be used. (2) When p_b has to send its own broadcast m', it sends control information of all messages like m received since its own previous broadcast. When p_c receives m', it knows not to deliver m' until the causally preceding m is received and delivered by it. As the system is asynchronous, p_c cannot verify with p_a in bounded time whether p_a actually sent m via a broadcast and until then p_c will not be able to deliver m'. If p_a is Byzantine and it never included p_c as a destination of m, i.e., did not send m via a broadcast but sent it to p_b via a unicast, the message m' from a correct p_b to a correct p_c is never delivered, resulting in a liveness violation.   □

In a multicast, a message is sent to a subset of processes and different send events can send to different multicast groups. We have the following result.

Theorem 3. *It is impossible to solve causal ordering (Definition 7) of multicast messages in an asynchronous message passing system with f Byzantine processes where $f < n$, as neither liveness nor strong safety is guaranteed.*

Proof. As neither strong safety nor liveness can be guaranteed for unicasts and broadcasts (Theorems 1, 2, respectively)—and unicasts and broadcasts are special cases of multicast—it follows that these guarantees cannot be provided for multicasts either.   □

4.2 Weak Safety and Liveness Without Cryptography

We now show a similar result to Theorem 1 with strong safety (Definition 7) defined in terms of the $\rightarrow$ relation replaced by weak safety (Definition 8) defined in terms of the $\xrightarrow{B}$ relation in the correctness criteria for causal ordering.

Theorem 4. *It is impossible to solve causal ordering (Definition 8) of unicast messages in an asynchronous message passing system with f Byzantine processes where $f < n$, as liveness cannot be guaranteed even though weak safety can be guaranteed.*

Proof. It was proved in [23, 25, 26] that liveness cannot be satisfied for the unicast mode of communication in a system with even one Byzantine process. Hence it cannot be satisfied in a system with f Byzantine processes where $f < n$.

Weak safety implies that in the space-time diagram of the execution, there is a path from a sender process p_a to a process p_c that goes through only correct processes. Along such a path, all processes including p_a and p_c are correct. It is straightforward to devise an algorithm in which causal dependencies on messages m sent by p_a (to their respective destinations) can be faithfully propagated along such paths as control information received on messages such as m' received by p_c. m belongs to $BCP(m')$, and if p_c is a destination of m as per the control information, p_c will wait for m to arrive and be delivered before delivering m'. m will certainly arrive as p_a is a correct process that must have sent m to p_c. Thus weak safety is guaranteed. □

Theorem 5. *It is impossible to solve causal ordering (Definition 8) of broadcast messages in an asynchronous message passing system with f Byzantine processes where $f < n$, as liveness cannot be guaranteed even though weak safety can be guaranteed.*

Proof. To prevent a liveness attack, it should not be possible for a Byzantine process p_b to insert a fake dependency on a fake message m sent (broadcast) by p_a. If p_b attempts to do so, it should be detectable by a correct process p_c so that p_c will not wait indefinitely to deliver a message m' carrying information about such a fake dependency. By sending broadcasts using Byzantine Reliable Broadcast (BRB) [3] which guarantees that exactly the same message is delivered to all or no correct process, p_c can wait for the arrival of m (if genuinely sent, it is guaranteed to eventually arrive at p_c if it arrived at p_b) before delivering m'; if m does not arrive, the dependency information is fake and the sender of m' that included information of the dependency on m is Byzantine. So there is no liveness violation. This guarantee of no liveness attack depends on the ability to do broadcasts satisfying the Agreement property of BRB – and a necessary condition for BRB is that $f < n/3$ [3]. It is well-known that the Agreement property cannot be satisfied when $f \geq n/3$ unless cryptography is used [19, 20]. As we are not allowing the use of cryptography in this theorem, when $f < n$, this necessary condition is not met and there is no way that a correct process p_c can verify that a reported causal dependency on a prior message is genuine. The prior message m may have been sent via a unicast by a Byzantine process p_a to correct process p_b. (Refer to the argument in the proof of Theorem 2.) Thus liveness attacks cannot be prevented when $f < n$.

The proof of weak safety being guaranteed in Theorem 4 for the unicast case applies almost identically for the guarantee of weak safety in this broadcast case. □

Theorem 6. *It is impossible to solve causal ordering (Definition 8) of multicast messages in an asynchronous message passing system with f Byzantine processes where $f < n$, as liveness cannot be guaranteed even though weak safety is guaranteed.*

Proof. Unicast and broadcast modes of communication are special cases of the multicast mode of communication. As it is not possible to guarantee liveness for unicasts (Theorem 4) and broadcasts (Theorem 5), it is not possible to guarantee liveness for multicasts.

The proof of weak safety being guaranteed in Theorem 4 for the unicast case applies almost identically for the guarantee of weak safety in the multicast case.
□

4.3   Strong Safety and Liveness Using Cryptography

Here we use the simple idea of relay broadcasts alongside cryptography. In a *relay broadcast*, when a process receives a message m for the first time, it broadcasts m to other processes.

Theorem 7. *It is impossible to solve causal ordering (Definition 7) of multicast messages in an asynchronous message passing system with f Byzantine processes where $f < n$, even with the use of cryptography as strong safety cannot be guaranteed even though liveness can be guaranteed.*

Proof. The logic that causal ordering (Definition 7) cannot be solved for the multicast mode of communication is as follows.

It was proved in [23, 25, 26] that strong safety cannot be satisfied in a system with even one Byzantine process. Hence it cannot be satisfied in a system with f Byzantine processes where $f < n$.

To prevent fake causal dependencies on send event e_a^x of message m sent by process p_a, p_a has to sign m with its private key K_a^- so that any other process that gets a copy of m can verify (using decryption by p_a's public key K_a^+) that m was truly sent by p_a. However to maintain confidentiality within the group, the message needs to be encrypted. When p_a has to multicast a message m to group G at event e_a^x, it creates the ciphertext C_m by encrypting m with the group key K_G and sends $K_a^-(C_m, e_a^x, G)$ signed by the private key K_a^- to all the processes in the system. When a correct process p_c receives $K_a^-(C_m, e_a^x, G)$ (whether from a or via a relay broadcast by some other process) for the first time, p_c does a relay broadcast of $K_a^-(C_m, e_a^x, G)$ to all the processes in the system. It then applies $K_a^+(X)$ to verify that m is a legitimate message sent/originated by p_a. If m passes this test and $p_c \in G$, p_c decrypts C_m using K_G and a local application receive event e_c^y of m can occur if causal dependencies are satisfied.

In the next message $K_c^-(C_{m'}, e_c^w, G')$ sent (multicast) by p_c, control information of the form $\langle K_c^-(K_a^-(C_m, e_a^x, G), e_c^y)\rangle$ (and for similar other messages application-received since the last multicast by p_c) is also included.

When $K_c^-(C_{m'}, e_c^w, G')$ (along with its control information of the form $\langle K_c^-(K_a^-(C_m, e_a^x, G), e_c^y)\rangle$) is received by p_d, if $p_d \in G'$, p_d first delivers messages (dependencies) of the form $K_a^-(C_m, e_a^x, G)$ in the control information if $p_d \in G$, before $C_{m'}$, i.e., m' is locally delivered. As all the dependencies can be verified as being genuine using the public key via $K_a^+(K_a^-(C_m, e_a^x, G))$, those messages C_m, i.e., m, must have been sent by p_a (directly or through relay broadcasts), and liveness is guaranteed.

Note that strong safety is not guaranteed, neither is a common order of delivery at the correct processes. However, a message from a correct process to another correct process will never be blocked waiting for fake causal dependencies to be satisfied, thus ensuring liveness. □

Theorem 8. *It is impossible to solve causal ordering (Definition 7) of unicast messages in an asynchronous message passing system with f Byzantine processes where $f < n$, even with the use of cryptography as strong safety cannot be guaranteed even though liveness can be guaranteed.*

Proof. The proof of Theorem 7 applies almost identically to unicasts with the observation that each multicast group is of size two – the sender and the receiver. □

Theorem 9. *It is impossible to solve causal ordering (Definition 7) of broadcast messages in an asynchronous message passing system with f Byzantine processes where $f < n$, even with the use of cryptography as strong safety cannot be guaranteed even though liveness can be guaranteed.*

Proof. The proof of Theorem 7 which is for multicast-based communication applies to the broadcast mode of communication which is a special case of group size equal to n. □

In the above three theorems, liveness can be guaranteed with $f < n$ because we simulate Byzantine Reliable Broadcast using cryptography and relay broadcasts, and overcome the limitation of $f < n/3$ without cryptography. BRB prevents equivocation, and we are able to meet all the specifications of BRB, and based on BRB, of BRM and BRU. For each multicast, there are up to n^2 point-to-point messages (or n broadcasts) sent.

4.4   Weak Safety and Liveness Using Cryptography

Theorem 10. *It is possible to solve causal ordering (Definition 8) of multicast messages in an asynchronous message passing system with f Byzantine processes where $f < n$, with the use of cryptography as weak safety and liveness can be guaranteed.*

Proof. Liveness can be guaranteed as shown in the proof of Theorem 7.

Weak safety can be guaranteed as shown in the proof of Theorem 6 (even without cryptography) – so the guarantee of weak safety holds with cryptography. Also, the algorithm described in the proof of Theorem 7 assuming cryptography guarantees weak safety because under the $\xrightarrow{B}$ relation, not just multicasts but also relay broadcasts by Byzantine processes are not considered. Relay

Table 3. Best-known bounds on f for solvability of causal ordering using deterministic algorithms in synchronous systems under different communication modes. SS = strong safety, WS = weak safety, L = liveness, $\overline{SS}, \overline{WS}, \overline{L}$ represent that strong safety, weak safety, liveness, respectively, cannot be guaranteed. For each entry in the table, the relation between f and n is explicitly indicated for properties that are satisfiable.

Mode of communic.	SS + L	SS + L with cryptography	WS + L	WS + L with cryptography
Unicasts	No, [28]	Yes, [24, 28]	Yes, [22, 25, 28]	Yes, Sect. 5.2
	$\overline{SS}, L(f < n)$	$SS, L, f < n/2$	$WS, L, f < n$	$WS, L, f < n$
Broadcasts	No, [28]	Yes, [28]	Yes, [28]	Yes, Sect. 5.2
	$\overline{SS}, L(f < n)$	$SS, L, f < n/2$	$WS, L, f < n$	$WS, L, f < n$
Multicasts	No, [28]	Yes, [28]	Yes, [28]	Yes, Sect. 5.2
	$\overline{SS}, L(f < n)$	$SS, L, f < n/2$	$WS, L, f < n$	$WS, L, f < n$

broadcasts by a Byzantine process can cause messages between the same sender-receiver correct process pair to be delivered out-of-order by selectively doing relay broadcasts or violate safety by selectively suppressing dependencies on messages sent by correct processes in the control information in a relay broadcast. However, a message chain containing a relay broadcast by a Byzantine process that does application-message forwarding is not valid under the $\xrightarrow{B}$ relation used in the definition of weak safety.   □

As unicasts and broadcasts are special cases of multicast, we have the following two results.

Corollary 1. *It is possible to solve causal ordering (Definition 8) of unicast messages in an asynchronous message passing system with f Byzantine processes where $f < n$, with the use of cryptography as weak safety and liveness can be guaranteed.*

Corollary 2. *It is possible to solve causal ordering (Definition 8) of broadcast messages in an asynchronous message passing system with f Byzantine processes where $f < n$, with the use of cryptography as weak safety and liveness can be guaranteed.*

5   Results for Synchronous Systems

Table 3 gives the best-known bounds on f for solvability of causal ordering using deterministic algorithms in synchronous systems under different communication modes for the same system assumptions and parameter values (strong safety/weak safety, liveness, with or without cryptography) as considered for asynchronous systems.

Rounds can be simulated in a synchronous system, where a process first sends messages within a round and then receives messages sent in that round. Thus all messages sent in a round are received by the end of that round. Rounds greatly simplify the design of synchronous distributed algorithms.

5.1 Strong Safety and Liveness Without Cryptography

As shown in [28], strong safety cannot be guaranteed with even a single Byzantine process when no cryptography is used.

Following the round-based cryptography-free algorithm in [28], liveness can be guaranteed with $f < n$ because all messages m' supposedly sent to a correct process in the causal past of message m would have been sent in earlier rounds than the round in which m is sent, and thus would have been delivered in an earlier round. Thus there is no need for m to wait on arrival at its destination for the arrival/delivery of m'. This result holds for multicasts, and by extension, to the special cases of unicast and broadcast.

5.2 Weak Safety and Liveness Without Cryptography

The two round-free algorithms – Sender-Inhibition and Channel Sync – given for unicasts in [22,25] guarantee weak safety and liveness, and work for $f < n$. The round-based cryptography-free algorithm for multicasts (and by extension, for the special cases of unicasts and broadcasts) given in [28], guarantees weak safety and liveness as shown in [28], and as per the logic for liveness outlined in Sect. 5.1.

5.3 Strong Safety and Liveness with Cryptography

The algorithm for unicasts given in [24] uses threshold encryption to guarantee strong safety and liveness. Threshold encryption requires $f < n/2$.

The cryptography-based algorithm in [28] for strong safety and liveness for multicasts (and for its special cases of unicasts and broadcasts) uses Byzantine Reliable Broadcast (BRB) in conjunction with threshold encryption. BRB requires that $f < n/3$ whereas threshold encryption requires that $f < n/2$. Thus the effective bound is $f < n/3$. In this algorithm, we propose replacing the BRB algorithm of Bracha by a module that implements Dolev-Strong authenticated Byzantine Agreement in synchronous systems [8], as done in [27]. Instead of doing a BRB broadcast of m, Dolev-Strong authenticated agreement is reached on m in exactly $f + 1$ rounds, tolerating up to $n - 1$ Byzantine processes, i.e., $f < n$. If the agreement value is a non-null message, then the m is processed further and the corresponding decryption shares are sent as per the rest of the algorithm in [28]; otherwise if the agreement value is the null message then m is ignored. Dolev-Strong algorithm authenticated Byzantine agreement requires $f < n$ whereas threshold encryption requires $f < n/2$. Thus the effective bound is $f < n/2$.

5.4   Weak Safety and Liveness with Cryptography

Weak safety and liveness can be guaranteed (for unicasts, multicasts, and broadcasts) in synchronous systems without the use of cryptography, for $f < n$ (see Sect. 5.2). So the result extends to systems with the use of cryptography.

6   Conclusions

Corresponding to the solvability and unsolvability results for various combinations of system assumptions shown in Table 1 for $f < n/3$ in asynchronous systems in [26], we showed results assuming only that $f < n$, thus considering systems with the highest degree of Byzantine fault-tolerance. The results for asynchronous systems are summarized in Table 2. We also gave the best-known bounds on f for solvability of causal ordering using deterministic algorithms in synchronous systems under the same combinations of system assumptions and parameters as for asynchronous systems. The results for synchronous systems are summarized in Table 3.

This study is significant because it is important to understand the limitations on what is solvable under the highest degree of Byzantine fault-resilience. In particular, when $f < n$ the results indicate what is solvable under Byzantine Sybil tolerance in peer-to-peer systems.

Some of our solvability results pertain to weak safety when $f < n$. We note that in practice, it is not possible to determine whether $m \xrightarrow{B} m'$ holds because this requires identifying the correct processes and the Byzantine processes. Therefore it is not possible to determine whether weak safety actually holds. The solvability results simply say that if weak safety holds then causal order can be satisfied.

Acknowledgments. We thank Hannes Hartenstein and Florian Jacob for suggesting the problem of studying Byzantine Sybil fault-tolerance for causal ordering. The paper has benefited from several discussions with Hannes and Florian.

References

1. Auvolat, A., Frey, D., Raynal, M., Taïani, F.: Byzantine-tolerant causal broadcast. Theoret. Comput. Sci. **885**, 55–68 (2021)
2. Birman, K.P., Joseph, T.A.: Reliable communication in the presence of failures. ACM Transactions on Computer Systems **5**(1), 47–76 (1987)
3. Bracha, G.: Asynchronous byzantine agreement protocols. Inf. Comput. **75**(2), 130–143 (1987)
4. Bracha, G., Toueg, S.: Asynchronous consensus and broadcast protocols. J. ACM **32**(4), 824–840 (1985). https://doi.org/10.1145/4221.214134
5. Cachin, C., Kursawe, K., Petzold, F., Shoup, V.: Secure and efficient asynchronous broadcast protocols. IACR Cryptol. ePrint Arch., p. 6 (2001). http://eprint.iacr.org/2001/006

6. Chandra, P., Gambhire, P., Kshemkalyani, A.D.: Performance of the optimal causal multicast algorithm: a statistical analysis. IEEE Trans. Parallel Distributed Syst. **15**(1), 40–52 (2004). https://doi.org/10.1109/TPDS.2004.1264784

7. Chandra, P., Kshemkalyani, A.D.: Causal multicast in mobile networks. In: 12th International Workshop on Modeling, Analysis, and Simulation of Computer and Telecommunication Systems (MASCOTS), pp. 213–220 (2004). https://doi.org/10.1109/MASCOT.2004.1348235

8. Dolev, D., Strong, H.R.: Authenticated algorithms for byzantine agreement. SIAM J. Comput. **12**(4), 656–666 (1983). https://doi.org/10.1137/0212045

9. Douceur, J.R.: The sybil attack. In: Druschel, P., Kaashoek, F., Rowstron, A. (eds.) IPTPS 2002. LNCS, vol. 2429, pp. 251–260. Springer, Heidelberg (2002). https://doi.org/10.1007/3-540-45748-8_24

10. Duan, S., Reiter, M.K., Zhang, H.: Secure causal atomic broadcast, revisited. In: 2017 47th Annual IEEE/IFIP International Conference on Dependable Systems and Networks (DSN), pp. 61–72. IEEE (2017)

11. Dwork, C., Lynch, N.A., Stockmeyer, L.J.: Consensus in the presence of partial synchrony. J. ACM **35**(2), 288–323 (1988). http://doi.acm.org/10.1145/42282.42283

12. Jacob, F., Beer, C., Henze, N., Hartenstein, H.: Analysis of the matrix event graph replicated data type. IEEE Access **9**, 28317–28333 (2021). https://doi.org/10.1109/ACCESS.2021.3058576

13. Kleppmann, M., Howard, H.: Byzantine eventual consistency and the fundamental limits of peer-to-peer databases. arXiv preprint arXiv:2012.00472 (2020)

14. Kshemkalyani, A.D., Singhal, M.: An optimal algorithm for generalized causal message ordering (abstract). In: Proceedings of the Fifteenth Annual ACM Symposium on Principles of Distributed Computing (PODC), p. 87. ACM (1996). https://doi.org/10.1145/248052.248064

15. Kshemkalyani, A.D., Singhal, M.: Necessary and sufficient conditions on information for causal message ordering and their optimal implementation. Distributed Comput. **11**(2), 91–111 (1998). https://doi.org/10.1007/s004460050044

16. Kshemkalyani, A.D., Singhal, M.: Distributed Computing: Principles, Algorithms, and Systems. Cambridge University Press (2011). https://doi.org/10.1017/CBO9780511805318

17. Lamport, L.: Time, clocks, and the ordering of events in a distributed system. Commun. ACM **21**(7), 558–565 (1978)

18. Lamport, L., Shostak, R.E., Pease, M.C.: The byzantine generals problem. ACM Trans. Program. Lang. Syst. **4**(3), 382–401 (1982). http://doi.acm.org/10.1145/357172.357176

19. Malkhi, D., Merritt, M., Rodeh, O.: Secure reliable multicast protocols in a WAN. In: Proceedings of the 17th International Conference on Distributed Computing Systems, pp. 87–94 (1997)

20. Malkhi, D., Reiter, M.K.: A high-throughput secure reliable multicast protocol. J. Comput. Secur. **5**(2), 113–128 (1997)

21. Mattern, F., Fünfrocken, S.: A non-blocking lightweight implementation of causal order message delivery. In: Birman, K.P., Mattern, F., Schiper, A. (eds.) Theory and Practice in Distributed Systems. LNCS, vol. 938, pp. 197–213. Springer, Heidelberg (1995). https://doi.org/10.1007/3-540-60042-6_14

22. Misra, A., Kshemkalyani, A.D.: Causal ordering in the presence of byzantine processes. In: 28th IEEE International Conference on Parallel and Distributed Systems, pp. 130–138. IEEE (2022). https://doi.org/10.1109/ICPADS56603.2022.00025

23. Misra, A., Kshemkalyani, A.D.: Solvability of byzantine fault-tolerant causal ordering problems. In: Koulali, M., Mezini, M. (eds.) 10th International Conference on Networked Systems. LNCS, vol. 13464, pp. 87–103. Springer (2022). https://doi.org/10.1007/978-3-031-17436-0_7

24. Misra, A., Kshemkalyani, A.D.: Byzantine fault-tolerant causal order satisfying strong safety. In: Proceedings of the 25th International Symposium on Stabilization, Safety, and Security of Distributed Systems (SSS), pp. 111–125 (2023). https://doi.org/10.1007/978-3-031-44274-2_10

25. Misra, A., Kshemkalyani, A.D.: Byzantine fault-tolerant causal ordering. In: 24th International Conference on Distributed Computing and Networking, pp. 100–109. ACM (2023). https://doi.org/10.1145/3571306.3571395

26. Misra, A., Kshemkalyani, A.D.: Byzantine-tolerant causal ordering for unicasts, multicasts, and broadcasts. IEEE Trans. Parallel Distributed Syst. **35**(5), 814–828 (2024). https://doi.org/10.1109/TPDS.2024.3368280

27. Misra, A., Kshemkalyani, A.D.: Detecting causality in the presence of byzantine processes: the case of synchronous systems. Inf. Comput. **301**, 105212 (2024). https://doi.org/10.1016/j.ic.2024.105212

28. Misra, A., Kshemkalyani, A.D.: Solvabiity of byzantine fault-tolerant causal ordering: synchronous systems case. In: Proceedings of the 39th ACM Symposium on Applied Computing (SAC), pp. 251–256 (2024). https://doi.org/10.1145/3605098.3636063

29. Pease, M.C., Shostak, R.E., Lamport, L.: Reaching agreement in the presence of faults. J. ACM **27**(2), 228–234 (1980). http://doi.acm.org/10.1145/322186.322188

30. Raynal, M., Schiper, A., Toueg, S.: The causal ordering abstraction and a simple way to implement it. Inf. Process. Lett. **39**(6), 343–350 (1991)

Byzantine Reliable Broadcast in Wireless Networks

Hao Lu, Jian Liu(✉), and Kui Ren

Zhejiang University, Hangzhou, Zhejiang 310027, China
{luhao,liujian2411,kuiren}@zju.edu.cn

Abstract. Byzantine reliable broadcast (BRB) is a fundamental primitive in distributed computing. BRB ensures that messages can be reliably broadcast among nodes, regardless of the presence of Byzantine nodes. Research on BRB in wireless networks receives relatively less attention, while there have been significant advancements in wired networks. Although current works have yielded remarkable results, they still have some limitations. Specifically, the forwarding frequency of each node is $O(n)$, and the fault-tolerant thresholds still have a gap with the thresholds in wired networks where less than one-third of nodes can be faulty. In this paper, we propose a new BRB protocol in an infinite grid (or finite toroidal) wireless network. With a forwarding frequency of $O(1)$, our BRB protocol achieves thresholds closer to the thresholds in wired networks. Moreover, we provide the derivations of new thresholds and formally prove the correctness of our BRB protocol.

Keywords: Byzantine fault-tolerant · Reliable broadcast · Wireless networks

1 Introduction

Byzantine reliable broadcast (BRB) is a fundamental primitive in distributed computing. The process begins with a source node that intends to disseminate messages to all other nodes in the system. Once a correct node reliably delivers a message, all other correct nodes will eventually reliably deliver the same message, even if some nodes (including the source node) are Byzantine. While noteworthy breakthroughs have been achieved, these endeavors predominantly concentrate on wired networks, overlooking the crucial aspect of wireless networks. However, BRB holds equal significance for applications in wireless (or radio) networks. We underscore two pivotal distinctions in achieving BRB in a wireless network, one being advantageous and the other detrimental:

- Each node has a broadcast radius r, and any message transmitted by a source node is received uniformly by all its neighbors (i.e., those within a distance of r from the source node). This property is advantageous for BRB because a node *cannot* equivocate to its neighbors.

S. Bonomi et al. (Eds.): SSS 2025, LNCS 16350, pp. 392–407, 2026.
https://doi.org/10.1007/978-3-032-11127-2_30

- When two nodes are beyond distance of r, their communication relies on the assistance of intermediary nodes for forwarding. Consequently, a source node colluding with an intermediary node retains the potential for equivocation.

These two characteristics highlight the unique opportunities and challenges in implementing BRB in wireless networks. Although the wireless channel is typically unreliable and prone to collisions, to better understand and address these aspects, starting with an idealized wireless channel is a practical approach. The idealized wireless channel exhibits FIFO (First-In-First-Out) characteristics, i.e., the messages broadcast by a source node will eventually be received by its neighbors in the sent order. In addition, it ensures that no message collisions will occur. The results in [3–5,11,16] are also based on the same assumption.

Remarkable results have been achieved in an infinite (or finite toroidal) wireless grid network where nodes are located on a grid consisting of 1×1 square units. In more detail, Koo [11] proves that achieving BRB in an asynchronous infinite grid wireless network is possible only when $f < \frac{1}{2}r(2r+1)$ in the L_∞ metric[1] where f denotes the number of Byzantine nodes in any single neighborhood. Bhandari et al. [4] introduce a BRB protocol that aligns with this established threshold and shows that BRB is achievable when $f < 0.23\pi r^2$ in L_2 metric[2]. Although these works achieve remarkable results, they still have some limitations. In these works, the forwarding frequency of each node is $O(n)$ (where n is the number of nodes within a neighborhood). The high forwarding frequency is costly in energy for wireless nodes, especially for energy-limited nodes. Furthermore, the fault-tolerant thresholds still have a gap with the thresholds in wired networks where less than one-third of nodes can be faulty.

Our Contribution. In this paper, we propose a new BRB protocol using the *aggregate signature* [6,15], which has $O(1)$ forwarding frequency and higher fault-tolerant thresholds than the state-of-the-art (SOTA) [4] presented in PODC '05. The aggregate signature is a cryptographic technique that allows multiple signatures from different nodes to be combined into a single signature, reducing the computational overhead required to verify signatures. In our BRB protocol, we use this technique to generate *certificates* that contain essential information enabling nodes to trust the content of messages, thus ensuring reliable message delivery. By using the aggregate signature technique, each node only needs to forward a certificate to convince others to reliably deliver the corresponding message, so our BRB achieves an $O(1)$ forwarding frequency. With low forwarding frequency demand, nodes do not need to have a large number of connections with others to receive and forward $O(n)$ messages. This reduced connection requirement means that a node can have more faulty neighbors. As a result, our BRB is more resistant to node failures. Specifically, we improve the fault-tolerant threshold to $f < \lfloor \frac{1}{2}(2r+1)(r+1) \rfloor$ in the L_∞ metric. We also present arguments suggesting that in the L_2 metric, when r is sufficiently large and $f < \lfloor 0.3\pi r^2 \rfloor$, BRB can be achieved. Compared to the SOTA [4] that has fault-tolerant thresholds of $f < \frac{1}{2}r(2r + 1)$ in the L_∞ metric and $f < 0.23\pi r^2$ in the L_2 metric,

---

[1] The distance between two nodes $\mathcal{N}_{(x_1,y_1)}$ and $\mathcal{N}_{(x_2,y_2)}$ is $\max\{|x_1 - x_2|, |y_1 - y_2|\}$.

[2] The distance between two nodes $\mathcal{N}_{(x_1,y_1)}$ and $\mathcal{N}_{(x_2,y_2)}$ is $\sqrt{(x_1 - x_2)^2 + (y_1 - y_2)^2}$.

our thresholds are closer to the thresholds in wired networks where one-third of nodes can be faulty. Finally, we provide a detailed derivation of new thresholds and a formal correctness proof of our BRB.

2 Preliminaries

2.1 Model

Wireless Grid Network. In this paper, we focus on an infinite (or finite toroidal) wireless grid network [11] where nodes are located on a grid consisting of 1×1 square units. Each node can be uniquely identified by its grid location (x, y), denoted as $\mathcal{N}_{(x,y)}$. All nodes have a broadcast radius of r, where $r \in \mathbb{Z}$. The messages broadcast by $\mathcal{N}_{(x,y)}$ can be received by all the nodes within a distance r from it. We refer to these nodes as the neighbors of $\mathcal{N}_{(x,y)}$. We remark that $\mathcal{N}_{(x,y)}$ and all its neighbors are termed as the *neighborhood* of $\mathcal{N}_{(x,y)}$, denoted as $nbd(x, y)$. For brevity, we may use $\mathcal{N}_i$ to denote $\mathcal{N}_{(x,y)}$ (i.e., use i to denote (x, y)).

In this paper, we consider two distance metrics, namely L_∞ metric and L_2 metric, which are induced by the L_∞ and L_2 norms [13]. In the L_∞ metric, the distance between two nodes $\mathcal{N}_{(x_1,y_1)}$ and $\mathcal{N}_{(x_2,y_2)}$ is $\max\{|x_1 - x_2|, |y_1 - y_2|\}$. In the L_2 metric, the distance is $\sqrt{(x_1 - x_2)^2 + (y_1 - y_2)^2}$. The analysis under the L_∞ metric is more manageable, which facilitates us in determining precise fault-tolerant thresholds. Moreover, it offers insightful intuitions that serve as the basis for our approximate reasoning regarding the L_2 metric of practical significance.

Communication Assumptions. We assume an asynchronous communication, where the sent message will eventually be received, but there is no time bound on the message delay. Following [3–5, 11, 16], we assume an *idealized shared wireless channel*, which exhibits FIFO (First-In-First-Out) characteristics and ensures that no message collisions will occur. Although real-world wireless channels are often unreliable and susceptible to collisions, it is beneficial to focus on an idealized wireless channel as a first step to investigate BRB. Specifically, this assumption allows us to first explore the fundamental principles and algorithms of BRB without the interference of complex real-world factors. Elements such as message conflicts and signal attenuation in real channels can then be gradually incorporated to refine and improve our proposed solutions, making them more practical and applicable.

Aggregate Signature Scheme. We assume a finite number of source nodes. Each node among the source nodes and their neighbors has a public/secret key pair. All nodes know all public keys, while nodes only know their own secret keys. A node can use its secret key to sign a message and output a signature. Any node can verify this signature by using the signer's public key.

A certain number of signatures usually serve as a *certificate* to identify that a message is signed by some nodes. The *aggregate signature* scheme can compress multiple signatures of a message into a single aggregate signature. Each node can verify the aggregate signature via an aggregate public key. As a result, this scheme significantly reduces the space and computation needed for verification compared to the naive approach when multiple signatures are required as certificates. How to achieve it is out of the scope of this paper; in fact, we can implement it based on Schnorr [15] or BLS [6]. The key operations of the scheme we used are as follows:

- $\text{Sign}(\text{SK}_i, m)$: Input: a secret key SK_i and a message m.
  Output: a signature σ of $\mathcal{N}_i$ for a message m.
- $\text{Aggre}(\sigma_j, ..., \sigma_k)$: Input: a set of signatures $\sigma_j, ..., \sigma_k$ of a common message.
  Output: an aggregate signature $\hat{\sigma}$ and a set of indexes of the input signatures $\sigma_j, ..., \sigma_k$.
- $\text{AggPK}(\text{PK}_j, ..., \text{PK}_k)$: Input: a set of public keys $\text{PK}_j, ..., \text{PK}_k$.
  Output: an aggregate public key $\hat{\text{PK}}$.
- $\text{Verify}(\text{PK}, m, \sigma)$: Input: a public key PK (could be an individual public key PK_i or an aggregate public key $\hat{\text{PK}}$, and a signature σ (could be an individual signature σ or an aggregate signature $\hat{\sigma}$) of a message m.
  Output: *true* when σ is a valid (individual/aggregate) signature of message m under PK; otherwise *false*.

The Aggre operation outputs a set of indexes that indicates the set of nodes $\hat{\mathcal{N}}$ that contribute to the aggregate signature $\hat{\sigma}$. This set enables nodes to compute the aggregate public key $\hat{\text{PK}}$ correspondingly by calling AggPK. Since the set has no impact on the core conclusions in the subsequent analysis, for the sake of brevity, we omit the set in subsequent sections. In our proposed BRB protocol, we use an aggregate signature to serve as a certificate to convince nodes to reliably deliver messages. More details are available in Sect. 3.

Adversary. In this paper, we consider Byzantine faults; a Byzantine (or faulty) node can behave arbitrarily, but cannot cause message collisions, and cannot forge a valid signature without access to the private key. We assume that up to f faulty nodes can exist in any neighborhood. Faulty nodes can collude with each other. For example, when a source node is faulty, it broadcasts two different messages m and m^*, respectively. Some faulty neighbors could forward m, while others forward m^* to cause equivocation.

2.2   Problem Definition

Byzantine Reliable Broadcast. A source node can reliably broadcast a message m by calling r_bcast(m, q) where $q \in \mathbb{N}$ is a sequence number. The sequence number q is used to differentiate the messages broadcast by a single node. A node can reliably deliver a message m with a sequence number q from $\mathcal{N}_i$ by outputting r_deliver$(m, q, \mathcal{N}_i)$. Formally, a BRB protocol satisfies the following properties:

- **Agreement:** if a correct node outputs r_deliver$(m, q, \mathcal{N}_j)$, then all other correct nodes will output r_deliver$(m, q, \mathcal{N}_j)$.
- **Integrity:** For each sequence number q and node $\mathcal{N}_i$, every correct node outputs r_deliver$(m, q, \mathcal{N}_i)$ at most once regardless of m.
- **Validity:** if a correct source node $\mathcal{N}_i$ calls r_bcast(m, q), then all correct nodes will eventually output r_deliver$(m, q, \mathcal{N}_i)$.

3   Proposed Byzantine Reliable Broadcast

In this section, we introduce our BRB protocol. We consider two types of nodes:

- *Nodes within the neighborhood of the source node.* They can reliably deliver the message directly because the source node cannot equivocate with its neighbors.
- *Nodes outside the neighborhood of the source node.* They reliably deliver a message upon receiving a valid *certificate* – that is, an aggregate signature contributed by $(f + 1)$ nodes within the neighborhood of the source node. Since at most f nodes can be faulty in any neighborhood, a node receiving this certificate is convinced that at least one correct node has reliably delivered the corresponding message.

These two types of nodes cover all nodes. Next, we introduce the handling mechanisms of our BRB protocol. The BRB protocol is initiated when a source node broadcasts a signed committed message. Fig. 1 shows the handling mechanisms of our BRB protocol:

1. Upon receiving a message from the source node, a neighbor of the source node reliably delivers the message and broadcasts a signed committed message (Fig. 1a).
2. Upon receiving a committed message from the source node or a neighbor of the source node, a node within the neighborhood of the source node waits for $f + 1$ committed messages from this neighborhood (including itself), combines their signatures into a certificate (an aggregate signature) and broadcasts the certificate (Fig. 1b).
3. When receiving a valid certificate for the first time, a node reliably delivers the corresponding message (if it has not delivered any message with the same sequence number) and forwards the certificate to others (Fig. 1c).

Figure 1a and Fig. 1b show the message handling of the nodes within the neighborhood of the source node. Fig. 1c shows the message handling of all nodes.

As we can see, each node in the system only needs to broadcast a certificate message (the source node and its neighbors also broadcast a committed message). Therefore, the forwarding frequency of each node is $O(1)$. The details of our BRB protocol are shown in Fig. 2. Notice that each node only processes messages with valid signatures. For brevity, we omit signature verification in Fig. 2.

Our BRB protocol works when $f < \lfloor \frac{1}{2}(2r + 1)(r + 1) \rfloor$ in the L_∞ metric and when $f < \lfloor 0.3\pi r^2 \rfloor$ in the L_2 metric. Next, we introduce how these thresholds are derived.

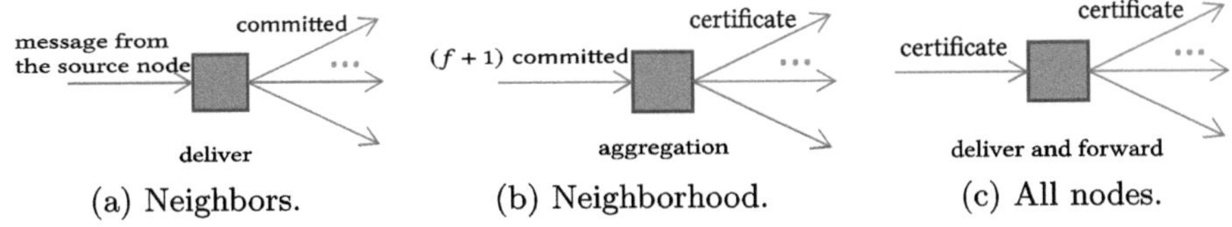

Fig. 1. The message handling mechanism of our reliable broadcast protocol.

Reliable Broadcast

By calling r_bcast(m, q), the source node $\mathcal{N}_s$ computes $\sigma_{q,s} := \mathrm{Sign}(\mathrm{SK}_s, M)$ and broadcasts $\langle \mathsf{committed}, s, m, q, \sigma_{q,s} \rangle$, where $M = \langle \mathsf{committed}, s, m, q \rangle$. Then, nodes run as the following:

1. **Node $\mathcal{N}_i$ within the neighborhood of the source node $\mathcal{N}_s$.**
   - Upon receiving the first valid **committed** message from $\mathcal{N}_s$ for m with a sequence number q:
     $\mathcal{N}_i$ outputs r_deliver$(m, q, \mathcal{N}_s)$;
     $\mathcal{N}_i$ computes $\sigma_{q,i} := \mathrm{Sign}(\mathrm{SK}_i, M)$, where $M = \langle \mathsf{committed}, s, m, q \rangle$;
     $\mathcal{N}_i$ broadcasts $\langle \mathsf{committed}, s, m, q, \sigma_{q,i} \rangle$.
   - Upon receiving $f + 1$ valid **committed** messages for m with a sequence number q:
     $\mathcal{N}_i$ computes $\hat{\sigma}_{q,m} := \mathrm{Aggre}(\{\sigma_*\}_{f+1})$;
     $\mathcal{N}_i$ broadcasts $\langle \mathsf{certificate}, s, m, q, \hat{\sigma}_{q,m} \rangle$.
   - When receiving a valid $\langle \mathsf{certificate}, s, m, q, \hat{\sigma}_{q,m} \rangle$ for the first time:
     $\mathcal{N}_i$ forwards this **certificate** message.

2. **Node $\mathcal{N}_i$ outside the neighborhood of the source node $\mathcal{N}_s$.**
   - When receiving a valid $\langle \mathsf{certificate}, s, m, q, \hat{\sigma}_{q,m} \rangle$ for the first time:
     $\mathcal{N}_i$ forwards this **certificate** message;
     $\mathcal{N}_i$ outputs r_deliver$(m, q, \mathcal{N}_s)$, if it has not delivered any message with a sequence number q.

Fig. 2. The details of our Byzantine reliable broadcast protocol.

3.1 Derivation of $f < \lfloor \frac{1}{2}(2r + 1)(r + 1) \rfloor$ in the L_∞ metric

In this section, we introduce how $f < \lfloor \frac{1}{2}(2r + 1)(r + 1) \rfloor$ is derived in the L_∞ metric. To better understand the derivation, we first explain some necessary information of Fig. 3. The source node $\mathcal{N}_{(a,b)}$ is located at (a, b) with a broadcast radius r. The nodes $\mathcal{N}_{(x,y)}$, where $\{(x, y) | a - r \leq x \leq a + r; b - r \leq y \leq b + r\}$, within the blue region are all nodes within the neighborhood of the source node $\mathcal{N}_{(a,b)}$. Node $\mathcal{N}_P$ is located at $(a + \beta r, b)$ where $0 \leq \beta \leq 1$. Nodes $\mathcal{N}_{(x,y)}$, where $\{(x, y) | |x - a| + |y - b| \leq \beta r\}$, within the yellow region in Fig. 3a are the more important nodes within the neighborhood of $\mathcal{N}_{(a,b)}$, which means that compared to $\mathcal{N}_P$, they have more (or the same number) shared neighbors with the source node $\mathcal{N}_{(a,b)}$. The yellow region contains $2\sum_{k=1}^{\beta r}(2k-1)+2\beta r+1 = (2r+1)((2-\beta)r+1)$ nodes. Nodes $\mathcal{N}_{(x,y)}$, where $\{(x,y)|a-(1-\beta)r \leq x \leq a+r; b-r \leq y \leq b+r\}$, within the dark green region in Fig. 3b are the shared neighbors of node $\mathcal{N}_P$ and the source node $\mathcal{N}_{(a,b)}$. The dark green region contains $(2r+1)((2-\beta)r+1)$ nodes. Nodes $\mathcal{N}_{(x,y)}$, where $\{(x,y)|a+1 \leq x \leq a+r; b-r \leq y \leq b+r\}$, within the red region in Fig. 3c are the neighbors on the right of the source node $\mathcal{N}_{(a,b)}$. The red region contains $(2r+1)r$ nodes. We summarize them in Table 1.

The intuition of this derivation is that "a valid certificate message will be constructed and received by all correct nodes".

We first consider how the neighborhood of a source node ensures that "a valid certificate message will be constructed". To construct a certificate message, a node must receive $f+1$ valid committed messages. Recall that only the neighbors of the source node broadcast committed messages. This implies that a node must have at least $2f+1$ shared neighbors with the source node because up to f nodes can be faulty. Obviously, compared to node $\mathcal{N}_P$, each node within the yellow region has more (or the same number) shared neighbors with the source node. This means that if node $\mathcal{N}_P$ can construct a valid certificate message, all nodes within the yellow region can also construct a valid certificate message.

Therefore, to have at least one correct node construct a valid certificate message, the number f of faulty nodes in the neighborhood of a source node should meet the following two rules:

1. The node $\mathcal{N}_P$ has at least $2f+1$ shared neighbors with the source node: $(2r+1)((2-\beta)r+1) \geq 2f+1$.
2. The yellow region contains at least $f+1$ nodes: $f < 2\beta r(\beta r+1)+1$.

Rule 1 ensures that all nodes within the yellow region can construct a valid certificate message. Rule 2 ensures that the yellow region contains at least one correct node. In summary, these two rules ensure that a valid certificate message can be constructed by at least one correct node.

Then, we consider how every neighborhood ensures that "a valid certificate will be received by all correct nodes". In order to spread a certificate message to all directions (that is, up, down, left, and right), each node should have at least one correct node in each direction. As shown in Fig. 3c, all neighbors on the right of node $\mathcal{N}_{(a,b)}$ are within the red region, which contains $(2r+1)r$ nodes. By symmetry, each node has $(2r+1)r$ nodes in every direction. Therefore, one of these $(2r+1)r$ nodes should be correct, which means that $f < (2r+1)r$.

In summary, the number f of faulty nodes in each neighborhood should meet:

1. $(2r+1)((2-\beta)r+1) \geq 2f+1$
2. $f < 2\beta r(\beta r+1)+1$
3. $f < (2r+1)r$

In fact, when $\beta = 0.8$, we get the maximum $f < \frac{1}{2}(2r+1)(1.2r+1)$. However, when $\beta = 0.8$, node $\mathcal{N}_P$ cannot be located at a grid point because $\beta r \bmod 1 \neq 0$.

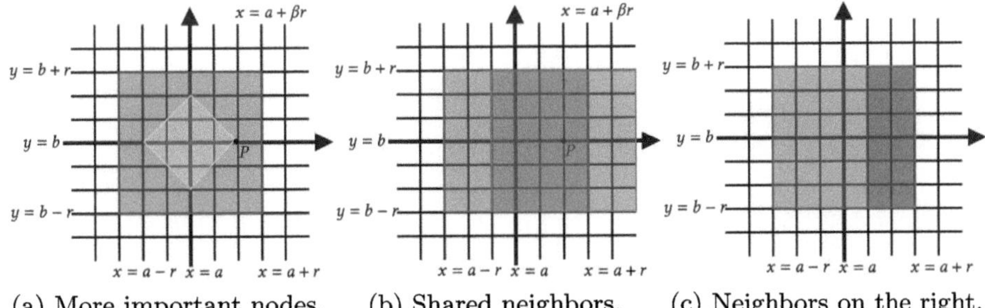

(a) More important nodes.   (b) Shared neighbors.   (c) Neighbors on the right.

Fig. 3. Derivation of f in an infinite grid under L_∞ metric.

Table 1. The summarized necessary information of Fig. 3.

Region	x-extent and y-extent	Description	Numbers of nodes
Blue	$a - r \leq x \leq a + r$ $b - r \leq y \leq b + r$	The neighborhood of $\mathcal{N}_{(a,b)}$, i.e., $nbd(a,b)$	$(2r + 1)^2$
Yellow	$\|x - a\| + \|y - b\| \leq \beta r$	The more important nodes within $nbd(a,b)$	$(2r + 1)((2 - \beta)r + 1)$
Dark green	$a - (1 - \beta)r \leq x \leq a + r$ $b - r \leq y \leq b + r$	The shard neighbors of $\mathcal{N}_P$ and $\mathcal{N}_{(a,b)}$	$2\beta r(\beta r + 1) + 1$
Red	$a + 1 \leq x \leq a + r$ $b - r \leq y \leq b + r$	The neighbors of $\mathcal{N}_{(a,b)}$ on the right	$(2r + 1)r$

Thus, we make $\beta = 1$ conservatively to allow node $\mathcal{N}_P$ to be located at a grid point and get $f < \lfloor \frac{1}{2}(2r + 1)(r + 1) \rfloor$. In conclusion, each neighborhood in our protocol can tolerate $f < \lfloor \frac{1}{2}(2r + 1)(r + 1) \rfloor$ faulty nodes.

The L_∞ metric simplifies the analysis, enabling precise fault-tolerance thresholds. It also provides intuitive insights that underpin an approximate argument for the practically significant L_2 metric. Next, we introduce derivation of $f < \lfloor 0.3\pi r^2 \rfloor$ in the L_2 metric.

3.2   Derivation of $f < \lfloor 0.3\pi R^2 \rfloor$ in the L_2 Metric

In this section, we introduce how $f < \lfloor 0.3\pi r^2 \rfloor$ is derived in the L_2 metric. Note that we provide an approximate argument based on [4], for sufficiently large r, the area is a good approximation for the number of nodes in these regions (it is increasingly accurate for large r). The intuition of this derivation is the same as that in the L_∞ metric, i.e., "a valid **certificate** message will be constructed, and received by all correct nodes". Similarly, to better understand this derivation, we first explain some necessary information of Fig. 4. The source node $\mathcal{N}_{(a,b)}$ is located at (a, b), and its broadcast radius is r. Nodes $\mathcal{N}_{(x,y)}$, where $\{(x,y)|(x - a)^2 + (y - b)^2 \leq r^2\}$, within the blue region are all nodes within the neighborhood of the source node $\mathcal{N}_{(a,b)}$. Node $\mathcal{N}_P$ is located at $(a + \alpha r, b)$ where $0 \leq \alpha \leq 1$. Nodes $\mathcal{N}_{(x,y)}$, where $\{(x,y)|(x - a)^2 + (y - b)^2 \leq (\alpha r)^2\}$, within the yellow region in Fig. 4a are the more important nodes within the neighborhood of $\mathcal{N}_{(a,b)}$, which means that compared to $\mathcal{N}_P$, they have more (or the same number) shared neighbors with the source node $\mathcal{N}_{(a,b)}$. The yellow

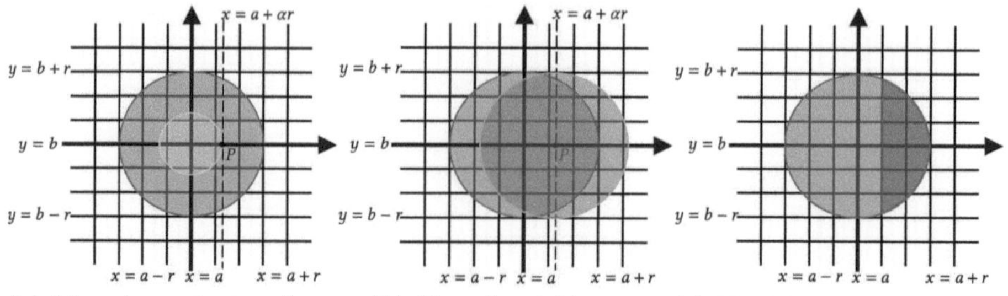

(a) More important nodes.   (b) Shared neighbors.   (c) Neighbors on the right.

Fig. 4. Derivation of f in an infinite grid under L_2 metric.

Table 2. The summarized necessary information of Fig. 4.

Region	x-extent and y-extent	Description	Numbers of nodes
Blue	$(x-a)^2+(y-b)^2 \leq r^2$	The neighborhood of $\mathcal{N}_{(a,b)}$, i.e., $nbd(a,b)$	πr^2
Yellow	$(x-a)^2+(y-b)^2 \leq (\alpha r)^2$	The more important nodes within $nbd(a,b)$	$\pi(\alpha r)^2$
Dark green	$(x-a)^2+(y-b)^2 \leq r^2$ $(x-a)^2+(y-b)^2 \leq (\alpha r)^2$	The shard neighbors of $\mathcal{N}_P$ and $\mathcal{N}_{(a,b)}$	$2r^2 cos^{-1}\frac{\alpha}{2} - \frac{1}{2}r^2\alpha\sqrt{4-\alpha^2}$
Red	$(x-a)^2+(y-b)^2 \leq r^2$ $a+1 \leq x$	The neighbors of $\mathcal{N}_{(a,b)}$ on the right	$0.5\pi r^2 - 2r - 1$

region contains $\pi(\alpha r)^2$ nodes. Nodes $\mathcal{N}_{(x,y)}$, where $\{(x,y)|(x-a)^2+(y-b)^2 \leq r^2; (x-a)^2+(y-b)^2 \leq (\alpha r)^2\}$, within the dark green region in Fig. 3b are the shared neighbors of node $\mathcal{N}_P$ and the source node $\mathcal{N}_{(a,b)}$. The dark green region contains $2r^2 cos^{-1}\frac{\alpha}{2} - \frac{1}{2}r^2\alpha\sqrt{4-\alpha^2}$ nodes. In more detail, let two circles of radius R and r and centered at $(0,0)$ and $(d,0)$ intersect in a region shaped like an asymmetric lens. The area of this lens is[3]: $r^2 cos^{-1}\left(\frac{d^2+r^2-R^2}{2dr}\right) + R^2 cos^{-1}\left(\frac{d^2+R^2-r^2}{2dR}\right) - \frac{1}{2}\sqrt{(-d+r+R)(d+r-R)(d-r+R)(d+r+R)}$. Thus, we can calculate the area of the dark green region in Fig. 4b by making $R = r$ and $d = \alpha r$, and get the area: $2r^2 cos^{-1}\frac{\alpha}{2} - \frac{1}{2}r^2\alpha\sqrt{4-\alpha^2}$. Nodes $\mathcal{N}_{(x,y)}$, where $\{(x,y)|(x-a)^2+(y-b)^2 \leq r^2; a+1 \leq x\}$, within the red region in Fig. 4c are the neighbors on the right of the source node $\mathcal{N}_{(a,b)}$. The red region contains $0.5\pi r^2 - 2r - 1$ nodes. We summarize them in Table 2.

Similarly to the derivation in Sect. 3.1, the number f of the faulty nodes in each neighborhood should meet:

1. $2r^2 cos^{-1}\frac{\alpha}{2} - \frac{1}{2}r^2\alpha\sqrt{4-\alpha^2} \geq 2f + 1$
2. $f < \pi(\alpha r)^2$
3. $f < 0.5\pi r^2 - 2r - 1$

When $\alpha = \sqrt{0.3}$, we get the maximum $f < 0.3\pi r^2$. Note that we do not select a strictly α to locate node $\mathcal{N}_P$ on a grid point because the threshold in this metric is an approximate argument. Therefore, in our protocol, each neighborhood should tolerate $f < \lfloor 0.3\pi r^2 \rfloor$ faulty nodes in the L_2 metric.

---

[3] https://mathworld.wolfram.com/Circle-CircleIntersection.html.

4 Correctness Proof

4.1 Correctness Proof When $f < \lfloor \frac{1}{2}(2r+1)(r+1) \rfloor$ in the L_∞ metric

In this section, we prove the correctness (validity, agreement, and integrity) of our proposed BRB protocol when $f < \lfloor \frac{1}{2}(2r+1)(r+1) \rfloor$ in the L_∞ metric. The intuition of our proof is that a valid certificate message will be constructed and received by all correct nodes. In addition, all valid certificate messages with the same sequence number certify the same message. We consider the worst case where the "important" nodes that are near the source node are faulty. A node near the source node has more shared neighbors with the source node and is more likely to influence the construction of certificate messages. As a result, a BRB protocol can handle all cases if it can handle the worst case. To better understand the proof, we first provide the necessary information of Fig. 5 in Table 3. Next, we first prove the agreement.

Lemma 1. *If a source node first calls* r_bcast(m, q), *then at least one correct node will construct a valid* certificate *for m with a sequence number q and broadcast it.*

Proof. Due to the nature of an idealized shared wireless channel, all neighbors of the source node will first receive m with q (if the source node is faulty, it may broadcast other messages with q after m). Then, each correct neighbor will broadcast a committed message for m.

Each of the nodes within the yellow region of Fig. 5a (including the boundary) has at least $(2r+1)(r+1)$ shared neighbors with the source node. This is because the node $\mathcal{N}_P$ within the yellow region has the least number of shared neighbors with the source node, i.e., the dark green region in Fig. 5b, which contains $(2r+1)(r+1)$ nodes. The nodes within the yellow region must receive the committed messages for m from the correct shared neighbors. As a result, each of them can receive at least $(2r+1)(r+1) - f \geq f + 1$ committed messages, which means that any of them can construct a valid certificate message. Then, we only need to prove that there is at least one correct node within this yellow region, which is straightforward because the number of nodes within this yellow region is $2r(r+1)+1 > \lfloor \frac{1}{2}(2r+1)(r+1) \rfloor > f$. Therefore, a valid certificate message will be constructed by at least one correct node.

Lemma 2. *If there is a valid* certificate *message for m with a sequence number q, then no valid* certificate *message for m' ($m' \neq m$) with q can be constructed.*

Proof. We prove this by contradiction. At least one correct node within the neighborhood of the source node broadcasts a committed message for m because there is a valid certificate message for m. Due to the nature of an idealized wireless channel, all correct nodes within the neighborhood of the source node broadcast committed messages for m. Suppose a valid certificate message for m' ($m' \neq m$) with a sequence number q is constructed. This implies that at least one correct node within the neighborhood of the source node broadcasts a committed message for m', which leads to a contradiction.

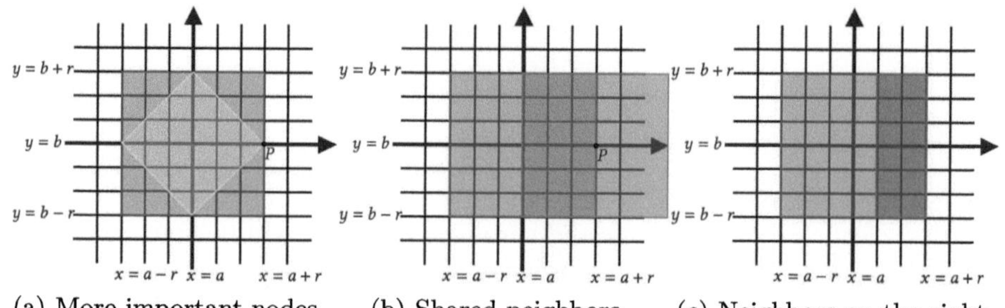

Fig. 5. Correctness proof in an infinite grid under L_∞ metric.

Table 3. The summarized necessary information of Fig. 5.

Region	x-extent and y-extent	Description	Numbers of nodes
Blue	$a - r \leq x \leq a + r$ $b - r \leq y \leq b + r$	The neighborhood of $\mathcal{N}_{(a,b)}$, i.e., $nbd(a,b)$	$(2r+1)^2$
Yellow	$\|x - a\| + \|y - b\| \leq r$	The more important nodes within $nbd(a,b)$	$(2r+1)(r+1)$
Dark green	$a \leq x \leq a + r$ $b - r \leq y \leq b + r$	The shard neighbors of $\mathcal{N}_P$ and $\mathcal{N}_{(a,b)}$	$2r(r+1)+1$
Red	$a + 1 \leq x \leq a + r$ $b - r \leq y \leq b + r$	The neighbors of $\mathcal{N}_{(a,b)}$ on the right	$(2r+1)r$

Lemma 3. *If a valid certificate message for m is constructed and broadcast by a correct node, all correct nodes will receive it and reliably deliver m.*

Proof. If a correct node $\mathcal{N}_{(a,b)}$ broadcasts a certificate message and at least one correct neighbors in each of four directions (i.e., up $\mathcal{N}_{(*,b+y)}$, down $\mathcal{N}_{(*,b-y)}$, left $\mathcal{N}_{(a-x,*)}$, right $\mathcal{N}_{(a+x,*)}$, $0 < x, y \leq r$) forwards it, then all correct nodes will receive this certificate message. Next, we prove that the certificate message can be propagated in four directions.

We prove this by contradiction. Suppose a correct node $\mathcal{N}_{(a,b)}$ broadcasts a certificate message, but no neighbor on the right forwards this certificate message. Then, all these $r(2r+1)$ neighbors on the right (within the red region of Fig. 5c) should be faulty. However, $r(2r+1) > f$, which leads to a contradiction. Therefore, if a correct node broadcasts a certificate message, at least one correct neighbor on its right will forward this certificate message. The same reasoning applies to the other three directions, enabling the message to spread in four directions. Thus, all correct nodes will receive this certificate message for m. By Lemma 2, with the same sequence number, no valid certificate message for m' can be constructed. Therefore, all correct nodes will first receive a certificate message for m and reliably deliver m.

Theorem 1 (Agreement). *If a correct node outputs* r_deliver$(m, q, \mathcal{N}_j)$, *then all other correct nodes will output* r_deliver$(m, q, \mathcal{N}_j)$.

Proof. Suppose a correct node $\mathcal{N}_i$ outputs r_deliver$(m, q, \mathcal{N}_j)$, we consider the following two cases:

- *Node $\mathcal{N}_i$ within the neighborhood of the source node $\mathcal{N}_j$*. In this case, due to the nature of an idealized shared wireless channel, all neighbors of the source node will output r_deliver$(m, q, \mathcal{N}_j)$. By Lemma 1, a valid certificate message for m with a sequence number q will be constructed and broadcast by a correct node. By Lemma 3, all correct nodes will receive this valid certificate message for m and output r_deliver$(m, q, \mathcal{N}_i)$.
- *Node $\mathcal{N}_i$ outside the neighborhood of the source node $\mathcal{N}_j$*. In this case, node $\mathcal{N}_i$ must receive a valid certificate message for m. By Lemma 3, all correct nodes will also receive this valid certificate message and output r_deliver$(m, q, \mathcal{N}_i)$.

Therefore, if a correct node outputs r_deliver$(m, q, \mathcal{N}_j)$, then all other correct nodes will output r_deliver$(m, q, \mathcal{N}_j)$. Next, we prove the integrity.

Theorem 2 (Integrity). *For each sequence number q and node $\mathcal{N}_i$, every correct node outputs* r_deliver$(m, q, \mathcal{N}_i)$ *at most once regardless of m.*

Proof. Suppose a correct node $\mathcal{N}_j$ outputs r_deliver$(m, q, \mathcal{N}_i)$, we consider the following two cases:

- *The node $\mathcal{N}_j$ within the neighborhood of the source node $\mathcal{N}_i$*. In this case, $\mathcal{N}_j$ outputs r_deliver$(m, q, \mathcal{N}_i)$ when it receives a first valid committed message with a sequence number q from $\mathcal{N}_i$.
- *The node $\mathcal{N}_j$ outside the neighborhood of the source node $\mathcal{N}_i$*. In this case, $\mathcal{N}_j$ outputs r_deliver$(m, q, \mathcal{N}_i)$ when it receives a valid certificate message for the first time.

Either case has only one rule of outputting. In addition, a node is either within or outside the neighborhood of the source node. Therefore, for each sequence number q and node $\mathcal{N}_i$, every correct node outputs r_deliver$(m, q, \mathcal{N}_i)$ at most once regardless of m.

Theorem 3 (Validity). *If a correct node $\mathcal{N}_i$ calls* r_bcast(m, q), *then all correct nodes will output* r_deliver$(m, q, \mathcal{N}_i)$.

Proof. By Lemma 1, a valid certificate message for m with a sequence number q will be constructed and broadcast by a correct node. By Lemma 3, all correct nodes will receive this valid certificate message and output r_deliver$(m, q, \mathcal{N}_i)$.

4.2   Correctness Proof When $f < \lfloor 0.3\pi R^2 \rfloor$ in the L_2 metric

In this section, we prove the correctness (validity, agreement, and integrity) of our proposed BRB when $f < \lfloor 0.3\pi r^2 \rfloor$ in the L_2 metric. In the L_2 metric, for sufficiently large r, the area can be used as a good approximation for the number of nodes in these regions [4], which serves as the basis for our deductions. The intuition of the proof in the L_2 metric is the same as that in the L_∞ metric. We still consider the worst case where the "important" nodes that are near the source node are faulty. In the L_2 metric, a node near the source node also has more shared neighbors with the source node and is more likely to influence the construction of certificate messages. As a result, a BRB protocol can handle all cases if it can handle the worst case. Similarly, we provide the necessary information of Fig. 6 in Table 4. Next, we first prove the agreement.

Lemma 4. *If a source node first calls* r_bcast(m, q), *then at least one correct node will construct a valid* certificate *message for m with a sequence number q and broadcast it.*

Proof. Similarly to the proof of Lemma 1 in the L_∞ metric, each correct neighbor of the source node will broadcast a committed message for m.

The node $\mathcal{N}_P$ within the yellow region in Fig. 6a has the least number of shared neighbors with the source node. Specifically, their shared neighbors are within the dark green region in Fig. 6b, and their number can be approximated as $N_1 = 2r^2 \cos^{-1} \frac{\sqrt{0.3}}{2} - \frac{1}{2}r^2\sqrt{1.11}$. Thus, each node within the yellow region has at least N_1 shared neighbors with the source node. When $r \geq 4$, each node within the yellow region can receive at least $N_1 - f \geq f + 1$ committed messages because each of them will receive the committed messages for m from the correct shared neighbors. This means that any of them can construct a valid certificate message for m. Furthermore, there is at least one correct node within this yellow region because $f < \lfloor 0.3\pi r^2 \rfloor$. Therefore, at least one correct node will construct a valid certificate message for m and broadcast it.

Lemma 5. *If there is a valid* certificate *message for m with a sequence number q, then no valid* certificate *for m′ (m′ ≠ m) with q can be constructed.*

Proof. The proof remains identical to that of Lemma 2.

Lemma 6. *If a valid* certificate *message is constructed and broadcast by a correct node, all correct nodes will receive it and reliably deliver it.*

Proof. Similarly to the proof of Lemma 3, we prove this by proving that a certificate message can be propagated in four directions.

We prove this by contradiction. Suppose that a correct node broadcasts a certificate message, but no correct neighbor on the right forwards this certificate message. Then, all these $r(2r+1)$ neighbors on the right (within the red region of Fig. 6c) should be faulty. However, $(0.5\pi r^2 - 2r - 1) - f \geq 1$ when $r \geq 4$, which leads to a contradiction. Therefore, if a correct node broadcasts a certificate message, at least one correct neighbor on its right will forward this certificate message.

The same reasoning applies to the other three directions, enabling the message to spread in four directions. Therefore, if a valid certificate message is constructed and broadcast by a correct node, all correct nodes will receive it. By Lemma 5, with the same sequence number, no valid certificate message for $m′$ can be constructed. Therefore, all correct nodes will first receive a certificate message for m and reliably deliver m.

Theorem 4 (Agreement). *If a correct node $\mathcal{N}_i$ outputs* r_deliver($m, q, \mathcal{N}_j$), *then all other correct nodes will output* r_deliver($m, q, \mathcal{N}_j$).

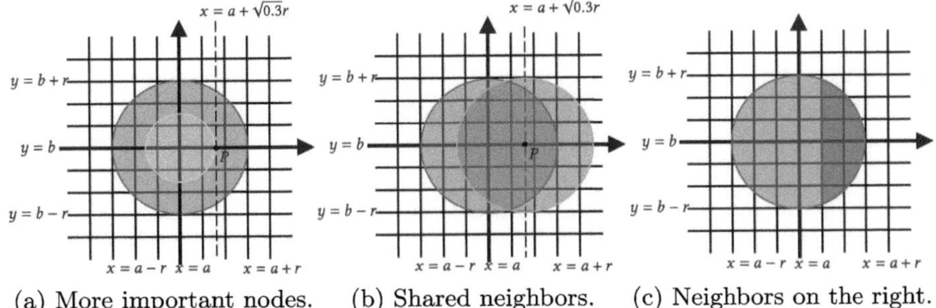

(a) More important nodes.   (b) Shared neighbors.   (c) Neighbors on the right.

Fig. 6. Correctness proof in an infinite grid under L_2 metric.

Table 4. The summarized necessary information of Fig. 6.

Region	x-extent and y-extent	Description	Numbers of nodes
Blue	$(x - a)^2 + (y - b)^2 \leq r^2$	The neighborhood of $\mathcal{N}_{(a,b)}$, i.e., $nbd(a,b)$	πr^2
Yellow	$(x - a)^2 + (y - b)^2 \leq 0.3r^2$	The more important nodes within $nbd(a,b)$	$0.3\pi r^2$
Dark green	$(x - a)^2 + (y - b)^2 \leq r^2$ $(x - a)^2 + (y - b)^2 \leq 0.3r^2$	The shard neighbors of $\mathcal{N}_P$ and $\mathcal{N}_{(a,b)}$	$2r^2 \cos^{-1} \frac{\sqrt{0.3}}{2} - \frac{1}{2}r^2 \sqrt{1.11}$
Red	$(x - a)^2 + (y - b)^2 \leq r^2$ $a + 1 \leq x$	The neighbors of $\mathcal{N}_{(a,b)}$ on the right	$0.5\pi r^2 - 2r - 1$

Proof. With Lemma 4 and Lemma 6, the agreement proof remains identical to that of Theorem 1.

Theorem 5 (Integrity). *For each sequence number q and node $\mathcal{N}_i$, every correct node outputs* r_deliver$(m, q, \mathcal{N}_i)$ *at most once regardless of m.*

Proof. The integrity proof remains identical to that of Theorem 2.

Theorem 6 (Validity). *If a correct node $\mathcal{N}_i$ calls* r_bcast(m, q), *then all correct nodes will output* r_deliver$(m, q, \mathcal{N}_i)$.

Proof. Based on the Lemma 4, and Lemma 6, the proof of this theorem remains identical to the Theorem 3.

5   Related Work

Byzantine Reliable Broadcast in Wired Networks. Byzantine reliable broadcast (BRB) is a fundamental problem in fault-tolerant distributed systems, introduced by Bracha and Toueg [7–9]. They show that BRB is achievable in wired asynchronous networks when less than one-third of nodes can be faulty. Then, hundreds of works such as [1,2,10,14] follow this result and aim to reduce the communication cost. However, these results cannot be directly adapted to wireless networks because they do not capture the spatially dependent connectivity of wireless networks. Specifically, in a wired network, the connection is in a peer-to-peer manner in which nodes can directly communicate with each other. In contrast, in a wireless network, a node can only communicate directly with the one within its broadcast radius.

Byzantine Reliable Broadcast in Wireless Grid Networks. In wireless networks, Koo [11] provides the first analysis of BRB in an infinite (or finite-toroidal) wireless grid network under the assumption of no message collisions and address spoofing. He shows how to achieve a BRB that tolerates roughly $\frac{1}{2}r^2$ Byzantine faults in the L_∞ metric. He also proves the impossibility of achieving BRB when more than $\lceil \frac{1}{2}r(2r+1) \rceil$ neighbors of any correct node are Byzantine nodes in the L_∞ metric. Vaikuntanathan [16] shows that the protocol proposed by Koo [11] indeed tolerates $\frac{1}{\sqrt{2}}r^2$ Byzantine faults. Bhandari et al. present a BRB algorithm [4] that matches the impossibility threshold in the L_∞ metric. They also present an approximate threshold $f < 0.23\pi r^2$ in the L_2 metric. Then, they present a simpler characterization [5] of a sufficient condition for BRB to augment [4]. In [12], Koo et al. allow Byzantine nodes to cause a known bounded number of collisions and address spoofing of arbitrary other nodes. They show that without the no-collision assumption, BRB is still achievable and has the same fault-tolerant threshold as the no-collision case of [4,11].

Although current works achieve remarkable results, the established thresholds in wireless grid networks still have a gap with those in wired networks. In this paper, we focus on the no-collision case and aim to explore the thresholds.

6   Conclusion

In this paper, we focus on the no-collision case of infinite (or finite-toroidal) wireless grid networks to explore the thresholds of the Byzantine reliable broadcast (BRB). We present new thresholds of BRB in the L_∞ and L_2 metrics that are closer to the thresholds in wired networks. Furthermore, we design a BRB protocol that matches these new thresholds with $O(1)$ forwarding frequency. Moreover, we provide detailed derivations of the new thresholds and formal correctness proofs of our BRB. In future work, it will be interesting to know whether tighter thresholds can be obtained. Additionally, we will explore how to adapt this idea for the bounded-collision case.

Acknowledgments. This work is funded by National Key Research and Development Program of China(2023YFB2704000) and the Fundamental Research Funds for the Central Universities 226-2025-00004.

References

1. Alhaddad, N., et al.: Balanced byzantine reliable broadcast with near-optimal communication and improved computation. In: Proceedings of the 2022 ACM Symposium on Principles of Distributed Computing, PODC 2022, pp. 399–417. Association for Computing Machinery, New York (2022). https://doi.org/10.1145/3519270.3538475

2. Alhaddad, N., Duan, S., Varia, M., Zhang, H.: Succinct erasure coding proof systems. Cryptology ePrint Archive, Paper 2021/1500 (2021). https://eprint.iacr.org/2021/1500

3. Bhandari, V., Vaidya, N.H.: Reliable broadcast in wireless networks with probabilistic failures. In: IEEE INFOCOM 2007 - 26th IEEE International Conference on Computer Communications, pp. 715–723 (2007) https://doi.org/10.1109/INFCOM.2007.89

4. Bhandari, V., Vaidya, N.H.: On reliable broadcast in a radio network. In: Proceedings of the Twenty-Fourth Annual ACM Symposium on Principles of Distributed Computing, PODC 2005, pp. 138–147. Association for Computing Machinery, New York (2005) https://doi.org/10.1145/1073814.1073841

5. Bhandari, V., Vaidya, N.H.: On reliable broadcast in a radio network: A simplified characterization. Tech. Rep, CSL, UIUC (2005)

6. Boneh, D., Lynn, B., Shacham, H.: Short signatures from the weil pairing. In: Boyd, C. (ed.) ASIACRYPT 2001. LNCS, vol. 2248, pp. 514–532. Springer, Heidelberg (2001). https://doi.org/10.1007/3-540-45682-1_30

7. Bracha, G.: Asynchronous byzantine agreement protocols. Inf. Comput. **75**(2), 130–143 (1987). https://doi.org/10.1016/0890-5401(87)90054-X

8. Bracha, G., Toueg, S.: Resilient consensus protocols. In: Proceedings of the Second Annual ACM Symposium on Principles of Distributed Computing. PODC 1983, pp. 12–26. Association for Computing Machinery, New York (1983) https://doi.org/10.1145/800221.806706

9. Bracha, G., Toueg, S.: Asynchronous consensus and broadcast protocols. J. ACM **32**(4), 824–840 (1985). https://doi.org/10.1145/4221.214134

10. Das, S., Xiang, Z., Ren, L.: Asynchronous data dissemination and its applications. In: Proceedings of the 2021 ACM SIGSAC Conference on Computer and Communications Security, CCS 2021, pp. 2705–2721. Association for Computing Machinery, New York (2021). https://doi.org/10.1145/3460120.3484808

11. Koo, C.Y.: Broadcast in radio networks tolerating byzantine adversarial behavior. In: Proceedings of the Twenty-Third Annual ACM Symposium on Principles of Distributed Computing, PODC 2004, pp. 275–282. Association for Computing Machinery, New York (2004). https://doi.org/10.1145/1011767.1011807

12. Koo, C.Y., Bhandari, V., Katz, J., Vaidya, N.H.: Reliable broadcast in radio networks: the bounded collision case. In: Proceedings of the Twenty-Fifth Annual ACM Symposium on Principles of Distributed Computing, PODC 2006, pp. 258–264. Association for Computing Machinery, New York (2006) https://doi.org/10.1145/1146381.1146420

13. Kreyszig, E.: Advanced engineering mathematics. John Wiley & Sons. Inc, Singapore (1993)

14. Nayak, K., Ren, L., Shi, E., Vaidya, N.H., Xiang, Z.: Improved extension protocols for byzantine broadcast and agreement. arXiv: 2002.11321 (2020)

15. Schnorr, C.P.: Efficient signature generation by smart cards. J. Cryptol. **4**(3), 161–174 (1991). https://doi.org/10.1007/BF00196725

16. Vaikuntanathan, V.: Brief announcement: broadcast in radio networks in the presence of byzantine adversaries. In: Proceedings of the Twenty-Fourth Annual ACM Symposium on Principles of Distributed Computing, PODC 2005, pp. 167. Association for Computing Machinery, New York (2005). https://doi.org/10.1145/1073814.1073844

Beyond Pairwise Comparisons: Unveiling Structural Landscape of Mobile Robot Models

Shota Naito[ID], Tsukasa Ninomiya[ID], and Koichi Wada[(✉)][ID]

Hosei University, Tokyo, Japan
`{shota.naito.5x,tsukasa.ninomiya.4w}@stu.hosei.ac.jp, wada@hosei.ac.jp`

Abstract. Understanding the computational power of mobile robot systems is a fundamental challenge in distributed computing. While prior work has focused on pairwise separations between models, we explore how robot capabilities, light observability, and scheduler synchrony interact in more complex ways.

We show that the **Exponential Times Expansion** (*ETE*) problem is solvable only in the strongest model–*fully-synchronous* robots with mutual lights ($\mathcal{LUMI}^F$). The **Hexagonal Edge Traversal** (*HET*) and *TAR(d)** problems illustrate how internal memory and lights interact with synchrony: under weak synchrony, memory alone is insufficient, while full synchrony can substitute for both.

In the *asynchronous* setting, we classify problems such as *LP–MLCv*, *VEC*, and *ZCC*, revealing fine-grained separations between $\mathcal{FSTA}$ and $\mathcal{FCOM}$ robots. We also analyze **Vertex Traversal Rendezvous** (*VTR*) and **Leave Place Convergence** (*LP–Cv*), showing the limitations of internal memory in symmetric scenarios.

Our results extend the known separation map of 14 canonical robot models, uncovering structural phenomena only visible through higher-order comparisons and providing new criteria for impossibility.

1 Introduction

1.1 Background and Motivation

The computational power of autonomous mobile robots operating through *Look-Compute-Move* (*LCM*) cycles has been a central topic in distributed computing. Robots are modeled as anonymous, uniform, disoriented points in the Euclidean plane. In each cycle, a robot observes the configuration (*Look*), computes a destination (*Compute*), and moves accordingly (*Move*). These agents can solve distributed tasks collectively, despite their simplicity.

The weakest standard model, $\mathcal{OBLOT}$, assumes robots are oblivious (no persistent memory) and silent (no explicit communication). Since its introduction [22], this model has been extensively studied for basic coordination tasks

This work was supported in part by JSPS KAKENHI Grant Number 25K03079.

such as Gathering [1–3,7–9,12,18,22], Pattern Formation [13,16,22–24], and Flocking [6,17,21]. The limitations imposed by obliviousness and silence have led to extensive investigation of stronger models. A widely studied enhancement is the luminous robot model, $\mathcal{LUMI}$ [10], where robots are equipped with a visible, persistent light (i.e., a constant-size memory) used for communication and state retention. The lights can be updated during *Compute* and are visible to others, enabling both memory and communication in each cycle.

To understand the individual roles of memory and communication, two submodels of $\mathcal{LUMI}$ have been introduced: $\mathcal{FSTA}$ (persistent memory without communication), and $\mathcal{FCOM}$ (communication without memory) [4,5,20]. Studying these four models–$\mathcal{OBLOT}$, $\mathcal{FSTA}$, $\mathcal{FCOM}$, and $\mathcal{LUMI}$–clarifies the contribution of internal robot capabilities to problem solvability.

External assumptions such as synchrony and activation schedulers also play a crucial role. The *semi-synchronous* (SSYNCH) model [22] activates an arbitrary non-empty subset of robots in each round, who perform one atomic LCM cycle. Special cases include the *fully-synchronous* (FSYNCH) model, where all robots are activated every round, and the *restricted synchronous* (RSYNCH) model [5], where activations in consecutive rounds are disjoint.

In contrast, the *asynchronous* (ASYNCH) model [11] introduces arbitrary but finite delays between phases of each robot's cycle, under a fair adversarial scheduler. This model captures high uncertainty and has proven computationally weaker.

Weaker variants of ASYNCH have been studied by assuming partial atomicity: for example, LC-**atomic** and CM-**atomic** models [20]. These restrict delays to only parts of the cycle and allow finer-grained modeling.

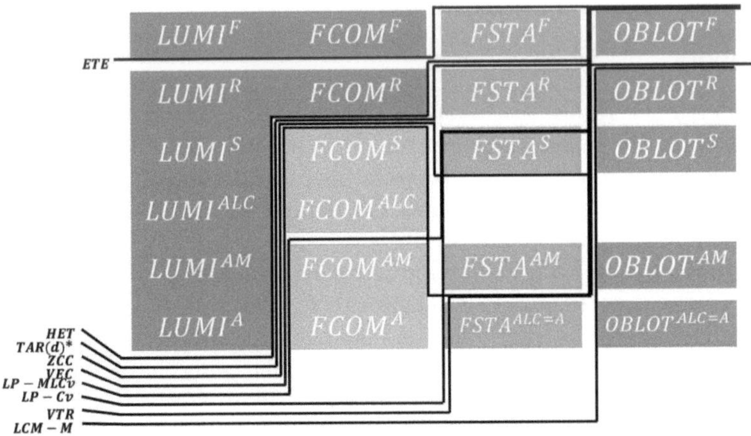

Fig. 1. Hierarchy of 14 robot models combining robot capabilities and scheduler classes. Also this figure captures newly established separations.

These internal models and external schedulers define a two-dimensional space of 14 canonical robot models. Previous work has explored their pairwise computational relationships by proving *separation* (i.e., a problem solvable in one model but not the other) and identifying *equivalence* or *orthogonality* between some models [5, 10, 15].

1.2 Our Contributions

This paper goes beyond pairwise comparisons and explores *triadic relationships*, revealing new structural properties in the model hierarchy.

- In Sect. 4, we show that some problems are solvable only in the strongest model $\mathcal{LUMI}^F$ (equivalent to $\mathcal{FCOM}^F$), but not in either $\mathcal{FSTA}^F$ or $\mathcal{LUMI}^R$, highlighting the role of synchrony in enabling luminous computation. We also introduce the problems HET and $TAR(d)^*$, which separate intermediate models such as $\mathcal{FCOM}^S$ and $\mathcal{FSTA}^R$ from weaker models like $\mathcal{OBLOT}^F$.
- In Sect. 5, we show that some problems require both memory and communication under *asynchronous* or *semi-synchronous* settings, while stronger synchrony can render these capabilities unnecessary.

These results reveal asymmetries and complex trade-offs between memory, communication, and synchrony that are not captured by pairwise analysis. They underscore the need for a global landscape perspective on distributed robot models.

2 Models and Preliminaries

2.1 Robots

We consider a set $R_n = \{r_0, \ldots, r_{n-1}\}$ of $n > 1$ mobile robots moving in the Euclidean plane $\mathbb{R}^2$, where they are treated as points. Each robot is *autonomous*, *identical*, and *homogeneous*, i.e., there is no central control, no identifiers, and they all execute the same algorithm. We denote by $r(t) \in \mathbb{R}^2$ the position of robot r at time t.

Each robot has its own local coordinate system centered at its position and can observe the others' positions relative to it. The robots are *disoriented*, i.e., their coordinate systems may differ and change over time. However, they have a common *chirality* (i.e., clockwise orientation of the plane is shared).

A robot repeatedly executes *Look-Compute-Move* (LCM) cycles:

- *Look*: takes a snapshot of the positions of all robots in its local coordinate system. Multiplicity detection is not assumed.
- *Compute*: computes a destination based on the snapshot.
- *Move*: moves in a straight line to the destination. Movements are rigid, i.e., not interruptible.

In the standard model $\mathcal{OBLOT}$, robots are *oblivious* (no memory between cycles) and *silent* (no explicit communication). In the $\mathcal{LUMI}$ model, robots are equipped with visible *lights* whose value (called *color*) is chosen from a finite set C and remains unchanged until explicitly updated. The snapshot includes the positions and colors of other robots. If $|C| = 1$, this coincides with the $\mathcal{OBLOT}$ model.

Two submodels of $\mathcal{LUMI}$ have been studied:

- $\mathcal{FSTA}$: Robots can read only their own light. They are *silent* but can remember a finite state.
- $\mathcal{FCOM}$: Robots can read only others' lights. They are *oblivious* but can communicate finite-state information.

2.2   Schedulers and Atomicity

The behavior of the system depends on the activation schedule of the robots, i.e., when each robot performs its *Look Compute Move* cycle. All schedulers we consider are *fair*, meaning each robot is activated infinitely often.

We assume each operation completes in finite time and that *Look* and *Compute* are instantaneous unless stated otherwise. Let $t_L(r, i)$, $t_C(r, i)$, $t_B(r, i)$, $t_E(r, i)$ denote the times when robot r performs its i-th *Look*, *Compute*, *Move-begin*(M_B), and *Move-end*(M_E), respectively. These satisfy:

$$t_L(r, i) < t_C(r, i) < t_B(r, i) < t_E(r, i) < t_L(r, i+1).$$

Synchronous Models.

- SSYNCH: In each round, an arbitrary non-empty subset of robots performs an *LCM* cycle in lockstep (*semi-synchronous*) [22].
- RSYNCH: Like SSYNCH, but after a possible finite *fully-synchronous* prefix, robots are activated in disjoint subsets [5].
- FSYNCH: All robots execute a full *LCM* cycle in every round (*fully-synchronous*).

Asynchronous Models. In ASYNCH [11], there are no assumptions on relative activation or durations, except fairness. Based on atomicity constraints, we define four subclasses:

- A_M (*M*-**atomic**): No robot performs *Look* while another is moving.

$$\forall r, s, \; i, j : \; t_L(r, i) \notin [t_B(s, j), t_E(s, j)]$$

- A_{CM} (*CM*-**atomic**): No robot performs *Look* during another's *Compute*-to-*Move* phase.

$$\forall r, s, \; i, j : \; t_L(r, i) \notin [t_C(s, j), t_E(s, j)]$$

- A_{LC} (*LC*-**atomic**): No robot performs *Look* during another's *Look*-to-*Compute* phase.

$$\forall r, s, \ i, j: \ t_L(r, i) \notin (t_L(s, j), t_C(s, j)]$$

- *LCM*-**atomic**-ASYNCH: Entire *LCM* cycle is atomic.

$$\forall r, s \in R, \ \forall i, j \in \mathbb{Z}_{>0}: \ t_L(r, i) \notin (t_L(s, j), t_E(s, j)].$$

This model is equivalent to SSYNCH [15].

In models with lights, light changes occur at the end of *Compute* and are immediately visible to others.

2.3   Problems and Computational Relationships

Let $\mathcal{M} = \{\mathcal{LUMI}, \mathcal{FCOM}, \mathcal{FSTA}, \mathcal{OBLOT}\}$ denote the models, and $\mathcal{S} = \{F, S, A, A_{LC}, A_M, A_{CM}\}$ the schedulers.

A *configuration* of n robots is a function $\gamma : \mathcal{R}_n \to \mathbb{R}^2 \times C$ giving each robot's position and light color in a fixed global coordinate system (unknown to the robots). When colors are omitted, we refer to it as a *geometric configuration* or *placement*. In $\mathcal{OBLOT}$, $|C| = 1$, and colors carry no information.

A *problem* is defined by a set of temporal geometric predicates specifying valid initial, intermediate, and terminal placements, as well as conditions on n. A terminal configuration is one in which the robots stop moving forever.

Let $M(K)$ be the set of problems solvable by model $M \in \mathcal{M}$ under scheduler $K \in \mathcal{S}$. For models $M_1, M_2 \in \mathcal{M}$ and schedulers $K_1, K_2 \in \mathcal{S}$, we write:

- $M_1^{K_1} \geq M_2^{K_2}$ if $M_1(K_1) \supseteq M_2(K_2)$ (not less powerful),
- $M_1^{K_1} > M_2^{K_2}$ if $M_1^{K_1} \geq M_2^{K_2}$ and $M_1(K_1) \neq M_2(K_2)$ (strictly more powerful),
- $M_1^{K_1} \equiv M_2^{K_2}$ if $M_1(K_1) = M_2(K_2)$ (equivalent),
- $M_1^{K_1} \perp M_2^{K_2}$ if each can solve problems the other cannot (incomparable).

3   Known Separation Landscape

Let $\mathcal{M} = \{\mathcal{LUMI}, \mathcal{FCOM}, \mathcal{FSTA}, \mathcal{OBLOT}\}$ be the set of models under investigation and $\mathcal{S} = \{F, S, A, A_{LC}, A_M, A_{CM}\}$ be the set of schedulers under consideration.

From the definitions,

Theorem 1. *For any $M \in \mathcal{M}$ and any $K \in \mathcal{S}$:*

(1) $M^F \geq M^S \geq M^{A_{LC}} \geq M^A$
(2) $M^F \geq M^S \geq M^{A_{CM}} \geq M^{A_M} \geq M^A$
(3) $\mathcal{LUMI}^K \geq \mathcal{FSTA}^K \geq \mathcal{OBLOT}^K$
(4) $\mathcal{LUMI}^K \geq \mathcal{FCOM}^K \geq \mathcal{OBLOT}^K$

As for $X \in \{\mathcal{FSTA}, \mathcal{OBLOT}\}$, since robots cannot observe the colors of the other robots, we have $X^{A_{CM}} \equiv X^{A_M}$ and $X^{A_{LC}} \equiv X^A$.

Let us also recall the following equivalences:

Theorem 2.   **(1)** $\mathcal{LUMI}^F \equiv \mathcal{FCOM}^F$ [14],
(2) $\mathcal{LUMI}^R \equiv \mathcal{LUMI}^S \equiv \mathcal{LUMI}^A \equiv \mathcal{FCOM}^R$ [5, 10],
(3) $\mathcal{FCOM}^S \equiv \mathcal{FCOM}^{A_{LC}}$ [15], and
(4) $\mathcal{FCOM}^{A_{CM}} \equiv \mathcal{FCOM}^{A_M} \equiv \mathcal{FCOM}^A$ [15].

Considering the equivalences above, we treat equivalent configurations as a single model represented by the strongest variant whenever possible. However, when there is a widely used conventional name for a weaker equivalent, we follow the standard notation for clarity. For example, although A_{CM} and A_M are equivalent for all robot models, we use the more familiar A_M to denote both. Similarly, when models are equal, only the strongest variant is counted in our classification, which reduces the total distinct configurations to 14, that is, $\mathcal{LUMI}^F$, $\mathcal{FSTA}^F$,

Table 1. Each cell indicates the known relation between two configurations: $>$ means that the left configuration strictly separates the right one, with the concrete problem given in parentheses. $\perp$ indicates orthogonality, again with the separation problem noted. $\leftarrow$ means the relation trivially follows from an inclusion by definition. The top row and leftmost column uses abbreviations: $\mathcal{LUMI} = \mathcal{L}$, $\mathcal{FCOM} = \mathcal{FC}$, $\mathcal{FSTA} = \mathcal{FS}$, $\mathcal{OBLOT} = \mathcal{O}$. Problems used for separation are described in the full paper [19]. The first table shows pairs for SSYNCH and stronger schedulers (*synchronous* part), and the second table covers SSYNCH and weaker schedulers (*asynchronous* part).

	$\mathcal{L}^F$	$\mathcal{FS}^F$	$\mathcal{O}^F$	$\mathcal{L}^R$	$\mathcal{FS}^R$	$\mathcal{O}^R$	$\mathcal{FC}^S$	$\mathcal{FS}^S$	$\mathcal{O}^S$
$\mathcal{L}^F$	-	$>$ (CYC)	$>$ (CGE)	$>$ ($\leftarrow$)	$>$ ($\leftarrow$)	$>$ ($\leftarrow$)	$>$ ($\leftarrow$)	$>$ ($\leftarrow$)	$>$ ($\leftarrow$)
$\mathcal{FS}^F$	-	-	$>$ (CGE)	$\perp$ (CGE, CYC)	$>$ (CGE)	$>$ ($\leftarrow$)	$\perp$ (CGE, CYC)	$>$ (OSP)	$>$ ($\leftarrow$)
$\mathcal{O}^F$	-	-	-	$\perp$ (CGE*, CYC)	$\perp$ (CGE*, OSP)	$>$ (CGE*)	$\perp$ (CGE*, CYC)	$\perp$ (CGE*, TAR(d))	$>$ (CGE*)
$\mathcal{L}^R$	-	-	-	-	$>$ (CYC)	$>$ ($\leftarrow$)	$>$ (OSP)	$>$ (CYC)	$>$ ($\leftarrow$)
$\mathcal{FS}^R$	-	-	-	-	-	$>$ (OSP)	$\perp$ (OSP, CYC)	$>$ (OSP)	$>$ ($\leftarrow$)
$\mathcal{O}^R$	-	-	-	-	-	-	$\perp$ (SRO, CYC)	$\perp$ (SRO, TAR(d))	$>$ (SRO)
$\mathcal{FC}^S$	-	-	-	-	-	-	-	$\perp$ (CYC, TAR(d))	$>$ (CYC)
$\mathcal{FS}^S$	-	-	-	-	-	-	-	-	$>$ (TAR(d))
$\mathcal{O}^S$	-	-	-	-	-	-	-	-	-

	$\mathcal{L}^S$	$\mathcal{FC}^S$	$\mathcal{FS}^S$	$\mathcal{O}^S$	$\mathcal{FC}^{A_M}$	$\mathcal{FS}^{A_M}$	$\mathcal{O}^{A_M}$	$\mathcal{FS}^A$	$\mathcal{O}^A$
$\mathcal{L}^S$	-	$>$ (OSP)	$>$ (OSP)	$>$ ($\leftarrow$)	$>$ ($\leftarrow$)	$>$ ($\leftarrow$)	$>$ ($\leftarrow$)	$>$ ($\leftarrow$)	$>$ ($\leftarrow$)
$\mathcal{FC}^S$	-	-	$\perp$ (TAR(d), CYC)	$>$ (CYC)	$>$ (RDV)	$\perp$ (TAR(d), CYC)	$>$ (CYC)	$\perp$ (SM, CYC)	$>$ (CYC)
$\mathcal{FS}^S$	-	-	-	$>$ (TAR(d))	$\perp$ (TAR(d), CYC)	$>$ (MLCv)	$>$ ($\leftarrow$)	$>$ (TAR(d))	$>$ ($\leftarrow$)
$\mathcal{O}^S$	-	-	-	-	$\perp$ (MLCv, CYC)	$\perp$ (TAR(d), MLCv)	$>$ (MLCv)	$\perp$ (SM, MLCv)	$>$ (MLCv)
$\mathcal{FC}^{A_M}$	-	-	-	-	-	$\perp$ (TAR(d), CYC)	$>$ (CYC)	$\perp$ (SM, CYC)	$>$ (CYC)
$\mathcal{FS}^{A_M}$	-	-	-	-	-	-	$>$ (TAR(d))	$>$ (TAR(d))	$>$ ($\leftarrow$)
$\mathcal{O}^{A_M}$	-	-	-	-	-	-	-	$\perp$ (SM, TF)	$>$ (TF)
$\mathcal{FS}^A$	-	-	-	-	-	-	-	-	$>$ (SM)
$\mathcal{O}^A$	-	-	-	-	-	-	-	-	-

$\mathcal{OBLOT}^F, \mathcal{LUMI}^R, \mathcal{FSTA}^R, \mathcal{OBLOT}^R, \mathcal{FCOM}^S, \mathcal{FSTA}^S, \mathcal{OBLOT}^S,$
$\mathcal{FCOM}^{A_M}, \mathcal{FSTA}^{A_M}, \mathcal{OBLOT}^{A_M}, \mathcal{FSTA}^A,$ and $\mathcal{OBLOT}^A.$

Table 1 summarizes all known results on the pairwise relations between models [5,14,15], clearly separated into the cases for SSYNCH and stronger schedulers (*synchronous*) and SSYNCH and weaker schedulers (*asynchronous*). Following these tables, we present a unified inclusion and separation diagram (Separation Map) that visually represents these results (Fig. 2).

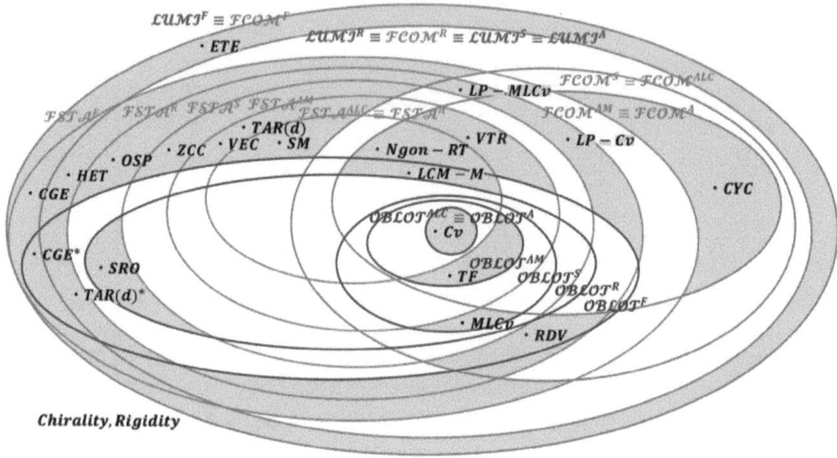

Fig. 2. Classification of computational problems based on their solvability across 14 canonical robot models. Each nested ellipse corresponds to a model class, with outer ellipses representing more powerful models. A dot represents a problem, and its placement indicates the set of models in which it is solvable. Problems highlighted in **yellow with bold labels** (e.g., *ETE*, *HET*, *TAR(d)**, etc.) are newly introduced in this paper. The shaded gray region represents model classes for which this paper identifies problems that separate them from strictly weaker models. That is, for each shaded region, at least one problem has been found that is solvable within that region but unsolvable in all strictly inner models. The remaining white areas correspond to model combinations for which the existence of such separating problems is still unknown, thus constituting open questions. (Color figure online)

4   Separation Under Synchronous Schedulers

This section presents the key separation results for SSYNCH and stronger schedulers. In particular, we introduce three new problems–*ETE* (Exponential Times Expansion), *HET* (Hexagonal Edge Traversal), and *TAR(d)** (Infinite Triangle Rotation with parameter d)–each designed to reveal non-trivial gaps that arise when varying the degree of synchrony or capability.

Figure 1 illustrates the known separation landscape for SSYNCH and stronger schedulers, highlighting where the *ETE, HET,* and *TAR(d)** problems contribute to revealing strict separations or orthogonal relations that cannot be derived by trivial inclusion alone. The problem *ETE* separates $\mathcal{LUMI}^F$ from both $\mathcal{FSTA}^F$ and $\mathcal{LUMI}^R$. *HET* separates $\mathcal{FCOM}^S$ from $\mathcal{FSTA}^R$, demonstrating a capability-scheduler trade-off. *TAR(d)** distinguishes $\mathcal{FSTA}^R$ and $\mathcal{FCOM}^S$ from $\mathcal{OBLOT}^R$, indicating the computational advantage of either communication or memory under strong synchrony.

Formal proofs for all separation results described in this paper are provided in the full paper [19] due to the space limitation.

4.1   The Most Difficult Problem (*ETE*)

Definition 1 *ETE (Exponential Times Expansion).* *Let $n \geq 3$, $k = 2^{n-1}$, and $d : \mathbb{N} \to \mathbb{R}$ be a non-invertible function. A sequence of patterns $C_0, \ldots, C_k$ is constructed as follows. C_0 is a circle of $n - 1$ robots around a central one. For $i > 0$, each non-central robot at position (x, y) is moved to a new position $(f(x, g_x), f(y, g_y))$, where $g_i = (g_x, g_y)$ is the center of gravity and*

$$f(a, b) = \lfloor b + d(i) \cdot (a - b) \rfloor.$$

This corresponds to scaling each robot by a factor of $d(i)$ from the center of gravity (Fig. 3).

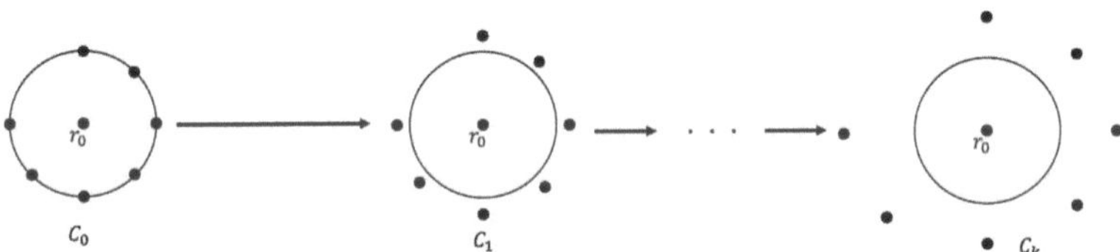

Fig. 3. Movement of the *ETE* problem.

Lemma 1. *Let $n \geq 3$. $ETE \notin \mathcal{LUMI}^{F'}$, where F' is any synchronous scheduler such that the first activation does not contain all robots.*

Lemma 2. *$ETE \notin \mathcal{FSTA}^F$.*

Lemma 3. *For $n \geq 3$, $ETE \in \mathcal{LUMI}^F$.*

Proof Sketch. (See Section 4.1 in [19] for the formal proof) The *ETE* problem combines two critical features: the expansion of distances from the Center of Gravity Expansion (*CGE**) and the step-wise transformation of configurations in a predefined cyclic order (*CYC*) [5].

First, the *CGE** component requires robots to correctly compute the distance from the center and move in synchrony to preserve the scaling factor $d(i)$. If robots are not activated simultaneously, local observations become inconsistent, making it impossible to guarantee correct expansion under any scheduler weaker than FSYNCH, even for the strongest $\mathcal{LUMI}$ model (Lemma 1).

Second, the *CYC* property requires robots to maintain a unique sequence of configurations. Without global synchrony, robots may diverge in phase, making the cyclic transition infeasible for any model with only internal memory, such as $\mathcal{FSTA}^F$ (Lemma 2).

In contrast, $\mathcal{LUMI}^F$ trivially solves *ETE* by coordinating simultaneous *Look-Compute-Move* cycles with full light observability, enabling exact distance computation and synchronized phase transitions (Lemma 3). $\square$

Theorem 3. *ETE is solved in* $\mathcal{LUMI}^F$, *but not in any of* $\mathcal{FSTA}^F$ *or* $\mathcal{LUMI}^R$.

4.2   Scheduler-Capability Interplay (*HET*)

Definition 2 *HET (Hexagonal Edge Traversal).* *On a regular hexagon with vertices* $V = \{v_0, \ldots, v_5\}$, *two robots,* r *and* q, *start at opposite vertices* v_i *and* $v_{(i+3) \bmod 6}$, *respectively. Each robot must traverse the edges to its opposite vertex and return (Fig. 4).*

The traversal is governed by two strictly increasing infinite time sequences:

$$0 \leq t_0 < t_1 < \ldots, \quad 0 \leq T_0 < T_1 < \ldots$$

These sequences must satisfy a cyclic condition for all $j \geq 0$:

$$t_0 = t_{13j \bmod 13}, \quad t_1 = t_{13j+1 \bmod 13}, \quad \ldots,$$

$$T_0 = T_{13j \bmod 13}, \quad T_1 = T_{13j+1 \bmod 13}, \quad \ldots$$

Lemma 4. $HET \in (\mathcal{FSTA}^F \cap \mathcal{LUMI}^R)$.

Lemma 5. $HET \notin (\mathcal{FSTA}^R \cup \mathcal{OBLOT}^F \cup \mathcal{FCOM}^S)$.

Proof Sketch. (See Section 4.2 in [19] for the formal proof) The *HET* problem requires each robot to traverse exactly one edge on a regular hexagon and then return to its initial position, under a symmetric configuration where all robots must break symmetry and remember whether they have completed exactly one round-trip.

In the $\mathcal{FSTA}$ model under a *fully-synchronous* scheduler (FSYNCH), robots can coordinate this operation using internal state to break symmetry and track

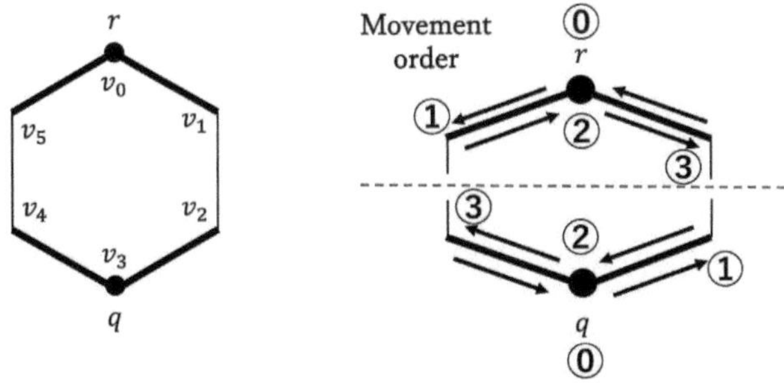

Fig. 4. Movement of the *HET* problem.

their traversal phase (Lemma 4,). However, if the scheduler is weakened to RSYNCH, robots may be activated in an interleaved order, causing inconsistencies in how their internal states align: some robots may advance to the next phase before others observe it, making it impossible to ensure exactly one traversal without external synchronization (Lemma 5).

Similarly, if the model is $\mathcal{OBLOT}$ (no internal memory, no lights), even under FSYNCH the robots cannot store whether they have started or finished the traversal, so they cannot guarantee a single round-trip (Lemma 5). If the model is $\mathcal{FCOM}$ under SSYNCH, the robots rely on external lights only, but partial synchrony allows phase divergence–robots may read inconsistent light states across partial activations, preventing reliable traversal (Lemma 5).

In contrast, $\mathcal{LUMI}^R$ can solve *HET* because robots have both internal state and external visible lights. Lights act as shared memory to align phases, compensating for interleaved activations and enabling the group to maintain exactly one round-trip traversal per robot, even under RSYNCH(Lemma 4).   □

Therefore, *HET* separates not only $\mathcal{FSTA}^F$ and $\mathcal{FSTA}^R$, but also $\mathcal{OBLOT}^F$, $\mathcal{FCOM}^S$ and $\mathcal{FSTA}^R$ from $\mathcal{LUMI}^R$.

Theorem 4. *HET is solved in both of $\mathcal{FSTA}^F$ and $\mathcal{LUMI}^R$, but not in any of $\mathcal{FSTA}^R$, $\mathcal{OBLOT}^F$ or $\mathcal{FCOM}^S$.*

4.3 Synchrony Can Substitute Capability Only Under Fsynch ($TAR(d)*$)

The $TAR(d)*$ problem generalizes $TAR(d)$ [14] by requiring robots to rotate a triangle configuration infinitely many times with parameter d (e.g., distance or rotation degree), rather than terminating after a single arrangement.

We can show that $TAR(d)*$ shows how perfect synchrony can substitute for memory and communication, but when synchrony is weakened, only models with both internal state and external observability–like $\mathcal{LUMI}^R$–can still solve it. Thus, the following theorem holds. The details are shown in Sect. 4.3 in [19].

Theorem 5. $TAR(d)*$ *is solved in both of $\mathcal{OBLOT}^F$ and $\mathcal{LUMI}^R$, but not in any of $\mathcal{FSTA}^R$ or $\mathcal{FCOM}^S$.*

5   Separation Under Asynchronous Schedulers

This section addresses separation results that involve *asynchronous* and partially *synchronous* schedulers.

Specifically, Sect. 5.1 and Sect. 5.2 present problems that can be solved by $\mathcal{LUMI}^A$, but cannot be solved by $\mathcal{OBLOT}$ under full synchrony ($\mathcal{OBLOT}^F$). These problems illustrate how $\mathcal{FCOM}$ and $\mathcal{FSTA}$ models depend critically on their scheduler assumptions and reveal non-trivial trade-offs between internal memory, external lights, and the degree of synchrony.

In contrast, Sect. 5.3 focuses on a unique problem that is solvable by any model other than $\mathcal{OBLOT}$ under the weakest scheduler (ASYNCH), but remains unsolvable by $\mathcal{OBLOT}$ even under RSYNCH. It becomes solvable for $\mathcal{OBLOT}$ only when perfect synchrony (FSYNCH) is assumed.

These results together clarify how the separation landscape extends beyond *synchronous* settings and how *asynchronous* schedulers highlight the minimal capabilities required for solving coordination problems.

5.1   Fully Asynchronous Mutual Light Trade-Off Problems (*LP–MLCv, VEC, ZCC*)

Figure 1 summarizes three representative problems–*LP–MLCv, VEC,* and *ZCC*– that each demonstrate how specific trade-offs between internal memory, external lights, and scheduler strength determine solvability.

Each line in the figure shows which pairs of configurations are strictly separated by these problems, clarifying how $\mathcal{LUMI}^A$ can solve them, while $\mathcal{OBLOT}^F$ cannot, and revealing which combinations of $\mathcal{FCOM}$ and $\mathcal{FSTA}$ depend critically on the scheduling assumptions.

Definition 3 *LP–MLCv (Leave the Place – MLCv). Two robots, r and q, start at distinct points. The LP–MLCv process consists of two phases. First (LP), they move away from each other along the line connecting them, vacating their starting points. Second (MLCv), they solve the collision-free linear convergence problem without increasing the inter-robot distance (Fig. 5).*

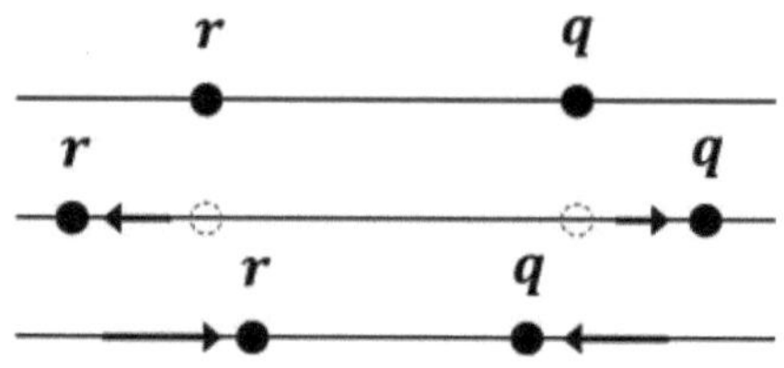

Fig. 5. Movement of the *LP–MLCv* problem.

Lemma 6. $LP\text{–}MLCv \notin (\mathcal{OBLOT}^F \cup \mathcal{FSTA}^S \cup \mathcal{FCOM}^S)$.

Lemma 7. $LP\text{–}MLCv \in (\mathcal{FSTA}^R \cap \mathcal{FCOM}^S)$.

Proof Sketch. (See Section 5.1 in [19] for the formal proof) The *LP–MLCv* problem combines two phases: first, both robots must vacate their initial points by moving away along the connecting line (the "Leave" phase), then they must solve the collision-free linear convergence problem (*MLCv*) without ever increasing their distance.

The second phase, *MLCv*, is well-known to be solvable even by $\mathcal{OBLOT}$ under a *fully-synchronous* scheduler (FSYNCH): robots can measure the distance and converge symmetrically along the line without collisions or internal state.

However, the first phase, *LP*, is crucially harder for $\mathcal{OBLOT}$. Vacating both initial positions requires symmetry breaking–both robots must decide to move away in opposite directions, ensuring they do not remain or move toward the same point. $\mathcal{OBLOT}$ robots, by definition, are completely oblivious and have no internal state or external lights. Even under a *fully-synchronous* scheduler (FSYNCH), they cannot remember whether they have already moved, nor can they observe any marker indicating which direction they should choose. Thus, they cannot coordinate a consistent "leave" action while preserving symmetry: each robot sees exactly the same local configuration and cannot deterministically choose to move away in a unique direction. This makes it impossible for $\mathcal{OBLOT}^F$ to guarantee that both robots will simultaneously leave their starting positions (Lemma 6).

Under $\mathcal{FSTA}^S$, where the scheduler is *semi-synchronous* and robots have only internal memory, robots cannot reliably guarantee that both start moving simultaneously, nor observe the other's intent before moving. This asymmetry means that one robot may stay while the other moves, failing to vacate both initial points as required (Lemma 6).

In contrast, $\mathcal{FSTA}^R$ (ROUNDROBIN after FSYNCH) can solve *LP–MLCv* because the deterministic activation order breaks the symmetry: each robot's internal state can record its phase, ensuring that both robots vacate their positions and then safely switch to *MLCv* (Lemma 7).

Similarly, $\mathcal{FCOM}^S$ can solve *LP–MLCv*: robots can use external lights to signal the intent to leave, ensuring the other robot sees the phase change and moves accordingly, even under *semi-synchronous* activation (Lemma 7).

Thus, combining Lemma 7 and Theorem 2 (2), *LP–MLCv* can be solved in $\mathcal{LUMI}^A$.

On the other hand, $\mathcal{FCOM}^A$ cannot solve it: the *fully-asynchronous* scheduler breaks any guarantee that light-based signaling will be observed consistently, and without internal state, the robots cannot resolve conflicts that arise from inconsistent observations (Lemma 6). $\square$

Therefore, *LP–MLCv* shows that the combination of the leave phase and collision-free convergence exposes the limits of purely oblivious robots ($\mathcal{OBLOT}^F$) and illustrates a non-trivial trade-off between memory, lights, and scheduler assumptions.

Theorem 6. *LP–MLCv is solved in* $\mathcal{FSTA}^R$ *and* $\mathcal{FCOM}^S$, *but not in any of* $\mathcal{OBLOT}^F$, $\mathcal{FSTA}^S$ *or* $\mathcal{FCOM}^S$.

Remark on VEC and ZCC. Due to space constraints, we omit the formal definitions of the problems VEC and ZCC, but we briefly discuss their separation role in contrast to that of $LP–MLCv$. All three problems are designed to reveal the distinct contributions of internal robot capabilities (memory vs. communication) under weak synchrony.

$LP–MLCv$ demonstrates that memory can be substituted by communication under fully *asynchronous* schedulers: the problem is solvable in $\mathcal{FCOM}^A$ but not in $\mathcal{FSTA}^A$. In contrast, VEC and ZCC highlight the opposite separation: they are solvable in $\mathcal{FSTA}^A$ but not in $\mathcal{FCOM}^{AM}$. This asymmetry indicates that the relative advantage of memory or communication depends not only on the task but also on the granularity of synchrony. Moreover, while $LP–MLCv$'s separation is aligned with the synchrony weakening hierarchy, VEC and ZCC expose limitations of communication models even under intermediate atomic schedulers. These results show that the trade-off between memory and communication is inherently problem-dependent and sensitive to scheduler assumptions. The details of VEC and ZCC are shown in Sect. 5.1 in [19].

Theorem 7. *VEC is solved in $\mathcal{FSTA}^{A_M}$, but not in any of $\mathcal{OBLOT}^F$ or $\mathcal{FCOM}^S$ or $\mathcal{FSTA}^{A_{LC}}$.*

Theorem 8. *ZCC is solved in $\mathcal{FSTA}^S$, but not in any of $\mathcal{OBLOT}^F$ or $\mathcal{FSTA}^{A_M}$ or $\mathcal{FCOM}^S$.*

5.2   Synchrony is Necessary for FSTA ($LP–Cv$, VTR)

Remark on $LP–Cv$. The $LP–Cv$ problem is a simplified variant of $LP–MLCv$ in which the second phase ($MLCv$) is replaced by the standard convergence problem (Cv). Then $LP–Cv$ requires only the visibility of other robots' communication states, while memory use is not essential.

Despite its reduced complexity, $LP–Cv$ still exhibits a strong separation: it is solvable in $\mathcal{FCOM}^A$ but not in $\mathcal{FSTA}^A$, showing that communication alone suffices in this case, whereas persistent memory does not. This reinforces the message of $LP–MLCv$, but under weaker assumptions and with simpler structure. The details of this problem are shown in Section 5.2 in [19].

Theorem 9. *$LP–Cv$ is solved in both of $\mathcal{FSTA}^R$ and $\mathcal{FCOM}^{A_M}$, but not in any of $\mathcal{OBLOT}^F$ or $\mathcal{FSTA}^S$.*

Definition 4 VTR (Vertex Traversal). *Three robots, r, q, s, occupy three vertices of a regular hexagon, with no two being opposite each other. The task is for each robot to travel to its opposite vertex and return to its start exactly once (Fig. 6).*

Lemma 8. $VTR \notin (\mathcal{OBLOT}^F \cup \mathcal{FSTA}^{A_{LC}})$.

Lemma 9. $VTR \in (\mathcal{FSTA}^{A_M} \cap \mathcal{FCOM}^S)$.

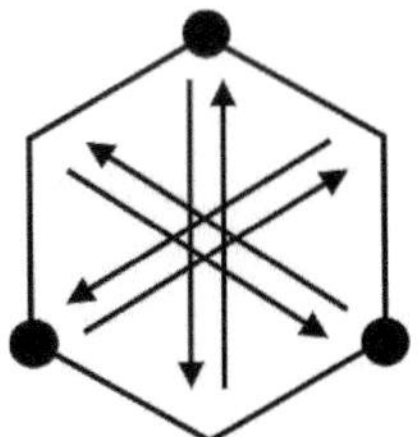

Fig. 6. Movement of the *VTR* problem.

Proof Sketch. (See Section 5.2 in [19] for the formal proof) The *VTR* problem requires each robot to traverse to its opposite vertex and return exactly once, under symmetric initial conditions and without persistent memory.

In $\mathcal{OBLOT}^F$, symmetry cannot be broken, and robots cannot remember past moves, making it impossible to ensure *exactly one round trip* per robot (Lemma 8).

In $\mathcal{FSTA}^{A_{LC}}$, although scheduling may help break symmetry over time, the lack of memory prevents distinguishing whether a robot is on its outward or return path, leading to ambiguity or infinite loops (Lemma 8).

In $\mathcal{FSTA}^{A_M}$, minimal memory (or visible state) allows robots to mark their traversal phase, enabling exact control over movement cycles (Lemma 9).

$\mathcal{FCOM}^S$ can use external lights to record traversal states, which is sufficient to ensure each robot performs one and only one round trip (Lemma 9).   □

Theorem 10. *VTR is solved in both of $\mathcal{FSTA}^{A_M}$ and $\mathcal{FCOM}^S$, but not in any of $\mathcal{OBLOT}^F$ or $\mathcal{FSTA}^{A_{ALC}}$.*

5.3   Full Synchrony or Light Is Necessary (*LCM–M*)

Definition 5 *LCM–M (Least Common Multiple – Movement).* *Place three robots A, B, C in this order on a straight line. In the initial configuration, the ratio of the distances between the robots is $AB : BC = 2 : 1$. In this problem, the following behavior is required of robots B and C.:*

- *Robot B shall perform the action "move away from A by the initial distance d_{AB} between A and B" two times.*
- *Robot C shall perform the action "move away from A by the initial distance d_{AC} between A and C" once.*

Here, let $0 < t_b, 0 < t_b' < t_c$. Let t_b be the time when robot B arrives at a location $2d_{AB}$ away from robot A, and t_b' be the time when robot B arrives at a location $3d_{AB}$ away from robot A. Let t_c be the time when robot C arrives at a location $2d_{AC}$ away from robot A (Fig. 7).

Lemma 10. $LCM\text{-}M \notin \mathcal{OBLOT}^R$.

Lemma 11. $LCM\text{-}M \in (\mathcal{OBLOT}^F \cap \mathcal{FSTA}^{A_{LC}} \cap \mathcal{FCOM}^{A_M})$.

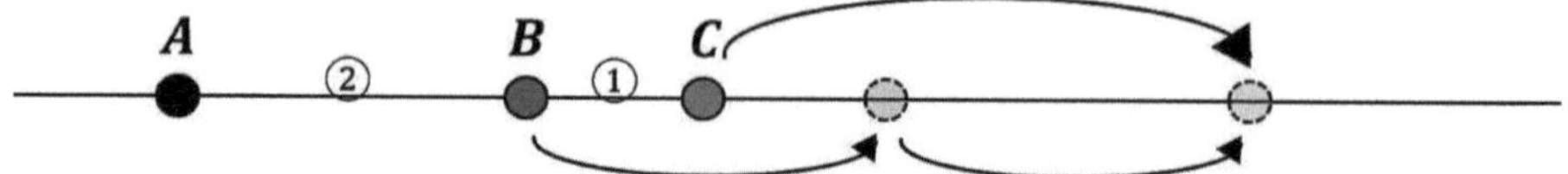

Fig. 7. Movement of the *LCM-M* problem.

Proof Sketch. (See Section 5.3 in [19] for the formal proof) The *LCM–M* problem requires robot B to perform two moves and robot C to perform one move, all at exact distances relative to robot A, and the key constraint is their **relative timing**: $t_b < t'_b < t_c$.

In $\mathcal{OBLOT}^F$, global synchronization ensures that robots can move deterministically in exact steps, maintaining the required timing and distance conditions. Thus, the task is solvable (Lemma 10).

In $\mathcal{OBLOT}^R$, although activations are sequential, the robots lack any memory or signaling capability. This makes it impossible to guarantee the required ordering of movements and distinguish the first move from the second for B. As such, the robots cannot coordinate the correct *LCM–M* behavior, and the task is unsolvable (Lemma 11).

In $\mathcal{FSTA}^{A_{LC}}$, despite being *asynchronous*, the ability to retain **internal state** (via local computation) allows each robot to count its own moves and maintain timing consistency, making the task solvable (Lemma 11).

In $\mathcal{FCOM}^{A_M}$, external visible lights enable even better coordination by encoding move-phase information visible to all robots. This allows the robots to correctly sequence their movements even under full asynchrony (Lemma 11). □

Theorem 11. *LCM–M is solved in* $\mathcal{OBLOT}^F$, $\mathcal{FSTA}^{A_{LC}}$ *and* $\mathcal{FCOM}^{A_M}$, *but not in* $\mathcal{OBLOT}^R$.

6   Concluding Remarks

This paper has extended the known landscape of mobile robot computation by identifying novel triadic separations and structural phenomena that emerge only when comparing three or more robot models simultaneously.

Figure 2 summarizes illustration of the problem landscape for the 14 major robot models, each defined by a pair of internal capability ($\mathcal{OBLOT}$, $\mathcal{FSTA}$, $\mathcal{FCOM}$, $\mathcal{LUMI}$) and scheduler class (Fsynch, Rsynch, Ssynch, Asynch or its variants). Each problem is positioned within the minimal model class that can solve it. The grey-shaded region represents the union of two areas: (1) the regions for which separation results were already known in previous work, and (2) the additional regions where new problems were introduced in this paper to complete the landscape. Problems highlighted in yellow (e.g., *ETE, TAR(d)*, LP–MLCv*) are newly defined in this work and serve to establish previously unknown separations. White regions remain unexplored or unresolved, and represent potential directions for future work.

References

1. Agmon, N., Peleg, D.: Fault-tolerant gathering algorithms for autonomous mobile robots. SIAM J. Comput. **36**(1), 56–82 (2006)
2. Ando, H., Osawa, Y., Suzuki, I., Yamashita, M.: A distributed memoryless point convergence algorithm for mobile robots with limited visivility. IEEE Trans. Robot. Autom. **15**(5), 818–828 (1999)
3. Bouzid, Z., Das, S., Tixeuil, S.: Gathering of mobile robots tolerating multiple crash faults. In: The 33rd International Conference on Distributed Computing Systems, pp. 334–346 (2013)
4. Buchin, K., Flocchini, P., Kostitsyna, I., Peters, T., Santoro, N., Wada, K.: Autonomous mobile robots: refining the computational landscape. In: APDCM 2021, pp. 576–585 (2021)
5. Buchin, K., Flocchini, P., Kostitsyna, I., Peters, T., Santoro, N., Wada, K.: On the computational power of energy-constrained mobile robots: algorithms and cross-model analysis. Inf. Comput. **303**, 105280 (2025)
6. Canepa, D., Potop-Butucaru, M.: Stabilizing flocking via leader election in robot networks. In: Proceedings of 10th International Symposium on Stabilization, Safety, and Security of Distributed Systems (SSS), pp. 52–66 (2007)
7. Cicerone, S., Di Stefano, G., Navarra, A.: Gathering of robots on meeting-points. Distrib. Comput. **31**(1), 1–50 (2018)
8. Cieliebak, M., Flocchini, P., Prencipe, G., Santoro, N.: Distributed computing by mobile robots: gathering. SIAM J. Comput. **41**(4), 829–879 (2012)
9. Cohen, R., Peleg, D.: Convergence properties of the gravitational algorithms in asynchronous robot systems. SIAM J. Comput. **34**(15), 1516–1528 (2005)
10. Das, S., Flocchini, P., Prencipe, G., Santoro, N., Yamashita, M.: Autonomous mobile robots with lights. Theoret. Comput. Sci. **609**, 171–184 (2016)
11. Flocchini, P., Prencipe, G., Santoro, N., Widmayer, P.: Hard tasks for weak robots: the role of common knowledge in pattern formation by autonomous mobile robots. In: 10th International Symposium on Algorithms and Computation (ISAAC), pp. 93–102 (1999)
12. Flocchini, P., Prencipe, G., Santoro, N., Widmayer, P.: Gathering of asynchronous robots with limited visibility. Theoret. Comput. Sci. **337**(1–3), 147–169 (2005)
13. Flocchini, P., Prencipe, G., Santoro, N., Widmayer, P.: Arbitrary pattern formation by asynchronous oblivious robots. Theoret. Comput. Sci. **407**, 412–447 (2008)
14. Flocchini, P., Santoro, N., Wada, K.: On memory, communication, and synchronous schedulers when moving and computing. In: Proceedings of 23rd International Conference on Principles of Distributed Systems (OPODIS), pp. 25:1–25:17 (2019)
15. Flocchini, P., Santoro, N., Sudo, Y., Wada, K.: On asynchrony, memory, and communication: separations and landscapes. In: 27th International Conference on Principles of Distributed Systems (OPODIS 2023). Leibniz International Proceedings in Informatics (LIPIcs), vol. 286, pp. 28:1–28:23 (2024)
16. Fujinaga, N., Yamauchi, Y., Ono, H., Kijima, S., Yamashita, M.: Pattern formation by oblivious asynchronous mobile robots. SIAM J. Comput. **44**(3), 740–785 (2015)
17. Gervasi, V., Prencipe, G.: Coordination without communication: the case of the flocking problem. Discret. Appl. Math. **144**(3), 324–344 (2004)
18. Izumi, T., et al.: The gathering problem for two oblivious robots with unreliable compasses. SIAM J. Comput. **41**(1), 26–46 (2012)
19. Naito, S., Ninomiya, T., Wada, K.: Beyond pairwise comparisons: unveiling structural landscape of mobile robot models (2025). https://arxiv.org/abs/2508.19805. arXiv:2508.19805

20. Okumura, T., Wada, K., Défago, X.: Optimal rendezvous $\mathcal{L}$-algorithms for asynchronous mobile robots with external-lights. In: Proceedings of 22nd International Conference on Principles of Distributed Systems (OPODIS), pp. 24:1–24:16 (2018)
21. Souissi, S., Izumi, T., Wada, K.: Oracle-based flocking of mobile robots in crash-recovery model. In: Proceedings of 11th International Symposium on Stabilization, Safety, and Security of Distributed Systems (SSS), pp. 683–697 (2009)
22. Suzuki, I., Yamashita, M.: Distributed anonymous mobile robots: formation of geometric patterns. SIAM J. Comput. **28**, 1347–1363 (1999)
23. Yamashita, M., Suzuki, I.: Characterizing geometric patterns formable by oblivious anonymous mobile robots. Theoret. Comput. Sci. **411**(26–28), 2433–2453 (2010)
24. Yamauchi, Y., Uehara, T., Kijima, S., Yamashita, M.: Plane formation by synchronous mobile robots in the three-dimensional euclidean space. J. ACM **643**, 16:1–16:43 (2017)

NestedBTO: A Timestamp-Based STM Protocol for Closed Nested Transactions with Opacity Guarantees

Nischay Ranjan[1,2]($\boxtimes$) , Rohit Kapoor[1]($\boxtimes$) , and Sathya Peri[1]($\boxtimes$)

[1] Indian Institute of Technology Hyderabad, Hyderabad, India
`rohitkapoor9312@gmail.com`, `sathya_p@cse.iith.ac.in`
[2] GEC Sheohar, Department of Science, Technology, and Technical Education,
Government of Bihar, Patna, Bihar, India
`nischayranjan.dstte@bihar.gov.in`, `cs21resch11012@iith.ac.in`

Abstract. Software Transactional Memory (STM) provides a high-level abstraction for managing concurrency, alleviating the complexity of traditional lock-based programming. While nesting support is critical for enabling composability and reducing the vulnerability window of long transactions, existing STM protocols for closed nesting often fall short in either offering rigorous correctness guarantees or demonstrating practical performance benefits. This paper introduces NestedBTO, a new STM protocol for closed nested transactions that uses timestamp ordering to allow child transactions to be executed in parallel while maintaining consistency according to the opacity criterion. NestedBTO satisfies opacity for all transactional histories, as demonstrated by our formal proof of correctness. We conduct extensive experimental evaluations using application-level workloads STAMP and concurrent red-black tree and hashtable benchmarks to assess its effectiveness. The results reveal that NestedBTO achieves substantial improvements in throughput and abort rates compared to other state-of-the-art STM, i.e., NesTM. These findings demonstrate that NestedBTO offers a scalable and practical solution for high-performance transactional memory systems.

Keywords: Software Transactional Memory · Opacity · Closed Nesting

1 Introduction

The multicore architecture has gained predominance in the design of modern computing systems. To gain the most out of it, concurrent programming has gained interest. However, developing a concurrent program is significantly more complex than developing its sequential counterpart. Locks have traditionally been used broadly for implementing parallelism. However, lock-based programs

This work is partly supported by Core Research Grant CRG/2022/009391.

pose complex issues with synchronization and are susceptible to various problems such as deadlocks and priority inversion. Moreover, the composition of larger software systems by integrating smaller components with locks creates serious complications [11]. A Software Transactional Memory (STM) system [12,23] enables concurrent processes to coordinate and communicate by performing transactions on shared memory. A transaction represents a sequence of operations that executes atomically–appearing as a single, indivisible action–even when other transactions are running concurrently. Within a transaction, memory operations can be performed on transactional objects (*t-obj*). If a transaction completes successfully, it commits, making all its updates visible to other transactions. If a transaction fails, it aborts, causing all its modifications to be undone, ensuring that no changes are observed by other transactions. Access to *t-obj* is permitted exclusively through transactions.

When, during the execution of a transaction, another transaction is invoked, the process is said to be a transaction nesting [10]. Nested transactions take the elementary transactional model to a multi-level structure, in which a group of sub-transactions can recursively call additional sub-transactions, forming a transactional tree. Different varieties of nesting exist, notably closed, open, and flat nesting [17]. In closed nesting, the failure (abort) of a child transaction does not directly affect the outcome of its parent transaction. However, it is significant to note that the updates of a child transaction become visible to other transactions only after a successful parent transaction is committed.

A primary reason for instituting transaction nesting into STM systems was to obtain composability requirements. **Composability** is the essential property of modular programming, allowing independent components to be assembled into a larger, more complex structure. In the case of transactional systems, this means placing together several subtransactions into one larger entity called a transaction. Another driving force is to mitigate the problems long-running transactions pose, particularly those with extended vulnerability periods. In this regard, the vulnerability period is when a transaction is still vulnerable to conflicts with other transactions that run concurrently. A practical remedy would be to decompose a long transaction into several smaller subtransactions, thereby executing them over multiple cores so that both execution time and vulnerability windows are minimized. Such a strategy is starting to gain traction in the context of transactional data structure library implementation [1,6,15,24].

Considering the widespread nested parallelism, such as nested parallel loops, recursive function calls, and calls to parallel libraries in real-world applications, these nested situations also motivate the developers of nested STM systems [25]. Nested calls are equally important in smart contract execution in the blockchain; that is, they aid in composability but also in interaction complexities. Designing an STM protocol which is able to support transactions nesting is really a nontrivial task; the difficulty of figuring out contention across several different levels of nesting, while at the same time making certain that the executions of such nesting are consistent in terms of sequential semantics, adds complexity to this problem [7,9,13,18]. When multiple transactions–nested or otherwise–access shared data, it is essential that their executions remain correct. Opacity has been widely accepted as the key correctness criterion for concurrent transactions in STM systems. Although several STM implementations have been proposed to support nested transactions, existing efforts focus mainly on two nesting models: linear nesting [5,11,16,22] and parallel nesting [2,3,14,20,25]. However, none of these models provides a formal guarantee of opacity. Although systems like HparSTM [14] offer an informal discussion on supporting opacity, they do not provide a rigorous proof. Addressing this gap, Peri et al. [18] formally extended the notion of opacity to nested transactions and identified a class of schedules that preserve opacity under closed nesting.

In concurrent transaction execution within software transactional memory systems, **Opacity** [8] has stood as a fundamental correctness criterion. In addition to ensuring serializability, Opacity requires that even the aborted transaction can have the illusion of no concurrency, i.e., the aborted transaction should also view the consistent state of the shared memory. The challenge of ensuring opacity becomes particularly sophisticated within the context of nested transactions, where an aborted parent transaction may have committed children and vice versa.

This paper introduces **NestedBTO**, a Software Transactional Memory system that supports parallel closed nesting based on timestamp ordering. It extends the timestamp-based approach introduced in our earlier study [21]. Our key contribution lies in proposing a parallel-nested STM system and formally proving that our protocol adheres to the class of schedules defined by Peri et al. for closed nested transactions. This proof of correctness establishes that **NestedBTO** satisfies opacity under arbitrary nesting depths–something not demonstrated in prior works.

Our Core Contributions (in Sect. 3) are as Follows:

1. We present a novel STM system that supports **closed nesting to arbitrary depth** while **formally guaranteeing opacity.**
2. The system enables **parallel execution of subtransactions**, supports **adaptive flattening** of shallow nests, and allows **work-stealing** across independent subtrees to improve performance.
3. We design a **new timestamp protocol** that ensures **serializability and correctness** across unrestricted levels of nested transactions.

In addition, we conduct an extensive performance evaluation of our STM system, comparing it against flat nesting and serial execution on benchmark data structures such as hash tables, counters, Red Black tree, and STAMP. Our results demonstrate that **NestedBTO** maintains theoretical correctness and delivers practical scalability.

Roadmap: The remainder of this paper is structured as follows: The preliminary data and the system model are presented in Sect. 2. The **NestedBTO** algorithm, as designed and with its operating semantics, is addressed in greater detail in Sect. 3. Section 4 presents an experimental algorithm evaluation that compares its performance with existing approaches. Finally, Sect. 5 concludes the paper and discusses potential directions for future research.

2   Preliminaries and System Model

2.1   Transaction Tree

A transaction consists of a sequence of read and write operations on transactional objects *t-obj*, and may invoke subtransactions that can themselves recursively spawn further subtransactions. This hierarchical execution naturally forms a transaction tree (see Fig. 1), where each node represents a transaction or an operation, with simple memory operations as leaves. Transactions perform two types of operations: *memory operations* and *transactional operations*, including *commit* and *abort*. When a transaction t_p invokes a subtransaction t_{pi}, a new child node is added under node t_p in the transaction tree. If t_p completes successfully, it issues a commit; otherwise, it ends with an abort. Aborted subtransactions (e.g., t_{qi}) lead to the removal of their subtree, while committed ones remain until garbage-collected. The root transaction t_0 sits at level 0 and manages globally shared *t-obj*. Transactions directly under t_0 are referred to as top-level transactions. Any transaction t_x with k children is denoted as t_{x1}, t_{x2}, ..., t_{xk}. Transactions use two local buffers: a *read_list* and a *write_list* . For a read $r_x(y)$, the transaction first checks its local *write_list* , then the read list, and finally queries ancestors up to t0 if necessary. Writes $w_x(y)$ add the object to the *write_list* . Each transaction begins with memory operations and concludes with subtransactions and terminal operations. A parent transaction executes its terminal operation only after all its subtransactions complete. On commit, a subtransaction's *write_list* merges into its parent's. On abort, its updates are discarded.

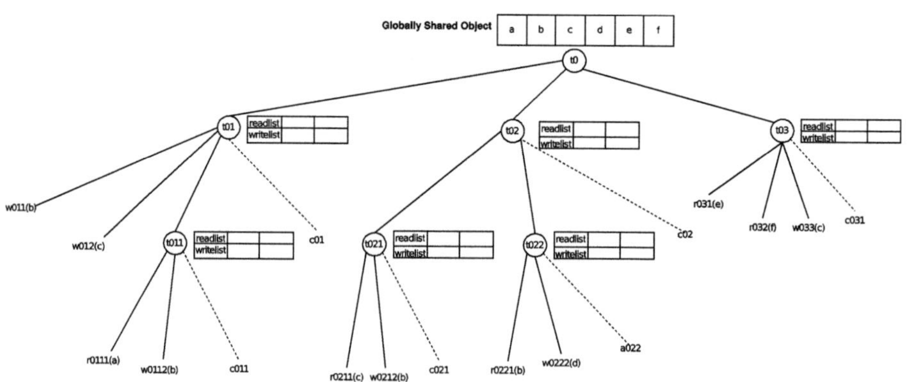

Fig. 1. Transaction Tree.

2.2   Computation Model

We have developed a system that enables a set of n processes (or threads), denoted as $p_1, p_2, \ldots, p_n$, to interact with a shared collection of transactional objects (t-obj) using atomic transactions. Each process can perform a predefined set of operations within a transactional context. A new transaction is started by the *begin* operation. The value of the object x is retrieved by the *read(x)* operation from either the local state of the current transaction or, in the event that it is absent, from one of its ancestor transactions. The transactional object x's value is locally updated to v within the transaction by the *write(x,v)* operation. To enable nested execution, the *assignthread(t_i)* operation creates a subtransaction as a child of transaction t_i. The *tryC()* operation is used to try to commit a transaction. If it succeeds, the updates from the transaction are sent to the buffer of its parent, and the operation returns a *commit* (in short, $\mathcal{C}$); if not, it returns an abort $\mathcal{A}$). The *tryA()* operation can also be used to explicitly abort a transaction, ending it and returning an abort.

It's important to note that both the *read* and *TryC* methods have the potential to return $\mathcal{A}$ which signifies that the underlying algorithm has forcefully aborted the transaction. However, in cases where $\mathcal{A}$ is not returned, it indicates that the operations have been executed successfully without any forced termination. Additionally, each transaction adheres to the following constraints: (1) a transaction executes one operation at a time, and (2) a parent transaction must wait until all of its children have terminated before entering the validation phase for its commit.

2.3   Histories and Base Formalism

A history is a sequence of events encompassing invocations and responses to transactional operations. The set of events is denoted as $H.evts$. In the context of Transaction history, considering an execution sequence, let H' represent the set of transactions invoked during that execution. The order relation between transactions, denoted as $\prec_{H'}$, is defined as follows:

1. $t_1 \prec_{H'} t_2$; t_1 terminates before t_2
2. $t_1 \nprec_{H'} t_2$; t_1 and t_2 are concurrent.

At the transaction level, the execution is defined as the partial order $\widehat{H'} = \langle H', \prec \rangle$, referred to as the **transaction history**. A transaction history is deemed *sequential* if no (sub)transactions within it occur concurrently. A sequential transaction history is *legal* if each read operation returns the last write on the same object.

Level-Wise History: In the context of closed nested STM, each transaction node t is associated with a local buffer, which functions as a shared memory for its descendants(i.e. its child and sub-child transactions). Consequently, every level or node in the transaction tree possesses a distinct, *level-wise history*, which reflects the operation performed within the scope of that transaction and its descendants. We denote this history as $\widehat{H_t'} = \langle H_t', \prec_{H_t'} \rangle$ representing a level-wise transaction history at t, where $\widehat{H_t'}$ represents the set of operations visible at node t and $\prec_{H_t'}$ captures the partial order among those operations. Similarly, a level-wise sequential history at node t, representing a total order consistent with the observed operations, is denoted as $\widehat{H'}_t^{seq} = \langle H_t'^{seq}, \prec_{H_t'^{seq}} \rangle$. Here, references to other transactions (e.g., parent or sibling) are restricted to those within the same transaction tree–i.e., sharing the same root and nested structure–not from separate threads or trees.

External Read: An external read operation within a transaction t denotes the reading of a transactional object x when it is not locally present within the transaction's buffer. Instead, t accesses x from the nearest ancestor (in the bottom-up hierarchy), possessing a local copy of x. In cases where x is not found in any ancestor's buffer, the transaction t resorts to the root transaction, where it would certainly find the t-obj x. Illustrated in Fig. 1, the operation $r_{0211}(c)$ exemplifies an external read for both t_{021} and t_{02}.

Propagated Read Objects: Propagated read objects refer to instances where a subtransaction t reads a transactional object x from the local buffer of an ancestor t', which is different from its immediate parent t_p. In executing the read operation, t propagates the read list of all its ancestors until t', ensuring that x is included in the read list. Consequently, the value of x retained in t_p and other ancestors' read lists aligns with the value obtained during t's read operation. This propagated value of x remains accessible to subsequent descendants of t_p and its ancestors. Illustrated in Fig. 1, the operation $r_{0211}(c)$ propagates the read buffer of t_{02}, thereby enabling its utilization by t_{022}.

Commit Write: In close nested transactions, the commit write operation occurs when a transaction t commits its operations. During this process, the *write_list* of t is merged with that of its parent transaction and updates all the t-obj, which is common in parent and child transactions. If a transactional object is absent in the parent's write list, it is added accordingly. Concerning Fig. 1, consider t_{021}; upon committing, the transactional object b in its write list integrates into t_{02}'s write list. Similarly, upon t_{011} committing, the transactional objects in its write list update the corresponding values in t_{021}'s write list.

Correctness of the Closed Nested Transactions: Guerraoui and Kapalka [8] introduced opacity as a correctness criterion for software transactions, ensuring schedule consistency. Opacity requires that an equivalent serial schedule maintains the original schedule real- time ordering and lastWrites of read operations from the original schedule, even accounting for aborted transactions treated as read-only in serial schedules. In nested transactions, complexity arises from aborted transactions that potentially have committed sub-transactions, impacting the correctness of subsequent operations. Committed sub-transactions of aborted ones can have commit-writes, affecting other transactions. The **Closed Nested Opacity (CNO)** definition [19] extends the traditional notion of opacity to system with closed nested transactions. *CNO* ensures that there exists a serial schedule preserving the order of operations, the partial order among transactions, and the lastWrite equivalence for read operations. Unlike standard opacity, *CNO* accounts for the added complexity of nested transactions by guaranteeing consistent reads across all transactions, including those within aborted subtransactions. This redefined definition is essential for maintaining correctness in systems that support nested transactional execution.

3 Nested Basic Timestamp Ordering Algorithm

This section presents a timestamp-based algorithm designed for the closed nested STM system, facilitating the parallel execution of nested transactions. Referred to as **Nested BTO**, this algorithm has been developed to enhance the efficiency of transactional processing within a closed nested environment. Subsequently, we establish our algorithm's robustness by providing proof demonstrating its adherence to the opacity correctness criteria. Graph characterization achieves this validation by showcasing the soundness and reliability of the Nested BTO algorithm in ensuring opacity.

The Working Principle

The algorithm employs an invariant approach for assigning unique identifiers (a string), which also function as timestamps, to transactions. For instance, if a transaction holds the identifier i, it will also be its timestamp. Intuitively, the timestamp tells about the "time" the transaction began. Adding numerical suffixes to the parent transactions' identifiers helps in the generation of string identifiers for their child transactions. Hence, this design enables the lexicographic comparison of transaction identifiers or timestamps, providing a structured means of considering their chronological order.

The invoking transaction's timestamp accompanies its every read and write operation within the system. Each transaction is associated with the #child, term_count, and status variables. The variable #child aids as an indicator of the number of children associated with a given transaction. The variable term_count accounts for the count of child transactions that have terminated either successfully committed or aborted. The variable status provides information about the current state of the transaction. Furthermore, every *t-obj* is associated with two variables: *max-r* and *max-w*. These variables store the timestamp of the most recent transaction that successfully executed a read and write operation on the respective transactional object.

Now, we will discuss the core concepts underlying the read, write, and tryC operations performed by a transaction T_i. The inspiration for these principles is drawn from the timestamp algorithms crafted by Bernstein and Goodman for databases [4]. The following operational principles are presented first informally; their full algorithmic realization (with timestamps and pseudocode) appears later.

1. **Read Principle:**When transaction T_i invokes a *read(x)* operation, it retrieves the value v from either its own local buffer or its ancestors' local buffers (say T_p). If x is found in the *write_list* of an ancestor T_p, the read principle checks whether the same object x has been concurrently updated in T_p's *write_list* by another transaction T_j, where T_j's timestamp is greater than that of T_i. If such a conflict exists, a recursive abort is triggered on T_i, which cascades to abort all of its descendants. Otherwise, the *max_r(x)* field in the T_p's *write_list* is updated with T_i's identifier. If x is instead found in the ancestor's *read_list* , the *max_r(x)* field in the local *read_list* is updated with T_i. It is important to emphasize that T_i and T_j are part of the same transaction (sub)tree at T_p.
2. **Write Principle:** Transaction T_i updates data in its local buffer, precisely to *write_list*.
3. **Commit Principle:** Before committing, a transaction T_i invokes a series of checks:
   (a) *Child Termination Check:* At first, T_i verifies whether all of its child transactions (if any) have terminated. It is done by comparing the invariants #child count with term_count. If not all children are done, T_i waits.
   (b) *Validation for Each t-objx in Its write_list :*
       i. If a transaction T_k (which is a sibling or a descendant of a sibling of T_i, and part of the same transaction tree rooted in parent(T_i)) has read x from parent(T_i), and the timestamp of T_i is less than that of T_k (that is, $i < k$), then T_i must abort.
       ii. If a sibling transaction T_k (i.e., a transaction with the same parent as T_i) has written to x in parent(T_i), and $i < k$, then again, T_i must abort.

   (c) *Commit Approval:* If none of the above conflict conditions are met, the transaction T_i is allowed to commit successfully.

The Protocol

Let T represent the set of transactions, X denotes the set of objects, and V signifies the possible values associated with these objects. Furthermore, we assume a thread pool comprises n threads, with each transaction invoked by a thread from this designated pool.

The STM system comprises the following operations or functions, each executed in response to a transaction's initiation, reading, writing, or attempt to commit.

initialize(): This operation initializes the STM system. In this process, it assigns an identifier of 0 to the root transaction and designates its parent as NULL. Furthermore, the STM system is presumed to possess knowledge of all *t-objs*ever accessed. During initialization, these *t-objs*are assigned a value of 0 by the root transaction and subsequently included in its writelist.

begin$_i$(): The operation allows a transaction to invoke other transactions involving a thread from the thread pool, invoking another (sub)transaction. Notably, all top-level transactions are invoked by t0. In the case of a (sub)transaction t_j invoked by another transaction t_i, its identifier is assigned by appending its position in the child list of t_i to the identifier of t_i. This operation also sets the status of any newly invoked transaction to ACTIVE.

read$_i$(): Through this operation, a transaction initiates a read operation; it first searches for the corresponding *t-obj* in its local buffer, *write_list* followed by the *read_list* It searches in its ancestor transactions if it fails to locate the required *t-obj* in its local buffer. This search process begins with its immediate parents and may extend upwards in the transaction hierarchy, eventually reaching the root transaction t_0. Once the transaction successfully locates the required *t-obj*, it traverses back down the order, populating the local read buffers of all the ancestor transactions it encountered with its timestamp information.

write$_i$(): The write operation creates and appends *t-obj*x to the *write_list* of the t_i, if it is not there. Otherwise, it updates the value of the *t-obj*x.

traverse_up$_i$(x): This operation examines whether the transactional objects *t-obj*x are present in the ancestors of transaction t_i. A transaction stack stores the ancestors of t_i, and the search may extend up to t_0.

traverse_down$_i$(): Upon locating the necessary *t-obj* in the ancestors' buffer, it is propagated downward into the local read buffer *read_list* of all the intermediate transactions until t_i.

TryC$_i$(): This operation is invoked when a transaction t_i intends to commit, aiming to update the content of *t-obj*s in its *write_list* to its parent's *write_list* The process involves several steps:

1. *Check for Children Termination:* Firstly, t_i checks whether all its children (if any) have terminated.
2. *Iterative Check for Updates*: For each *t-obj*x in the *write_list* of t_i, the following checks are performed:
   Suppose t_i and t_k are siblings with t_p as their parent-
   (a) *read check:* Suppose t_k has read x from t_p, *max-r(x)* in t_p's *read_list* or *max-w(x)* in t_p's *write_list* (depending on the source of the read) stores the timestamp of t_k. If the timestamp of t_i is less than the timestamp of t_k, t_i is considered too late for the update, leading to its abortion.
   (b) *update check:* If t_k has already performed an update on t_p's *write_list* for *t-obj*x, *max-w(x)* in the *write_list* stores the timestamp of t_k. If the timestamp of t_i is less than the timestamp of t_k, t_i is deemed too late for the update, resulting in its abortion.
3. *Reflect Updates to Parent*: If the checks are successful for all *t-obj*s in t_i's *write_list* , the proposed updates by t_i are reflected in its parent's *write_list*.

Note that for Root Transaction t_0, the tryC protocol ensures all child transactions have terminated, commits without conflict validation (no siblings/parent), and merges its *write_list* into the global state.

Abort$_i$(): This operation is responsible for aborting the current transaction. It is invoked when the validation within the *TryC* operation for a specific transaction t_i fails. In such cases, the system triggers this operation to handle the abortion of the transaction t_i.

Rec_abort$_i$(): In the event of a transaction aborting at a higher level of the hierarchy, all its children must also be aborted. This scenario unfolds when a transaction traverses upward to read a particular *t-obj* say x, and encounters an ancestor, say t_p, that undergoes abortion. The entire subtree rooted at t_p is also aborted in this case. It's essential to note that this mechanism differs from the standalone Abort() operation, where a transaction independently aborts itself without affecting its children or ancestors.

Consider the following scenario depicted in the Fig. 2 to explain the functionality of the algorithm: Suppose t_p is a top-level transaction with t_{p1} and t_{p2} as its two children. Following our assumption that t_p invokes memory operations first and then subtransaction operations, it initiates a read operation on *t-obj* y and z, adds them to its *read_list* The *max-r* of these variables is then set to t_p. Subsequently, when t_p attempts to invoke t_{p1} and t_{p2}, and if threads are available in the thread pool, t_{p1} and t_{p2} are concurrently invoked by different threads. Assuming enough threads are available, t_{p1} and t_{p2} are invoked concurrently. In this context, consider the scenario where t_{p1} has a descendant t_i that initiates a *read(x)*. The read operation of t_i traverses upward, with each ancestor attempting to locate the transactional object x in its parent's buffer. Concurrently, t_{p2} is engaged in a *write(x)* operation on its local *write_list*.

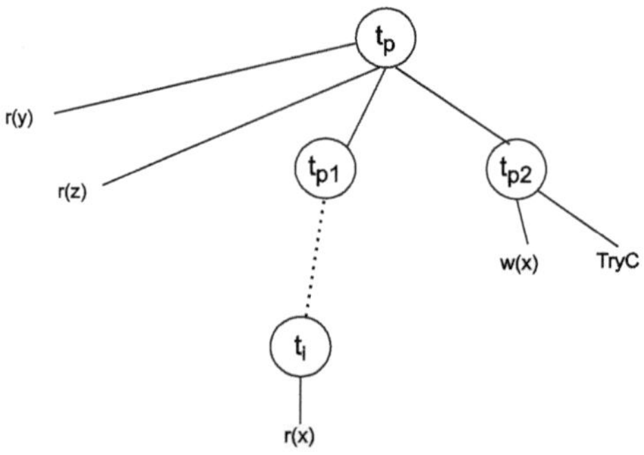

Fig. 2. Transactional Operations.

In this context, let's examine different cases:

1. **TryC - read conflict**: If t_{p2} successfully updates before t_{p1} completes its search for x in t_p's buffer, x is added to the *write_list* of t_p *max-w(x)* = t_{p2}. When t_{p1} attempts to locate x in t_p's buffer (checking *write_list* first and then *read_list*), it discovers that a later transaction (t_{p2} in this) case has appended x to t_p's *write_list* Consequently, *Rec_abort()* of t_{p1} is invoked, leading to the abortion of t_{tp1} along with all its descendants.

2. **Read - TryC conflict**: If t_{p1}'s search for x precedes t_{p2}'s update operation, t_{p1} can traverse upward until t_0 where it definitively finds x. Subsequently, it traverses down, adding x to the *read_list* of t_p, while t_{p2} appends x to the *write_list* of t_p. Any subsequent *read(x)* operations by descendants of t_{p1} retrieve the propagated value in t_{p1}'s and its descendant's *read_list*

3. **TryC - TryC conflict**: As updates from (sub)transactions are reflected in the parent's *write_list* , consider a scenario where a descendant of t_{p1} executes a *write(x)*, leading to x being appended to t_{p1}'s *write_list* Consequently, when t_{p1} invokes the *tryC* operation and conflicts with t_{p2}, t_{p1} has to invoke *Rec_abort()* if t_{p2} appends x to t_{p1}'s *write_list* first. Alternatively, both can terminate successfully if t_{p1} commits first, with t_{p2} overwriting t_{p1}.

Theorem. *The history generated by the NestedBTO algorithm satisfies the* ***Conflict preserving close nested opacity(CP-CNO)*** [19].

4 Evaluation

Experimental Setup: To evaluate the efficacy of NestedBTO, we compared its performance with NesTM using the STAMP benchmark suite and also against microbenchmarks such as concurrent red-black tree and concurrent hashtable. We have chosen the NesTM as it is the closest point of comparison. The experiments were carried out on a 16-core Intel Xeon processor (32 threads with hyperthreading) with 64GB of RAM. All implementations were compiled using GCC 11.3 with -O3 optimization flags. The complete pseudocode and a reference implementation of the NestedBTO protocol are publicly available[1].

4.1 STAMP

For the **STAMP** benchmark, the transactional workload was designed with 3–5 operations per nesting level, reflecting fine-grained parallelism common in realistic applications. For top-level transactions, NestedBTO consistently exhibited a 5–10% reduction in normalized performance difference (NPD) compared to NesTM. This improvement is attributed to NestedBTO's use of timestamp-based conflict resolution, which avoids lock contention during commit phases. In contrast, NesTM's commit-lock mechanism and repeated read-set validation introduce additional overheads, particularly in benchmarks with frequent barriers, such as intruder and vacation.

When evaluating scalability with increased nesting levels (NL $\geq$ 1), NestedBTO demonstrated superior performance. Its lexicographic timestamp ordering led to fewer redundant aborts by enforcing a deterministic serialization order. NesTM, on the other hand, experienced a performance bottleneck due to its lock-based commit protocol, especially at nesting levels of 3 or more,

---

[1] https://github.com/rohit-kapoor/STM-BTO-nested/tree/main.

Algorithm 1. NestedBTO Protocol

---

enum State ABORT = -1, ACTIVE = 0, COMMIT = 1
SUCCESS = 1, FAILURE = -1

State of the transaction object
1: $oid \in I$
2: $val \in V$
3: $lock\ obj$
4: $max\_r, max\_w \leftarrow$ **NULL**
State of the transaction
5: $tid \leftarrow$ **NULL**
6: $status \leftarrow$ **ACTIVE**
7: $lock\ trans$
8: $\#child,\ term\_count \leftarrow 0$
9: $read\_list, write\_list \subseteq X$
10: $parent's\ id \leftarrow$ NULL
11: $child\_list \subset T$
STM_initialize()
12: $t0.id \leftarrow 0$
13: $t0.parent \leftarrow$ NULL
14: $objcounter \leftarrow 0$
15: **for all** $x \in X$ used by the STM system **do**
16:     $x.id \leftarrow ++objcounter$
17:     add x to $t0.wl$
18: **end for**
STM_begin$_i$()
19: new trans t
20: $t_j.parent \leftarrow t_i$
21: $t_j.id \leftarrow t_i.id.append(++\#child)$
22: $t_j.status \leftarrow$ **ACTIVE**
STM_read$_i$(x)
23: **if** $t_i.status \leftarrow$ **ABORT then**
24:     return **FAILURE**
25: **end if**
26: search locally, $write\_list$, and $read\_list$
27: **if** $x\ is\ present\ locally$ **then**
28:     return $x.val$
29: **else**
30:     $val \leftarrow traverse\_up_i\ ()$
31:     return val
32: **end if**
STM_write$_i$(x,val)
33: **if** $t_i.status \leftarrow$ **ABORT then**
34:     return **FAILURE**
35: **end if**
36: **if** $x\ is\ in\ the\ writelist$ **then**
37:     $x.val \leftarrow val$
38: **else**

39:     $create\ t\text{-}obj\ x\ in\ the\ writelist$
40:     $x.val \leftarrow val$
41:     $x.max\_w \leftarrow t_i.id$
42: **end if**
STM_TryC$_i$()
43: **if** $t_i.status \leftarrow$ **ABORT then**
44:     return **FAILURE**
45: **end if**
46: **while** $t_i.\#child < t_i.term\_count$ **do**
47:     wait
48: **end while**
49: **for all** d in $ti.write\_list()$ **do**
50:     /* Lock the $t\text{-}obj$s in predefined order to avoid deadlocks */
51:     $lock\ obj(d)$ in $t_i.parent\ writelist$ $and\ readlist$
52: **end for**
53: **for all** d in $t_i.write\_list()$ **do**
54:     **if** $t_i.parent.readlist[d].max\_r > t_i.id$ OR $t_i.parent.writelist[d].max\_r > t_i.id$ OR $t_i.parent.writelist[d].max\_w > t_i.id$ **then**
55:         **for all** d in $ti.write\_list()$ **do**
56:             $unlock\ obj(d)$ in $t_i.parent$ $writelist\ and\ readlist$
57:         **end for**
58:         $Try\_abort(t_i)$
59:         return **FAILURE**
60:     **end if**
61: **end for**
62: **for all** d in $ti.write\_list()$ **do**
63:     update $t_i.parent.writelist[d].val \leftarrow d.val$
64:     update $t_i.parent.writelist[d].max\_w \leftarrow t_i.id$
65: **end for**
66: **for all** d in $ti.write\_list()$ **do**
67:     $unlock\ obj(d)$ in $t_i.parent\ writelist$ $and\ readlist$
68: **end for**
69: $t_i.parent.term\_count++$
70: $t_i.status \leftarrow COMMIT$
71: $t_i.parent.term\_child\_list.remove(t_i)$
72: delete t_i
73: return **COMMIT**

---

where throughput dropped by 15–20% in high-contention scenarios. Furthermore, NestedBTO's smaller transactional footprints resulted in better cache efficiency, reducing L1/L2 cache misses by approximately 12–18% based on simulation traces. NesTM's approach of merging metadata accesses adversely affected temporal locality. In benchmark-specific evaluations, both protocols performed similarly in low-conflict workloads such as the labyrinth, with less than a 2% difference in NPD. However, in high-conflict workloads, such as intruders, NestedBTO significantly reduced abort rates by 25 to 30% through early conflict resolution using recursive timestamp checks. The aggressive validation strategy of NesTM in such workloads led to more frequent rollbacks.

The Fig. 3 summarizes the normalized performance difference (NPD) for three representative STAMP benchmarks, measured using 16 threads:

The evaluation results affirm that NestedBTO's design, including timestamp-based conflict resolution and lightweight metadata, offers measurable performance gains over NesTM in systems with fine-grained transactional workloads. However, there are two key limitations to note. First, while NestedBTO performs well for nesting depths up to two, deeper nesting levels may introduce timestamp comparison overheads that require further optimization. Second, the current results are derived from simulations; validating the findings using real hardware performance counters, such as cycle counts and cache miss rates, would further strengthen the analysis.

4.2   Red Black Tree

Experimental Setup: We evaluated both protocols using a concurrent red-black tree microbenchmark under the following conditions: a thread pool containing 128 threads and a Non-uniform memory access (NUMA) architecture. The workload consisted of 20% writes (insert and delete operations) and 80% reads (lookup) per transaction. Performance metrics comprise throughput (in transactions per millisecond), abort rates and latency distributions.

Performance Analysis

1. **Throughput scaling with nesting depth:** NestedBTO showed significantly better scalability than NesTM as nesting levels increased. For example, with a nesting depth of 2, NestedBTO completed the benchmark in 9.8 ms compared to 11.4 ms by NesTM, resulting in a performance gain of 16.3%. At levels 3 and 5, the performance improvements were 21. 6% and 24. 8%, respectively. This performance gain can be attributed to NestedBTO's timestamp-based conflict detection, which minimizes synchronization overhead between parent and child transactions. By contrast, NesTM's commit-locking mechanism suffers from increasing validation complexity as the nesting depth grows, roughly scaling quadratically with the nesting level.

2. **Abort Behaviour:** NestedBTO produced more predictable and lower abort rates than NesTM, primarily due to its deterministic timestamp ordering strategy. At a nesting depth of 3 with 64 threads, abort rates for NestedBTO were 6%, 14%, and 19% for levels 1, 2, and 3, respectively. NesTM showed

greater abort rates of 8%, 15%, and 22% for the same levels. The differences arise from the different design decisions: NestedBTO avoids unnecessary aborts by early conflict resolution, whereas NesTM's locking and validation procedures might lead to cascading aborts amid a dispute.

3. **Latency Feature:** NestedBTO was superior in terms of latency performance, especially in terms of high-tail-latency reduction. At the 50th percentile, NestedBTO had a transaction latency of about 29.1 ms; NesTM had 34.6 ms. At the 99th percentile, the latencies were 42.3 and 53.6 ms for NestedBTO and NesTM, respectively. NestedBTO's non-blocking validation, which prevents

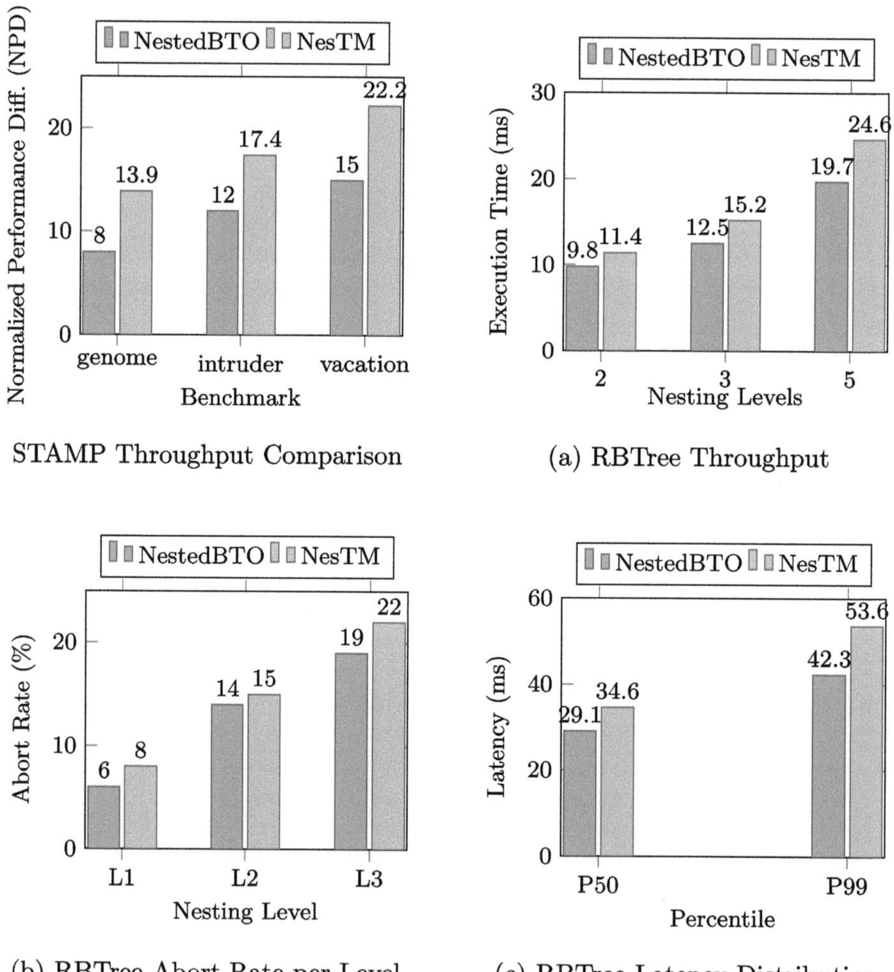

STAMP Throughput Comparison

(a) RBTree Throughput

(b) RBTree Abort Rate per Level

(c) RBTree Latency Distribution

Fig. 3. Performance comparison of NestedBTO and NesTM across STAMP and Red-Black Tree benchmarks. (a)–(c) illustrate throughput, abort rates, and tail latencies in red-black tree workloads. (Color figure online)

commit-phase bottlenecks, is the fundamental reason for this improvement. On the other hand, NesTM's locking during the commit phase introduces contention, especially during tree rebalancing, which led to observed latency variance of 15%, compared to only 8% for NestedBTO.

The experimental results demonstrated two advantages of NestedBTO. First, NestedBTO achieves a reasonably good throughput, even with deep nesting levels, with a capacity throughput of about 25% at nesting level 5 compared to a high degradation of up to 38% with NesTM. Second, NestedBTO's conflict resolution reduces cascading aborts by 20–30% for heavily nested transactions. However, there are two points to keep in mind: First, with a nesting level beyond five, the timestamp comparisons have overhead issues, and there will be a runtime penalty of around 12%. Second, for shallow nesting depths (levels 1–2), NesTM at times outperformed NestedBTO up to 5%-due mainly to its lock granularity being more suitable for hardware.

5   Conclusion and Future Work

The paper presents **NestedBTO**, which is considered a new STM protocol for closed nesting. Under the opacity consistency model, the algorithm permits parallel execution of nested transactions while guaranteeing correctness. **NestedBTO** is demonstrated to have major performance advantages over NesTM via extensive benchmarking with STAMP, concurrent Red Black Tree, and concurrent hashtable. One possible direction is to extend **NestedBTO** to allow multi-versioned transactional objects. The extension would bring about more concurrency by permitting object versions to coexist, hence allowing more flexible and efficient transactional operations. One can also delve deeper into optimizations for specific application scenarios and do further benchmarking on various workloads to gather further insights into the performance characteristics of **NestedBTO**. On the other hand, pondering integration with emerging hardware architectures and tapping into possible distributed implementations make for interesting areas for future work. Moreover, investigating the integration of **NestedBTO** with emerging hardware architectures and exploring its applicability in distributed systems are intriguing areas for future exploration. Overall, the versatility and robustness of **NestedBTO** lay a solid foundation for continued research and innovation in transactional memory systems.

References

1. Assa, G., Meir, H., Golan-Gueta, G., Keidar, I., Spiegelman, A.: Using nesting to push the limits of transactional data structure libraries. arXiv preprint arXiv:2001.00363 (2020)
2. Baek, W., Bronson, N., Kozyrakis, C., Olukotun, K.: Implementing and evaluating nested parallel transactions in software transactional memory. In: Proceedings of the Twenty-Second Annual ACM Symposium on Parallelism in Algorithms and Architectures, pp. 253–262 (2010)

3. Barreto, J., Dragojević, A., Ferreira, P., Guerraoui, R., Kapalka, M.: Leveraging parallel nesting in transactional memory. ACM Sigplan Notices **45**(5), 91–100 (2010)

4. Bernstein, P.A., Goodman, N.: Multiversion concurrency control—theory and algorithms. ACM Trans. Database Syst. (TODS) **8**(4), 465–483 (1983)

5. Dalessandro, L., Spear, M.F., Scott, M.L.: Norec: streamlining STM by abolishing ownership records. ACM Sigplan Notices **45**(5), 67–78 (2010)

6. Dickerson, T.D., Gazzillo, P., Herlihy, M., Koskinen, E.: Proust: a design space for highly-concurrent transactional data structures. arXiv preprint arXiv:1702.04866 (2017)

7. Doherty, S., Groves, L., Luchangco, V., Moir, M.: Towards formally specifying and verifying transactional memory. Formal Aspects Comput. **25**, 769–799 (2013)

8. Guerraoui, R., Kapalka, M.: Opacity: a correctness condition for transactional memory. Technical report (2007)

9. Guerraoui, R., Kapalka, M.: On the correctness of transactional memory. In: Proceedings of the 13th ACM SIGPLAN Symposium on Principles and Practice of Parallel Programming, pp. 175–184 (2008)

10. Haines, N., Kindred, D., Morrisett, J.G., Nettles, S.M., Wing, J.M.: Composing first-class transactions. ACM Trans. Program. Lang. Syst. (TOPLAS) **16**(6), 1719–1736 (1994)

11. Harris, T., Marlow, S., Peyton-Jones, S., Herlihy, M.: Composable memory transactions. In: Proceedings of the Tenth ACM SIGPLAN Symposium on Principles and Practice of Parallel Programming, pp. 48–60 (2005)

12. Herlihy, M., Moss, J.E.B.: Transactional memory: architectural support for lock-free data structures. In: Proceedings of the 20th Annual International Symposium on Computer Architecture, pp. 289–300 (1993)

13. Kobus, T., Kokocinski, M., Wojciechowski, P.T.: The correctness criterion for deferred update replication. Program TRANSACT **15** (2015)

14. Kumar, R., Vidyasankar, K.: Hparstm: a hierarchy-based STM protocol for supporting nested parallelism. In: The 6th ACM SIGPLAN Workshop on Transactional Computing (TRANSACT 2011) (2011)

15. Lebanoff, L., Peterson, C., Dechev, D.: Check-wait-pounce: increasing transactional data structure throughput by delaying transactions. In: Pereira, J., Ricci, L. (eds.) DAIS 2019. LNCS, vol. 11534, pp. 19–35. Springer, Cham (2019). https://doi.org/10.1007/978-3-030-22496-7_2

16. Moravan, M.J., et al.: Supporting nested transactional memory in logtm. ACM SIGARCH Comput. Archit. News **34**(5), 359–370 (2006)

17. Moss, J.E.B., Hosking, A.L.: Nested transactional memory: model and architecture sketches. Sci. Comput. Program. **63**(2), 186–201 (2006)

18. Peri, S., Vidyasankar, K.: Correctness of concurrent executions of closed nested transactions in transactional memory systems. Theoret. Comput. Sci. **496**, 125–153 (2013)

19. Peri, S., Vidyasankar, K.: Correctness of concurrent executions of closed nested transactions in transactional memory systems. In: Aguilera, M.K., Yu, H., Vaidya, N.H., Srinivasan, V., Choudhury, R.R. (eds.) ICDCN 2011. LNCS, vol. 6522, pp. 95–106. Springer, Heidelberg (2011). https://doi.org/10.1007/978-3-642-17679-1_9

20. Ramadan, H., Witchel, E.: The xfork in the road to coordinated sibling transactions. In: 4th ACM SIGPLAN Workshop on Transactional Computing (TRANSACT 2009) (2009)

21. Ranjan, N., Kapoor, R., Peri, S.: Short paper: an efficient framework for supporting nested transaction in STMs. In: International Conference on Networked Systems, pp. 204–210. Springer (2024)
22. Saha, B., Adl-Tabatabai, A.R., Hudson, R.L., Minh, C.C., Hertzberg, B.: MCRT-STM: a high performance software transactional memory system for a multi-core runtime. In: Proceedings of the Eleventh ACM SIGPLAN Symposium on Principles and Practice of Parallel Programming, pp. 187–197 (2006)
23. Shavit, N., Touitou, D.: Software transactional memory. In: Proceedings of the Fourteenth Annual ACM Symposium on Principles of Distributed Computing, pp. 204–213 (1995)
24. Spiegelman, A., Golan-Gueta, G., Keidar, I.: Transactional data structure libraries. ACM SIGPLAN Notices **51**(6), 682–696 (2016)
25. Volos, H., Welc, A., Adl-Tabatabai, A.R., Shpeisman, T., Tian, X., Narayanaswamy, R.: Nepaltm: design and implementation of nested parallelism for transactional memory systems. In: Proceedings of the 14th ACM SIGPLAN Symposium on Principles and Practice of Parallel Programming, pp. 291–292 (2009)

Gathering of Asynchronous Robots on Circle with Limited Visibility Using Finite Communication

Avisek Sharma[1]($\boxtimes$) iD, Satakshi Ghosh[2] iD, and Buddhadeb Sau[1] iD

[1] Department of Mathematics, Jadavpur University, Kolkata, West Bengal, India
`{aviseks.math.rs,buddhadeb.sau}@jadavpuruniversity.in`
[2] Department of Computer Science and Engineering, Sister Nivedita University, Kolkata, West Bengal, India
`satakshi.g@snuniv.ac.in`

Abstract. This work addresses the gathering problem for a set of autonomous, anonymous, and homogeneous robots with limited visibility operating in a continuous circle. The robots are initially placed at distinct positions, forming a rotationally asymmetric configuration. The robots agree on the clockwise direction. In the θ-visibility model, a robot can only see those robots on the circle that are at an angular distance $< \theta$ from it. Di Luna *et al.* [DISC'20] have shown that, in $\pi/2$ visibility, gathering is impossible. In addition, they provided an algorithm for robots with π visibility, operating under a semi-synchronous scheduler. In the π visibility model, only one point, the point at the angular distance π is removed from the visibility. Ghosh *et al.* [SSS'23] provided a gathering algorithm for π visibility model with robot having finite memory ($\mathcal{FSTA}$), operating under a special asynchronous scheduler.

If the robots can see all points on the circle, then the gathering can be done by electing a leader in the weakest robot model under a fully asynchronous scheduler. However, previous works have shown that even the removal of one point from the visibility makes gathering difficult. In both works, the robots had rigid movement. In this work, we propose an algorithm that solves the gathering problem under the π-visibility model for robots that have finite communication ability ($\mathcal{FCOM}$). In this work the robot movement is non-rigid and the robots work under a fully asynchronous scheduler.

Keywords: Limited visibility · Finite communication · Asynchronous robots · Continuous circle · Distributed algorithms

1 Introduction

Swarm robotics has been well-studied over the past two decades in distributed computing. Here a collection of simple and inexpensive robots (mobile computing units) collaboratively perform a task. In this line of research, we investigate to find the minimal capability required for the robots to complete a certain

S. Bonomi et al. (Eds.): SSS 2025, LNCS 16350, pp. 442–457, 2026.
https://doi.org/10.1007/978-3-032-11127-2_33

task. Generally, robots are considered dimensionless computational entities on the Euclidean plane. They are expected to complete a given task by coordinating with each other. The robots operate through Look-Compute-Move (LCM) cycles. On activation, a robot takes a snapshot of its surroundings in its vicinity (Look). Then taking information in the snapshot it runs an inbuilt algorithm and outputs a position to move (Compute). Next, it moves to that computed position (Move). The robots are generally autonomous (having no central control), anonymous (no unique identifier), identical (physically indistinguishable), and homogeneous (all robots run the same distributed algorithm). Depending upon the time of activation there are different schedulers. In the fully synchronous setting (FSYNC), the time is divided into equal rounds and simultaneously executes the phases of their LCM cycles. The semi-synchronous model (SSYNC) is the same as the FSYNC model, except for the fact that not all robots are necessarily activated in each round. The most general model is the fully asynchronous model (ASYNC) where there are no assumptions regarding the synchronization and duration of the robot actions.

There are different robot models depending on the capabilities of the robots. The weakest robot model is $\mathcal{OBLOT}$. In this model, the robots are oblivious, i.e., robots have no persistent memory to remember their past actions or past configurations, and robots are silent, i.e., robots have no explicit communication ability. Next, robots may be equipped with a persistent light that can take colors from a predefined pallet consisting of a finite number of colors. If this light is visible to itself then it serves as a finite memory. This model is called $\mathcal{FSTA}$ ([5]). If this light is visible only to other robots, then this serves as a communication architecture. This model is called $\mathcal{FCOM}$ ([14]). If this light is visible to all robots including itself then the model is called $\mathcal{LUMI}$ ([2]). In the $\mathcal{LUMI}$ model, the robots have finite memory and also have communication ability.

In this work, we are concerned with the most fundamental coordination task, Gathering ([1,18]). Here, the robots are asked to meet at a point, not known beforehand. In general, robots have full visibility, but in practice, we may not have this luxury due to hardware limitations. In the limited visibility model ([1,7,8,15]), a robot can see only those robots that are at a distance less than a constant $R > 0$. The R is called the visibility radius of the robots. In this work, the robots are operating in a continuous circle [4]. Initially, the robots are placed at distinct positions in a circle. All robots agree on the clockwise direction. The initial forming configuration by the robots is rotationally asymmetric and has no multiplicity point, i.e., there is no location on the circle that is occupied by more than one robot. A configuration is rotationally asymmetric, which means no non-trivial rotation with respect to the center will keep the configuration unchanged. In the θ-visibility model, the robots can see only those robots that have an angular distance less than θ from it (See Fig. 1).

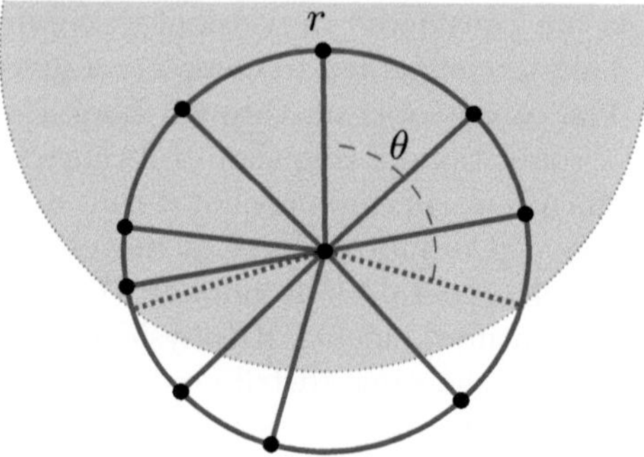

Fig. 1. Visibility for robot r: black solid circles denote robots. Robot r can see the robots that are at an angular distance θ from it.

2   Related Work and Our Contribution

The gathering problem by a group of robots has been studied extensively in different terrains and models (See the survey in [6]). Here we shall restrict our discussion to the continuous terrain when robots have limited visibility. In [1], Ando *et al.* first time considered gathering in the plane by robots with limited visibility. But there is a lighter version of gathering, called *Point Convergence* which has been studied in the literature. The point convergence problem asks to design a distributed algorithm such that for any given $\epsilon > 0$ there is a time t_ϵ such that for any time $t \geq t_\epsilon$ distance between any pair of robots is less than ϵ. In works [3,4,7,8,10–12,15,17], authors considered gathering under limited visibility and in the works [1,11,12], authors considered point convergence with limited visibility. In [4,10] gathering on continuous circles has been considered by robots with limited visibility. In [13], Mondal *et al.* considered the arbitrary pattern formation problem on continuous circles.

Next, we only focus on the details of the previous works where the gathering on continuous circles has been considered. In [4], Di Luna *et al.* considered first the gathering of robots on a continuous circle with limited visibility. They considered the $\mathcal{OBLOT}$ robot model. They showed that rotationally asymmetric configuration at the initial is necessary. They first showed that if robots have $\pi/2$ visibility then the gathering is impossible. Then they proposed an algorithm to gather robots if robots have π visibility. The π visibility is a minimal limitation because it only removes one point from the visibility, the point on the circle at the angular distance π. We call the position at the angular distance π from a robot as *antipodal* position of it. While in full visibility, there is a gathering algorithm by electing a unique leader in a rotationally asymmetric configuration. Next, since robots are assumed to have multiplicity detection ability, so formation of one multiplicity almost solves the problem. However, the removal of one point from the visibility makes it challenging. First, the election of one unique leader is not possible and also there is a chance of formation of two multiplicity points at antipodal positions. In [4], authors considered robots operating under a semi-synchronous scheduler. Also, they assumed the robot's movements are rigid, i.e., when a robot decides to move at a point then the adversary cannot stop its movement midway. They allow a robot to cross another robot while moving.

Next, in [10] Ghosh *et al.* attempted to solve it under the same model but in an asynchronous scheduler. However, authors could solve it for a special type of asynchronous scheduler. In addition, the authors traded synchrony with memory. They used the $\mathcal{FSTA}$ robot model. In both works, the rigidity of robots' movement remains a crucial and necessary assumption. So, in a practical scenario where maintaining rigidity is challenging, both of the previous solutions do not work.

In this work, the authors have proposed a gathering algorithm under a fully asynchronous scheduler. Here, authors trade synchrony with communication. The robot model considered here is $\mathcal{FCOM}$. In addition, the authors removed the assumption of rigidity. The robots' movements are non-rigid. Since $\mathcal{LUMI}$ is an equally or more strong model than $\mathcal{FCOM}$, it also gives a solution for the $\mathcal{LUMI}$ model. Furthermore, to validate our contribution from the work in [10], where the $\mathcal{FSTA}$ model has been used, in Sect. 6, we have shown that for non-rigid movements $\mathcal{FCOM}$ is not a stronger model than $\mathcal{FSTA}$[1]. We have defined a problem in the same terrain (continuous circle) and the same visibility model that is solvable in model $\mathcal{FSTA}$ but not solvable in $\mathcal{FCOM}$.

Our Contribution: In this work, a set of robots are initially deployed at distinct locations on a continuous circle, initially forming a rotationally asymmetric configuration. The robots can freely move on the perimeter of the circle. The robots cannot see all points of the circle. The robots are operating under the π visibility model, i.e., each robot cannot see the point at an angular distance π from it. The robots agree on the clockwise direction. Robots have a weak multiplicity detection ability, i.e., from a snapshot, they can detect a multiplicity point but cannot count the number robots occupying that point. The robot model is $\mathcal{FCOM}$, i.e., each robot is equipped with a persistent external light that can take colors from a predefined palette consisting of finitely many colors. The robot's movements are non-rigid, i.e., a robot's movement can be stopped by an adversary midway. Although, for the sake of fairness, every time a robot attempts to move, it must move a constant distance $\delta > 0$ towards its destination location. The robots are operating under a fully asynchronous scheduler. In addition, to justify our contribution, we prove that model $\mathcal{FCOM}$ is not as strong as $\mathcal{FSTA}$ (a model used in previous work) when robot movements are non-rigid. This work first proposes a gathering algorithm for the π visibility model under a fully asynchronous scheduler and with non-rigid robot movements. The Table 1 compares our contribution with the other relevant works. Proofs of most results are omitted due to space constraint (See full version [16] for the proofs).

---

[1] See [9] for better understanding of model comparisons.

Table 1. Comparison table

Work	Model	Scheduler	Rigidity
Di Luna *et al.* [4]	$\mathcal{OBLOT}$	Semi-synchronous	Yes
Ghosh *et al.* [10]	$\mathcal{FSTA}$	A special asynchronous scheduler	Yes
This work	$\mathcal{FCOM}$	Fully asynchronous Scheduler	No

3   Model Definition and Preliminaries

Let a set of k robots be initially placed on a circle C on the Euclidean plane. The robots look identical from the outside. Robots are autonomous and homogeneous. All robots are governed by the same algorithm. The robots agree on the clockwise direction. The robots can freely move in the circle. To widen the scope of the applicability, we assume that the robots cannot cross each other on the circle. If a robot meets another robot in the middle of its move, its movement is paused until it is activated again. The robot's movements are not rigid. That is, an adversary can stop a robot's movement at any point. But for the sake of fairness, we assume that once a robot starts moving it will at least make a constant $\delta > 0$ amount of movement towards its computed destination. Each robot is equipped with an external light that can take finitely many predefined colors from the palate P. The colors of the light are persistent, i.e., they remain unchanged unless it is changed by the robot according to the algorithm.

We assume that initially, each robot is occupying a distinct position on C. Let a and b be two distinct points on C. The clockwise angular distance between a and b, denoted as $\alpha(a, b)$ is the angle subtended at the center of C by a robot while traversing from a to b on C in a clockwise direction. Similarly, $\overline{\alpha}(a, b)$ $(= 2\pi - \alpha(a, b))$ denotes the counterclockwise angular distance between a and b. For two robots r_1 and r_2 located on C at points a and b respectively, we denote the clockwise (resp, anticlockwise) angular distance between them as $\alpha(r_1, r_2)$ (resp, $\overline{\alpha}(r_1, r_2)$) which is equal to $\alpha(a, b)$ (resp, $\overline{\alpha}(a, b)$).

We assume the robots have limited visibility. Each robot can see all robots at an angular distance $< \pi$. The point at the angular distance π is not visible to a robot. The robots operate through Look-Compute-Move cycles. Upon activation in the look phase, the robots scan the visible part of the circle and obtain the position and color of the light of the visibility. In the compute phase, the robots run the inbuilt algorithm and obtain a position and a color c from the palate P. In the move phase first, it sets the color of its light at c and then moves the computed position.

The robots are operating under a fully asynchronous scheduler. The robots activate independently and execute their phases of the LCM cycle independently. The duration of the phases can take an unbounded amount of time. For a robot, the duration of the phases can be different from other robots and also from its own earlier cycle.

A finite set of ordered pairs $(p, \omega(p))$ is said to be a *configuration of robots* (or, simply a *configuration*) where the first quadrant of each pair p is a unique point on C occupied by a robot and $\omega(p) = \{1, mult\}$. The $\omega(p) = 1$ denotes that p is occupied by only one robot and $\omega(p) = mult$ denotes that p is occupied by more than one robot. A point p having $\omega(p) = mult$ is said to be a *multiplicity* point. If a configuration has no multiplicity point, then we can ignore the second quadrant of the members of C. For such a case, a configuration is equivalent to a finite set of points on C.

In a snapshot of a robot r, it receives ordered pairs $(p, \omega(p))$, where p is a point visible to r occupied by a robot. Thus, robots have weak multiplicity detection ability, i.e., a robot can detect whether a point p is occupied by one robot or more than one robot from the value of $\omega(p)$, but it cannot count the number of robots occupying a multiplicity point.

Let C be a configuration with no multiplicity point. Let S be a finite set of points on C occupied by the robots in C. The C is said to be *rotational symmetric* if there is a nontrivial rotation of points of S with respect to the center which leaves the configuration unchanged. A configuration with no multiplicity point is said to be *rotationally asymmetric* if it is not rotationally symmetric. A robot r is said to be an antipodal robot if there exists a robot r' on the angular distance π of the robot. In such a case, r and r' are said to be *antipodal* robots to each other.

Let r be a robot in a given configuration with no multiplicity point and let $r_1, r_2, \ldots, r_n$ be the other robots on the circle in clockwise order. Then the angular sequence for robot r is the sequence $(\alpha(r, r_1), \alpha(r_1, r_2), \alpha(r_2, r_3), \ldots, \alpha(r_n, r))$. We denote this sequence as $\mathcal{S}(r)$. Further, $\mathcal{S}(r, r_i)$ denote the following subsequence of $\mathcal{S}(r)$: $(\alpha(r, r_1), \alpha(r_1, r_2), \alpha(r_2, r_3), \ldots, \alpha(r_{i-1}, r_i))$. Further, we call $\alpha(r, r_1)$ as the leading angle of r. We denote the leading angle of a robot r as $\lambda(r)$.

Note that as the configuration is initially rotationally asymmetric, by results from the paper [4] we can say that all the robots have distinct angle sequences.

Definition 1 (Lexicographic Ordering). *Let $\tilde{\alpha} = (\alpha_1, \ldots, \alpha_n)$ and $\tilde{\beta} = (\beta_1, \ldots, \beta_n)$ be two finite sequences of reals of same length. Then $\tilde{\alpha}$ is said to be lexicographically strictly smaller sequence than $\tilde{\beta}$ if $\alpha_1 < \beta_1$ or there exists $1 < k < n$ such that $\alpha_i = \beta_i$ for all $i = 1, 2, \ldots, k$ and $\alpha_{k+1} < \beta_{k+1}$. $\tilde{\alpha}$ is said to be lexicographically smaller sequence than $\tilde{\beta}$ if either $\tilde{\alpha} = \tilde{\beta}$ or $\tilde{\alpha}$ is lexicographically strictly smaller sequence than $\tilde{\beta}$.*

Definition 2 (True leader). *In a configuration with no multiplicity point, a robot with the lexicographically smallest angular sequence is called a true leader.*

If the configuration is rotationally asymmetric and contains no multiplicity point, there exists exactly one robot that has strictly the smallest lexicographic angle sequence. Hence there is only one true leader for such a configuration. Since a robot on the circle cannot see whether its antipodal position is occupied by a robot or not. So a robot can assume two things: 1) the antipodal position is empty, let's call this configuration $C_0(r)$, 2) the antipodal position is nonempty, let's call this configuration $\mathcal{C}_1(r)$. So, a robot r can form two angular sequences. One considering $C_0(r)$ configuration and another considering $C_1(r)$. The next two definitions are from the viewpoint of a robot. If the true leader robot can confirm itself as the true leader, we call it a *cognizant* leader. If the true leader or some other robot has an ambiguity of being a true leader depending on the possibility of $\mathcal{C}_0$ or $\mathcal{C}_1$ configuration, then we call it *undecided* leader. For a robot r, there may be the following possibilities.

- Possibility-1: $\mathcal{C}_0(r)$ configuration has rotational symmetry, so $\mathcal{C}_1(r)$ is the only possible configuration.
- Possibility-2: $\mathcal{C}_1(r)$ configuration has rotational symmetry, so $\mathcal{C}_0(r)$ is the only possible configuration.
- Possibility-3: Both $\mathcal{C}_0(r)$ and $\mathcal{C}_1(r)$ has no rotational symmetry, so both $\mathcal{C}_0(r)$ and $\mathcal{C}_1(r)$ can be possible configurations.

Definition 3 (Cognizant leader). *A robot r in a rotationally asymmetric configuration with no multiplicity point is called a cognizant leader if r is the true leader in any possible configurations.*

Note that, the cognizant leader is definitely the true leader of the configuration. Thus, if the configuration is asymmetric and contains no multiplicity point, there is at most one cognizant leader.

Definition 4 (Undecided leader). *A robot r in a rotationally asymmetric configuration with no multiplicity point is called an undecided leader if both $C_0(r)$ and $C_1(r)$ are possible configurations and r is a true leader in one configuration but not in another.*

Definition 5 (Follower robot). *A robot in an asymmetric configuration with no multiplicity point is said to be a follower robot if it is neither a cognizant leader nor an undecided leader.*

Definition 6 (Expected leader). *A robot in an asymmetric configuration with no multiplicity point is said to be an expected leader if it is not a follower robot. That is, an expected leader is either a cognizant leader or an undecided leader.*

Note that the above definitions are set in such a way that the cognizant leader (or, an undecided leader or, a follower robot) can recognize himself as the cognizant leader (or, an undecided leader or, a follower robot).

Definition 7. *For two robots r and r' situated at different positions on the circle with $\alpha(r, r') = \theta$, we define $[r, r']$ as the set of points x on the circle such that $0 \leq \alpha(r, x) \leq \theta$ and (r, r') as the set of points x on the circle such that $0 < \alpha(r, x) < \theta$.*

Definition 8. *Let r be a robot in a configuration, then a robot r_1 is said to be situated at the left of r if $\alpha(r, r_1) > \pi$ and said to be at right if $\alpha(r, r_1) < \pi$.*

Note that, the true leader of the configuration can become an undecided leader and a robot other than the true leader can also become an undecided leader. The next results lead us to find the maximum possible number of expected leaders in a rotationally asymmetric configuration with no multiplicity points.

Proposition 1. *Let $(\alpha_1, \ldots, \alpha_k)$ be the angular sequence of the true leader, say r_0, in a rotationally asymmetric configuration with no multiplicity point. Then there cannot be another robot r' with the following properties.*

1. r' is at left side to r_0,
2. $\mathcal{S}(r', r_0) = (\alpha_1, \alpha_2, \ldots, \alpha_i)$.

Next, we state a simple observation in the following Proposition 2.

Proposition 2. *Suppose there is a rotationally asymmetric configuration with no multiplicity point with the true leader, r_0 (say), then on including a robot, say r, on the circle at an empty point without bringing any rotational symmetry, the true leader of the new configuration must be in $[r_0, r]$.*

For an undecided leader r, there may be two possibilities. The first one is when r is a true leader in $\mathcal{C}_0(r)$ configuration but not in $\mathcal{C}_1(r)$. The second one is when r is a true leader in $\mathcal{C}_1(r)$ configuration but not in $\mathcal{C}_0(r)$. We show that the second possibility can not occur. We formally state the result in the following Proposition.

Proposition 3. *If a robot r is an undecided leader in asymmetric configuration with no multiplicity point, r is the true leader in $\mathcal{C}_0(r)$ configuration but r is not the true leader in $\mathcal{C}_1(r)$ configuration.*

From Proposition 3 one can observe that if the true leader of an asymmetric configuration with no multiplicity point is an undecided leader then its antipodal position must be empty. Also if a robot, which is not the true leader of the configuration, becomes an undecided leader then its antipodal position must be non-empty. We record these observations in the following Corollaries.

Corollary 1. *In a rotationally asymmetric configuration with no multiplicity point if the true leader of the configuration is an undecided leader then its antipodal position must be empty.*

Corollary 2. *In a rotationally asymmetric configuration with no multiplicity point if a robot other than the true leader becomes an undecided leader then its antipodal position must be non-empty.*

In Proposition 4, we record an important property regarding the locations of the expected leaders, and proposition 5 determines the total number of possible expected leaders in a rotationally asymmetric configuration without multiplicity points.

Proposition 4. *Let $\mathcal{C}$ be a rotationally asymmetric configuration with no multiplicity point and L be the true leader of the configuration. Then another expected leader r of $\mathcal{C}$ must satisfy $\alpha(L, r) \geq \pi$.*

Proposition 5. *For any given rotationally asymmetric configuration with no multiplicity point, there can be at most one undecided leader other than the true leader.*

Let $\mathcal{C}$ be a rotationally asymmetric configuration with no multiplicity point. Then $\mathcal{C}$ will have a true leader and from above Proposition 5, there can be at most one more undecided leader. Hence we can have the following four exhaustive cases for $\mathcal{C}$.

1. $\mathcal{C}$ has exactly one expected leader and that is a cognizant leader.
2. $\mathcal{C}$ has exactly one expected leader and that is an undecided leader.
3. $\mathcal{C}$ has exactly two expected leaders and both are undecided leaders.
4. $\mathcal{C}$ has exactly two expected leaders. One of them is a cognizant leader and another one is an undecided leader.

The Fig. 2 gives the existence of all four above cases. For the third case, we observe two properties in the following Propositions.

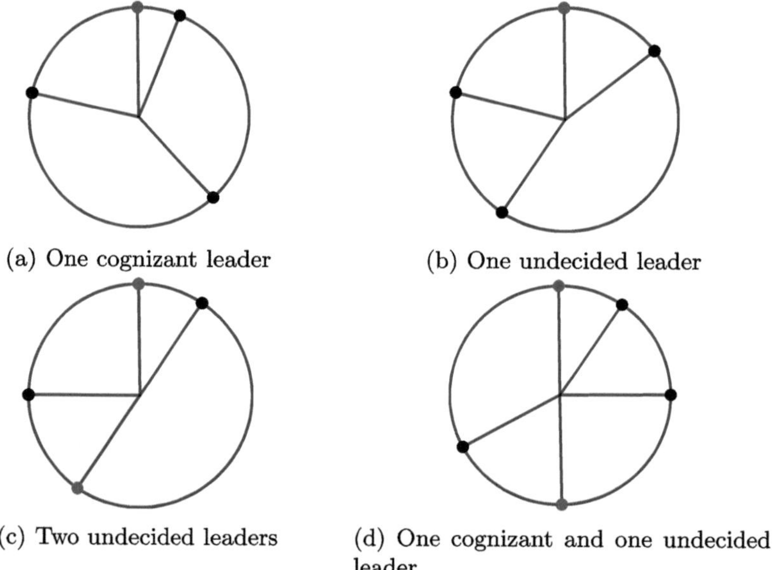

(a) One cognizant leader   (b) One undecided leader

(c) Two undecided leaders   (d) One cognizant and one undecided leader

Fig. 2. Different possibilities of expected leaders; Red-colored and blue-colored solid circles respectively denote a cognizant leader and an undecided leader. (Color figure online)

Next, we record more results in Proposition 6 and Proposition 7 regarding the positions of the expected leaders, which guides the design of the proposed algorithm.

Proposition 6. *If for a rotationally asymmetric configuration with no multiplicity point, there are two undecided leaders then they can not be antipodal of each other.*

Proposition 7. *If for a rotationally asymmetric configuration with no multiplicity point, there are two undecided leaders then their clockwise neighbors cannot be antipodal to each other.*

4   The Proposed Algorithm

Here we present an algorithm in Algorithm 1 to gather the robots. Each robot has the palate consisting the colors: `leader_present`, `leader_absent`, `verify`, `cognizant`, `undecided`, `off`. Initially, all colors of the robots' are set at `off`. If a robot cannot see any robot then it moves $\pi/2$ distance clockwise. If there is no multiplicity point in its visibility, then the robot verifies whether it is a cognizant leader or not. If it is a cognizant leader then it changes its color to `cognizant` and moves to its clockwise neighbor. If the robot is an undecided leader then it verifies that its clockwise neighbor is safe to move or not. Before further we give the definition of a *safe* clockwise neighbor.

Definition 9 (Safe neighbor). *Suppose r is an undecided leader and s is the first clockwise neighbor of r. The robot s is said to be a safe neighbor of r if the first clockwise neighbor of the true leader of $C_1(r)$ configuration is not antipodal to s.*

If the clockwise neighbor of the undecided leader is a safe neighbor then the robot changes its light's color to `undecided` and moves to the clockwise neighbor. If the clockwise neighbor is not a safe neighbor then it verifies whether its C_0 configuration consists of another undecided leader or not. If its C_0 does not consist of another undecided leader, it signals the neighbor to verify whether its antipodal position is occupied or not. It changes its color to `verify`.

Suppose a follower robot r sees its counter-clockwise neighbor, say r_0 with color `verify`. Let s and s_0 be the antipodal positions of the r and r_0 respectively. The follower robot verifies whether there is a robot in the interval $[s_0, s)$. If there is a robot then it changes its color to `leader_present` otherwise, changes its color to `leader_absent`. Now if a robot sees its clockwise neighbor has color `leader_absent`, then it changes its color to `cognizant` and moves to its clockwise neighbor. If it sees its clockwise neighbor has color `leader_present`, then it changes its color to `off`.

Now, suppose there is a multiplicity point and a robot r can see the multiplicity point. If r is not at the multiplicity point and its clockwise or counterclockwise neighbor is a multiplicity point then the robot moves to the closer multiplicity point. Suppose r is at a multiplicity point and there is another multiplicity point visible. In this case, if the clockwise distance of the multiplicity point from r is less than π, then r moves to another multiplicity point.

Algorithm 1: Gathering algorithm for π-visibility; executed by a generic robot r with initial light color **off**;

1 **if** *there is a robot visible* **then**
2   **if** *there is no multiplicity point* **then**
3   **if** *the robot r is a cognizant leader* **then**
4   Change the color to **cognizant**;
5   Move to the clockwise neighbor;
6   **else if** *the robot r is an undecided leader* **then**
7   **if** *the clockwise neighbor of r is safe* **then**
8   Change the color to **undecided**;
9   Move to the clockwise neighbor;
10   **else if** *$C_0(r)$ configuration does not have another undecided leader other than r* **then**
11   **if** *the clockwise neighbor has color off* **then**
12   Change the color to **verify**;
13   **else if** *the clockwise neighbor of r has color leader_absent* **then**
14   Change the color to **cognizant**;
15   Move to its clockwise neighbor;
16   **else if** *the clockwise neighbor of r has color leader_present* **then**
17   Change the color to **off**;
18   **else if** *the robot r is a follower robot* **then**
19   **if** *the counter-clockwise neighbor has color verify* **then**
20   Let r_0 be the robot with color **verify**;
21   Let s and s_0 be the antipodal positions of r and r_0 respectively;
22   **if** *there is a robot in $[s_0, s)$* **then**
23   Turn the light to **leader_present**;
24   **else**
25   Turn the light to **leader_absent**;
26   **else if** *there is a multiplicity point but the robot r is not at any multiplicity point* **then**
27   **if** *its clockwise or counter-clockwise neighbor is a multiplicity point* **then**
28   Move to the closer multiplicity point;
29   **else if** *there is another visible position that is a multiplicity point* **then**
30   **if** *clockwise angular distance from the multiplicity point is $< \pi$* **then**
31   Move to the multiplicity point;
32 **else**
33   Move $\pi/2$ distance in clockwise direction;

5   Correctness of the Proposed Algorithm

First, we categorize a rotationally asymmetric configuration with no multiplicity points in the following:

❖ *Configuration-A:* Only the expected leader is the true leader of the configuration. If the expected leader is an undecided leader then its clockwise first neighbor is safe.

❖ *Configuration-B:* There are two expected leaders in the configuration.

● *Configuration-BI:* One cognizant leader and one undecided leader.

● *Configuration-BII:* Two undecided leaders.

❖ *Configuration-C:* One expected leader which is an undecided leader sees that its clockwise first neighbor is not safe.

Lemma 1. *If the initial configuration is type configuration-A, then after finite-time execution of Algorithm 1 at least one and at most two multiplicity points will form or, it will turn into a configuration-C. If two multiplicity points are formed then they will be non antipodal to each other.*

Lemma 2. *If the initial configuration is type configuration-C, then after finite time execution of Algorithm 1 only one multiplicity point will form.*

Lemma 3. *If the initial configuration is type configuration-B, then after finite time execution of Algorithm 1 at least one and at most two multiplicity point will form. If two multiplicity points are formed then they will be non antipodal to each other.*

Theorem 1. *From any rotationally asymmetric configuration with no multiplicity point, by finite time execution of Algorithm 1 the robots can form at least one and at most two multiplicity points, and then all robots gather at a point on the circle.*

Proof. Let $\mathcal{C}$ be a rotationally asymmetric configuration with no multiplicity point. Then there can be three exhaustive possible configurations, Configuration-A, Configuration-B, and Configuration-C. From lemma 1, lemma 2, and lemma 3, we can say that in any type of configuration, at least one and at most two multiplicity points will form. On the formation of the multiplicity point, if a robot sees a multiplicity point, from line 27 of the algorithm robots one by one move to their closest multiplicity point. If only one multiplicity is formed, then all robots will eventually gather there except one robot at the antipodal position of the multiplicity point. If that robot does not move at all, then eventually it will not see any other robot. Then according to line 33 of the algorithm, it will move $\pi/2$ in a clockwise direction. After this, it will be able to see the multiplicity point. If a robot is at a multiplicity point and can not see any other multiplicity point then it does not move anywhere further. If there are two multiplicity points created, then from Lemma 1 and Lemma 3. the multiplicity points are non antipodal. Thus, each robot will be able to see at least one of the multiplicity points. Thus, eventually, all robots will gather at those two multiplicity points. After this according to the line 29-31, robots of the one multiplicity point will move to another multiplicity point. If two robots simultaneously start moving towards another multiplicity point, then a robot might see three multiplicity points in its view. But in such a scenario the robot does nothing. Hence, after a finite time, all robots will gather at a point.   □

Hence, we can conclude the following theorem.

Theorem 2. *There exists a gathering algorithm that gathers any set of robots with finite communication and π visibility from any initial rotationally asymmetric configuration under an asynchronous scheduler.*

6 $\mathcal{FCOM}$ Is Not as Powerful as $\mathcal{FSTA}$

In this section show that model $\mathcal{FCOM}$ is not as strong as $\mathcal{FSTA}$ under limited visibility. For the sake of fairness, we keep the underlying topology and visibility model the same as concerned in the proposed work. We define a problem 3TO7 in the following definition.

Definition 10. *Let there be two robots r_1 and r_2 placed on a circle such that $\alpha(r_1, r_2) = \pi/2$. The problem 3TO7 asks only the robot r_2 to move clockwise to occupy the position such that $\alpha(r_1, r_2)$ becomes $7\pi/6$ (See Fig. 3). The robots have π visibility model and the robot's movements are non-rigid. The robots agree on the clockwise direction.*

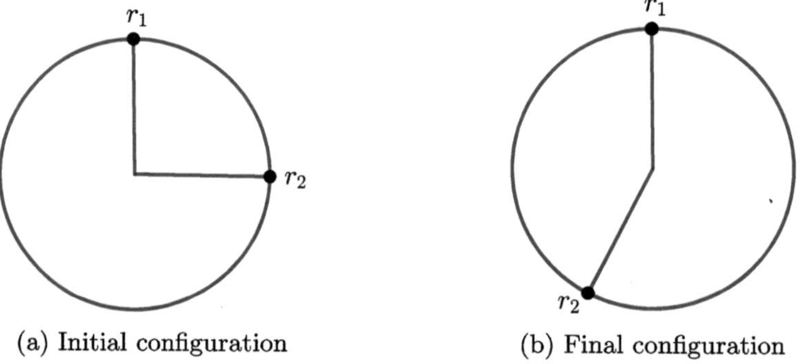

(a) Initial configuration   (b) Final configuration

Fig. 3. Configuration related to problem 3TO7.

We show that this problem is solvable in $\mathcal{FSTA}$ under an asynchronous scheduler but not solvable in $\mathcal{FCOM}$ under a fully asynchronous scheduler. We propose the following Algorithm 2 to solve the problem with two robots having two states off and done.

Next, we show that Algorithm 2 solves Problem 3TO7 under a fully asynchronous scheduler.

Algorithm 2: Executed by robot r; state of r initially set to off

---

```
 1  if state.r = off then
 2  │   if there is a robot r' visible and α(r, r') = 3π/2 then
 3  │   │   state.r ← done;
 4  │   │   Move 2π/3 angular distance clockwise;
 5  │   else
 6  │   │   Do nothing;

 7  if state.r = done then
 8  │   if there is another robot r' visible then
 9  │   │   if α(r, r') ≠ 5π/6 then
10  │   │   │   Move clockwise to make α(r, r') = 5π/6;
11  │   else
12  │   │   Move π/6 angular distance clockwise;
```

Lemma 4. *There is an algorithm that solves the problem* 3TO7 *with two robots having model* $\mathcal{FSTA}$ *under a fully asynchronous scheduler.*

In the next Lemma 5, we show that the problem 3TO7 is not solvable by robots with $\mathcal{FCOM}$ model even under a fully synchronous scheduler.

Lemma 5. *There is no algorithm that solves the problem* 3TO7 *with two robots having model* $\mathcal{FCOM}$ *even under a fully synchronous scheduler.*

Hence, from Lemma 4 and Lemma 5, there is a problem that is solvable in $\mathcal{FSTA}$ but not solvable in $\mathcal{FCOM}$ under limited visibility. This proves that $\mathcal{FCOM}$ is not as powerful as $\mathcal{FSTA}$.

7 Conclusion

This work deals with the gathering of a set of robots with limited visibility deployed over a continuous circle. The robots are autonomous, anonymous, identical, and homogeneous. The robots operate through the Look-Compute-Move cycle. Initially, robots are at distinct positions forming a rotationally asymmetric configuration. The robots are required to meet at a point not known beforehand. The robots have the π visibility model. Each robot cannot see the point at an angular distance π from it. This work intends to extend the gathering problem under a limited visibility model as full visibility is not practical due to hardware limitations.

In this work, the robots agree on the clockwise direction. Each robot is equipped with a persistent external light ($\mathcal{FCOM}$) that can take colors from a predefined palette consisting finitely many colors. The robots are operating under a fully asynchronous scheduler. The robot's movements are non-rigid. This is the first work for the π visibility model under the fully asynchronous scheduler with non-rigid robots' movement. In addition, to validate our contribution, we attach a section showing that the model $\mathcal{FCOM}$ is not as powerful as model $\mathcal{FSTA}$. As a future work, there are a few questions listed below that are still open.

1. Is it impossible to gather the robots with $\mathcal{OBLOT}$ in π visibility under the fully-asynchronous scheduler?
2. Is it possible to gather the robots with $\theta(< \pi)$ visibility model?

Acknowledgment. The first author would like to thank the University Grants Commission (UGC), the Government of India for support. The third author is supported by the Science and Engineering Research Board (SERB), India.

References

1. Ando, H., Oasa, Y., Suzuki, I., Yamashita, M.: Distributed memoryless point convergence algorithm for mobile robots with limited visibility. IEEE Trans. Robot. Autom. **15**(5), 818–828 (1999). https://doi.org/10.1109/70.795787
2. Das, S., Flocchini, P., Prencipe, G., Santoro, N., Yamashita, M.: Autonomous mobile robots with lights. Theoret. Comput. Sci. **609**, 171–184 (2016). https://doi.org/10.1016/j.tcs.2015.09.018
3. Degener, B., Kempkes, B., Langner, T., Meyer auf der Heide, F., Pietrzyk, P., Wattenhofer, R.: A tight runtime bound for synchronous gathering of autonomous robots with limited visibility. In: Proceedings of the Twenty-third Annual ACM Symposium on Parallelism in Algorithms and Architectures, pp. 139–148 (2011)
4. Di Luna, G.A., Uehara, R., Viglietta, G., Yamauchi, Y.: Gathering on a circle with limited visibility by anonymous oblivious robots. Theoret. Comput. Sci. **1025**, 114974 (2025). https://doi.org/10.1016/j.tcs.2024.114974
5. Flocchini, P., Santoro, N., Viglietta, G., Yamashita, M.: Rendezvous with constant memory. Theoret. Comput. Sci. **621**, 57–72 (2016). https://doi.org/10.1016/j.tcs.2016.01.025
6. Flocchini, P.: Gathering, pp. 63–82. Springer International Publishing, Cham (2019). https://doi.org/10.1007/978-3-030-11072-7_4
7. Flocchini, P., Prencipe, G., Santoro, N., Widmayer, P.: Gathering of asynchronous oblivious robots with limited visibility. In: Ferreira, A., Reichel, H. (eds.) STACS 2001. LNCS, vol. 2010, pp. 247–258. Springer, Heidelberg (2001). https://doi.org/10.1007/3-540-44693-1_22
8. Flocchini, P., Prencipe, G., Santoro, N., Widmayer, P.: Gathering of asynchronous robots with limited visibility. Theoret. Comput. Sc. **337**(1), 147–168 (2005). https://doi.org/10.1016/j.tcs.2005.01.001
9. Flocchini, P., Santoro, N., Sudo, Y., Wada, K.: On Asynchrony, Memory, and Communication: separations and landscapes. In: Bessani, A., Défago, X., Nakamura, J., Wada, K., Yamauchi, Y. (eds.) 27th International Conference on Principles of Distributed Systems (OPODIS 2023). Leibniz International Proceedings in Informatics (LIPIcs), vol. 286, pp. 28:1–28:23. Schloss Dagstuhl – Leibniz-Zentrum für Informatik, Dagstuhl, Germany (2024). https://doi.org/10.4230/LIPIcs.OPODIS.2023.28

10. Ghosh, S., Sharma, A., Goswami, P., Sau, B.: Brief announcement: Asynchronous gathering of finite memory robots on a circle under limited visibility. In: Dolev, S., Schieber, B. (eds.) Stabilization, Safety, and Security of Distributed Systems - 25th International Symposium, SSS 2023, Jersey City, NJ, USA, October 2-4, 2023, Proceedings. LNCS, vol. 14310, pp. 430–434. Springer (2023).https://doi.org/10.1007/978-3-031-44274-2_32
11. Katreniak, B.: Convergence with limited visibility by asynchronous mobile robots. In: Kosowski, A., Yamashita, M. (eds.) SIROCCO 2011. LNCS, vol. 6796, pp. 125–137. Springer, Heidelberg (2011). https://doi.org/10.1007/978-3-642-22212-2_12
12. Kirkpatrick, D., Kostitsyna, I., Navarra, A., Prencipe, G., Santoro, N.: On the power of bounded asynchrony: convergence by autonomous robots with limited visibility. Distrib. Comput. **37**(3), 279–308 (2024)
13. Mondal, B., Goswami, P., Sharma, A., Sau, B.: Arbitrary pattern formation on a continuous circle by oblivious robot swarm. Theoret. Comput. Sci. **1021**, 114882 (2024). https://doi.org/10.1016/j.tcs.2024.114882
14. Okumura, T., Wada, K., Défago, X.: Optimal l-algorithms for rendezvous of asynchronous mobile robots with external-lights. Theoret. Comput. Sc. **979**, 114198 (2023). https://doi.org/10.1016/j.tcs.2023.114198
15. Poudel, P., Sharma, G.: Time-optimal gathering under limited visibility with one-axis agreement. Information **12**(11) (2021). https://doi.org/10.3390/info12110448
16. Sharma, A., Ghosh, S., Sau, B.: Gathering of asynchronous robots on circle with limited visibility using finite communication (2025). https://arxiv.org/abs/2509.04004
17. Souissi, S., Défago, X., Yamashita, M.: Using eventually consistent compasses to gather memory-less mobile robots with limited visibility. ACM Trans. Auton. Adaptive Syst. (TAAS) **4**(1), 1–27 (2009)
18. Terai, S., Wada, K., Katayama, Y.: Gathering problems for autonomous mobile robots with lights. Theoret. Comput. Sci. **941**, 241–261 (2023). https://doi.org/10.1016/j.tcs.2022.11.018

Space-Time Trade-Off in Bounded Iterated Memory

Guillermo Toyos-Marfurt[1,2]($\boxtimes$) and Petr Kuznetsov[1]

[1] Télécom Paris, Institut Polytechnique de Paris, Palaiseau, France
`{guillermo.toyos,petr.kuznetsov}@telecom-paris.fr`
[2] Inria Saclay, Télécom SudParis,Institut Polytechnique de Paris, Palaiseau, France

Abstract. The celebrated asynchronous computability theorem (ACT) characterizes tasks solvable in the read-write shared-memory model using the *unbounded full-information protocol*, where in every round of computation, each process shares its complete knowledge of the system with the other processes. Therefore, ACT assumes shared-memory variables of unbounded capacity. It has been recently shown that *bounded* variables can achieve the same computational power at the expense of extra rounds. However, the exact relationship between the bit capacity of the shared memory and the number of rounds required in order to implement one round of the full-information protocol remained unknown.

In this paper, we focus on the asymptotic *round complexity* of bounded iterated shared-memory algorithms that simulate, up to isomorphism, the unbounded full-information protocol. We relate the round complexity to the number of processes n, the number of iterations of the full information protocol r, and the bit size per shared-memory entry b. By analyzing the corresponding *protocol complex*, a combinatorial structure representing reachable states, we derive necessary conditions and present a *bounded* full-information algorithm tailored to the bits available b per shared memory entry. We show that for $n > 2$, the round complexity required to implement the full-information protocol satisfies $\Omega((n!)^{r-1} \cdot 2^{n-b})$. Our results apply to a range of iterated shared-memory models, from regular read-write registers to atomic and immediate snapshots. Moreover, our bounded full-information algorithm is asymptotically optimal for the iterated collect model and within a linear factor n of optimal for the snapshot-based models.

Keywords: Theory of computation · Distributed Computing Models · Iterated Models · Combinatorial Topology · Communication Complexity · Round Complexity

1 Introduction

In distributed systems, one 'rof the key characteristics is the system's computational power, that is, the range of tasks it allows for solving. Models of distributed computing are defined by a plethora of parameters and their comparative task computability analysis remains a significant challenge. The celebrated

S. Bonomi et al. (Eds.): SSS 2025, LNCS 16350, pp. 458–474, 2026.
https://doi.org/10.1007/978-3-032-11127-2_34

Asynchronous Computability Theorem (ACT) [11] characterizes wait-free task computability for read-write shared-memory models.

Combinatorial topology [9] has been proven essential in characterizing such iterative shared-memory models [6,11,15]. Specifically, the set of executions of a given model can be represented as a *simplicial complex*: a combinatorial structure that enables analysis of the model's key properties. In this context, a distributed *task* is defined as a tuple $(\mathcal{I}, \mathcal{O}, \Delta)$, where $\mathcal{I}$ and $\mathcal{O}$ represent the *input* and *output* complexes, defining the task's inputs and outputs, and $\Delta : \mathcal{I} \to 2^{\mathcal{O}}$ is a mapping that associates each input assignment with the set of *allowed* output assignments.

ACT [11] states that a task is *solvable* if and only if there exists a *chromatic* map from $\mathcal{I}$ to $\mathcal{O}$ that *respects* Δ.[1] This characterization assumes the *full-information protocol*, in which the processes share their *complete* (rapidly growing) state in every round of computation, which relies on shared-memory variables of *unbounded* size.

It has been proved that, strictly speaking, in the *wait-free* read-write iterated model, unbounded memory is not necessary for preserving computational power [4]. Indeed, variables of minimal (one-bit) capacity can be used to implement an iteration of the full-information protocol, at the expense of running through multiple rounds. But this raises a fundamental question: What is the cost of bounded memory? Specifically, while bounded memory may be theoretically sufficient for simulating full-information, what practical costs, such as round complexity, are required to achieve the same computing power?

This paper characterizes a family of iterated shared-memory algorithms [17], called *ITER*, that implement the full-information protocol. We asymptotically determine the necessary number of rounds an *ITER* algorithm should go through so that its protocol complex becomes isomorphic to r rounds of the full information protocol.

Our approach uses the concept of process *distinguishability* to map shared memory values in a way that transmits full-information within bounded bits. By mapping shared memory entries to states of the full-information protocol, we derive the number of rounds required to emulate r iterations of the full-information protocol, as a function of the available registers and the number of processes n.

Our contributions are as follows. For a given input complex $\mathcal{I}$, we establish the relationship between round complexity and available bits b per shared-memory entry required to simulate r iterations of a full-information protocol. Specifically, for systems with three or more processes, $(n > 2)$, we show that the round complexity (i.e., the asymptotic number of iterations required by the algorithm) has a lower bound of $\Omega((n!)^{r-1} \cdot 2^{n-b})$. For two processes, the round complexity is 1, achievable with just 2-bit registers [19]. We also present a construction to obtain bounded full-information algorithms that are asymptotically optimal in round complexity for the iterated collect model, and execute at

---

[1] In the special case when the task is *colorless*, the characterization boils down to the existence of a continuous map from $|\mathcal{I}|$ to $|\mathcal{O}|$ (geometric realizations of $\mathcal{I}$ and $\mathcal{O}$).

most $O((n!)^{r-1} \cdot 2^{n-b} \cdot n)$ rounds in the atomic and immediate snapshot models. These results apply to a broad class of iterated shared-memory algorithms, *ITER*, which include well-known iterated models such as the Iterated Immediate Snapshot (IIS) [2], Iterated Atomic Snapshot (IAS) [3,16]. and Iterated collect (IC) [1]. An extended version of this paper with full proofs and additional details is available at [21].

Roadmap. The paper is organized as follows. We overview the related work in Sect. 2. In Sect. 3, we recall the combinatorial topology tools and concepts used throughout this work. In Sect. 4, we describe the iterated family of algorithms *ITER*. Then in Sect. 5, we give the necessary conditions that algorithms in *ITER* must satisfy for having a protocol complex equivalent to the full-information protocol. Next, in Sect. 6, we introduce Greedy Star, a construction for obtaining bounded full-information protocols based on the combinatorial condition of Sect. 5. Finally, in Sect. 7, we discuss the final results.

2   Related Work

This paper builds on the research started in [19], and further extended in [20], which characterized the iterated immediate snapshot model (IIS) [2,7] by analyzing its protocol complex, which is equivalent to the iterated standard chromatic subdivision [14]. The authors determine the necessary and sufficient conditions on the amount of shared-memory to simulate the full-information protocol *in the same number of rounds* as its unbounded counterpart. Our work extends this line of research by allowing an iterated algorithm to perform multiple rounds for a single iteration of the unbounded protocol, examining the relationship between round complexity and bit complexity. We further generalize these ideas to a broader class of protocols, which we refer to as *ITER*.

The Asynchronous Computability Theorem (ACT), introduced by Herlihy and Shavit [11], showed the impossibility of wait-free set agreement, solving the long-standing problem at the time [2,18]. Herlihy, Kozlov and Rajsbaum [9] offered a comprehensive treatment of combinatorial topology as a tool in distributed computing. Hoest and Shavit [12] proposed an alternative IIS model to study its execution time complexity. Several extensions considered adversarial models and dynamic networks [3,5,15].

Recently, Delporte-Gallet et al. [4] characterized the computational power of bounded registers in the wait-free and t-resilient models. Their focus was on (task) *computability*, not addressing the relationship between bounded shared-memory size and *round complexity*. In this paper, we fill this gap.

3   Preliminaries

In this section, we briefly review the key definitions and concepts, introduced in [19], which are essential for this work.

Simplicial Complex. We represent the states of iterated shared-memory protocols as topological-combinatorial spaces, called *simplicial complexes* [9]: a set of *vertices* and an inclusion-closed set of vertex subsets, called *simplices*. We call the simplicial complex of $n + 1$ vertices with its power set the n-dimensional simplex Δ^n. The leftmost triangle in Fig. 1 corresponds to Δ^2. The *dimension* of a simplex Δ, denoted as $dim(\Delta)$, is its number of vertices $V(\Delta) - 1$. The dimension of a simplicial complex is equal to the dimension of the largest simplex it contains. We call *faces* the simplices contained in a simplicial complex, where a 0-face is a vertex. The set $Faces(\mathcal{A})$ consists of all faces of the simplicial complex $\mathcal{A}$. As an example, $Faces(\Delta^2)$ has three 0-faces (vertices), three 1-faces (edges), and one 2-face (the simplex itself). We call *facets* the simplices which are not contained in any other simplex. We denote the set of vertices of a simplicial complex $\mathcal{A}$ as $V(\mathcal{A})$. Two vertices are *adjacent* if there is a 1-face (an "edge") in the simplicial complex containing both vertices. We denote $\deg(\mathcal{A}, v)$ the number of adjacent vertices to a vertex v in $\mathcal{A}$. We call a simplicial complex *chromatic* when it comes with a *coloring function* $\pi : V(\mathcal{A}) \to \Pi$ that assigns every vertex to a process identifier. We consider that all simplicial complexes are chromatic. Given a simplicial complex $\mathcal{A}$, the *star* of $\mathcal{S} \subseteq \mathcal{A}$, $\mathrm{St}(\mathcal{A}, \mathcal{S})$, is a subcomplex made of all simplices in $\mathcal{A}$ containing a simplex of $\mathcal{S}$ as face.

Subdivisions. In general, a subdivision operator $\tau : \mathcal{A} \to \mathcal{B}$ is a map that "divides" the simplices in $\mathcal{A}$ into smaller simplices. For a rigorous definition see [9]. We say that a simplicial complex $\mathcal{B}$ subdivides $\mathcal{A}$ if there exists a subdivision operator τ such that $\mathcal{B} = \tau(\mathcal{A})$. Subdivisions are intersection preserving: for any subcomplexes $\mathcal{A}, \mathcal{B} \subseteq \mathcal{I}$, $\tau(\mathcal{A}) \cap \tau(\mathcal{B}) = \tau(\mathcal{A} \cap \mathcal{B})$. The *standard chromatic subdivision* of a simplicial complex $\mathcal{A}$, denoted as $\mathrm{Ch}\,\mathcal{A}$, is a complex whose vertices are tuples (c, σ) where c is a color and σ is a face of $\mathcal{A}$ containing a vertex of color c. A set of vertices of $\mathrm{Ch}\,\mathcal{A}$ defines a simplex if for each pair (c, σ) and (c', σ'), $c \neq c'$ and either $\sigma \subseteq \sigma'$ or $\sigma' \subseteq \sigma$. For Δ^n, the vertices (c, Δ^n) define a *central simplex* in $\mathrm{Ch}\,\mathcal{A}$, these vertices are called *central vertices*. Note that Ch is itself a subdivision operator: for any simplicial complex $\mathcal{A}$, the complex $\mathrm{Ch}\,\mathcal{A}$ subdivides $\mathcal{A}$. For instance, if $\mathcal{A} = \Delta^2$, then $\mathrm{Ch}\,\Delta^2$ subdivides Δ^2 as shown in Fig. 1 (top right).

In this paper, we characterize a necessary condition for full-information protocols using the combinatorial growth of vertex degrees under iterative chromatic subdivisions of an input complex. Specifically, we use Theorem 5 of [20], originally stated in terms of the number of faces in the local neighborhood of a vertex. In our setting, we consider the case corresponding to the number of 1-dimensional faces adjacent to a vertex, that is, its degree.

Lemma 1 (Bound on the iterative chromatic subdivision [20]). *Let $r >$ 1, $\mathcal{K}$ a n-dimensional simplicial complex and $v \in V(\mathcal{K})$. The following is a tight asymptotic bound on the degree of v in $\mathrm{Ch}^r\,\mathcal{K}$:*

$$\deg(\mathrm{Ch}^r\,\mathcal{K}, v) \in \Theta\left((n!)^{r-1} 2^n n \right)$$

Tasks. A *distributed task* is defined as a tuple $(\mathcal{I}, \mathcal{O}, \Delta)$, where $\mathcal{I}$ and $\mathcal{O}$ are, resp., the *input complex* and the *output complex*, describing the task's input and output configurations, and $\Delta : \mathcal{I} \to 2^{\mathcal{O}}$ is a map relating each input assignment to all output assignments allowed by the task. In this work we consider general *colored* tasks, where $\mathcal{I}$ and $\mathcal{O}$ are chromatic simplicial complexes, allowing processes to have different sets of inputs and outputs. To solve a task, a distributed system uses a (communication) *protocol* to share information between processes. The set of configurations reachable by the protocol from a given initial configuration are expressed via a carrier map, denoted by Ξ. A carrier map is a function from a simplicial complex to its power set that preserves face inclusion: for all simplices $\tau, \sigma \in Faces(\mathcal{I})$ with $\tau \subseteq \sigma$, it holds that $\Xi(\tau) \subseteq \Xi(\sigma)$. We refer to this function as the *protocol map*. From the protocol complex $\Xi(\mathcal{I})$, a simplicial map (i.e. a function that maps simplices to simplices) determines the outputs of processes by mapping the simplices of the protocol complex to an output complex, respecting the specification of the problem given by Δ: $\delta \circ \Xi(\mathcal{I}) \subseteq \Delta(\mathcal{I})$. We refer to δ as the *decision function*.

4　System Model

4.1　Iterated Memory

We introduce *ITER*, a family of iterated memory models. Algorithm 1 shows the pseudo-code of a model in *ITER*. The motivation behind this family is to provide a *generic* representation that covers all iterated read-write shared-memory protocols, allowing us to characterize them.

Computation is structured in a sequence of *shared-memory layers* $M[r]_{1 \leq r \leq R}$, where each layer is an array of n SWMR registers (one per process), initialized with $\bot$. Processes then can write to their dedicated registers and read the registers of the layer. The r-th round of algorithm A, denoted $A(r)$, consists of each process performing a write to its register on $M[r]$, reading all registers of the layer, and updating its local state accordingly (i.e., one iteration of the loop in Algorithm 1).

The models in *ITER* are *wait-free*: each process reaches an output in a finite number of its own steps, regardless of the other processes. Each process is initially assigned an *input*, and maintains its state in a local variable s. In each iteration, the process writes a representation of its current state using an *encode* function (line 2). Then, it reads the whole memory layer (line 3) and changes the internal process state according to a *next_state* transition function. Finally, after a fixed number of rounds R, the process returns a value according to its task-specific decision function δ_i according to its internal state.

Note that if both the encoding and the state function are the identity – that is, *encode* : *state* $\mapsto$ *state* and *next_state* : (*view*, *state*) $\mapsto$ *view* – the protocol corresponds to the *unbounded* full-information protocol and the bit complexity is n^r in the r-th iteration. Here, only the *read-write pattern* characterizes the algorithm. We denote this particular set of algorithms as FI_P, where P is the pattern it uses. For example, FI_{IIS} refers to the well-studied full-information iterated

---

Algorithm 1: The family of iterated memory algorithms. Code for process $p_i \in \Pi$, $R > 0$ iterations and write-read pattern P.

Shared: $M[R, |\Pi|]$ array of R arrays of shared memory, each with $|\Pi|$ entries.
Initial: $v = input(i)$ ▷ What the process sees. At first, its input.
Initial: $s \leftarrow next\_state(v, \perp, 0)$ ▷ Initial state of the process.
1 **for** $r := 1$ *to* R **do**
    write-read pattern:
2      $M[r, i] \leftarrow write_P(encode(s, r))$;
3      $v \leftarrow read_P(M[r])$;
4      $s \leftarrow next\_state(s, v, r)$;
5 **end**
6 **return** $\delta_i(s)$;

---

immediate snapshot (IIS) algorithm, FI_{IAS} to its atomic snapshot counterpart, and FI_{IC} to the iterated collect variant [8].

However, if we impose restrictions on the size of outputs in the encoding function we can obtain algorithms that can be implemented using *bounded-size* registers. For instance, by using the immediate snapshot access pattern and a bounded encoding function, we get the Bounded Iterated Immediate Snapshot (B-IIS) model studied in [19]. Thus, the objective of this work is to characterize algorithms in *ITER* that are equivalent to FI_P but use bounded memory at each shared-memory layer.

For brevity, given an algorithm $A \in ITER$, let $\mathcal{S}$ be the set of possible states of the variable s, V the set of possible views obtained by the read primitive and E the image of *encode* (also called *encoding set*). An algorithm $A \in ITER$ is equipped with a sequence of functions $\{\omega_k\}_{k \in [R]}$ and $\{\sigma_k\}_{k \in [R]}$ such that at round r, $\sigma_r \in \{\sigma_k\}_{0 < k \leq R}, \sigma_r : \mathcal{S} \times V \mapsto \mathcal{S}$ and $\omega_r \in \{\omega_k\}_{0 < k \leq R}, \omega_r : \mathcal{S} \mapsto E$ are used as the *encode* and *next_state* functions respectively.

4.2 Topological Model

We consider a system of n asynchronous processes, which can fail arbitrarily by crashing. Processes communicate using an algorithm A from the family of iterated shared memory algorithms *ITER* described in Sect. 4.1.

We denote by Ξ_A the protocol map of an algorithm $A \in ITER$. Since A uses a different pair of functions (ω_r, σ_r) at each iteration r, it induces a distinct intermediate protocol map $\Xi^*_{A(r)}$ for each round r. We write $\Xi_{A(k)}(\mathcal{I})$ to refer to the protocol complex obtained after applying the sequence of intermediate protocol maps from round 1 to k: $\Xi_{A(k)}(\mathcal{I}) = \Xi^*_{A(k)} \circ \Xi^*_{A(k-1)} \circ \cdots \circ \Xi^*_{A(1)}(\mathcal{I})$. Accordingly, we have that $\Xi_A = \Xi_{A(R)}$, were R is the last round of A (as defined in Algorithm 1).

Figure 1 highlights the resulting protocol maps of the well-known full information protocols that use different shared memory primitives. We look for bounded memory protocol maps that yield the same output complex as the unbounded full information protocols FI_P. That is, given a simplicial input complex $\mathcal{I}$, and an algorithm $A \in ITER$ using a read write pattern P. We want that the protocol complex $\Xi_A(\mathcal{I})$ to be *isomorphic* to the protocol complex yielded by FI_P. Particularly, if A uses immediate snapshot, we want that $\Xi_A(\mathcal{I}) \cong \Xi_{IIS}(\mathcal{I}) \cong \mathrm{Ch}(\mathcal{I})$

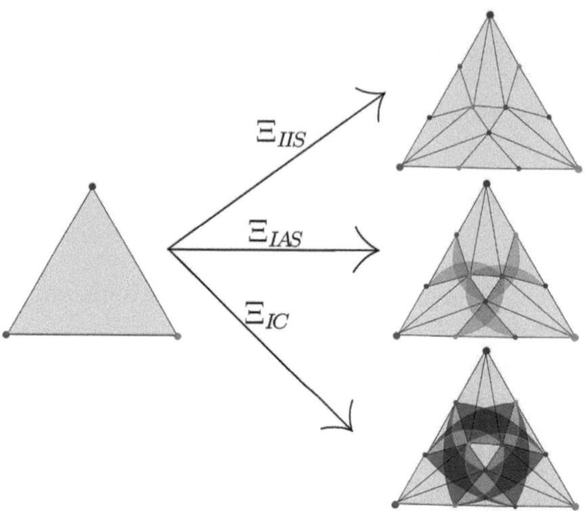

Fig. 1. Application of the *immediate snapshot, atomic snapshot* and *iterated collect* protocol maps to Δ^2 (shown on the right, from top to bottom). Starting from $\Xi_{IIS}(\Delta^2)$, the new possible configurations in the resulting protocol complexes of Ξ_{IAS} and Ξ_{IC} are highlighted in orange and purple, respectively. Note that $\Xi_{IIS}(\Delta^2) \subset \Xi_{IAS}(\Delta^2) \subset \Xi_{IC}(\Delta^2)$ and $\Xi_{IIS} \cong \mathrm{Ch}$.

5   A Necessary Condition for Full-Information Protocols

In this paper, we aim to identify a condition that an algorithm $A \in ITER$ using a write-read pattern P must satisfy in order to achieve an equivalence, up to isomorphism, with the (unbounded) full-information algorithm FI_P. More precisely, we derive a necessary condition on the protocol complex $\Xi_A(\mathcal{I})$, given an input complex $\mathcal{I}$.

5.1   Properties of FI_P

Our first objective is to establish some properties of the full-information protocol map of FI_P. For example, we know that $\Xi_{FI_{IIS}}(\mathcal{I}) = Ch$ as established in [14]. However, this does not hold for all write-read patterns.

In Fig. 1, we depict the protocol maps corresponding to one iteration of the iterated atomic snapshot (FI_{IAS}) and iterated collect (FI_{IC}) in 3-process systems.

These protocol complexes do not preserve planarity, implying they are not subdivisions of the input complex. Nevertheless, they share some key properties with subdivision operators. Specifically, Lemma 2 and Lemma 3 demonstrate that for any write-read pattern P, FI_P induces a mesh-shrinking protocol map [9]. Proofs are deferred to the extended version [21].

Therefore, any algorithm $A \in ITER$ that is equivalent, up to isomorphism, to a full-information protocol map must also satisfy the properties stated in Lemmas 2 and 3. These lemmas are applied in the proofs of Lemma 6 and Theorem 1.

Lemma 2. *Let $\mathcal{I}$ be an input complex, FI_P an unbounded full information algorithm using the write-read pattern P and Ξ_{FI_P} its protocol map. Then: $\forall v, w \in V(\mathcal{I}), \{v, w\} \notin \Xi_{FI_P}(\mathcal{I})$ (proof in [21]).*

Lemma 3. *Let $\mathcal{I}$ be an input complex and $\mathcal{A}, \mathcal{B} \subseteq \mathcal{I}$, FI_P an unbounded full information algorithm using the write-read pattern P and Ξ_{FI_P} its protocol map: $\Xi_{FI_P}(\mathcal{A}) \cap \Xi_{FI_P}(\mathcal{B}) = \Xi_{FI_P}(\mathcal{A} \cap \mathcal{B})$ (proof in [21]).*

5.2 The Necessary Condition

We now show which conditions $A \in ITER$ must satisfy in order to have the same properties as the unbounded full information protocol FI_P. The key concept in achieving this equivalence is *process distinguishability*. The goal is for a process to determine the state of another by reading the value it has written in shared memory. Importantly, it is not necessary for processes to write all of their knowledge into shared memory; a small, decodable, value may suffice to uniquely identify a state within the protocol complex. This allows us to define encoding functions that use a bounded number of shared-memory bits.

The following definitions define the notion of distinguishability for subcomplexes.

Definition 1 (Vertex distinguishability). Let $\mathcal{K}$ be a simplicial complex and $\omega : V(\mathcal{I}) \to E$. We say that $v \in V(\mathcal{K})$ is *distinguishable* in $\mathcal{K}$ *under* ω if no adjacent vertex to v has an adjacent vertex that uses the same encoding as v and has the same color as v: $\forall u \in V(\mathcal{K}), \{v, u\} \in \mathcal{K} \implies \nexists x \in V(\mathcal{K}) \setminus \{v\} : \{u, x\} \in \mathcal{K} \wedge \omega(v) = \omega(x) \wedge \pi(x) = \pi(v)$.

Definition 2 (Simplicial subcomplex distinguishability). Let $\mathcal{K}$ be a simplicial complex, $\mathcal{B} \subseteq \mathcal{K}$ subcomplex, and $\omega : V(\mathcal{K}) \to E$. We say that $\mathcal{B}$ is distinguishable in $\mathcal{K}$ under ω if every vertex $v \in V(\mathcal{B})$ is distinguishable in $\mathcal{K}$ under ω.

The primary challenge, and the key difference from [19], is that algorithm A can take an arbitrary number of rounds to obtain the same output complex as r iterations of the unbounded full-information protocol. Figure 2 presents a commutative diagram that illustrates this concept. The bounded algorithm A may require multiple rounds to achieve a protocol isomorphic to a single iteration of FI_P.

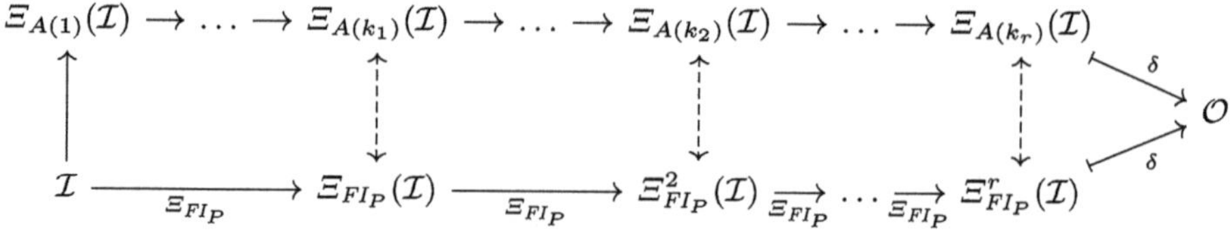

Fig. 2. Commutative diagram illustrating the equivalence between the unbounded full-information protocol map Ξ_{FI_P} and the bounded shared-memory iterated protocol map Ξ_A, using the write pattern P, over an input complex $\mathcal{I}$. $\Xi_A(k)$ denotes the associated protocol map at the k-th iteration of algorithm A. Dashed arrows indicate the existence of an isomorphism. The bounded algorithm A is able to solve the same set of tasks as the unbounded full information protocol ($\delta \circ \Xi_A(\mathcal{I}) \cong \delta \circ \Xi_{FI_P}(\mathcal{I})$).

Theorem 1 gives a necessary condition for a bounded algorithm A to be equivalent to its unbounded full information counterpart FI_P. In a nutshell, each face of the input complex must be made distinguishable in $\mathcal{I}$ at some round of A. To make a face σ distinguishable in an input complex $\mathcal{I}$, the encoding function has to assign distinct values to each vertex in σ so that, if the input configuration is σ, all processes can effectively map the observed values to the corresponding states in σ. The encoding functions of A, ω_A must account for each global input configuration and convey the same information as in the full information protocol. We consider each face $\sigma \in Faces(\mathcal{I})$ as the subcomplex it induces in $\mathcal{I}$ when reasoning about distinguishability.

Theorem 1. *Let $\mathcal{I}$ be an input complex, $r > 0$, $A \in ITER$ be an algorithm using the write-read pattern P, and $\{\omega_k\}_{k \leq r}$ the sequence of encoding functions of A. Then:*

$$\Xi_{A(r)}(\mathcal{I}) \cong \Xi_{FI_P}(\mathcal{I}) \Rightarrow \forall \sigma \in Faces(\mathcal{I}) \; \exists k \leq r : \sigma \text{ is distinguishable in } \mathcal{I} \text{ under } \omega_k$$

Proof. By contradiction, suppose that there exists a face $\sigma \in \mathcal{I}$ which is not distinguishable in $\mathcal{I}$ under any encoding function in $\{\omega_k\}_{k \leq r}$. As $\Xi_{A(r)}(\mathcal{I}) \cong \Xi_{FI_P}(\mathcal{I})$, by Lemma 3 we have that for any subcomplex $\mathcal{A} \subseteq \mathcal{I}$, $\Xi_{A(k)}(\sigma \cap \mathcal{A}) = \Xi_{A(k)}(\sigma) \cap \Xi_{A(k)}(\mathcal{A})$. We want to show that this requires that σ is distinguishable in $\mathcal{I}$ under some $\omega_k \in \{\omega_k\}_{k \leq r}$.

Consider the case where $\mathcal{I}$ is a simplicial complex made of two simplices α and β of dimension n such that $\alpha \cap \beta = \sigma$ and $dim(\sigma) = n - 1$. Note that both α and β have a vertex which is not in σ. Let a and b be such vertices respectively and p the respective process associated to both vertices: $p = \pi(a) = \pi(b)$. Let p write the same value for both states, that is: $\omega_k(a) = \omega_k(b)$. As a and b are adjacent to any vertex in σ, then α is not distinguishable in $\mathcal{I}$ under ω_k. Analogously, β is neither distinguishable in $\mathcal{I}$ under ω_k. By contradiction, suppose $\Xi_{A(k)}(\sigma) = \Xi_{A(k)}(\alpha) \cap \Xi_{A(k)}(\beta)$. Consider the execution where the global input configuration is α and a process $p' \neq p$ reads the value written by p and p reads nothing but its own input. Let $c_{p'}$ be the resulting final state of process p'. Because $\omega_k(a) = \omega_k(b)$, from the perspective of p', the execution is *indistinguishable* from the execution where process p is in state b. Thus, the vertex $c_{p'}$ is both in $\Xi_{A(k)}(\alpha)$ and $\Xi_{A(k)}(\beta)$. However, $a \notin \sigma$ and $b \notin \sigma$. Thus, $c_{p'} \notin \Xi_A(\sigma)$, a contradiction.

Therefore, there exists a $\omega_k \in \{\omega_k\}_{k \leq r}$ such that σ is distinguishable in $\mathcal{I}$ under ω_k, which contradicts the initial hypothesis. $\square$

Theorem 1 demonstrates that in an iterative protocol, each facet is made distinguishable at some round k and remains so until the final round R. This is a defining property of the iterative protocol maps Ξ_A: they must subdivide the facets of the input complex at some step. Indeed, making a facet distinguishable is necessary for the protocol to make progress.

Let b denote the number of bits available in shared memory for each process entry per round in A. The number of possible encodings is then limited by this bit count: $|\operatorname{Im}\omega_{A(k)}| \leq 2^b - 1$. Thus, the task of designing a bounded iterative protocol A that matches the unbounded full-information protocol's behavior becomes a problem of identifying suitable encoding functions $\omega_{A(k)}$, constrained by $|\operatorname{Im}\omega_{A(k)}| \leq 2^b - 1$, so that A can distinguish all facets in $\mathcal{I}$ within as few rounds as possible. However, achieving this optimally is non-trivial: in fact, the problem is NP-hard [13]. This is shown by finding an optimal sequence is NP-hard via reduction from Set Cover (proof deferred to the extended version [21]).

First, we define the *distinguishable simplicial subcomplex* of an encoding function ω as the subset of faces of an input complex $\mathcal{I}$ that are *distinguishable* under ω. Then, we can denote the distinguishable simplicial subcomplex of a sequence of encoding functions as follows:

Definition 3. Let ω_A be a sequence of encoding functions $\{\omega_{A(k)}\}_{0 < k \leq R}$ and $\mathcal{I}$ an input complex. We define $\mathcal{D}(\omega_A, \mathcal{I})$, the *distinguishable subcomplex of $\mathcal{I}$* under ω_A, as follows:

$$\mathcal{D}(\omega_A, \mathcal{I}) = \bigcup_{0 \leq k < R} \{\sigma \in \operatorname{Faces}(\mathcal{I}) : \sigma \text{ is distinguishable in } \mathcal{I} \text{ under } \omega_{A(k)}\}$$

Observe that $\mathcal{D}(\omega_A, \mathcal{I})$ is a well-defined simplicial subcomplex, as if a face $\sigma \in \mathcal{D}(\omega_A, \mathcal{I})$ is distinguishable under some $\omega_{A(k)}$, all its faces $\sigma' \subseteq \sigma$ are also distinguishable under the same encoding function. Using this definition we can restate Theorem 1 as follows:

Corollary 1. *Let $\mathcal{I}$ be an input complex, $A \in ITER$ using write-read pattern P, FI_P the unbounded full-information algorithm, ω_A the corresponding sequence of encoding functions of A and $\mathcal{D}(\omega_A, \mathcal{I})$ its distinguishable subcomplex.*

$$\Xi_A(\mathcal{I}) \cong \Xi_{FI_P}(\mathcal{I}) \Rightarrow \mathcal{D}(\omega_A, \mathcal{I}) = \mathcal{I}$$

Lemma 4. *Let $\mathcal{I}$ be an input complex, $b > 0$, $A \in ITER$ taking R rounds and $\{\omega_{A(k)}\}_{0 < k \leq R}$ its corresponding sequence of encoding functions such that $|\operatorname{Im}\omega_{A(k)}| \leq 2^b - 1 \ \forall 0 < k \leq R$ and $\mathcal{D}(\omega_A, \mathcal{I}) = \mathcal{I}$. Finding a sequence ω_A such that there is no shorter sequence satisfying the former conditions is NP-Hard (proof available in [21]).*

5.3 Combinatorial Characterization of the Necessary Condition

We now link the necessary condition of Lemma 6 to a combinatorial property of the input complex and subsequently present a result on its asymptotic growth with respect to the number of full-information rounds and available bits of shared-memory.

Lemma 5. *Let $\mathcal{I}$ be an input complex, and let $A \in ITER$ be an algorithm using the write-read pattern P such that $\Xi_A(\mathcal{I}) \cong \Xi_{FI_P}(\mathcal{I})$. Let b denote the number of bits per entry in the iterated memory of A. Then the number of iterations required by A – that is, the length of its sequence of encoding functions ω_A – satisfies the following lower bound:*

$$|\omega_A| \geq \left\lceil \frac{\max_{v \in V(\mathcal{I})} \deg(\mathcal{I}, v)}{n(2^b - 1)} \right\rceil$$

Proof (sketch). Consider a vertex v of maximum degree D. Distinguishing all D adjacent states requires that each encoding function separate as many neighbors as possible. Since a function over b bits yields at most $2^b - 1$ usable values, one function can cover only part of the neighborhood. Repeating this across $\lceil D/(2^b - 1) \rceil$ functions suffices to distinguish every state, establishing the bound. Full details appear in the extended version [21].

Lemma 5 gives a topological condition on the number of rounds required by A to yield a full-information protocol. It is of interest understanding how the value grows asymptotically. To achieve this, we will take advantage of the results and tools provided in [19]. We recommend the reader to check [19] if it is interested in the tools used, such as the f-vector, a common tool in polyhedral combinatorics [22], to perform the asymptotic analysis.

First, we show that the degree of a vertex in the input complex, after r iterations of the full-information protocol map, will be bigger than applying the standard chromatic subdivision r times.

Lemma 6. *Let $\mathcal{I}$ be an input complex, FI_P the unbounded full information protocol, $v \in V(\mathcal{A})$, $r > 0$, and Ch the standard chromatic subdivision operator. Then the following inequality holds:*

$$\deg(\Xi^r_{FI_P}(\mathcal{I}), v) \geq \deg(\mathrm{Ch}^r(\mathcal{I}), v)$$

Proof (sketch). The result follows by induction on r from the properties of the full information protocol and chromatic subdivisions. The full proof can be found in [21].

By applying the asymptotic bound from Lemma 1 to the construction in Lemma 6, we derive a lower bound on the round complexity as a function of the number of available bits b.

Theorem 2. *Let $\mathcal{I}$ be an n-dimensional input complex, with $n > 2$ and let $r > 0$. Let FI_P denote the unbounded full information protocol that uses a write-read pattern P. Let $A \in ITER$ be an algorithm that uses a write-read pattern P, but uses at most b bits per process per round. Any such algorithm A satisfying $\Xi_A(\mathcal{I}) \cong \Xi^r_{FI_P}(\mathcal{I})$ must perform $\Omega\left((n!)^{r-1} \cdot 2^{n-b}\right)$ rounds.*

Corollary 2. *Under the same context of Theorem 2, if we want A to perform a single round per iteration of the full-information protocol, that is $|\omega_A| = 1$, then $\Omega(rn \log n)$ is a lower bound on the bit complexity of A. Conversely, if we want $\Xi_A(\mathcal{I}) \cong \Xi^r_{FI_P}(\mathcal{I})$ using 2 bits, we have that $\Omega((n!)^{r-1}2^n)$ is a lower bound on the round complexity of A.*

6   Greedy Star: An Asymptotically Optimal Algorithm

In this section, we present Algorithm 2, an algorithm that leverages the inequality of Lemma 5 to construct a sequence of encoding functions ω_S. We devise an algorithm to construct a sequence of encoding functions that yield *distinguishable subcomplexes* forming a cover of the input complex. We call this algorithm Greedy Star, as it proceeds by iteratively selecting vertices in $\mathcal{I}$ and ensuring that each selected vertex is distinguishable by assigning it a unique encoding within its star. To maintain distinguishability, the algorithm enforces that the star of each selected vertex has an empty intersection with the stars of previously selected vertices within an encoding function.

Algorithm 2: Greedy Star encoding algorithm: $GS(\mathcal{I})$

Input: Chromatic input complex $\mathcal{I}$.

Output: Sequence of encoding functions $\{\omega_r\}_{r \geq 0}$ such that $\mathcal{D}(\{\omega_r\}_{r \geq 0}, \mathcal{I}) = \mathcal{I}$.

1 $\mathcal{A} \leftarrow \emptyset; V \leftarrow V(\mathcal{I}); W \leftarrow \emptyset; r \leftarrow 0$

2 **while** $\mathcal{A} \neq \mathcal{I}$ **do**

3   let $\omega_r : V(\mathcal{I}) \to \mathbb{N} \cup \{\bot\}, \ \omega_r(x) = \bot$

4      $\mathcal{U} \leftarrow \emptyset; V_u \leftarrow \emptyset$

5   **foreach** $v \in V$ **do**

6          $\mathcal{S} \leftarrow \mathrm{St}(\mathcal{I}, v)$

7   **if** $\forall w \in V(\mathcal{S}), \mathrm{St}(\mathcal{I}, w) \cap \mathcal{U} = \emptyset$ **then**

8              $\mathcal{U} \leftarrow \mathcal{U} \cup \mathcal{S}$

9              $V_u \leftarrow V_u \cup V(\mathcal{S})$

            `// redefine `ω_r` for vertices in the star.`

10   let $\forall x \in V(\mathcal{S}), k \in \mathbb{N}, \ \omega_r(x) = k : \nexists y \in V(\mathcal{S}), \ x \neq y \wedge \omega_r(y) = k \wedge \pi(y) = \pi(x)$

11   **end**

12   **end**

13      $\mathcal{A} \leftarrow \mathcal{A} \cup \mathcal{U}; V \leftarrow V \setminus V_u; W \leftarrow W \cup \{\omega_r\}; r \leftarrow r + 1$

14 **end**

15 **return** W

Lemma 7. *The Greedy Star Cover algorithm outputs a sequence of encoding functions such that $D(\omega_S(\mathcal{I}), \mathcal{I}) = \mathcal{I}$.*

Proof (sketch). Intuitively, the claim follows from the fact that each star eventually covers all simplices, and stars are disjoint by construction. The full proof is available in [21].

Observe that each encoding function returned by Greedy Star requires a number of distinct encodings proportional to the maximum number of facets in the star of some vertex. However, the number of available encodings is bounded: $|\operatorname{Im}\omega_{A(k)}| \leq 2^b - 1$. Therefore, a bounded algorithm $A \in ITER$ must simulate each encoding function $\omega_S(k)$ over multiple iterations in order to assign unique encodings to all the facets in a star. In particular, each $\omega_S(k)$ must be further decomposed into multiple encoding functions, ensuring that all facets in the corresponding star remain distinguishable across these iterations. Furthermore, any process p at a vertex v such that $\omega_S(k)(v) = \bot$ ignores the contents of shared memory during such iteration and preserves its local state.

Lemma 8. *Let $\mathcal{I}$ be an input complex, $b > 0$, and let $\omega_S(\mathcal{I})$ be the sequence of encoding functions constructed by the Greedy Star algorithm such that for each $\omega_k \in \omega_S(\mathcal{I})$, $|\operatorname{Im}\omega_k| \leq 2^b - 1$.*

$$|\omega_S| \leq 4 \left\lceil \frac{\max_{v \in V(\mathcal{I})} \deg(\mathcal{I}, v)}{n(2^b - 1)} \right\rceil$$

Proof (sketch). At each iteration, the algorithm assigns unique encodings to a subset of facets in disjoint vertex stars. Since each encoding function has at most $2^b - 1$ distinct values, multiple iterations are needed to encode each star. A careful argument shows that after at most 4 iterations, all facets are assigned unique encodings (see the extended version for details [21]).

Using Greedy Star, we construct algorithms $A \in ITER$ with a deterministic sequence of encoding functions ω_A, computable locally by each process. The next-state function $\sigma_{A(k)}$ then reconstructs process states from these encodings. We show that an algorithm using the iterated collect pattern, along with the encoding functions produced by Greedy Star, faithfully simulates the unbounded full-information iterated collect protocol.

Theorem 3. *Let $\mathcal{I}$ be an input complex, and let $GS \in ITER$ be an algorithm that uses the same read-write pattern as Iterated Collect, with a sequence of encoding functions $\omega_{GS} = GS(\mathcal{I})$ produced by the Greedy Star algorithm. Then:*

$$\Xi_{GS}(\mathcal{I}) \cong \Xi_{FI_{IC}}(\mathcal{I}),$$

Proof. First, we will show that $\Xi_{GS}(\mathcal{I}) \subseteq \Xi_{FI_{IC}}(\mathcal{I})$. We want to show that any global configuration of $GS(\mathcal{I})$ is also a valid global configuration in FI_{IC}. By the pseudocode of Algorithm 1, there is an execution of GS where every process writes and reads only its own input. Such execution is present at $\Xi_{FI_{IC}}(\mathcal{I})$.

Now consider executions where processes may read, at some round in GS, the inputs of additional processes. Since processes retain all previously read inputs, it suffices to show that any execution of iterated collect can be extended by allowing a process to read additional inputs. Consider an execution α of iterated collect and a process p which has not read the input of some other process q in α. We can construct an execution α' identical to α except that p additionally reads the input of q. It suffices to rearrange the read operation of p on the register of process q in α such that it occurs afterwards the write operation by q. Note that the views of all other processes remain unchanged. Therefore, any global configuration reachable in GS is also reachable in FI_{IC}, and we have $\Xi_{GS}(\mathcal{I}) \subseteq \Xi_{FI_{IC}}(\mathcal{I})$.

It remains to show that $\Xi_{FI_{IC}}(\mathcal{I}) \subseteq \Xi_{GS}(\mathcal{I})$. Consider a final configuration $\sigma \in \Xi_{FI_{IC}}(\mathcal{I})$. Because GS covers the entire input complex, there exists a round k in GS where the input configuration is distinguishable under $\omega_{GS}(k)$. Hence, there exists an execution β in GS such that the view of all process at round k is exactly the same as in σ. We now construct an extension β' of β in which all processes terminate while preserving the same final configuration σ. Because the model allows arbitrary asynchrony, we can extend the execution by scheduling all remaining steps of processes such that each read after round k returns a value already observed by the reading process. Consequently, no process obtains new inputs beyond what it had at round k. Thus, no process gains additional information and their local state remains unchanged Thus, the final configuration of β' is equal to σ. Therefore, $\Xi_{FI_{IC}}(\mathcal{I}) \subseteq \Xi_{GS}(\mathcal{I})$ and $\Xi_{FI_{IC}}(\mathcal{I}) \cong \Xi_{GS}(\mathcal{I})$. $\square$

From a bounded implementation of the iterated collect model, we can derive bounded algorithms with stronger communication patterns, such as atomic and immediate snapshots. In particular, Algorithm 5 from [4] simulates a single round of immediate snapshot using n rounds of iterated collect. This simulation can be applied in a black-box manner to our bounded iterated collect implementation, yielding bounded full-information algorithms for snapshot-based models. As a result, if the bounded iterated collect algorithm terminates in R rounds, the corresponding derived algorithm terminates in $O(n \cdot R)$ rounds. Therefore, the asymptotic round complexity is preserved up to a linear factor in the number of processes.

7 Final Results and Conclusions

Combining Theorem 2 and Theorem 3, we derive asymptotic bounds that characterize the fundamental trade-off between the bit complexity (b bits available per process per round) and the round complexity (i.e., the number of rounds) of bounded full-information algorithms.

Theorem 4 (Tight bound on the full-information iterated collect). *Let $\mathcal{I}$ be an $(n-1)$-dimensional input complex, with $n > 2$ and let $r > 0$. Let FI_{IC} denote the unbounded full-information protocol based on iterated collect. Let*

$A \in ITER$ be an algorithm that uses the same read-write pattern as iterated collect, but uses at most b bits per process per iteration. Let $|\omega_A|$ denote the number of encoding functions used by A, i.e., the number of rounds performed. Then there exists such an algorithm A satisfying $\Xi_A(\mathcal{I}) \cong \Xi^r_{FI_{IC}}(\mathcal{I})$, and the number of rounds satisfies: $|\omega_A| \in \Theta\left((n!)^{r-1} \cdot 2^{n-b}\right)$. Moreover, no such algorithm exists that uses asymptotically fewer rounds.

Theorem 5 (Bounds on the full-information iterated snapshot).

Let $\mathcal{I}$ be an $(n-1)$-dimensional input complex, with $n > 2$ and let $r > 0$. Let FI_S denote the unbounded full-information protocol that uses atomic or immediate snapshots as its read-write pattern S ($S \in \{IAS, IIS\}$). Let $A \in ITER$ be an algorithm that uses the same read-write pattern as FI_S, but with at most b bits per process per iteration. Let $|\omega_A|$ denote the number of encoding functions used by A, i.e., the number of rounds performed. Then any such algorithm A satisfying $\Xi_A(\mathcal{I}) \cong \Xi^r_{FI_S}(\mathcal{I})$ must perform at least $|\omega_A| \in \Omega\left((n!)^{r-1} \cdot 2^{n-b}\right)$ rounds, and there exists such an algorithm that performs at most $|\omega_A| \in O\left((n!)^{r-1} \cdot 2^{n-b} \cdot n\right)$ rounds.

The Greedy Star algorithm introduced in this work provides what appears to be the first efficient construction for simulating unbounded full-information protocols using a bounded number of bits per round. Notably, our results apply to several iterated memory models studied in the literature, including iterated collect, atomic snapshot, and immediate snapshot. This contribution highlights the power of combinatorial topology not only as an analytical tool for distributed systems, but also as a constructive tool for designing algorithms that exploit the topological structure of computation.

Several research directions remain open. One promising avenue is to adapt the techniques developed here to problems beyond full-information, such as approximate agreement [9]. Additionally, an important open question is whether the upper bound for bounded implementations of snapshot-based models can be improved. Specifically, given b-bit snapshot entries per process per round, it remains to determine whether an unbounded full-information snapshot protocol can be implemented using fewer than $O\left((n!)^{r-1} \cdot 2^{n-b} \cdot n\right)$ rounds. Finally, one can also consider this question in the context of bounded-memory *adversarial* models [10,15].

Acknowledgments. This work was supported by CHIST-ERA (grant ANR-23-CHR4-0009).

References

1. Attiya, H., Welch, J.L.: Distributed computing - fundamentals, simulations, and advanced topics, 2nd edn. Wiley, Wiley series on parallel and distributed computing (2004)

2. Borowsky, E., Gafni, E.: Immediate atomic snapshots and fast renaming. In: PODC, pp. 41–51. ACM Press, New York (1993). https://doi.org/10.1145/164051. 164056
3. Bouzid, Z., Gafni, E., Kuznetsov, P.: Strong equivalence relations for iterated models. In: Aguilera, M.K., Querzoni, L., Shapiro, M. (eds.) OPODIS 2014. LNCS, vol. 8878, pp. 139–154. Springer, Cham (2014). https://doi.org/10.1007/978-3-319-14472-6_10
4. Delporte-Gallet, C., Fauconnier, H., Fraigniaud, P., Rajsbaum, S., Travers, C.: The computational power of distributed shared-memory models with bounded-size registers. In: PODC, p. 310–320. Association for Computing Machinery, New York (2024). https://doi.org/10.1145/3662158.3662789
5. Delporte-Gallet, C., Fauconnier, H., Rajsbaum, S.: Communication complexity of wait-free computability in dynamic networks. In: Richa, A.W., Scheideler, C. (eds.) SIROCCO 2020. LNCS, vol. 12156, pp. 291–309. Springer, Cham (2020). https://doi.org/10.1007/978-3-030-54921-3_17
6. Gafni, E., Kuznetsov, P., Manolescu, C.: A generalized asynchronous computability theorem. In: PODC (2014)
7. Gafni, E., Rajsbaum, S.: Distributed programming with tasks. In: Lu, C., Masuzawa, T., Mosbah, M. (eds.) OPODIS 2010. LNCS, vol. 6490, pp. 205–218. Springer, Heidelberg (2010). https://doi.org/10.1007/978-3-642-17653-1_17
8. Guerraoui, R., Kuznetsov, P.: Algorithms for Concurrent Systems. EPFL Press (2018)
9. Herlihy, M., Kozlov, D.N., Rajsbaum, S.: Distributed Computing Through Combinatorial Topology. Morgan Kaufmann (2014)
10. Herlihy, M., Rajsbaum, S.: The topology of shared-memory adversaries. In: PODC, pp. 105–113 (2010)
11. Herlihy, M., Shavit, N.: The topological structure of asynchronous computability. J. ACM 46(2), 858–923 (1999)
12. Hoest, G., Shavit, N.: Toward a topological characterization of asynchronous complexity. SIAM J. Comput. 36(2), 457–497 (2006). https://doi.org/10.1137/S0097539701397412
13. Karp, R.M.: Reducibility among combinatorial problems. Complexity Comput. Comput. 85–103 (1972)
14. Kozlov, D.N.: Chromatic subdivision of a simplicial complex. Homology, Homotopy Appli. 14(2), 197–209 (2012)
15. Kuznetsov, P., Rieutord, T., He, Y.: An asynchronous computability theorem for fair adversaries. In: Proceedings of the ACM Symposium on Principles of Distributed Computing, PODC 2018, pp. 1–10. (2018)
16. Rodrigues, H., Jones, R.: A cyclic distributed garbage collector for network objects. In: Babaoğlu, Ö., Marzullo, K. (eds.) WDAG 1996. LNCS, vol. 1151, pp. 123–140. Springer, Heidelberg (1996). https://doi.org/10.1007/3-540-61769-8_9
17. Rajsbaum, S.: Iterated shared memory models. In: López-Ortiz, A. (ed.) LATIN 2010. LNCS, vol. 6034, pp. 407–416. Springer, Heidelberg (2010). https://doi.org/10.1007/978-3-642-12200-2_36
18. Saks, M., Zaharoglou, F.: Wait-free k-set agreement is impossible: the topology of public knowledge. In: Proceedings of the 25th ACM Symposium on Theory of Computing, pp. 101–110 (May 1993)
19. Toyos-Marfurt, G., Kuznetsov, P.: On the bit complexity of iterated memory. In: Structural Information and Communication Complexity: 31st International Colloquium (SIROCCO 2024), pp. 456–477. Springer-Verlag, Berlin, Heidelberg (2024). https://doi.org/10.1007/978-3-031-60603-8_25

20. Toyos-Marfurt, G., Kuznetsov, P.: On the bit complexity of iterated memory (2024). https://doi.org/10.48550/arXiv.2402.12484
21. Toyos-Marfurt, G., Kuznetsov, P.: Space-time trade-off in bounded iterated memory (2025). https://doi.org/10.48550/arXiv.2509.13157
22. Ziegler, G.M.: Shellability and the Upper Bound Theorem, pp. 231–290. Springer New York (1995)

Author Index